IFIP Advances in Information and Communication Technology 420

IFIP – The International Federation for Information Processing

IFIP was founded in 1960 under the auspices of UNESCO, following the First World Computer Congress held in Paris the previous year. An umbrella organization for societies working in information processing, IFIP's aim is two-fold: to support information processing within its member countries and to encourage technology transfer to developing nations. As its mission statement clearly states,

> IFIP's mission is to be the leading, truly international, apolitical organization which encourages and assists in the development, exploitation and application of information technology for the benefit of all people.

IFIP is a non-profitmaking organization, run almost solely by 2500 volunteers. It operates through a number of technical committees, which organize events and publications. IFIP's events range from an international congress to local seminars, but the most important are:

- The IFIP World Computer Congress, held every second year;
- Open conferences;
- Working conferences.

The flagship event is the IFIP World Computer Congress, at which both invited and contributed papers are presented. Contributed papers are rigorously refereed and the rejection rate is high.

As with the Congress, participation in the open conferences is open to all and papers may be invited or submitted. Again, submitted papers are stringently refereed.

The working conferences are structured differently. They are usually run by a working group and attendance is small and by invitation only. Their purpose is to create an atmosphere conducive to innovation and development. Refereeing is also rigorous and papers are subjected to extensive group discussion.

Publications arising from IFIP events vary. The papers presented at the IFIP World Computer Congress and at open conferences are published as conference proceedings, while the results of the working conferences are often published as collections of selected and edited papers.

Any national society whose primary activity is about information processing may apply to become a full member of IFIP, although full membership is restricted to one society per country. Full members are entitled to vote at the annual General Assembly, National societies preferring a less committed involvement may apply for associate or corresponding membership. Associate members enjoy the same benefits as full members, but without voting rights. Corresponding members are not represented in IFIP bodies. Affiliated membership is open to non-national societies, and individual and honorary membership schemes are also offered.

Daoliang Li   Yingyi Chen (Eds.)

Computer and Computing Technologies in Agriculture VII

7th IFIP WG 5.14 International Conference, CCTA 2013
Beijing, China, September 18-20, 2013
Revised Selected Papers, Part II

 Springer

Volume Editors

Daoliang Li
Yingyi Chen
China Agricultural University
China-EU Center for Information
and Communication Technologies in Agriculture (CICTA)
17 Tsinghua East Road, Beijing 100083, P.R. China
E-mail: {dliangl, chenyingyi}@cau.edu.cn

ISSN 1868-4238   e-ISSN 1868-422X
ISBN 978-3-662-52554-8   ISBN 978-3-642-54341-8 (eBook)
DOI 10.1007/978-3-642-54341-8
Springer Heidelberg New York Dordrecht London

Typesetting: Camera-ready by author, data conversion by Scientific Publishing Services, Chennai, India

Printed on acid-free paper

Springer is part of Springer Science+Business Media (www.springer.com)

Preface

First of all, I must express my sincere thanks to all authors who submitted research papers to support the 7th International Conference on Computer and Computing Technologies in Agriculture (CCTA2013) held in Beijing, China, during September 18–20, 2013.

The conference was hosted by the China Agricultural University; the IFIP TC5 Work Group (WG) on Advanced Information Processing for Agriculture (AIPA); the Agricultural Engineering Information Committee, Chinese Society of Agricultural Engineering. It was organized by the China-EU Centre for Information & Communication Technologies (CICTA).

Proper scale management is not only a necessary approach for agro-modernization and agro-industrialization but it is also required for the development of agricultural productivity. Thus, the application of different technologies in agriculture has become especially important and "informatized agriculture" and the "Internet of Things" have been sought out by many countries recently in order to scientifically manage agriculture so as to achieve low costs and high income. CICTA aims at boosting research on advanced and practical technologies applied in agriculture and promoting international communication and cooperation, and has successfully held seven international conferences since 2007.

The topics of CCTA2013 cover a wide range of interesting theory and applications of all kinds of technology in agriculture, including: the Internet of things and cloud computing; simulation models and decision-support systems for agricultural production; smart sensor, monitoring, and control technology; traceability and e-commerce technology; computer vision, computer graphics, and virtual reality; the application of information and communication technology in agriculture; and universal information service technology and service system development in rural areas.

We selected the 115 best papers among all those submitted to CCTA2013 for these proceedings, and all the papers are divided into two thematic sections. In this volume, creative thoughts and inspirations could be discovered, discussed and disseminated. It is always exciting to have experts, professionals, and scholars getting together with creative contributions to share inspiring ideas and hopefully accomplish great developments in these technologies of high demand.

Finally, I would like to express my sincere thanks to all the authors, speakers, session chairs, and attendees, both local and international for their active participation and support of this conference.

January 2014   Daoliang Li
                                                                    Chair of CCTA2013

Conference Organization

Sponsors

- China Agricultural University
- The IFIP TC5 Work Group (WG) on Advanced Information Processing for Agriculture (AIPA)
- Agricultural Engineering Information Committee, Chinese Society of Agricultural Engineering

Organizer

- China-EU Center for Information & Communication Technologies in Agriculture (CICTA)

Chair

- Daoliang Li

Conference Secretariat

Lihong Shen

Table of Contents – Part II

Table of Contents – Part I

Maize Seed Embryo and Position Inspection
Based on Image Processing

Yingbiao Wang, Liming Xu[*], Xueguan Zhao, Xingjie Hou

Key Laboratory of Soil-Machine-Plant System Technology, China Agricultural University,
Beijing, China, 100083
{wybjob,zhaoxueguan.com}@163.com,
{xlmoffice,hou19890809}@126.com

Abstract. Orientational planting cannot only make maize leaf growing consistently, but also improve the leaf photosynthetic capacity of unit area and yield of maize. The "tip" direction was detected and the angle of deflection was measured by using contour curvature analysis method of maize seed, the S component of HSV channel was preprocessed by Otsu method, the automatic extracting maize seeds embryo and orientation information have been realized through image color channel conversion, segmentation, preprocessing, and its contour characteristic analysis. A rectangular region ROI in the image was defined and counts up pixels within the region, the region-specific positive and negative ROI pixels were compared by T_M threshold and the embryo side towards was identified. This paper adopted Zheng-958, Jundan-20 and Zhongke-11 maize seed for research object, each variety was repeated three times by using the above methods. The results showed that the average accuracy of embryo inspection was more than 95%; the direction average angle was 2.2 °.

Keywords: Image processing, Maize seed, Orientational planting, Position inspection, Embryo.

1 Introduction

Maize is an important crop in the world. The orientational planting cannot only make maize leaf growing consistently, but also improve the maize leaf photosynthetic capacity of unit area and yield of maize. Thereby, enhancing unit area of maize leaf photosynthesis is able to increase yield [1-2]; there are many researchers to try to put maize seeds manually to be fixed and achieve orientational sowing, but the seeding efficiency is too low to suit for large-scale operations. In order to improve the orientational seeding efficiency and effect, it has to find a way for the seed orientation and fixed in the soil to adopt the combination of mechanization and automation, so the key technology is to achieve orientation of maize seeds. Because of its unique and irregular appearance characteristics of maize seed, it is more difficult to achieve

---

[*] Corresponding author.

D. Li and Y. Chen (Eds.): CCTA 2013, Part II, IFIP AICT 420, pp. 1–9, 2014.

orientation of maize seeds. Therefore, the image processing technology can be used to identify the embryo of the maize seed.

The image processing technology has been widely used in agricultural products inspection and grading, such as cereal grain color, grain shape and type of identification, fruit shape and defects [3-9] and other exterior quality inspection, but the application of maize seed orientation for mechanized seeding research rarely reported. In this paper, the embryo and position of maize were obtained by the image processing technology to provide technical support for subsequent mechanized orientational planting.

2   Image Processing and Inspection Methods

According to the situation of the maize sowing into the soil, there are 13 kinds of position and orientation relative to the row direction [10], but only four kinds of that are easy to implement mechanized planting, which are perpendicular or parallel to the row for embryo up and down (Fig.1). To determine the position and direction of maize seeds, each seed has a fixed orientation and position. This paper adopted Zheng-958, Jundan-20 and Zhongke-11 maize seed for research object, the direction radicle of the maize seed deflection angle was taken for the positioning standard in long axis direction, the embryo surface and obverse were distinguished through the embryo color image processing, this paper mainly located the maize seed classify feature and color feature detection.

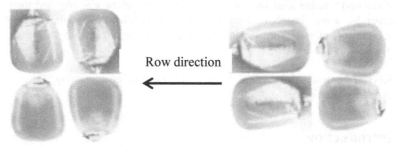

Row direction

(a) Embryo up and down, perpendicular to the row (b) Embryo up and down, parallel to the row

Fig. 1. Maize seed position and direction

2.1   Image Preprocessing

The greyscale image is divided into two kinds of A and B type based on Otsu segmentation method, the probability of occurrence and mean gray level of A and B type can be calculated, after that the inter-class variance can be calculated also, so the maximum variance was chosen as the best threshold. The background gradient of RGB are removed based on Otsu method, the noise was suppressed by using image smoothing operators, the result of applying a 3×3 median filter to the image and converting from RGB mode to HSV in Fig.2(b), the gray level distribution range was transform to [0-255] [11], and the H, S, V three single channel were extracted from HSV channel in Fig.2(b), otherwise, it were preprocessed with binarization processing and the morphological noise reduce processing.

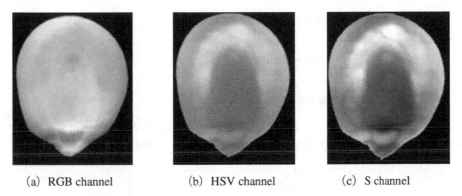

(a) RGB channel   (b) HSV channel   (c) S channel

Fig. 2. Image color channel analysis of maize seed

2.2 The Direction Inspection Method for Maize Seed

Most crop seeds have end "tip" in the direction of the long axis, and the other end is relatively flat, for example: garlic, pumpkin, and sunflower. For such seed direction discrimination method, the image edge of garlic clove pointed position corner was found using SUSAN detection algorithm, and to identify garlic seed direction [12]. The same method has been used in Reference [13] and Reference [14], the seed orientation is judged by scanning seed edge on fixed area and comparing both directions of the seed edge pixels number within the region. This method has the advantages of high success rate of identification easy to control, and can distinguish the general orientation, but it can't accurately quantify the deflection angle of the seed direction.

This paper adopted the best aspects of several algorithms, the seed head was distinguished by the area scanning method, the maize seeds and deflection angle were analyzed and distinguished with the contour curvature. The edge of preprocessed image was detected by using Candy operator, the maximum outer contours of the edges image of maize seed to be found and saved in $\{S_i\}$, the contour point of maximum curvature was found by the $[S_i, S_{i-k}]$ and $[S_i, S_{i+k}]$ of the inner product of two vectors, this point is the expectation, k is the precision of calculations, the definite steps is shown in Fig.3. Because the maximum curvature was on the terminal of contours, the minimal result of inner product is satisfactory. The maximum curvature point of the contour through round-robin comparison was turned into point of maize cusp. The straight line that connected to the maximum curvature point P_1 and corn contour centroid point P_0, that was regarded as the axis of the corn, the angle θ between the axis and the vertical line was calculated as a reference in a vertical direction, it is shown by formula (1), the direction will been determined by judging the size of θ, the inner product of vectors is shown in Fig.4 and the detection result is shown in Fig.5.

$$\theta = \cos^{-1} \frac{\overline{S_i S_{i+1}} \cdot \overline{S_i S_{i-1}}}{\left|\overline{S_i S_{i+1}}\right| \left|\overline{S_i S_{i-1}}\right|} \qquad (1)$$

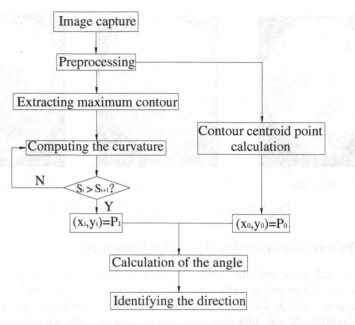

Fig. 3. Flow chart of Seeds direction detection algorithm

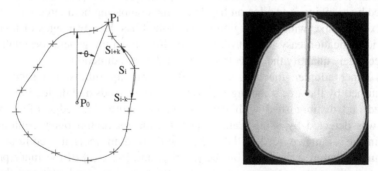

Fig. 4. Inner product of vectors   **Fig. 5.** Detection result

2.3   The Embryo Inspection Method for Maize Seed

As shown in Fig.6, the mass center of maize seeds for the grey value was obviously on the high side after binarization in the S channel for embryo up, but it has no obvious change for the embryo down, due to the influence that the surface has no embryo. The collected images were transformed from RGB to HSV channel, and its S channel was extracted to conduct gradation processing [15]. Binarization processing was done according to the threshold determined by Otsu method. A rectangular ROI region was defined, which central was the maize contour centroid, the length of the half major axis was regarded as the long side of the rectangle, the length of the half minor axis was regarded as the short side, so the ROI region was scanned progressively and calculated the number of the pixels which gray value was 255, The result P was tested according to the preset threshold T_M, the calculating process was shown in Fig.7; P is calculated by formula (2):

$$P = \begin{cases} 1 \ T = \left(\sum_{i=i_0}^{i=i_k} \sum_{j=j_0}^{j=j_k} M\left(i, j\right) = 255 \right) / S > T_M \\ 0 \ T = \left(\sum_{i=i_0}^{i=i_k} \sum_{j=j_0}^{j=j_k} M\left(i, j\right) = 255 \right) / S \le T_M \end{cases} \tag{2}$$

M (i, j) is the grey value of point (i, j), T is the rectangular area ratio of black pixels inside the rectangular area, S is the area of the rectangle, the point (i_0, j_0) and the point (i_k, j_k) are the diagonal point of the rectangle.

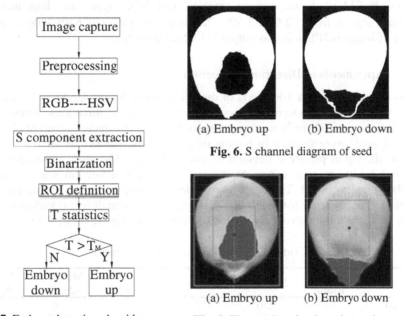

(a) Embryo up　　　　(b) Embryo down

Fig. 6. S channel diagram of seed

(a) Embryo up　　　　(b) Embryo down

Fig.7. Embryo detection algorithm　　　**Fig. 8.** The results of embryo inspection

If the P = 1, the maize embryo is upward; else P = 0, maize embryo is downward. Each 100 seeds of three varieties were selected: Zheng-dan958, Jun-dan20 and Zhong-ke11 for the experiment test. The T value will be to calculate according to the above method under the maize embryo was upward and downward, and the distribution of T value was obtained as shown in Fig.9. The T value was more than 0.45 when maize embryo was upward, while the T value is less than 0.15 in commonly for downward, The T is 0.4 to be the threshold of T_M and judge the orientations of embryo surface.

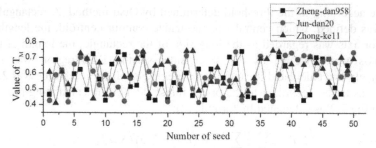

Fig. 9. TM scatter diagram

3   Experiments and Result Analysis

The position and orientation of maize seed are determined, which needed to the full surface contours and color information. The image has been taken by automatic Exmore R CMOS digital camera (SONY, DSC-WX7 type), the lens maximum aperture range is from F2.6 to F6.3, The image resolution is 640×480pix, and the format of image is JPEG, it was output via USB2.0 interface.

3.1   Experiments of Direction Inspection

In this experiment, each 100 seeds of the Zheng-dan958, Jun-dan20 and Zhongke11 maize varieties were chosen and divided into two kinds of directions. Among them, the radicle tips of 50 seeds were pointed to the top of the picture, the radicle tips of other 50 seeds were pointed to bottom of picture. The number of seed with radicle point to the top of picture is decided to qualified index. The result of inspection was achieved by image processing. At the same time, the deflection angle value was calculated and recorded. The test was repeated for three times, the results are shown in table 1. The accuracy for the direction inspection of three different variety seed was above 98%, the minimum average value of the deflection angle was 1.8 degree.

Table 1. The result of direction inspection

Factor	Time/ Result	Upward	Downward	Maximum (degree)	Minimum (degree)	Average (degree)
Manual		50	50	0	0	0
Zheng-dan958	1 time	50	50	3.5	1.8	2.5
	2 time	49	50	3.4	2.2	2.7
	3 time	50	49	3.5	1.8	2.2
	Accurate rate	99%	99%			
Jun-an20	1 time	50	50	3.8	2.1	2.8
	2 time	50	50	3.7	2.4	2.6
	3 time	49	50	3.5	1.9	2.4
	Accurate rate	99%	100%			
Zhong-ke11	1 time	49	49	3.3	1.7	2.3
	2 time	50	50	3.4	2.2	2.8
	3 time	49	50	3.6	2.4	2.9
	Accurate rate	98%	99%			

3.2 Experiments of Embryo Orientation Inspection

In the same way, each 100 seeds of the Zheng-dan958, Jun-dan20 and Zhongke11 maize varieties were chosen and divided into two kinds of embryo up and down, each test were chosen 50 seeds. It was tested same with the above method, the results were shown in table 2, the accurate rate of Zheng-dan958 for embryo up was 98%, the Jun-dan20 was 93%, the Zhong-ke11was 93%. In addition, the accurate rates for embryo down respectively were 99%, 98% and 95% respectively.

Table 2. The result of embryo orientation inspection

Factor	Time/Result	Embryo up	Embryo down
Manual		50	50
	1 time	47	49
	2 time	45	50
Zheng-dan958	3 time	50	50
	Accurate rate	94%	99%
	T average value	0.68	0.12
	1 time	48	50
	2 time	49	48
Jun-dan20	3 time	50	50
	Accurate rate	98%	98%
	T average value	0.59	0.13
	1 time	45	50
	2 time	46	47
Zhong-ke11	3 time	49	46
	Accurate rate	93%	95%
	T average value	45	50

4 Conclusions

Maize is the second large crop in China. The orientational planting cannot make maize leaf growing consistently, but also improve the maize leaf photosynthetic capacity of per unit area and yield of maize. So the extraction of maize embryo and orientation method based on image processing has been conducted in this paper. The automatic extracting maize seeds embryo and orientation information has been realized through image color channel conversion, segmentation, preprocessing, and its contour characteristic analysis. Through three maize varieties experiments, the conclusions are shown in follows:

A contour curvature analysis method was put forward to determine the maize seed "tip" direction, the maize seed "tip" direction was detected and the angle of deflection was measured using contour curvature analysis method.

The S channel of HSV image was preprocessed by Otsu method, a rectangular region ROI in the picture was defined and the pixels within the region had been calculated, the region-specific ROI pixels of the positive and negative maize seed were compared by T_M threshold, so the maize seed embryo side towards had been distinguished.

This paper adopted Zheng-958, Jundan-20 and Zhongke-11 maize seed for research object, each variety was repeated three times by using the above methods, the results showed that the average accuracy of embryo inspection was more than 95%; the direction average angle was 2.2 °.

Acknowledgment. The fund for this research was provided by Specialized Research Fund for the Doctoral Program of Higher Education (20120008110045) and the Fundamental Research Funds for the Central Universities (2012YJ105).

References

1. Pandey, A.K., Khatoon, S.: Effect of orientation of seed placement and depth of sowing on seedling emergence in Sterculia urens Roxb. Indian Forester 125(7), 720–724 (1999)
2. Hou, Y., Xu, L., Chen, L.: The Current Situation and Development Trend of Corn Mechanization Oriented Seeding Technology. Journal of Agricultural Mechanization Research (2), 10–14 (2012) (in Chinese)
3. Yarnia, M., Tabrizi, E.F.M.: Effect of Seed Priming with Different Concentration of GA3IAA and Kinetin on Azarshahr Onion Germination and Seedling Growth. J. Basic. Appl. Sci. Res. 2(3), 2657–2661 (2012)
4. Ming, S., Yiming, W., Yun, L., et al.: A hue based detecting approach to yellow rice kernel. Transanctions of the Chinese Society for Agricultural Machinery 36(8), 78–81 (2005) (in Chinese)
5. Liao, K., Marvin, R., Paulsen, M.R., et al.: Rea-l time detection of color and surface defects of maize kernels using machine vision. Journal of Agricultural Engineering Research 59(4), 263–271 (1994)
6. Cheng, H., Shi, Z., Yao, W., et al.: Corn breed recognition based on support vector machine. Transactions of the Chinese Society for Agricultural Machinery 40(3), 180–183 (2009) (in Chinese)
7. Yang, S., Ning, J., He, D.J.: Identification of corn breeds by BP neural network. Journal of Northwest Sc-I Tech. University of Agriculture and Forestry 32(Supp.), 189–192 (2004) (in Chinese)
8. Ying, Y., Cheng, F., Ma, J.: Rea-l time size inspection of citrus with minimum enclosing rectangle method. Journal of Biomathematics 19(3), 352–356 (2004) (in Chinese)
9. Ying, Y., Jing, H., Ma, J., et al.: Application of machine vision to detecting size and surface defect of Huanghua pear. Transanctions of the Chinese Society of Agricultural Engineering 15(1), 197–200 (1999) (in Chinese)
10. Chen, Y., Liao, T., Lin, C., et al.: Grape inspection and grading system based on computer vision. Transactions of the Chinese Society for Agricultural Machinery 41(3), 169–172 (2010) (in Chinese)
11. Jiang, G., Han, Y., Wang, Y., et al.: Directional and Precision Sowing Techniques of Corn. Agricultural Engineering 2(2), 17–20 (2012) (in Chinese)

12. Xu, L.: Randomized Hough transforms (RHT): Basic mechanisms, algorithms, and computational complexities. CVGIP: Image Understanding (57), 131–154 (1993)
13. Ning, J., He, D., Yang, S.: Identification of tip cap and germ surface of corn kernel using computer vision. Transactions of the CSAE 20(3), 117–119 (2004) (in Chinese)
14. Yang, Q., Li, J., He, R.: Direction identification of garlic seeds based on image processing. Acta Agriculturae Zhejiangensis 22(1), 119–123 (2010) (in Chinese)
15. Wang, H., Sun, Y., Zhang, T., et al.: Appearance Quality Grading for Fresh Corn Ear Using Computer Vision. Transactions of the Chinese Society for Agricultural Machinery 41(8), 156–158 (2011) (in Chinese)

Greenhouse Irrigation Optimization Decision Support System

Dongmei Zhang[1], Ping Guo[1], Xiao Liu[1], Jinliang Chen[1], and Chong Jiang[2]

[1] College of Water Resources & Civil Engineering,
China Agriculture University, Beijing, 100083, China
[2] School of Computer Science and Technology, Harbin Institute of Technology,
Harbin, Heilongjiang, 150001, China
{azhang_d_m,c0319xiaoxiao}@163.com, bguop@cau.edu.cn,
dchenjinliang1123@126.com

Abstract. Greenhouse irrigation optimization decision support system (GDSS) is developed for the irrigation management of greenhouse crops in the northwest arid area of China. GDSS forecasts on relative yield and aids to develop irrigation schedule in terms of growth periods, comprehensively considering soil, crops and water supply conditions. The system consists of three modules. The database module stores all kinds of data using the Access database. The model module includes optimization models of greenhouse crops under insufficient irrigation based on uncertainty. These models are programmed by lingo, which can be invoked through internal interface. The man-machine dialogue module is designed with the principal of user control, user-friendly, visuality, usability, conciseness and uniformity. The GDSS can provide the decision makers the alternative decision making under uncertainty.

Keywords: Decision support systems, Greenhouse irrigation optimization, Man-machine dialogue, Uncertainty.

1 Introduction

In the northwest of China, the available water resources is less than 2200 m^3 per capita, only one quarter of the world average level[1]. The agricultural water consumption accounts for approximately 70% of the total water uses. Improving irrigation management is most likely the best option in most agricultural systems for increasing the efficiency of water using so as to mitigate the shortage of water resources [2]. At present, an effective method to improve management efficiency is applying the modern technology particularly Computer and Database in agricultural irrigation management.

As efficient tools for improving management efficiency, Decision support systems (DSSs) are relatively new disciplines that have emerged from the development of earlier management information systems (MIS) which are data oriented and, for the most part, simply a means of retrieving data from large databases grounded on selected queries. This new discipline focuses on the design and development of DSSs,

D. Li and Y. Chen (Eds.): CCTA 2013, Part II, IFIP AICT 420, pp. 10–23, 2014.

while at the present time there is a solid conceptual footing and increasing number of applications that demonstrate their importance and efficiency in aiding management [3]. DSSs involve computer software and hardware, information theory, artificial intelligence, management science, and many other disciplines. DSSs are used to solve semi-structured and un-structured problems that cannot normally be expressed in unambiguous formulas [4]. Besides, DSS can effectively improve the decision-making ability of managers, as well as enhance the scientific of decision-making and the degree of information, especially for complicated management systems such as sustainable planning systems of rural area and irrigation management system.

A series of decision support systems for resource management have been developed, regarding the comprehensive planning of socioeconomic development and eco-environment protection, [3,5,6]. With the extensive application of uncertainty methods in optimization management, more and more uncertainty models have been put into above management systems, for example, UREM-IDSS[5] has been developed based on an inexact optimization model to aid decision makers in planning energy management systems. In the field of irrigation, [6] developed an integrated scenario-based multi-criteria decision support system (SMC-DSS) for planning water resources management in the Haihe River Basin. [7] developed the software for water-saving irrigation management and decision support system in the light of the complexity and real-timeliness of water use management in farmland irrigation.

In the northwest arid area of china Greenhouse has been widely applied because of the characters as energy-saving and manageable, hence, study on DSSs for optimal management of greenhouse irrigation is significant to alleviate water scarcity. [8]designed an expert system for mini-watermelon culture management in greenhouse based on the growth model was developed, the system was designed to help agronomists and famers to make strategic and tactical decisions. [9] introduced a decision support system for greenhouse constructed with data warehouse and data mining technology considering most of expert knowledge in agriculture is descriptive and experiential.

In the aspect of greenhouse irrigation optimization, A lot of experiments and research have been carried out, the theories of water consumption efficiency pattern, water use and optimal irrigation schedule of tomato[10], watermelon[11],muskmelon and hot pepper[12]of greenhouse in arid northwest China are pretty mature. However the application of these results is relatively limited because the knowledge is too complex for most of decision makers and a large part of theories are aimed at a given area and specific crops.

The above-mentioned research results indicated applications of DSSs and uncertainty methods in the field of water resource management are significant and commendable. A majority of useful and valuable theories regarding irrigation water management and allocation should be applied more widely. In this paper, consequences in the form of interval would recommended to decision makers as a result of applying uncertainty methods to a crop-water production function-Jensen model. Moreover, a greenhouse irrigation decision support system developed in this study is developed which is a further application of existing associated irrigation theories of water consumption efficiency pattern, water use and optimal irrigation schedule ,and is able to provide alternative decision makings scientifically and comprehensively for decision makers. Through reducing less important variables and keep relative significant variables as well as updating experimental data of representative areas, GDSS is applicable for the northwest arid area of china including where lack measured data.

The aim of this study is to build a common uncertainty based irrigation optimization model for greenhouse crops on the basis of existing associated irrigation theories, use the actual greenhouse irrigation experiment data and corresponding analysis results as references, and develop a greenhouse irrigation decision support system to provide alternative irrigation schedules for decision makers through comprehensive scenario analysis and an user-friendly graphical user interface (GUI).

2   System Design

Greenhouse irrigation optimization decision support system (GDSS) is developed for the irrigation management of greenhouse crops in the northwest arid area of china. GDSS forecasts relative yield and aids to develop irrigation schedule in terms of growth periods, comprehensively considered soil, crops and water supply conditions. The system contains meteorological data outside the greenhouse and all kinds of experimental data inside the greenhouse in Shiyanghe River Basin from 2008 to 2011.GDSS is based on the interval optimization model of greenhouse crop under insufficient irrigation, which figure out relative yield interval and water consumption interval of each growth period, and recommend an optimal irrigation procedure according to the soil and crop parameter and water supply conditions input by the users. Furthermore, scenario analysis of a series of information include the relative yield, water consumption and irrigation schedule according to the experimental data and User's input will presented thus to provides decision makers a clear comparison between all the circumstances so as to aid decision making efficiently. Fig1 demonstrates the system architecture.

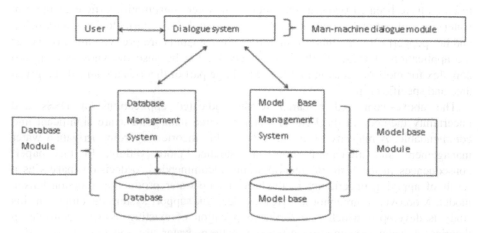

Fig. 1. System architecture

3 Database Module

The database module is responsible for the retrieval, updating and visualization of the information required, and is composed of a database and a database management system. The database is designed and implemented to represent all the relevant data, while the data management system comprises the software required to create, access and update the database [13].

All these data can be divided into two classes, one is static data from Shiyanghe River Basin, which is stored in Access database, and exit in the form of data table of data set. Consequently, database operations can be performed even disconnecting from the database. We use ReportViewer control to display chart, line graph and histogram in graphic user interface. Another data class is dynamic data consist of user input and calculated results. Data structure is showed as Fig2.

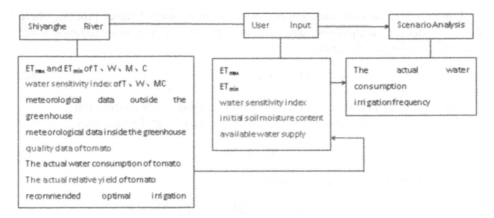

Fig. 2. Data structure

4 Model Module

4.1 Overview of Model

The optimization model of greenhouse crop under insufficient irrigation(OMGII) is based on the crop-water production function-Jensen model, which is structure logical and has been widely used in arid area in China [14]. With the consideration of uncertainty exit in formulating the insufficient irrigation schedule of greenhouse crops, the maximal evapotranspiration(ET_{max})and minimum evapotranspiration(ET_{min}) are set as interval value, therefore, the consequences of this model are interval number, which provides users with a reference range. The final model is as follows:

Object function :

$$\max f^{\pm} = \frac{Y^{\pm}}{Y_{\max}^{\pm}} = \prod_{i=0}^{n} \left(\frac{ET_i^{\pm}}{ET_{\max i}^{\pm}} \right)^{\lambda_i} \tag{1}$$

The object of OMGII is to maximum the relative yield. Where Y is actual yield, Y_{max} is the yield under sufficient irrigation, n is growth stages, a stage variable, ET_i is the actual water consumption of each growth stage, mm, ET_{maxi} is the maximal water consumption, mm, λ_i is water sensitivity index,

Subject to:

$$W = 1000\gamma H_i \left(\theta_i - \theta_{min} \right) \qquad (2)$$

Where W_i is the effective water available for crop at stage i, a state variable, mm, H_i is the depth of design root zone, mm, θ_i is average soil water content, g/g, θ_0 is the initial soil moisture content θ_{min} is the lower limit of soil water content, g/g, γ is the soil dry density, g/cm^3.

$$ET_i^{\pm} = W_i - W_{i+1} + m_i^{\pm} + P_i + G_i \qquad (3)$$

Equation (3) is a state transition equation indicating water balance process, where P_i is the effective precipitation, mm, G_i is irrigation quota, mm, m_i is irrigation water amount, decision variable, mm.

$$Q_{i+1}^{\pm} = Q_i^{\pm} - m_i^{\pm} \qquad (4)$$

Equation (4) is a state transition equation indicating water distribution process, where Q_i is increment of ground water, mm.

$$m_i^{\pm} \leq Q_i \qquad (5)$$

Equation (5) is an irrigation quota upper limit equation.

$$ET_{min\,i}^{\pm} \leq ET_i^{\pm} \leq ET_{max\,i}^{\pm} \qquad (6)$$

Equation (6) indicates the bound of actual water consumption.

$$\theta_{min} \leq \theta_i \leq \theta_f \qquad (7)$$

Equation (7) indicates the bound of average soil water content.

$$m_i^{\pm} \geq 0, \forall i \qquad (8)$$

Equation (8) is the variable nonnegative constraint.

4.2 Model Linkage

Above model is programmed by lingo, which supports programming in the form of dynamic linking library. On the basis of different number of growth period, two lingo programs are written. After sufficient case tests, by changing the file input output function to @pointer(n) and corresponding address defined in C#, it is able to transmit data from shared memory directly.

5   Man-Machine Dialogue Module

5.1   System Flow

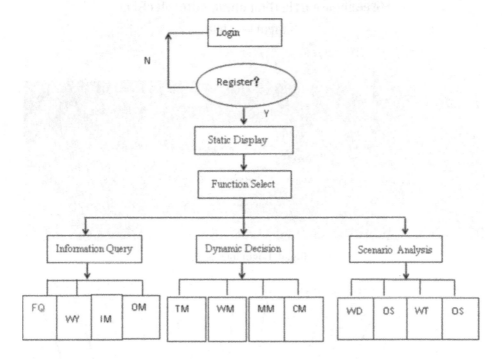

Fig. 3. System flow

5.2   GUI and Functions

By double-clicking the icon of the GDSS, the user can enter the system's login interface (Fig 4). After register an ID and check username and password, the user can enter Static display interface (Fig 5). Technical route, system function, framework, system flow, model can be displayed by clicking on buttons in the left side of static display interface.

Fig. 4. Login interface

Fig. 5. Static display interface

Users can access a function selection interface (Fig 6 by clicking on Function Selection button, this interface consists of three buttons corresponding three important functions of GDSS.

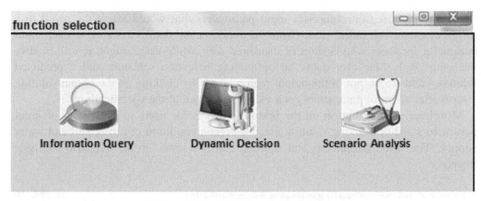

Fig. 6. Function selection interface

(1) Information Query

With the selection of Information Query, there entrance an information query interface (Fig 7), there are four parts as showed in Fig7, each of them includes two areas. The left area provides user Listbox for select time or district and checkbox for multiple selection of correlation parameters, as well as buttons for select display format of data on the report area on the right side. In addition, the report area on the right side has export capabilities for export useful data.

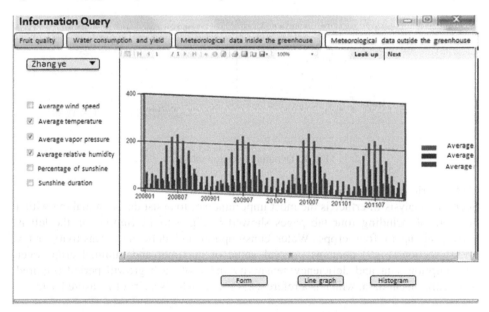

Fig. 7. Information query interface

(2) Dynamic Decision

The dynamic decision interface is able to help formulating optimizing irrigation schedule of four crops as Fig 8 showed after above dynamic linking. For each crop, the upper input region comprises input parameters that would affect the result to a different extent. An initial value (from the experimental data)is assigned to each input parameter for users who is short of measured data, while users might as well modify the input with their owe data. An optimizing irrigation schedule and a predicted relative yield are output in the output region below by clicking on the obtain solution button after all input parameters get a value conformed to the specification.

Moreover, a Save button on the bottom panel is for users to save a set of input parameters and optimizing solution under a certain condition of crops, soil and water supply. The saved parameters and solution can be viewed in the Scenario Analysis interface.

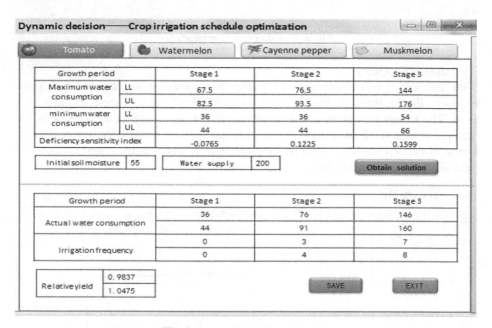

Fig. 8. Dynamic decision interface

(3) Scenario Analysis

Scenario analysis interface is the most important part to assist decision making with a tab control including four tab pages showed in Fig 9 to 12 Buttons on the left is corresponding to four crops. Water consumption and deficiency sensitivity index (WD) section (Fig9) contains interval value of maximal and minimal crop water consumption data and deficiency sensitivity index of each growth period acquired from experiment data, which is a reference for users who is short of measured data.

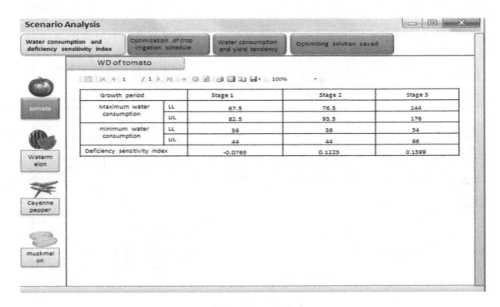

Fig. 9. WD section

Optimization of crop irrigation schedule (OS) (Fig 10)and water consumption and yield tendency (WT) section (Fig 11) consist of optimizing solutions calculated from the reference data. Three ReportViewers have been added to both of them, the upper one is used to display chart of optimizing irrational schedule and optimizing relative yield, while the lower two is for display line graph and histogram. In OR section, the lower report will display the range of actual water consumption at a certain water supply condition by choosing water supply and clicking show chart button, and the variation range of actual water consumption with the increase of water supply at a certain growth period will be revealed after choosing a growth period and clicking above button in the WT section.

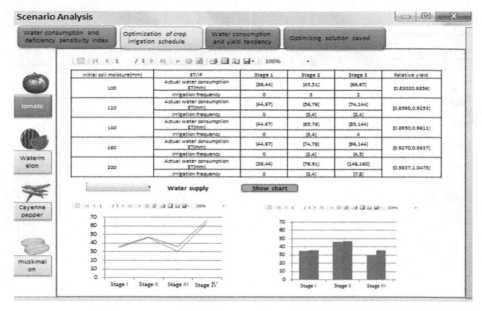

Fig. 10. OR section

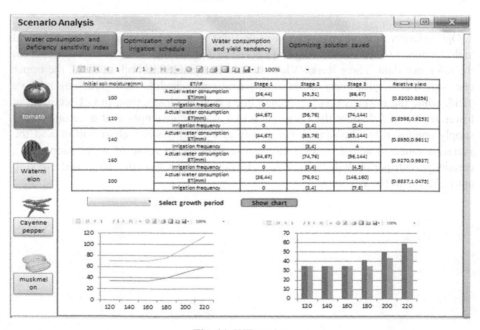

Fig. 11. WT section

The last section (Fig12) is for user to check the data that user saved before in the dynamic decision interface. User can contrast all the saved optimizing solution and delete useless data, thereby, GDSS helps decision maker to make a wise decision.

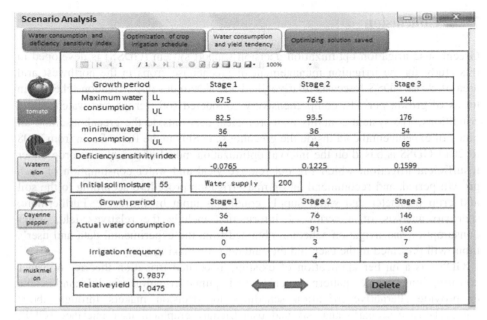

Fig. 12. OS section

6   Result Analysis and Discussions

Generally, the developed GDSS could be used to support greenhouse irrigation management in the northwest arid area of china. Through incorporating advanced database system, optimization technique, uncertainty method and User-friendly graphic user interface, a series of practical and scientific irrigation schedule under various circumstances could be recommended. The basic fruit quality, water consumption and yield, and meteorological data, compiled in the database system were collected from many greenhouse experiments in Shiyanghe River Basin. These first-hand data reflected the actual relation between irrigation schedule and fruit parameters and provided valuable references to districts lack of measured data. A dynamic decision function is developed for obtaining a recommended solution presented in the form of interval through a dynamic linking with lingo program of OMGII. The input parameters should be assigned according to the local actual data. In the last stage of the whole decision, users enter scenarios analysis interface, compare all the alternative solutions intuitively.

In future research, more valuable data such as experiment results from other district should be added into the database to provide more references. Moreover, the irrigation optimization model under uncertainty should consider more uncertainty factors besides water consumption thereby obtaining more precise solution.

7   Conclusion

Greenhouse irrigation optimization decision support system (GDSS) is developed in this study for the irrigation management of greenhouse crops in the northwest arid area of china. GDSS forecasts relative yield and aids to develop irrigation schedule in terms of growth periods, comprehensively considered soil, crops and water supply conditions. The system contains meteorological data outside the greenhouse and all kinds of experimental data inside the greenhouse in Shiyanghe River Basin from 2008 to 2011.GDSS is based on the interval optimization model of greenhouse crop under insufficient irrigation, which Fig out relative yield and water consumption of each growth period, and recommend an optimal irrigation procedure according to the soil and crop parameter and water supply conditions input by the user. Furthermore, scenario analysis of a series of information include the relative yield, water consumption and irrigation schedule obtained from the experimental data and user's input will presented to the user and thus aid decision making.

GDSS is a further application of existing associated irrigation theories of water consumption efficiency pattern, water use and optimal irrigation schedule ,and is able to provide alternative irrigation schedules for decision makers through above comprehensive scenario analysis and user-friendly graphical user interface (GUI). GDSS is convenient for users even those who are lack of computer programming or system modeling knowledge. Thus, users can concentrate on developing and comparing alternative irrigation schedule.

Acknowledgment. This research was supported by the National Natural Science Foundation of China (No. 41271536,), Government Public Research Funds for Projects of Ministry of Water Resources (No.201001061), International Science & Technology Cooperation Program of China (No. 2013DFG70990).

References

1. Li, W., Li, Y.P., Li, C.H., Huang, G.H.: An inexact two-stage water management model for planning agricultural irrigation under uncertainty. Agr Water Manage 97(11), 1905–1914 (2010)
2. Jensen, C.R., Battilani, A., Plauborg, F., et al.: Deficit irrigation based on drought tolerance and root signalling in potatoes and tomatoes. Agr Water Manage 98(3), 403–413 (2010)
3. Huang, G.H., Qin, X.S., Sun, W., Nie, X.H., Li, Y.: An optimisation-based environmental decision support system for sustainable development in a rural area in China. Civ. Eng. Environ. Syst. 26(1), 65–83 (2009)
4. Ji, X., Kang, E., Chen, R., Zhao, W., Xiao, S., Jin, B.: Analysis of water resources supply and demand and security of water resources development in irrigation regions of the middle reaches of the Heihe River Basin, Northwest China. Agricultural Sciences in China 5(2), 130–140 (2006)
5. Cai, Y.P., Huang, G.H., Lin, Q.G., Nie, X.H., Tan, Q.: An optimization-model-based interactive decision support system for regional energy management systems planning under uncertainty. Expert Syst. Appl. 36(2), 3470–3482 (2009)

6. Weng, S.Q., Huang, G.H., Li, Y.P.: An integrated scenario-based multi-criteria decision support system for water resources management and planning–A case study in the Haihe River Basin. Expert Syst. Appl. 37(12), 8242–8254 (2010)
7. Chen, F.Z., Song, N., Wang, J.L.: Water-saving irrigation management and decision support system. Transaction of the CSAE (S2), 1-6 (2009)
8. Xu, G., Guo, S., Zhang, C., et al.: Expert system for mini-watermelon culture management in greenhouse based on the growth model. Transactions of the Chinese Society of Agriculture Engineering (04), 157–161 (2006)
9. Wang, C., Li, M., Wang, L., et al.: Decision support system for greenhouse based on data warehouse and data mining. Transactions of the Chinese Society of Agriculture Engineering (11), 169–171 (2008)
10. Wang, F., Kang, S., Du, T., Li, F., Qiu, R.: Determination of comprehensive quality index for tomato and its response to different irrigation treatments. Agr Water Manage 98(8), 1228–1238 (2011)
11. Hu, Z., Tian, X., Ma, Z., Bao, X., Zhang, J.: Research on High Yield and Efficient Water-saving Planting Mode in Shiyang River Basin. Water Saving Irrigation 1, 11 (2011)
12. Chen, P., Du, T., Wang, F., Dong, P.: Response of Yield and Quality of Hot Pepper in Greenhouse to Irrigation Control at Different Stages in Arid Northwest China. Scientia Agriculture Sinica (09), 3203–3208 (2009)
13. Chen, Y., Jiang, Y., Li, D.: A decision support system for evaluation of the ecological benefits of rehabilitation of coal mine waste areas. New. Zeal J. Agr. Res. 50(5), 1205–1211 (2007)
14. Jiao, Y., Luo, Y., Li, Y.: Effect of Stochastic Error of Sensitivity Indexes of Jensen's Model on Optimal Water Consumption. Chinese Agricultural Science Bulletin (2011)

Study on Pear Diseases Query System Based on Ontology and SWRL

Qian Sun and Yong Liang

School of Information Science and Engineering,
Shandong Agricultural University, Taian, 271018, China
{applesq,yongl}@sdau.edu.cn

Abstract. This paper studied the construction of Pear Diseases Domain Ontology (PDDO), and the realization of query system based on PDDO and SWRL. First, an approach to build PDDO based on SWRL was proposed, which consists of confirming core concepts, adding the properties of concepts and the relationships between concepts, adding the instances of concepts, representing domain ontology, adding SWRL rules and reasoning. Then the query system model and implementation algorithm were given. The query system, which integrates SWRL with Jess reasoning engine based on the SWRL rules, realized disease query, instance query, and diagnosis query by Protégé-OWL API. The query system realized excavating implicit relationships and renewing PDDO relative to previous system. Through the query system based on reasoning pear diseases knowledge can be obtained from PDDO according to user needs, furthermore, new and inferred knowledge are written to the PDDO owl file.

Keywords: Protégé, SWRL, Protégé-Owl API, pear; ontology, query, JESS.

1 Introduction

With the development of owl ontology language, more and more knowledge systems based on domain ontology are developed. The development of knowledge system includes knowledge representation, storage, reasoning, query and so on, in which query and reasoning are the key technologies, through them knowledge can be obtained from ontology [1]. In order to realize pear diseases knowledge query based on reasoning, the query system based on PDDO and SWRL is studied in this paper.

2 Technologies and Tools on Ontology

2.1 Protégé

Protégé is a open source ontology editor, which allows user to model ontology. The Protégé platform provides two main ways of modeling ontology via the protégé-frames and protégé-owl editors. Protégé ontology can be exported into a variety of formats including OWL [2] ,RDF(S) [3], and XML Schema. Further more, Protégé

D. Li and Y. Chen (Eds.): CCTA 2013, Part II, IFIP AICT 420, pp. 24–33, 2014.

can be extended by way of a plug-in architecture and a Java-based Application Programming Interface (API) for building knowledge-based tools [4].

2.2   Protégé-OWL API

The Protégé-OWL API is an open source Java library for the Web Ontology Language and RDF(S). The API provides classes and methods to load and save OWL files, to query and manipulate OWL data models, and to reason. Furthermore, the API is optimized for the implementation of graphical user interfaces [5].

2.3   SWRL

SWRL is a Semantic Web Rule Language based on a combination of the OWL DL and OWL Lite sublanguages of the OWL Web Ontology Language with the Unary/Binary Datalog RuleML sublanguages of the Rule Markup Language. SWRL includes a high-level abstract syntax for Horn-like rules in both the OWL DL and OWL Lite sublanguages of OWL. A model-theoretic semantics is given to provide the formal meaning for OWL ontology including rules written in this abstract syntax [6].

2.4   JESS

Jess is a Java-based rule engine. Jess system consists of a rule base, fact base, and an execution engine. It has been used in Protégé-based tools, e.g. SWRLJessTab[7], SweetJess, JessTab.

3   Construction of Pear Diseases Domain Ontology Based on SWRL

At present, ontology construction methodologies have not been standardized, there are numerous frequently quoted approaches. In this paper, an approach for building domain ontology based on SWRL is proposed. The process of it consists of the following phases:

3.1   Confirming Core Concepts

In this case, core concepts of pear diseases domain are collected according to features needed by diseases diagnosis and pear diseases. Firstly, "Pear-tree" is selected as the first concept, and then the concepts relating to "Pear-tree" are selected. Table1 gives the names and explanations of these concepts.

Table 1. The general concepts of PDDO

Name		Explanation
Disease		Refers to the categories of the pear diseases
Growing-period		Reflects the time that pear elapses from seeding to mature
Part		Refers to the diseased parts of pear, instances: root, fruit
Pathogen	Feature	Reflects the features of pathogens. Instance: black small point
	P-kind	Refers to the kinds of the pathogens that cause pears fall ill
Pear-tree		Refers to pears
Symptom	Color	Reflects the variety of colors that the diseased parts of pears change to
	Shape	Reflects the variety of shapes of the spots.
	Dynamic-symptom	Reflects the symptoms of the diseased pears , instance: rotting.

3.2 Adding the Properties of Concepts and the Relationships between Concepts

In this case, every core concept has different properties to be added, for example: "describe" is added as the property of "disease" to describe the features of diseases. In addition, the relationships between concepts are added. Table2 gives the relationships between "Pear-tree" and other concepts.

Table 2. The relationships between "Pear-tree" and other concepts

Name	Explanation
At-part	Reflects the relationship between "Pear-tree" and "part"
Has-ds	Reflects the relationship between "Pear-tree" and "Dynamic-symptom"
Has-c	Reflects the relationship between "Pear-tree" and "Color"
Has-disease	Reflects the relationship between "Pear-tree" and "Disease"
Has-shape	Reflects the relationship between "Pear-tree" and "Shape"
Has-feature	Reflects the relationship between "Pear-tree" and "Feature"
Has-pathogen	Reflects the relationship between "Pear-tree" and "P-kind"
At-period	Reflects the relationship between "Pear-tree" and "Growing-period"

3.3 Adding the Instances of Concepts

It is necessary for building domain ontology to supply the instances of concepts. For example: in this case, "change color" , "die", "dry-rot", " falling-off", " putrescence" , "rotting", "spotting", "wilting " are added as instances of "Dynamic-symptom". "branch", " fruit", "fruit-stem", "leaf", and "root" are added as instances of "Part". Names of pear diseases are added as instances of "Disease", for example:. "Steptomyces-scabis", "Pear-Brown-blight", "Pear-Rust", "Pear-Valsa-Canke",

"Pear-anthracnose", "Pear-blackPedicle-disease",   Pear-black-shank", " Pear-black-spot", "Pear-blight" and so on.

3.4   Representing Pear Diseases Domain Ontology

In this case, protégé 3.4.8 is selected as the developing tool, so the PDDO can be represented by OWL. Further illustrate below:

Firstly, according to the confirmed core concepts, corresponding classes of PDDO are created by using protégé. Classes structure of PDDO can be shown by OWLVizTab in protégé (see Fig.1). Secondly, data properties and object properties of classes are added, relationships between concepts can be represented by way of adding object properties. For example: Fig.2 represents the relationships between "Pear-tree" and other classes by JambalayaTab. Finally, instances of classes are added. After that, an owl file (tree.owl) is created by protégé.

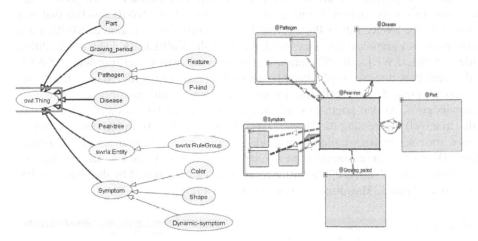

Fig. 1. Classes structure of PDDO   **Fig. 2.** Relationships between "Pear-tree" and other classes

3.5   Adding SWRL Rules and Reasoning

3.5.1. SWRL Editor

The SWRL Editor is an extension to Protégé-OWL, which supports the interactive editing of SWRL rules. The editor can be used to create, edit, read and write SWRL rules .It is accessible through the SWRLTab within Protégé-OWL [8] (see Fig.3).

3.5.2 Building SWRL Rules Library

A rule axiom consists of an antecedent (body) and a consequent (head), each of which consists of a set of atoms. Atoms can be of the form C(x), P(x,y), sameAs(x,y) or differentFrom(x,y), where C is an OWL description, P is an OWL property, and x,y are either variables, OWL individuals or OWL data values. where both antecedent and consequent are conjunctions of atoms written a1 ∧ ... ∧ an. Variables are indicated using the standard convention of prefixing them with a question mark (e.g., ?x) [9].

For example: the symptoms of "rust of pear" are as follow: it mainly destroys leaves, makes leaves show circle yellow scabs, further more some yellow acicular points on scabs. The description of this symptom by SWRL rule is as follow:

Pear-tree(?x) ∧ At-part(?x, leaf) ∧ Has-ds(?x, spotting) ∧ Has-c(?x, yellow) ∧ Has-shape(?x, circle) ∧ Has-feature(?x, yellow-acicular-small-point) → Has-disease(?x, Pear-Rust)

Using this syntax, some rules relating to pear diseases diagnosis are created by the SWRL Editor, so that SWRL rules library based on PDDO is built. (see Fig.3).

3.5.3   Reasoning Based on SWRL Rules

The SWRL Editor itself does not perform any inference. However, a bridge mechanism is provided to allow interoperation with rule engines [10]. At present, the Jess rule engine is supported, and is accessible through the SWRLJessTab that is a plug-in to the SWRLTab in Protege-OWL. It supports the execution of SWRL rules and provides a graphical interface to interact with the SWRLJessBridge. After editing rules, "OWL+SWRL->JESS" button can be pressed, which can transfer all SWRL rules and Pear diseases OWL knowledge to the Jess rule engine. When "Run Jess" button is pressed, Jess will run its inference engine and possibly generate new knowledge[11]. At that point, this inferred knowledge can be passed back to the owl file (tree.owl) by pressing the "Jess->OWL" button (see Fig.3). In this case, reasoning based on rules realizes building new relationships between instances of "Pear-tree" and "Disease". For example: a new relationship between "pear3" the instance of "Pear-tree" and "Pear-Rust" the instance of "Disease" is inferred by the above SWRL rule ,that is "pear3" Has-disease " Pear-Rust" (see Fig.4).

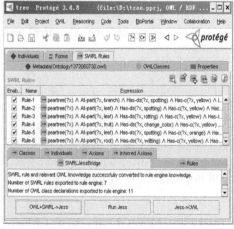
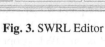
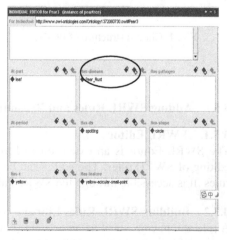

Fig. 3. SWRL Editor   **Fig. 4.** New relationship inferred

4   Design and Realization of Query System Based on PDDO and SWRL

4.1   Architecture of Query Model

In order to design query system, the model of it is built at first .The model consists of six modules, which are shown in the Fig.5. Further illustrate below: By using interactive query interface, users can customize query conditions, which are sent into query processor. The functions of the query processor are executing corresponding algorithms and calling Protégé-OWL API methods according to the given query conditions. Parsing ontology file, it is a way to access and draw information from ontology file. Jess rule engine can transfer all SWRL rules and Pear diseases OWL knowledge to the engine, run its inference engine, and pass inferred knowledge back to the owl file .Finally, the query results can be outputted in the visual interface.

4.2   Parsing the Pomology Domain Ontology

In this study Protégé-OWL API is used to parse the PDDO. An OWLModel is created through the ProtegeOWL.createJenaOWLModel(), which can load an OWL files, then the resources can be created, queried, deleted through the methods of OWLModel .In this case , owlModel1 is created to load "tree.owl" file. The methods of Protégé-OWL API that are used in this study are listed below: the certain OWLClass can be obtained by calling of getOWLNamedClass(), the certain RDFIndividual can be obtained by calling of getRDFIndividual(), the certain RDFProperty can be obtained by calling of getRDFProperty(), calling getUserDefinedOWLObjectProperties() can get all object properties collection.

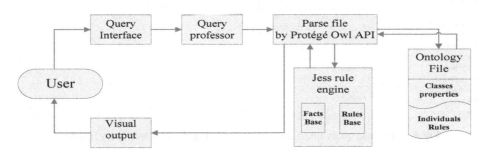

Fig. 5. Flow-process diagram of query model

4.3   Realization of Query System Based on PDDO and SWRL

This query system realizes three kinds of query, which are disease query, instance query, and diagnosis query. Further illustrate below:

1) Disease Query

The instances of "Disease" class are listed on the left side of visual query interface, and the TextArea is on the right side. While user selects one from the list, the introduction of the certain pear disease is displayed in the TextArea (see Fig.6). Details of the method are as follows:

Firstly, the instances list of "Disease" class are realized by using JList. GetOWLNamedClass() is called by the defined owlModel1 to obtain the OWLClass "Disease", and then getInstances() is called by it to obtain all instances of OWLClass "Disease", finally, the instances are saved into a collection, iterated, and added into the listModel of JList. The algorithm is as follows:

```
Collection ins=Disease.getInstances();
DefaultListModel listModel = new DefaultListModel();
for(Iterator i4=ins.iterator();i4.hasNext();)
      {    OWLIndividual in=(OWLIndividual ) i4.next();
           String t=in.getLocalName();
           listModel.addElement(t);  }
```

Secondly, addMouseListener () is added to answer that the user clicks one from the list. Calling GetRDFIndividual() that OWLModel provides to change the instance selected by user into the RDFIndividual, and then getPropertyValue () is called by the certain RDFIndividual to obtain the value of the "describe" data property.

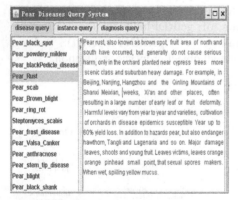

Fig. 6.　Disease query interface

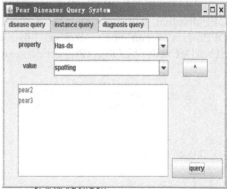

Fig.7.　Instance query interface

2) Instance Query

Interactive instance query supports query on the instances of classes based on SWRL rule. User can select a property from properties of classes through the "property" JComboBox, and then select a value of the certain property through "value" JComboBox. If the "∧" button is pressed , another property and it's value can be selected. While user presses the "query" button, the query results are displayed in the TextArea. For example : if the instances of "Pear-tree" class that show yellow circle spots on the leaves would be queried , user can select "At-part", "leaf" , "Has-ds", "spotting" , "Has-c", "yellow", "Has-s", "circle" and press "query" button at last (see Fig.7).Details of the method are as follows:

Firstly, obtaining properties and corresponding values to generate a string used by SQWRL queries. For example: with regard to the above query, the string generated is as follows : Pear-tree(?x) ∧At-part(?x, leaf) ∧ Has-ds(?x, spotting) ∧ Has-c(?x, yellow) ∧ Has-shape(?x, circle).

Secondly, creating an instance of SQWRL query engine for owlModel1.

Finally, running SQWRL queries. Calling the `runSQWRLQuery()that` SQWRL query engine provides to run SQWRL queries.

3) Diagnosis Query

Interactive diagnosis query supports adding instances of "Pear-tree" class into the PDDO and diagnosis based on reasoning. User can input the name of a instance in the "name" JTextField, select properties from properties of "Pear-tree" class through the "property" JComboBox, and then select values of properties through "value" JComboBox, after that, pressing "add" button can realize adding a new instance, it's properties, and values, furthermore, pressing "diagnosis" button can realize diagnosing the disease that the certain instance gets by reasoning based on SWRL(see Fig.8). Details of the method are as follows:

Firstly, adding a new instance, properties and values. GetOWLNamedClass() is called by the defined owlModel1 to obtain the OWLClass "Pear-tree", and then createOWLIndividual() is called by this OWLClass to create a new instance in owlModel1, in the end setPropertyValue (OWLProperty, OWLIndividual) is called by the certain instance to add   it's properties and values.

Secondly, reasoning based on SWRL. Getting all SWRL rules (SWRL rules library) using the SWRLFactory, and then creating an instance of SWRLRuleEngine for owlModel1 by SWRLRuleEngineFactory.create(), finally using the infer( ) that SWRLRuleEngine interface provides to load rules and knowledge from owlModel1 into a rule engine, run the rule engine, and write any inferred knowledge back to owlModel1.

Thirdly, inferred knowledge are written to the owl file. Fig.9 shows the modified file, the blue block represents the instance, it's properties and values that are added in Fig.8, "<Has-disease rdf:resource="#Pear_Rust"/>" of blue block is a new relationship obtained by reason. The algorithm is as follows:

```
FileOutputStream   outFile1=new   FileOutputStream(uri);//
uri file path
Writer out=new OutputStreamWriter(outFile1,"UTF-8");
OWLModelWriter omw=new
OWLModelWriter(owlModel1,owlModel1.getTripleStoreModel().
getActiveTripleStore(),out);omw.write();out.close();
FileInputStream file1 = null;file1 = new
FileInputStream(uri1);
```

Fig. 8. Disease query interface   **Fig. 9.** Modified owl file

5   Conclusion

In this paper, the modeling of PDDO based on SWRL is studied, design and realization techniques of the query system based on PDDO and SWRL are proposed. This query system based on reasoning realizes disease query, instance query, and diagnosis query by using Protégé-Owl API, furthermore, it supports interactive query. Through it pear diseases knowledge can be obtained from PDDO according to user needs, new and inferred knowledge are written to the PDDO owl file. Query results show that the query system based on PDDO and SWRL is practical for users to query information from PDDO.

Acknowledgment. I would like to express my gratitude to all those who have helped me during the writing of this thesis. I acknowledge the help of Professor Liang Yong. I do appreciate his professional instructions. Last but not the least, my gratitude also extends to my family who have been assisting, supporting and caring for me all of my life.

References

1. Li Hua, Q.: Ontology Storage and Querying Technology: Master's Thesis, Beijing University of Posts and Telecommunications (2007)
2. Harmelen, F., Hendler, J., Horrocks, I., et al.: OWL Web Ontology Language Reference. World Wide Web Consortium (February 10, 2004), http://www.w3.org/tr/owl-ref
3. http://www.w3.org/RDF/
4. http://protege.stanford.edu/
5. http://protege.stanford.edu/plugins/owl/api/
6. http://www.w3.org/Submission/SWRL/#1

7. Golbreich, C., Imai, A.: Combining SWRL rules and OWL ontologies with Protégé OWL Plugin, Jess, and Racer. In: 7th International Protégé Conference, Bethesda, MD (2004)
8. http://protege.cim3.net/cgi-bin/wiki.pl?SWRLEditorFAQ
9. http://www.daml.org/2003/11/swrl/abstract.html#2.1
10. : Knowledge Representation and Semantic Reasoning of Mandarin Fish Disease Diagnosis Based on Ontology and SWRL. Journal of Library and Information Sciences in Agriculture 21(06) (June 2009)
11. Excavating ImplicitRelation Based on SWRL. Information Analysis and Research (2011)

Applications and Implementation of Decomposition Storage Model (DSM) in Paas of Agricultural

Shuwen Jiang, Tian'en Chen[*], Jing Dong, and Cong Wang

Department of Information Engineering, NERCITA
Beijing, 100097, China
jiangsw@nercita.org.cn

Abstract. With involvement and the popularization of the Internet of things and cloud computing technology in the modern information agriculture, RDBMS has been difficult to resolve perception storage and analysis of mass data in Internet of things.big data storage and computing are a hotspot and difficulty of research on agriculture cloud computing in recent years. This article is based on agricultural cloud Paas platform, rural areas and farmers services to designe and implemente a distributed storage of large-scale data of perception using the DSM and distributed file system. While providing services of vast agricultural data of perception storage and analysis through agriculture Paas platform, while implementing agricultural cloud computing. DSM is designed based on Hbase as a mass of NoSql database which provides real-time efficient reading and writing, high scalability and high availability. DSM is deploied on the distributed file system of Hadoop which is based on HDFS. While MapReduce of hadoop can provide high-speed large file analysis and processing. Experiments indicate that the DSM technology based on application of agricultural Paas can meet requirement of the perceptive big data, strong scalability and so on.

Keywords: DSM,cloud computing, HDFS, hadoop, Hbase, Paas.

1 Introduction

With the industrial upgrading of agricultural informationization in China, the Internet of things technology and cloud computing is also more and more been used in agricultural products. Big data problem have been followed in the Internet of things technology[1] and cloud computing[2]. Because of the regional distribution is more and more widely, characteristics of wireless sensor network scale is more and more big[3],Sensors' data as the basis most important data of Internet of things technology appeared explosive growth. agricultural big data problem directly appear in front of us. For example, if the sensors' frequency is 5 s, then a sensor got data quantity is 17280 one day, 1000 sensors got data which are nearly 20 million for a day, the amount of data that there are 6.3 billion reasons for a year. Now these big data have

---

[*] Corresponding author.

D. Li and Y. Chen (Eds.): CCTA 2013, Part II, IFIP AICT 420, pp. 34–41, 2014.

saved based on relation or object model database and a large number of small files.But using a relational database and log files storage cost and conventional tool to analyze these data have been faced with resources and operational problem in whole system.

recent years , big data problem is emerging continuously,cloud computing technology also had further development on the analysis of big data. Google first distributed HDFS and storage model. Have developed out of the many open source framework based on Google.which represented by apache Hadoop of distributed computing technology and HBase's DSM. Current agricultural sensors' data type is single, semi-structured and oriented to single data record, therefore DSM technology and distributed computing is suitable for agricultural data processing mode. According to agricultural sensory data are extensively distributed and big, single structure and so on, which used Hadoop to agricultural Paas cloud platform, through DSM technology and HBase open-source database construct storage of Paas. Agricultural cloud Paas platform would put sensors'data storage as a service of platform and opening interface to the outside. the users through the interface to use big data storage services while agricultural Paas cloud also can provide big data analytics service at the same time.

2   Overall Design of DSM on the Paas

2.1   DSM and Hbase

The article "A decomposition storage model" proposed the DSM (decomposition storage model) detail concept at SIGMOD conference in 1985, and the Sybase had DSM Sybase IQ database system 2004 years or so which is mainly used for on-line analysis, query intensive applications such as data mining. DSM compared with the NSM (N - ary storage model), the main difference is that the DSM will all records in the same field and NSM is aggregation of all the fields in each record[4].

The HBase is an open source implementation of DSM. HBase is distributed and the columns of the storage system which provides real-time, speaking, reading and writing and random access to big data sets. HBase table automatically cut into different region, each area contains a list of all the lines of a subset. HBase is composed of a master node server coordination of one or more area (region server). HBase implementation depends on the Zookeeper to coordinate, zookeeper select a node as the Master, the rest of the nodes in the region server. HBase table consists of rows and columns[5], by default, HBase automatically assigns the timestamp when inserted in the cell.content of tables' cells is a byte array which is not explained. Each row of the column are grouped to form columns as column families, all the columns of the cluster members have the same prefix, columns in the group by qualifier.as a result, each column means for the column family: the qualifier.

Agricultural sensory data characteristics is a single structure, large amount of data. Use relational database to build storge in the early stages what did not consider the characteristic of large scale and distributed of sensors'data.and relational database in the bottleneck with the development of the Internet of things technology. general solution is done by copying or partition distributed storage, but the installation and maintenance cost is very high. Use of distributed storage technology, the expansion characteristics can dynamically add or delete storage nodes without changing the existing data storage way, distribution of the data on the server cluster cleverly. one of the important advantage that agricultural Sensory data used stored in columns is that the entire database is automatically indexed via selection rules are defined by column in the query.when query by only a few fields that greatly reduce the read of big data via storing gathering data of each field for columns storage.while it is more easily to design better compression/decompression algorithm for this storage of gathered.

2.2   Overall Structure of DSM on the Paas

Design a sense data storage service on agriculture Paas platform according to sensors' data type is single, large amount of data.architecture is shown in figure 1. It contains three layers:

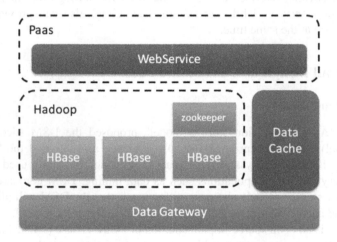

Fig. 1. Structure of DSM on the Paas

(1) the data gateway and the cache: the sensors'data which were got form the Internet of things access to distributed storage cluster via the gateway.the gateway classify the sensors'data accroding to different times and different regions of the data.the gateway also is called data access entrance.If all the area sensors'data are collected to the cluster database at a time, because of the characteristic of the Hadoop cluster to handle large files[6] that will query real time data and analysis take a long time. therefore sensors'data will be collected and stored in the cache from different regions, the cache data is used by the relational database, depending on the time to cache a particular interval data, has been used to carry out real-time query and analysis.

(2)Distributed storage cluster : distributed storage of the sensory data, the hadoop cluster obtain real-time sensors' data from the data gateway and the cache, and have a persistence. the distributed cluster also manage these metadata and distribution with cluster scheduling.

(3) the WebService service: this mainly provided sensors' data storge service, real-time query and big data analytics services for the agricultural Paas platform. Developers can also use the the sensors'data services which Paas platform provides to customize special application of agricultural data storage and analysis.

3   Application of DSM in the of Paas

3.1   Construct of Hbase on Paas

HBase is a distributed storage application which built on a Hadoop cluster.So first to set up a Hadoop cluster and configure Hadoop cluster. Node number of hadoop's DataNode is 5, the NameNode is 1, Jobtracker node is 1 and a Master node.Each node's CPU is the Intel i3 and 2 Gb ram while have a 500 gb hard drive and the operating system for Ubuntu 10. Use Hadoop version is cloudare open-source, built-in HBase version corresponding to a zookeeper. Set the data block is 64 MB, replications is 3. This Hadoop cluster as a Paas platform is distributed structures for computing platform, data storage. The construction of the Hbase specific steps are as follows:

(1) deployment of zookeeper.zookeeper as Hbase's management scheduling, which is deployed separately on a node.It is used to control the Hbase distributed storage cluster. Zookeeper determine to monitor specific information of Hbase through extracting configuration files.

(2) deployment of Hbase. First to config Hbase on Hadoop NameNode, Hbase is introducesd to the Hadoop cluster; Second to config the associated files of zookeeper and Hbase on Hadoop cluster; Finally, deployment of Hbase in the DataNode and start the Hbase cluster to realize the construction of distributed storage[7].

3.2   Structure Designe of DSM

A Storage architecture is designed by this paper is two structure tables, It respectively contain the sensor group table and sensors'data tables. Sensor group table is mainly used to store sensor group information. Sensorgroupid as a row with the info as a column family.Info provide key/value of key-value pairs to define sensor group information such as name, address, description for the column.Which is including (info: gid, info: IP, info: port, info: region servers, info: avaCapacity, info: the location). Sensors'data table mainly storage sensors' data of sensor group.data table will be sensorgroupid and reverse timestamp as a row, sensordata as a column family and column family contains (sensordata: tem, sensordata: co2, sensordata: soiltem, sensordata: hum, sensordata: soilhum, sensordata: sun)[8].

Efficient reading is the key design of DSM for big data of Sensors'data. For sensor group table, using sensor group id as a row, because it is usually queried as a keyword in the entire sensor group.while the sensory data table row key used a combination of sensor group id and reverse time stamp. so that we can observe real-time data according to a sensor group data by time stamp. The latest sensory data will be put in front of the row. Table design is illustrated below as Table 1.

Table 1. Design of sensordata table

RowKey	times tamp	timeColumnFamliy : sensordata			
	time1	sensordata:tem	sensordata:hum	sensordata:sun	sensordata:co2
sensorGroupid	time2	temValue1	humValue1	sunValue1	co2Value1
	time3	temValue2	humValue2	sunValue2	co2Value2

3.3   Sensors'data Storage Service

DSM applications of Paas in agricultural is mainly embodied in as a service provided to developers or users.the user encapsulate specific applications by calling service. And sensors' data storage as a service use a three layer scheduling method. First having a persistence at the bottom of the infrastructure.Using Hbase to sensors'data storage.Second processing a the OO storage operation on the Paas layer. The operation of each table in the Hbase needs to acquire the objects of HTable, then use the put and get methods to complete the data operation of insertting and reading. Using HBaseAdmin object to complete the operation of creating and deleting table. Finally At the application level, store operation is encapsulated on the Paas and form the service interface while releasing a version of the WebService.Developers and users can invoke the WebService of storage service to storage sensors'data via the Paas. storage service on Paas as shown in the figure2 below.

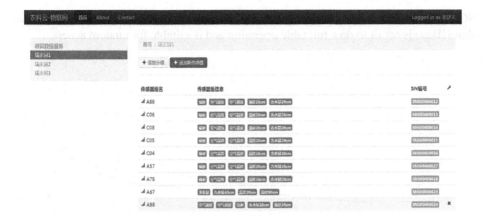

Fig. 2. Storage service on Paas

4   Experimental Results

To verify efficiency of storage and query of HBase and analysis ability of large data, comparing Hadoop cluster which used 20 nodes with single computer in operation of GET and SCAN data's time from the 100 sensor groups for a month. Figure 3 shows the time relationship of GET data between single machine's sensor nodes and cluster.

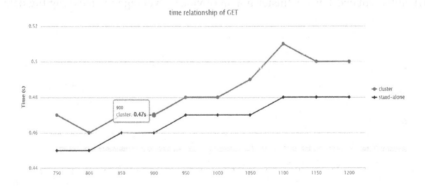

Fig. 3. Time relationship of GET data between stand-alone and cluster

Can be Seen from the Diagram. The gap of standalone and cluster is not big when the amount of data is small. The amount of data increases with the passage of time, the advantage of cluster began to highlight.The growth of reading time is lower than that of single. HBase cluster had an visible obvious advantage for storing big data. Figure 4 shows the time relationship of SCAN data between single machine's sensor nodes and cluster.Can be seen from Fig.4. When cluster and stand-alone execute operation of

SCAN at the same time, efficiency of cluster is lower than stand-alone's. This shows that HBase is not fit to do a full table operation and is suitable for random access.

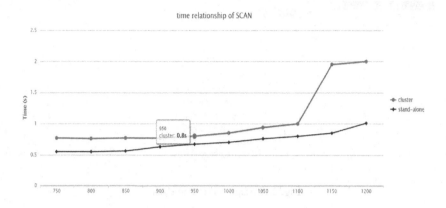

Fig. 4. Time relationship of SCAN data between stand-alone and cluster

For example, analysising nearly the highest temperature for a month. Hbase cluster and single machine carries on a comparison while analyzing by MapReduce in Hbase and analyzing by traditional method in stand-alone.As shown in the figure5 below, the gap of single and cluster is not big when the data volume is less than 1G. Cluster's advantage began to highlight with data volume reaches G level. Analyzing time significantly reduced.HBase cluster got an obvious advantage in analyzing big data .

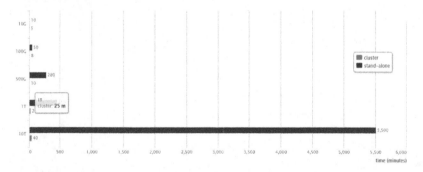

Fig. 5. Time relationship of analyzing between stand-alone and cluster

5   Conclusions

This paper analyzed the problem of sensors'data storage and analysis under the background of big.So putting forward a distributed storage based on DSM architecture and combined with agriculture Paas platform to provide a service. Distributed cluster using Hadoop and Hbase cluster database while putting real-time sensors' data fast persist in distributed file system through reasonable design DSM.

Experiments show that using DSM in the big data persistence has high storage efficiency[9]and having a speed improvements by using computing of distributed cluster.

Acknowledgment. This work was supported in part by National science and technology support project(2013BAD15B05). It also has been done under the help of the team of the Information engineering.

References

1. Conti, J.P.: The internet of things. Communications Engineer 4(6), 20–25 (2006)
2. Yin, G.-M., Liu, Y.-J., Guo, G.-X.: Cloud Computation Applied in the Distributed
3. Yick, J., Mukherjee, B., Ghosal, D.: Wireless sensor networksurvey. Computer Networks 52(12), 2292–2330.6 (2008)
4. Yan, Q.-L., Sun, L., Wang, M., Le, J.-J., Liu, G.-H.: Heuristic Mechanism for Query Optimization in Column-Store Data Warehouse.Chinese Journal of Computers 10 (2011)
5. Chang, F., Dean, J., Ghemawat, S., et al.: BigTablel:A distributedstorage system for structured data. ACM Trans. on Computer Systems 26(2), 1–26 (2008)
6. Hadoop, http://hadoop.apache.org/
7. He, Y., Tang, Y., Lin, Y.: Analysis of Construction Profiling Based on Multi-tenant Architecture Paas. Digital Communication, 3 (2012)
8. Zhou, L.-Z., Chen, Q.-K.: HBase-Based Storage System for Wireless Sensor Information of Agriculture. Computer Systems & Applications, 8 (2012)
9. Wang, S., Wang, H.-J., Qin, X.-P., Zhou, X.: Architecting Big Data:Challenges,Studies and Forecasts. Chinese Journal of Computers, 10 (2010)

Estimation of Pig Weight by Machine Vision: A Review

Zhuo Li[1], Cheng Luo[1], Guanghui Teng[1], and Tonghai Liu[2]

[1] College of Water Conservancy and Civil Engineering,
China Agricultural University, Beijing, China
[2] Department of Computer Science and Information Engineering,
Tianjin Agricultural University, Tianjin, China
{lizhuo_lin,Luocheng06,Tonghai_1227}@163.com, futong@cau.edu.cn

Abstract. In pig production, food conversion ratio and profit can be evaluated by real time detection of pig live weight. Traditional pig weight detections usually require direct contact with pigs, which are limited by its low efficiency and result in a lot of stresses even to death. The non-contact detection of pig body weight has become a challenge in pig production for decades. Digital image analysis and machine vision method enable the real time estimation of pig live weight by detecting pig critical body dimensions without any contact. This article elucidated the advantages and limitations of each detection method of pig body weight by comparing the system framework and estimation models. The research trends of contactless pig weight estimation were analyzed as well.

Keywords: Pig, weight, machine vision, estimation, reviews.

1   Introduction

Weight is an important index in pig rearing. Pig's daily gain and nutritional status can be evaluated immediately through getting pig's weight timely [1]. Even the feed utilization efficiency can be detected by combined with automatic feeder [2]. Then pigs in good or bad nutrient status could be raised separately to meet uniform marketing weight standard.

In traditional weight measurements, pigs need to be moved to weighing devices such like mechanical scale and electronic balance. The whole process is time consuming and laborious, it often requires at least two stockmen to spend 3-5 minutes for each pig [3]. This process brings lot stress to pig and even leads to sudden death; feed intake is going down on the day of weighing compared to before and after the weigh day [4]. Recent years, some manufacturers added weighing sensors inside the automatic feeder to weigh and record pig weight real-time. Most of these apparatuses are expensive and prone to erosion by sewage. In addition, they need to remold original piggery.

Due to numerous problems of contact weight measurement, the non-contact measurement attracted attentions to measure or estimate pig weight. Early in 1988, digital image analysis were proposed to have about 90 kinds of potential applications in stock farming, which can be used to estimate pig weight [5]. Many researches have

D. Li and Y. Chen (Eds.): CCTA 2013, Part II, IFIP AICT 420, pp. 42–49, 2014.

proven that there is being significant correlation between animal body sizes and their body weight. Using image analysis and machine vision technology, key sizes or back area of pig could be obtained. Combining with the relation model of body sizes and body weight, pig live weight can be estimated accurately. The measurement technique based on machine vision presents many merits, such as non-contact, fast and labor saving. The non-contact measurement has not been widely applied in practical engineering, although it was researched for decades. In this article, the application of machine vision technique in estimating pig weight has been reviewed from two respects of system frame work and prediction model according to the form of machine vision.

2   System Framework

A complete machine vision system mainly includes following aspects: light source, lens, camera, image capture card, image processing platform, machine vision software, I/O (Input/ Output) devices and execution control mechanism. The control mechanism is not needed in pig weight estimation system, and the only output is pig weight information. Thus this chapter summaries as the following aspects: camera location, light source and image trigger method.

2.1   Camera Location

In order to get pig body sizes as many as possible, cameras were not only set at the top of pigs, but also at the side of pigs. The camera on the top of pig could get the pig back area and the one at the side of pig is set to get pig's body height [6]. To reduce the number of cameras, pig was put in a weight cage, and a mirror was installed on the top canted 45 degree. Camera was fixed in the flank of pig. Consequently pig's black area and its body height could be attained from one shoot [7]. There are still some disadvantages. Cameras are prone to contaminated by smudginess, and the pig being measured is possibly covered by other pigs.

For the protection of cameras, most researchers only set up the camera on the top of pig. The image from top view can provide the pig body width and body length rather than the pig body height. Binocular vision technology could get depth information with two cameras fixed on the top of pig through image matching. The pig body height information could be figured out by 3d coordinates of points taken from the binocular vision [8].

The location of camera should be convenient for image capture and analysis, so that image with only one pig with straight body and static could be taken. Camera is commonly set up at the top of the feeding station or drinker, because pig's body is straight and motionless when it is drinking or eating. If only one pig is allowed to enter into feeding station, which is too narrow for pig to turn round or bend. High quality images could be attained when camera was put on the top of feeding station [9].

Similarly, cameras were also fastened upper the drinker for the high quality image [10]. Most time pig's body will be straight when it is drinking. But when the temperature is high in summer, pig will scramble for the drinker, and like to lie in the drinking area. This brings a lot of troubles for image collection and pig contour extraction.

With the development of digital camera, shutter speed advance greatly and reduce the requirement for pig state of motion. Even fast moving pig can be snapped clearly. A gallery with a camera on the top is built to let pigs go through [11, 12]. System is always running to detect the appearance of pig and catch appropriate images. But the shortcoming of this method is that pig tend to remain and explore in the aisle, which will influence image capture. And this method need to remold pig house which increase the investment and difficulty of system implementation.

On the whole, when pig house is already built, putting the camera on the top of feeder or drinker area is more practical way, automatic feeding station that allowed only one pig enter one time is the idealist way.

2.2   Light Source

Light source is the most essential part of machine vision system. A good light environment not only enhances the contract between foreground and background, and simplify image processing algorithm, but also increases system speed and robustness. In the pig body size detection and body estimation system, light source also plays a very important role.

Light is scarcely taken into account in the early research. Only the relationship between illumination intensity and binaryzation threshold is discussed briefly. Two filament lamps or two strip fluorescent tubes for illumination were used in most researches. In consideration of field application the semi-closed structure was used, where only one pig is allowed to go through or entrance. In the semi-closed structure, the sun light is warded off to avoid impacting pig contour extraction, and the background was painted to black artificially to enhance the contrast [11].

In a real house the background is much complex. The illumination of image acquisition area is inhomogeneous, because most pig houses have window for day lighting. This brings lots of challenges to image analysis. Many researchers used automatic exposure to response the change and inhomogeneous of environment light, but this seems inadequate. It is easy to get the wrong part when extracting pig image, with the present of water spots or sunshine illuminate area in the field of view. It is inevitable for pigs to stain dirt whose color is close to the ground. From the top view, the dirt located at pig outlines will bring errors to contour extraction.

Different color lights were tested to simplify pig image analysis, one machine vision system use a projector as light source and a clearer pig image suitable for image analysis was obtained when using a red slide film printed with yellow mesh lines fixed before projector [13]. This resulted from the higher reflection ability of red light compared to that of the blue and green light when the color of pig body is present as white and slightly red.

A systemic study of light for pig weight estimation system was made. Several light sources' characteristics and price were compared, such as fluorescent lamp, halogen lamp, LED (Light Emitting Diode). The strip fluorescent lamp was suggested to use for uniform light. Limited by the piggery condition, the only location of light source is the top of pigs. Three different color filters were used to transform white light of fluorescent lamp to three monochromatic lights (red, green and blue) as light source, then image quality differences were analyzed. Red light is the best among three kinds of lights. Slant red light is recommended in practical to help the pig extraction [14].

2.3   Image Trigger Method

After camera is installed, system needs signal to trigger snap image. Some systems have no trigger hardware, and systems are processing image all the time no matter whether there is a pig or not in the view. Hot point and cross hair method were used to monitor some points gray value to judge whether pig is appeared in image [15, 16]. And the system is always in a state of computing and waiting, leading to a low efficiency. A new method was invented to snap image [17], where a flow switch installed in water line is used as a trigger. The flow switch change state when pig is drinking, then the connected camera is triggered to capture image. This method does not need system detect and process image all the time, and has higher efficiency. Combining with RFID (radio frequency identification devices) reader, the designated pig can be snapped. But this method needs to remodel water line with the increase of system cost.

Optoelectronic switch was used to detect the appearance of cow [18, 19], with the image snap controlled by MCU (Microprogrammed Control Unit). Sensor and peripheral circuit are included in this methods, resulting in the increase of system complexity and reduction of stability.

RFID reader is a good way to control snapshot. A RFID reader is fastened near the feeder or drinker. When pig is feeding or drinking, RFID reader fetch the ID of pig, then send a break to system to take a shoot. This way combines with pig identification and needs no more additional equipment to achieve image trigger snap conveniently.

3   Estimation Model

It is found that some pig's size have fine correlation relation with pig's weight. Studies on the selection of measurement parameters are necessary to estimate the pig's weight accurately.

The pig's weight was intended to be calculated by pig volume multiplying the pig's density. But pig's shape is irregular and difficult to be measured. In the previous research, pig was simplified to a cylinder and a cone. Then body length and chest circumference were used to compute volume to build relevant model with pig weight and the average relative error is about 2.8% [20]. However chest circumference can't be obtained by image technology, so this method is not practical. There is an obvious error in the estimation of pig's volume, because it's very hard to define regular shape to match every body part accurately.

The relationship of two dimensional area and one dimensional body measurements with weight were built respectively. Due to the structure of piggery, camera have to be put on the top of pig to catch the pig's back image for body size. Camera coordinate system could be built through calibration which helps the transform of image area to real area, and image length to real length. The relation of average number of pixel of pig back area with body weight was established. The relation varied among different breeds. The average error of estimation weight is below 5%. Within the range of 60-90kg weight, some breeds' error are under 2% [9]. A few body measurements were used to establish relationships with weight. The better model among those relationships is power equation between pig back area and weight, whose estimation error was less than 3.7% [15]. Those relationships built between number of pig back pixel and weight seems not universal. If any factors such as camera height and resolution ratio are changed, the model is no longer suitable and need to be rebuilt. If pig back area and body size are real dimensions, so that the model is not influenced by system configuration.

Pig height information is prerequisite for transform image area to real area. The object's coordinates obtained by single camera cannot compute the height information without other devices. Pig is assumed as a cylinder with certain radius (r) and height (of the ground, h). The values of h and r are gained from a lookup table of ASAE [16]. The model using back area (A4) without ear and neck with weight is as following.

$$W = -15.56 + 411.3A_4 \tag{1}$$

If each pig's model is corrected by 75 days' weight, estimation error and the relative error are within 1kg and 1.25%, respectively. But this method cannot be used in practical application obviously, owing to the need of the lookout table.

A calibration scale was put on 0.5m height to get pig body size and area information [3]. This method doesn't calibrate the camera, so when the pig's height does not equal to 0.5m, pig's width and length will be incorrect.

A technology similar to the structured light was used to measure pig height [10]. A slide projector was fixed to the ceiling above the drinker to project shadow lines of lattice pattern. The difference in pixel of shadow line length between on the shoulder of the pig and on the floor was measured, then pig height was obtained using a geometric relationship between pixel difference and height. The mathematical model is established as followed.

$$W = 5.68 \times 10^{-4}A^{1.16}H^{0.522} \tag{2}$$

Where W is the pig weight; A is the pig back area without pig ear; and H is the pig height. Twelve pigs were estimated for five times between 81 to 98 days old, the relative error is 2.1%. But this method is easy to be influenced by piggery light environment.

The distance from the pig's back to camera was recorded manually for every photo to get pig body height. And a ruler was put on the pig back to get the real size [21]. Using Matrox image process software to extract pig's size. Back area was found having well relevant (r=0.96) with weight compared to several body sizes. Body width also has better relevance (r=0.95). The best body measurement was extracted by the image process to build mathematical model. In order to estimate pig weight, the model was verified to be appropriate by regression analysis. The relative error is 6%. Artificial neural network was used to make model with 3% relative error later.

The pig's body size detect and weight estimation system was designed by Fu Weisen. The coordinates of key point in two pictures were found by image matching, and three-dimensional space coordinates could be computed through stereo vision principle. The relative error of pig's body size is around 1%. Eleven body measurements including pig length, width and height were used to build an estimation model through principal component analysis. Field experiment showed the average relative error of pig estimation weight is 0.77% [8], which indicated a certain practical value.

Back area has higher correlation coefficient with pig weight than body size, and the prediction model of area is also better than body size's model. It was found that both central and vertical projection area have very high dependency with weight, but the former is better than later [10]. Central projection area is the area of image range of pig projection to camera CCD sensor. Vertical projection area is the pig projection area without height influence, and it is a real two dimensional area. The central projection area have positive correlation with height, and it could be taken as a three dimensional body measurement. This also could explain why central projection area has better correlation with weight than vertical projection area.

In extraction of area, many researches demonstrated that pig's head, ear and tail have large range of motion and influence on the stability of area extraction. It should be removed out of the statistics. It is found that body length have higher correlation with weight, partly because the magnitude of length is larger than width and changes significantly over time. In contrast, width and height change little over time. But in actual measurement process, pig is always moving especially its head, neck and fore hoof. The uncertainty get high when the pig's body height and body width are tested which lower the correlation with weight. Therefore, parameters which are little influenced by the pig motion should be chosen. In addition, it is crucial to keep pig stand with a straight and head rising posture when it is snapped.

Varied breeds need different estimation models. Estimation model is affected rarely by the weight interval or genders. In Schofield's linear model, every pig has different intercept. The manual measuring weight was used to calibrate the model, with the drop of the relative error from 5% to 1.25% [16].

4   Summary

Machine vision technology has been widely used in the industry, but the applications in agriculture are mainly concentrated in the field of plant and food classification such as nondestructive testing. Noncontact testing of live animals is limited by many factors such as harsh environment and poor light condition, the estimation system could accommodate the environment of pig house and run stably for a long time. The image analysis algorithm should adapt to uneven illumination environment to get a right pig contour. Due to the moving nature of animals, it's a great challenge to snap a straight and head raised pig's image, the location of camera and image trigger mode can deal with this challenge. In the same time, how the estimation system combine with auto feeder and maximizing precision and repeatability should be considered as well.

Acknowledgment. This work was supported by the Special Fund for Agroscientific Research in the Public Interest (201003011), and supported by Chinese Universities Scientific Fund.

References

1. Doeschl-Wilson, A.B., Whittemore, C.T., Knap, P.W., Schofield, C.P.: Using visual image analysis to describe pig growth in terms of size and shape. Animal Science 79(Part 3), 415–427 (2004)
2. Huang, R., Zhong, C., Li, H., Geng, W.: The research of intelligent swine measurement system. Modern Agricultural Equipment 2012 (Z1), 64–66 (2012) (in Chinese)
3. Brandl, N., Jørgensen, E.: Determination of live weight of pigs from dimensions measured using image analysis. Computers and Electronics in Agriculture 15(1), 57–72 (1996)
4. Augspurger, N.R., Ellis, M.: Weighing affects short-term feeding patterns of growing-finishing pigs. Canadian Journal of Animal Science 82(3), 445–448 (2002)
5. DeShazer, J.A., Moran, P., Onyango, C.M., Randall, J.M., Schofield, C.P.: Imaging systems to improve stockmanship in pig production. Silsoe: AFRC Institute of Engineering Research (1988)
6. Yan, Y., Teng, G., Li, B., Shi, Z.: Measurement of pig weight based on computer vision. Chinese Society of Agricultural Engineering 2006 (02), 127–131 (2006) (in Chinese)
7. Schofield, C.P.: Evaluation of image analysis as a means of estimating the weight of pigs. Journal of Agricultural Engineering Research 47(C), 287–296 (1990)
8. Fu, W.: Study of Pig's Body Dimensions Detection and Weight Estimation Based-on Binocular Stereovision, China Agricultural University, Beijing (2011) (in Chinese); Doctor: 118
9. Schofield, C.P., Marchant, J.A., White, R.P., Brandl, N., Wilson, M.: Monitoring Pig Growth using a Prototype Imaging System. Journal of Agricultural Engineering Research 72(3), 205–210 (1999)
10. Minagawa, H., Murakami, T.: A hands-off method to estimate pig weight by light projection and image analysis. In: Livestock Environment VI: Proceedings of the 6th International Symposium, Louisville, Kentucky, USA, pp. 72–79 (2001)
11. Banhazi, T.M., Tscharke, M., Ferdous, W.M., Saunders, C., Lee, S.: Using Image Analysis and Statistical Modelling to Achieve Improved Pig Weight Predictions. In: Society for Engineering in Agriculture (Australia); Agricultural Technologies In a Changing Climate: The 2009 CIGR International Symposium of the Australian Society for Engineering in Agriculture, Brisbane, Queensland. Engineers Australia, pp. 69–79 (2009)

12. Wang, Y.S., Yang, W., Winter, P., Walker, L.: Walk-through weighing of pigs using machine vision and an artificial neural network. Biosystems Engineering 100(1), 117–125 (2008)
13. Minagawa, H., Taira, O., Nissato, H.: A color technique to simplify image processing in measurement of pig weight by a hands-off method. In: Proceedings of Swine Housing II, pp. 166–73. ASAE Publication, American Society of Agricultural Engineers, ST Joseph (2003)
14. Fu, W., Teng, G., Zong, C.: Study on Illumination Mode of Pig Growth Inspecting System Base on Binocular Stereovision Technology. In: 2009 ASABE Annual International Meeting, Reno, Nevada (2009)
15. Tscharke, M.J., Banhazi, T.M.: Growth recorded automatically and continuously by a machine vision system for finisher pigs. In: SEAg 2011: Diverse Challenges, Innovative Solutions, Surfers Paradise, Queensland, Australia. Engineers Australia, pp. 454–464 (2011)
16. Marchant, J.A., Schofield, C.P., White, R.P.: Pig growth and conformation monitoring using image analysis. Animal Science 68(1), 141–150 (1999)
17. Teng, G., Fu, W., Xie, Z.L., Huang, W.: An image collection trigger method and system based on livestock drinker (2009). CN101144705 (in Chinese)
18. Tasdemir, S., Urkmez, A., Inal, S.: Determination of body measurements on the Holstein cows using digital image analysis and estimation of live weight with regression analysis. Computers and Electronics in Agriculture 76(2), 189–197 (2011)
19. Tasdemir, S., Urkmez, A., Inal, S.: A fuzzy rule-based system for predicting the live weight of Holstein cows whose body dimensions were determined by image analysis. Turkish Journal of Electrical Engineering and Computer Sciences 19(4), 689–703 (2011)
20. Fu, W., Teng, G., Yang, Y.: Research on three-dimensional model of pig's weight estimating. Chinese Society of Agricultural Engineering 2006 (S2), 84–87 (2006) (in Chinese)
21. Wang, Y.S., Yang, W., Winter, P., Walker, L.T.: Non-contact sensing of hog weights by machine vision. Applied Engineering in Agriculture 22(4), 577–582 (2006)

Agricultural Field Environment
High-Quality Image Remote Acquisition

Fu Junqian, Xiao Deqin[*], and Deng Xiaohui

College of Informatics, South China Agricultural University, Guangzhou 510642, China
deqinx@scau.edu.cn

Abstract. In order to realize the acquisition of high-resolution and high-precision image in agricultural field environment, the agricultural field environment high-quality image remote acquisition system was designed, based on Canon DSLR camera. This paper made detailed design about core algorithm, including remote acquisition system architecture, camera access, image transmission and command response. The scheme used Fit-PC to control Canon 550D DSLR camera for image acquisition and WIFI wireless transmission module to transmit collected images to the remote server. Meanwhile, the server could also remotely operate camera to collect images in different ways through wireless transmission command, according to different needs. The experimental results showed that the system had three main advantages as high resolution, precision and image transmission rate in collecting images. The resolution of images collected in each node was up to 5184×3456, and the effective breadth of Canon camera image was 17 times as the Logitech camera. In addition, at a distance of 16 meters, image captured by Logitech camera could not see the target while the same one captured by Canon camera could still clearly display. Moreover, the average speed of wireless images transmission to the server was 0.97MB/S. Therefore, this system could meet the requirements of high-quality image acquisition and transmission in agricultural field environment.

Keywords: agricultural field environment, image acquisition, wireless transmission.

1 Introduction

In recent years, with the rapid development of computer technology, computer image processing technology has matured and has been widely used in agricultural field, its main application focused on crop seed resources detection, grading of agricultural products, agricultural machine vision, growth monitoring of agricultural products, precision irrigation, and assessment of agricultural machinery. Image information of farmland crops was important agricultural basic information, playing a key role in the early warning and monitoring field of crop growth status, environment and field insect pests in large area [1].

During these years, many scholars have studied on agriculture image acquisition. Fan Fengyi, etc, designed the agricultural remote monitoring system based on ARM

---

[*] Corresponding author.

D. Li and Y. Chen (Eds.): CCTA 2013, Part II, IFIP AICT 420, pp. 50–60, 2014.

and CMOS camera [2], using the wired way to transmit images. Xiong Yingjun, etc, combined ZigBee and GPRS, designed the system of acquisition and wireless transmission for farmland image [3]. Meanwhile the highest theoretical speed of GPRS network built by China Mobile was 171.2kb/s, but in fact current users' access speed was about 20kbps-40kbps, due to the transmission speed of actual data affected by network encoding and terminal support [4]. Therefore, it needed about 30 seconds to transmit a resolution of 640 × 480 BMP formatted image, and several minutes were needed for high-quality images. Zheng Yehan, etc, designed the wireless image acquisition system based on ARM and wireless LAN [5]. However, it would never meet the requirements of agricultural image analysis as 640 × 480 resolution collected by all these image acquisition systems. For example, the characterization of crop diseases and insect pests is usually not obvious in the low-resolution images, so it needs to magnify them by large multiples to identify. In a word, high-quality images are urgent needed in agricultural field.

This paper designed and realized an agricultural field environment high-quality image acquisition scheme based on Canon SLR camera. The scheme based on the Fit-PC-Windows-XP platform, using Canon EOS 550D DSLR (Digital Single Lens Reflex) cameras to get the high-quality images. Moreover, it used Fit-PC by controlling Canon camera to collect images and WIFI wireless transmission module to realize transmitting high-quality images transmission to the remote server.

2   The Architecture of Agriculture Field Environment High-Quality Image Remote Acquisition System

Fig.1 showed the sketch map of the system architecture of the scheme. Each node consisted of a Canon EOS550D DSLR camera, a Fit-PC and a WIFI module. Canon DSLR camera and Fit-PC physically connected through USB connection line. The image acquisition system running on the node host established an application-level link with Canon DSLR camera through Canon EDSDK API [6], and it used WIFI module to connect to the Internet network in order to transmit the collected images and accepted remote control commands. The server consisted of a PC server and a WIFI module. Through WIFI module, software running on PC connected to the Internet network, aiming to receive images from nodes and sent remote control commands to control nodes.

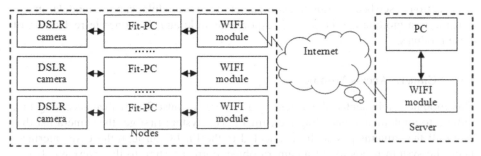

Fig. 1. System Architecture

In order to implement functions of timing automatic shooting, remote control manual shooting and automatic transmission in agricultural field environmental high-quality image acquisition this paper designed camera access module, shooting control module, command response module and image transmission module. The camera access module implemented application-level link to the camera, so that users could remotely control the camera through the system. The shooting control module included remote control manual shooting sub-module and timing shooting sub-module, respectively implementing remote control manual shooting function and timing automatic shooting function. The command response module implemented that sending remote control commands from server to the nodes, receiving commands from sever and executing them. Remote control commands were consist of the shooting command, the timing command and the canceled shooting command. The image transmission module included camera to Fit-PC images transmission sub-module and nodes to server images transmission sub-module, respectively implementing images local backup and remote image transmission. Fig.2 showed the cooperation relationship between the system modules.

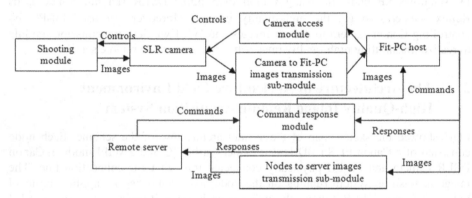

Fig. 2. System overall collaboration diagram

3   The Core Algorithm Design of Agriculture Field High-Quality Image Acquisition System

This section described the functions of camera access module, capture module, image transmission module and command response module and the concrete realization of these modules.

3.1   Camera Access Module

This module implemented the access to a camera. It must first realize accessing to a camera to implement controlling the camera. The system first get the camera list that listed all the cameras physically connected to the Fit-PC through the USB interface. Then the system needs to get a camera object corresponding to the index number in the camera list. After successfully getting a camera object, the system should attempt to establish a session with the camera. If succeed, then it had realized an application-level link with Canon DSLR camera. When the system exited, it must close the session with the camera and release the camera object [7]. Fig.3 showed the process of accessing the camera.

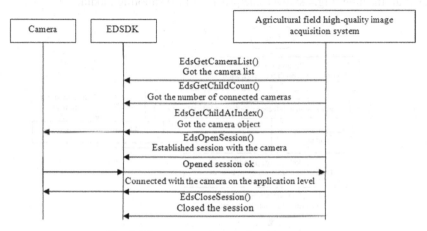

Fig. 3. The process of accessing the camera

3.2   Shooting Module

This module controlled cameras to capture images. There were two ways to realize the shooting function, according to the needs in agricultural field environment high-quality image acquisition system. One was remote control manual shooting, another was timing automatic shooting.

(1) Remote Control Manual Shooting
When the user pressed the "shooting" button in the interface, the server would send a message including shooting command to nodes. The node specified by the user controlled its camera to capture the high-quality image after receiving the message. Then the image would be transmitted to Fit-PC. Then the Fit-PC would send the image to the remote server.

(2) Timing Automatic Shooting
"Start timing" button and "end timing" button were in the interface, when user inputted interval and pressed "start timing" button, the server would send a message including timing command and time interval to nodes. After receiving the command, the node specified by the user would control the camera capture high-quality images automatically according to the time interval. When the user pressed the "end timing" button, the server would send a message including canceled timing command to nodes. The node specified by the user no longer periodically controlled its camera to capture images. The server first checked whether the time interval set by the user was greater than the system's maximum allowed value, or was less than the response time

of the system. If there were the above-described two situations, the system would prompt the user that time interval was unreasonable and required the user to reset the time interval. The system would not start the timing automatic shooting function before the time interval was set correctly. If the time interval were suitable, the system would enter timing automatic shooting mode. The user could change the time interval all the time. Fig.4 showed the process of the shooting module.

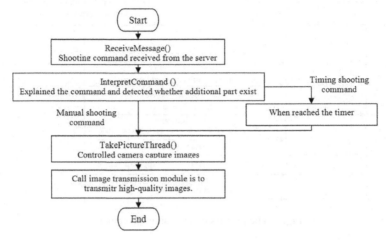

Fig. 4. The process of the shooting module

3.3   Image Transmission Module

This module implemented the transmission of images. In this paper, we divided image transmission into image transmission between camera and Fit-PC, image transmission between nodes and server. Both methods are explained below.

(1) Image Transmission between the Camera and Fit-PC
When the camera finished capturing high-quality images, the system needed to transmit the images to the Fit-PC. Image transmission between camera and Fit-PC was mainly relied on the callback function. The system registered the callback function after its initialization. Every time the camera captured an image, the callback function would be involved and send a transmit request to the system. When the system received the transmit request, it would check the local directory whether there was a subfolder named by the date when the system captured the image, this subfolder would be fill with images whose captured date was the same to the subfolder's name. Then the system would add an index after the date when the image was capture to identify the image's index in the subfolder. When everything was ready, the system downloaded images from the camera and saved them into Fit-PC's local hard disk. When the transmission was completed, system would inform the camera that the image transmission success so that the camera could continue to carry out other operations. If failed it would return other error codes prompting the corresponding error [8].

(2) Image Transmission between the Nodes and the Server
Every time the image transmission between the camera and Fit-PC finished, Fit-PC need to transmit the image to remote server by using WIFI module. The remote server collected images from all nodes and recorded their information into database for the subsequent data analysis. The system would first get information of the image, such as file name, file size. At the same time, the system would create a listening port waiting for the server's connection. The server connected to this port and then the system sent image data to server through this port. The system would send a transmit-message to the remote server about the node ID, the information of the image and the information of listening port through the connection which had already established when the system start. The remote server created a new file prepared for receiving the data of the image according to the information received from the node and tried to connect to the listening port. The format of this message was as table1.The system started reading data from the image and sent it if succeed. Remote server received these data and wrote them to the file. When data transmission finish, the system closed the connection and closed the listening port, the server closed the connection port and inserted the information of the image in the database. Fig.5 showed the process of image transmission module.

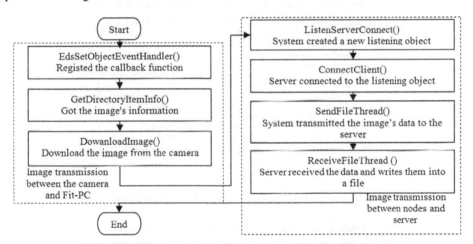

Fig. 5. The process of image transmission module

Table 1. The format of transmit-message messages

	Length(byte)
Node ID	8
Image name	24
File size	4
Listening port	4
In total	40

3.4   Command Response Module

This module implemented that sending remote control commands from server to the nodes, receiving commands from sever and executing them. After nodes' startup, they would try to establish connections with the server and the remote server would save all these connections. When users pressed buttons in the interface which corresponding to the functions they want the specified node achieved, it would trigger the corresponding button events. Then the server would broadcast a command-message to nodes included the node ID declared which node should response this message, controlling commands which included shooting command, timing command and canceled timing command, and additional information (example: time interval). The format of the message was as table2. When nodes received the command-message, they firstly checked whether their node IDs were match to the message's. If matched, the node would response differently based on the contents of the command. If not matched, other nodes would do nothing. Fig.6 showed the process of the command response module.

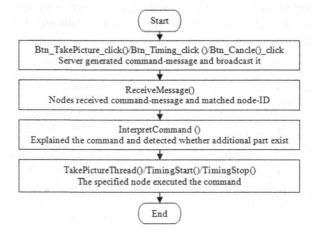

Fig. 6. The process of the command response module

Table 2. The format of command-message messages

	Length(byte)
Node ID	8
Command	8
Optional part	variable
In total	<=40

4   System Testing and Running

Users could be smoothly to capture images by remotely controlling nodes and transmit them to the server on display by using the above agricultural field environment high-quality image acquisition system. Fig.7 showed its timing automatic shooting interface. This paper also carried out comparative analysis for shooting accuracy, shooting distance and transmission performance of this scheme.

Fig. 7. System interface diagram

4.1   Shooting Accuracy Analysis

This section used Logitech pro9000 camera and Canon EOS 550D camera to capture images and then compared them. Firstly, we captured images for different targets at the same distance by using these two cameras and compared them, and then we captured images for the same target at different distance and compared them.

4.1.1   Different Targets at the Same Distance

This test used the above cameras to capture images for vegetable leaves, corn leaves, and tomato fruit and magnified them to compare. All images were captured from the target at a distance of 0.5m.Each image was five times magnified. These images were showed as Fig.8.The left side images were captured by Logitech pro9000 camera and the right side images were captured by Canon EOS 550D camera. In the first group of images contrast, the image captured by Logitech camera showed the vegetable leaves' macula lutea hazily while the image captured by Canon camera showed the leaves' macula lutea clearly. In the second group of images contrast, the image captured by Logitech camera could only showed lesions on the corn leaves hazily while the image captured by Canon camera could show a clear reduction of lesions on the corn leaves. In the third group of images contrast, people could not surely determine whether the tomato exist insect pests through the image captured by Logitech camera but they could surely confirm the tomato existed insect pests through the image captured by Canon camera. From the effect of contrast, the scheme could better satisfy the demand of agricultural the image analysis.

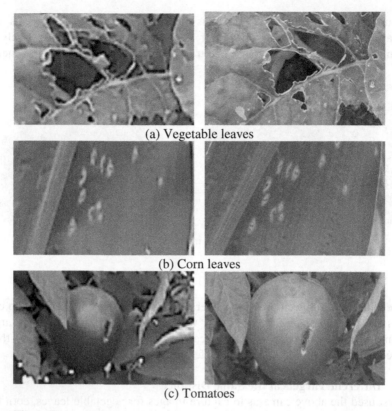

(a) Vegetable leaves

(b) Corn leaves

(c) Tomatoes

Fig. 8. The comparison of images for different targets at the same distance

4.1.2 Same Target in Different Distance

This test used Logitech pro9000 camera and Canon EOS 550D camera to capture images for the same target tomato at deferent distance respectively. Firstly, this test used the cameras to capture images at a distance of 0.5m to the target, then 1m, 2m, 4m, 8m, 16m. Then we used red rectangle to identify the target. Figure9 (a) showed images captured by Logitech pro9000 camera zooming to the original images of 17% and Figure9 (b) showed images captured by Canon camera zooming to the original images of 1%. However, they had the same display size in Fig.9. Therefore, the effective range of Canon camera image was 17 times as the Logitech camera.

| 0.5m | 1m | 2m | 4m | 8m | 16m |

Fig. 9. (a) Images captured by Logitech pro9000 camera (b) Images captured by Canon EOS 550D camera

Furthermore, Fig.10 showed the situation that each image in Figure9 amplified five times in target area. As we can see, at a distance of 0.5m, image captured by each camera could clearly show the target. However, at a distance of 1m, image captured by Logitech camera began to appear some noise to influence us to observe the target clearly. The farther the distance, the more noise generated, and we would be more and more difficult to observe the target. On the other hand, images captured by Canon camera could still clearly show the target at a distance of 1m, 2m and 4m, even 8m. Finally, at the distance of 16m, image captured by Logitech camera could not see the target while the one captured by Canon camera could still show the target clearly.

0.5m	1m	2m	4m	8m	16m

Fig. 10. (a) Magnified images captured by Logitech pro9000 camera (b) Magnified images captured by Canon EOS 550D cameras

4.2   Transmission Performance Analysis

Agricultural field environment high-quality image acquisition system transmitted images by using WIFI module on nodes. Data send by WIFI module passed through the router, eventually received by the server. The effective radius of a single high-power router was about 1.5km. With cascades of routers, it could improve the transmission range so that we could achieve long distance agricultural image acquisition.

This test captured images periodically through remote controlling camera and transmitted them to the remote server. The time interval was 10 seconds and this test totally captured 150 images. All images were successfully transmitted and the image transmission speed was from 0.72MB/S to 1.18MB / S, and the average speed of the transmission was 0.97MB / S. Fig.11 showed the result of this test.

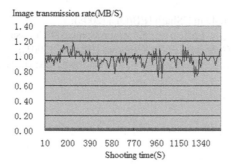

Fig. 11. Transmission performance analysis test result diagram

5   Conclusions

In this paper, aiming at the demand of current agricultural field environments on high-quality image acquisition, we designed and implemented an agricultural field environment high-quality image remote acquisition system based on Canon DSLR camera. And it used the system to conduct a series of experiments to test the feasibility of the system, and carried out comparative analysis for shooting accuracy, shooting distance and transmission performance. The scheme could achieve real-time high-quality image acquisition of agricultural field environment, to a certain extent, meeting the demand of agricultural high-definition image analysis. It was worth to mention that this scheme based on Fit-PC was a relatively high cost at the present stage. Therefore, the next step we should take was to migrate this system to a low-cost embedded platform in order to carry out large-scale applications

Acknowledgment. The authors would thank the project of the National Key Technology R&D Program (Project Name: Key Technologies for Agricultural Field Information Comprehensive Sensing and Rural Extension, Contact Number: 2011BAD21B01); the Natural Science Foundation of China (Contact Number: U0931001).

References

1. Wang, J., Guan, T.: Farmland image acquisition system based on GPRS and ARM. Journal of Agricultural Mechanization Research 10, 195–198 (2012)
2. Xiong, Y., Shen, M., Sun, Y., et al.: Design on system of acquisition and wireless transmission for farm land image. Transactions of the Chinese Society of Agricultural Engineering 42(3), 184–187 (2011)
3. Fan, F., He, D.: Agricultural remote monitoring system based on embedded linux. Computer Engineering 37(1), 249–253 (2011)
4. Wu, S.: The rate of transmission of the GPRS data system. China Data Communications 11, 20–23 (2002)
5. Zheng, Y., Tian, X.: Design and implementation of wireless image capturing system. Computer Engineering and Design 32(1), 110–113 (2011)
6. Canon, Canon EOS Digital SDK EDSDK2.10 API Programming Reference.2.10, Canon, 1-144 (2011)
7. John M.: Live View Sample in CSharp [EB/OL] (2008), http://tech.groups.yahoo.com/ group/CanonSDK/message/1176
8. MicroSoft, Still Image Connectivity for Windows [EB/OL] (2001), http://msdn.microsoft.com/en-us/windows/hardware/ gg463507.aspx

Study on Consultative Agricultural Knowledge Service System

Wang Xiguang

Agricultural information institute of CAAS, Beijing 100081, China
wangxiguang@hotmail.com

Abstract. Agricultural informatization has entered a rapid development period. With the improvement of basic rural information infrastructure and the Agricultural informatization resource increase, the information service models also need reform and innovation. The Consultative knowledge service system is people-oriented and support the user interactive information exchange, and it can also provide the remote technical consultation, Remote diagnosis and the final solution for the farmers. In this paper, we used open-source system OpenMeetings to design and secondary develop, and realize remote video communication among agricultural users, agricultural Technicians and agricultural professors. This system has been integrated in China Agricultural Technology Extension informatization platform, and support serving agriculture, rural areas and peasant and agricultural informatization. The actual test result shows that the system has effectively promoted Agricultural Technology Extension and information resources integration.

Keywords: Agricultural Technical knowledge service, Remote diagnosis, OpenMeetings, Agricultural Informatization.

1   Introduction

In recent years, the research on knowledge service has flourished in China and abroad, but the knowledge service does not have much progress in the agricultural industry. China is a large agricultural country, and to accelerate the popularization and promotion of modern agricultural technology is one of the key solutions to solve food security [1, 2]. Entering 21st century, the transformation of information technology in China's agriculture came into a stage of rapid development. The No. 1 Documents of the central government of China continued to focus on rural areas, agriculture and farmers for many years. With the explosive growth of agricultural information from the Internet and digital environment, the problem of how to enrich and extent the agricultural technicians knowledge and thereby enhancing the contribution of science and technology in China's agriculture and increasing farmers' income has become a significant part of the three dimensional rural issues. In order to improve the traditional agricultural technology extension methods, elevating the level of grassroots agricultural technology services, the state government attaches great importance to the reform and construction of the grassroots agricultural technology

D. Li and Y. Chen (Eds.): CCTA 2013, Part II, IFIP AICT 420, pp. 61–69, 2014.

extension system. The No. 1 Document of the central government in 2012 clearly stated that we should make full use of modern information technology and innovate the management and tools of grassroots agricultural technology extension. In order to focus on the farmers and effectively solve the various problems during agricultural production, to provide targeted agricultural technology consulting services, and to improve the quality of agricultural technology extension services, the new model of agricultural information technology is indispensable[3,4]. The knowledge service model of expert consultation establishes the interaction between agricultural experts, agricultural technicians, and farmers. Based on the online / offline communication, the one-on-one personalized consulting service can effectively narrow the distance between experts and grassroots farmers and agricultural technicians [5]. It can directly address the practical problems that occur in the production and provide corresponding solutions.

Remote video conference system originated from 1990s, gradually spreading nationwide since 2000. It is generally realized by audio/video compression and multimedia communication technology, supporting remote real-time information exchange and sharing as well as remote collaborate. Internet is utilized to satisfy geographically dispersed people's demand for audio and video communication in different regions. Currently it is mainly applied in international conference and remote teaching etc. Domestic farmers have the feature of wide geographical distribution, large user base and low computer skill. If hardware/software video conference systems are massively adopted, many arrangement and cost problems would occur. However, video conference system based on B/S framework is widely applied since it has low cost and easy deployment and usage.

At present there are many video conference systems based on Web browsers, of which OpenMeetings is a popular open-source video conference system with abundant features and multiple language packages. It is widely applied in many fields, e.g. Ma Guodong[6] (2011) proposed to engage reference service via video conference in library combining actual business of the library. As a result, favorable effect has been achieved through application of such system. Statistics shows that 97% townships and 80% administrative villages have basic Internet access in 2011. Such good network basic environment offers convenient conditions for expert consultation of knowledge service system designed for large number of peasants. In this case, the system is fully operational, laying solid foundation for agricultural development and scientific benefits of farmers. In this paper, based on sufficient research of OpenMeetings, design and secondary development are carried out, and all functions of agricultural expert consultation of knowledge service system have been released. Furthermore, through observation and actual test results, the feasibility of constructing consulting knowledge service system using OpenMeetings is confirmed.

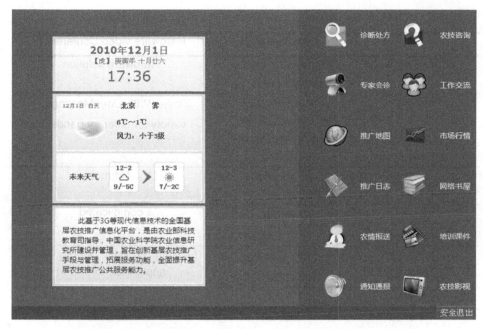

Fig. 1. China Agricultural Technology Extension informatization platform

2   OpenMeetings Introduction

As an open-source remote video conference system based on B/S framework, OpenMeetings can use current Internet access and limited hardware resources (PC, webcam, and microphone) to provide remote video, voice call, teaching board, real-time sharing and reading of documentary resources, instant message, interactive operation and etc. In this case, farmers' hardware requirements for video conference will be sharply reduced, simplifying users' operation. Furthermore, OpenMeetings' requirement for network bandwidth is also low, while user video default setting is 30 FPS with a video window of 320*240, which only requires an upstream bandwidth of about 50KB for user with stable network. Most ADSL network nationwide could satisfy OpenMeetings' need to maintain appropriate video call with other user. There are no limitations on usage and number of users [7].

Thus, compared to the hard wired video conference, OpenMeetings has more advantages in functionality, speed, network adaptability, usability and cost. Though OpenMeetings is not as good as video conference in respect of performance and image clarity, it can be deeply integrate with other network applications.

Table 1. OpenMeetings compare with Hardware video conference

	OpenMeetings	Hardware video Conference
Core Technique	streaming media service Red5	Hardware solution
Function	Text Message, Files Share, White broad, Video Record and others	Similar With OpenMeetings, variety for different bands
Speed	Fast after optimization	Normal
Deploy、 operation complexity	simply	complicated
Hardware Requirement	PC、 Camera, Microphone	Customized video Server, Client application for PC
Cost	Low	High
Period	Medium	Medium
Network applicability	Any	Customized Network

3   Consultative Knowledge Service System Design

Consultative knowledge service system, a key part of the information platform for grassroots agricultural extension, primarily provides remote expert consultative services to farmers and agricultural workers throughout the country, aimed to establish timely, accuracy and effective agriculture technology knowledge service. Through the expert consultative service system, face-to-face communication between the terminal user and experts is implemented and provide the direct and visualized remote agriculture consultative knowledge services.

The system compose with three modules real-time remote technical consulting services, remote diagnosis services for animal and plant pests and remote agricultural technology classroom. Real-time remote technical consulting service offers real-time technical answering services and technical advice for farmers, enabling experts and farmers communicate face-to-face on the Internet [8]. Remote-diagnosis services for animal and plant pests refer to an expert learning the symptom and diagnosing in the consultation room, through direct pictures, audio and video information captured by farmers in production fields with their 3G mobile terminals. Remote agricultural technology classroom enables the agricultural experts to start agricultural science and technology classes at any places with Internet access. The implementation process of consultative service system studied in this paper is as follows:

3.1   User Interface Reformation

User interface redesign is to realize by lzx script and second develop based on the OpenMeetings. In order to make the operation more convenient, some of the widgets move to the toolbar on the left. Some frequent used buttons are put on the top, such as white board, application, document upload, video record and screen sharing. Compared with default OpenMeetings interface, the video conference panel replaces the attribute panel, the reformatted interface as the Figure 2:

Fig. 2. Consultative knowledge service system User Interface

3.2   SSO (Single Sign On)

OpenMeetings executes 2 major steps to realize the SSO. First OpenMeetings use its integrated service to acquire SessionKey, and then generate a Hash string according the SessionKey to get the video conference room entrance URL address, which is similar with the external invitation URL. Meanwhile, the conference room type will be confirmed and Session will be initialed, and navigation menu resource will be released [9, 10].

3.3   Calendar Management

Administrators have the privilege to manage the calendar and assign the timesheet. The professor can provide the Consultative knowledge service and remote diagnosis services after his successful authentication. Then he can also invite some farmers or agricultural technicians to join the conference and start their face-to-face communication. Meanwhile the farmers or agricultural technicians consults the professors or ask him to on-line examine their production. If all participators sign-out and conference room is vacant, the conference session will be destroyed automatically by the system.

Farmer can check the calendar details by clicking the calendar link after he enters the conference room. The calendar displays corresponding professor's name and their timesheet for one week. The detail information includes sequence Number, professors' name, professors' description (specialty, resume and etc.) and timesheet period.

3.4   Video Conference Invitation

The invitation must be send out by the conference room builder, and the invited user must have been the authenticated. The users can invite the same professor at the same time, and the professor can accept more than one invitation. You can search your target users by name location and other parameter. The Process as the following:

3.4.1   Send out the Invitation

According to the users' parameter (Such as invitation user ID, the authenticated user ID, video conference room ID and etc.) generate an invitation record in the Database, and set the label as 1(1 means has been invited).

3.4.2 Query and Process the Invitation

If the user has been authenticated, the system will scan the database to check the invitation record. If the record is found, the user will be listed in the online user list which is located in the left side of the video conference interface.

3.4.3 Accept or Deny the Invitation

If the video conference room status is 1, the dialog will be pop up to ask the user to accept the invitation. If the user has been invited in the system and accepted the invitation, he will not be invitation again. While the users accept the invitation, he can enter the conference room directly and start the video communication [10].

3.5   Algorithms Optimization

For the agricultural users often use the low bandwidth connection to access our knowledge service system, we must consider such situation and try to obtain high-quality video communication. To enhance video communication representation in low-bandwidth network environment, 3 major solutions has been applied in the system, video frame reduction, compressed transmission, and bicubic interpolation algorithm.

3.5.1   Frame Rate Reduction

To ensure sound effects in video connections, our system will reduce 30 fps to 10 fps or 5 fps, while the client using low bandwidth connection. Coupled with the compression and transmission algorithm, the real-time traffic produced by video calls can decrease from about 100KB to about 20KB.

3.5.2   Compression and Transmission

With low bandwidth Internet connection, H.264 is the most efficient in video and voice coding and decoding, but it significantly raises requirements for the processing capacity of CPU. Through overall pre-research and testing, this platform finally adopts H.264 coding at low bandwidth and H.263 coding at high bandwidth, which is much flexible and provide better user experience.

3.5.3   Application of Bicubic Interpolation Algorithm

In order to reduce video transmission traffic and cope with the packets drop, Bicubic interpolation algorithm has been integrated. However, it takes in consideration of not only impact of gray value of directly adjacent point on the sampling point, but also impact from rate of gray value change between adjacent points. Thus, the bicubic interpolation algorithm can provide more accurate gray value of sampling points to be measured, with the disadvantage of large, complex and slow computation.

4   Test Result and Analysis

So far, expert consulting knowledge service system has implemented for half a year and has severed more than 1110 times for actual uses. In order to verify the application effect and enhance our system, appraisal system will be active at the end of video call each time, recording the call quality feedback of each time. The total effective appraisal we have received amounted to 286 times, of which positive feedback rate is more than 90%. Other knowledge service system is most based on library management system or some Management Information System. All of them are lack of information exchange and user interaction and cannot provide effective and accurate solutions. This system resolves such issues in an economic and practical way. The continuous strength of communication and interaction between agricultural members, farmers and agricultural specialists make it possible for specialists to provide face-to-face onsite training and service of solving problems step by step for agricultural members through distance video function. As a result, agricultural members unnecessary to speed large quantity of time on unfamiliar question but still not solving it, nor do they need to work hurry for fear of missing the suited weather for agricultural production. Through this way, level of serves of agricultural technology will be largely enhanced, making farmers satisfied truly on site. While providing consulting service, full play of guidance in agricultural fields is given to specialists to guide agricultural members and to cultivate their abilities of solving problems. Besides, functions such as work communication, knowledge sharing, independent learning and business service are also provided for the purpose of promoting agricultural members' active participation, self-directed learning and independent creation, making agricultural members learn from each other and progress together during the process of communication with specialists and other agricultural members[11]. In addition to this, the system also makes it possible for agricultural specialists to learn about the problems faced by farmers timely and to get the key point and development direction of agricultural scientific research, enabling specialists to work on significant issues of agricultural production and rural development and hot issues concerned by farmers.

Moreover, owing to the inconsistency of internet development across the country and high geographical heterogeneity, high-speed channel and low-speed channel are installed to ensure that users with different network speed can all obtain good voice quality and technical service. To testify the conversation delay and fluency in a multi-user environment, tests are conducted under the condition of concurrent users of 5, 10, 20 and 50 respectively. The specific data of test as in Table 2, the tested system is Ali Cloud Server Standard C and bandwidth is 10Mbps.

Table 2. Consultative knowledge service system test result for different Current Users

Current Users	Bandwidth for per-user(KB/s)	Server CPU Usage	Network Delay	Fluency
5 Users	85	34%	Less than 0.1s	Good
10 Users	105	36%	Less than 0.1s	Good
20 Users	129	40%	Less than 0.1s	Good
50 Users	152	45%	Less than 0.8s	Good

From the actual test result, CPU usage is much less than 50% when the current user is added to 50, we can see that the system does not rely on the CPU and consume much CPU resources. On the other hand, the server bandwidth increases with the current user number. During the test, it increases from 85kb/s (5 current users) to 152 kb/s (50 current users), Due to user exchange information and some control package. Limited by the test server bandwidth limitation, we did not execute the test cases for more than 50 current users. To solve such issues, we consider using the more servers and setup load balancing cluster.

5   Conclusions

Starting from the introduction of OpenMeetings' functions, this paper presents the basic structure and features of this system, all functions of consultative knowledge-based service system have been realized through applying the secondary development and integration of this system to basic agricultural technology extension information platform. The relatively low-cost development and deployment of this system make it possible for a full use of current network resource to shorten communication distance, which provides to agricultural production a direct, vivid and visual long-distance expertise. Consultative knowledge-based service system supports all kinds of platform and is available to most of browsers; it's so easy and simple that it effectively meets the need of long-distance consultation service. The test result of actual network operation has proved that OpenMeetings video conference system has a fine adaptability to current network in our country, even working in relatively slow networking bandwidth. In addition, basic network environment in rural area also offers a convenient condition for the coverage of expert consultative and knowledge-based service system to rural residents in our country, and in this way, it helps to give full play to the function of the system, which in turn lays a foundation for rural residents' obtaining of the achievement and development of agriculture. Though current system still has its flaws, for instance, large occupancy of server resource while recording a video, disconnection when the amount of users is oversized, but these problems will be studied and solved later in work.

Acknowledgment. Funds for this research was provided by the National Science and Technology Support Program(2011BAD21B01), that is Key Technologies for Agricultural Field Information Comprehensive Sensing and Rural extension.

References

1. Wu, Z.: Study on the New Knowledge Service System of Agricultural Science and Technology Based on the Farmers. Chinese Academy of Agricultural Sciences (2012)
2. Tan, C.-P., et al.: Research of Origin and Development of Agriculture Knowledge Service in China. Journal of Anhui Agricultural Sciences 12, 7440–7441 (2011)
3. Liu, J.: Thinking about innovation in the construction of agricultural knowledge service system. Agric. Network Inform. 10, 4–6 (2008)
4. Lu, Q.C., Li, Y.: Research on promotion process of knowledge service in the development of informatization in rural region of China. Technoecon. Manage. Res. 2, 92–95 (2009)
5. Zhou, G.-M., Qiu, Y., et al.: Design and Realization of Agriculture Knowledge Service System Based on Internet. Journal of Library and Information Sciences in Agriculture 02, 238–240 (2005)
6. Ma, G.: Applications of the Video-Conference Based on Open Meetings in Library. Journal of Modern Information 01, 146–149 (2011)
7. Open Meetings Home Page (May 10, 2013),
   http://openmeetings.apache.org/
8. Li, J., Wang, J., Wu, W., et al.: Design and Implementation of Web Video Conferencing System Based on Red5. Radio Communications Technology 06, 56–58+76 (2012)
9. Peng, L., Zhao, R.: The Research of Video Conference System on OpenMeetings. Computer Knowledge and Technology 12, 2909–2911+2922 (2011)
10. Lu, M., Wang, X.-D.: Design of Video Conference System Based on Techniques of FLV Streaming Media. Journal of Jilin University (Information Science Edition) 02, 186–190 (2010)
11. Tong, Y., Hu, W., et al.: A Review on the Video Quality Assessment Methods. Journal of Computer-Aided Design & Computer Graphics 05, 735–741 (2006)

Stochastic Simulation and Application of Monthly Rainfall and Evaporation

Nana Han and Yang-ren Wang

Department of Hydraulic Engineering, Tianjin Agricultural University, Tianjin 300384
{hnn23144,wyrf}@163.com

Abstract. Statistic was done for the precipitation and evaporation monitoring data of Yuncheng from 1971 to 2007. The first-order seasonal autoregressive models were set up, considering separately for the normal and skewed distribution of rainfall and evaporation. Long series of monthly precipitation and evaporation sequences were generated. Compared the average, standard deviation and other parameters of simulation sequences with measured sequence, the results of skewed simulation had better agreement with the measured one. Then it proved that skewness model maintains the main statistical characteristics of the measured sequence. 20 groups of equal length with the measured sequence of precipitation and evaporation sequence could be generated through the skewness model. The sequence of irrigation water of the winter wheat multiple corns planting type had been calculated, using the "α" value method. And comparative analysis with the measured result was performed. The simulation computed result was found in good coincidence with the measure computed result.

Keywords: precipitation, evaporation, seasonal autoregressive model, irrigation water.

1 Introduction

Crop irrigation water can be determined by the field water balance method, using the precipitation and evaporation data. Based on the long series of data, the scale of irrigation projects can be determined. Thus many of the data needed, actually the measured data were limited. Aiming at the randomicity of precipitation and evaporation, the precipitation and evaporation were randomly treated, and simulated using random hydrology methods. Then Long series of irrigation water sequence can be calculated in order to develop irrigation schemes more in line with actual crop water demand. Azhar [1] considered evapotranspiration, precipitation randomness, introduced a minimum of irrigation water requirement of rice cumulative probability distribution, and used to guide irrigated rice irrigation. Huan-Jie Cai [2] simulated the daily precipitation and studied the irrigation problems of wheat and maize. Zuo Xiaoxia [3] generated long series of precipitation and evaporation by the stochastic hydrology method. Then a long series of rice irrigation process were calculated using the cross-correlation of rainfall evaporation. Wen Ji [4] simulated the evapotranspiration,

D. Li and Y. Chen (Eds.): CCTA 2013, Part II, IFIP AICT 420, pp. 70–78, 2014.

and the results derived stochastic simulation model of winter wheat irrigation. Zai Song-mei [5] simulated the precipitation and evapotranspiration using random hydrology methods based on time-series, and then the simulated values were used into the water balance equation to determine the irrigation time and amount. Wen Ji [6] analyzed the stochastic simulation techniques of crop irrigation and pointed out that it has strong universality. This paper simulated the sequences of precipitation and evaporation by the seasonal autoregressive model, and calculated the sequence of crop irrigation water in Yuncheng.

2 Materials and Methods

2.1 Model

Seasonal autoregressive model is used to simulate monthly precipitation and evaporation. The general form of seasonal autoregressive model is as follows [7]:

$$Z_{t,\tau} = \varphi_{1,\tau} z_{t,\tau-1} + \varphi_{2,\tau} z_{t,\tau-2} + ... + \varphi_{p,\tau} z_{t,\tau-p} + \varepsilon_{t,\tau} \tag{1}$$

Where $Z_{t,\tau}$ is standardized sequence, $\varphi_{1,\tau}$、$\varphi_{2,\tau}$、 ... $\varphi_{p,\tau}$ are respectively the auto regression coefficients of the τ season from 1 to p orders, and $\varepsilon_{t,\tau}$ is the independent random sequences of the t year of the τ season.

Standardized sequence which is to eliminate the seasonal effects of average、 variance can be obtained by standardizing the hydrological series. Formula is as follows.

$$z_{t,\tau} = \frac{x_{t,\tau} - u_\tau}{\sigma_\tau} \tag{2}$$

Where u_τ is average, σ_τ is standard deviation, and $x_{t,\tau}$ is seasonal hydrological sequence(t=1,2, ..., n, n is the number of years ; τ =1,2, ..., w, w is the number of seasons), which can be represented through matrix.

$$
\begin{bmatrix}
x_{1,1} & x_{1,2} & x_{1,3} & \cdots & x_{1,w} \\
\vdots & \vdots & \vdots & \vdots & \vdots \\
x_{t,1} & x_{t,2} & x_{t,3} & \cdots & x_{t,w} \\
\vdots & \vdots & \vdots & \vdots & \vdots \\
x_{n,1} & x_{n,2} & x_{n,3} & \cdots & x_{n,w}
\end{bmatrix} = \left\{ x_{t,\tau} \right\}_{n \times w}
$$

Studies show that a first or second order model will be able to meet the requirements. Thus this article uses a first order seasonal autoregressive model. When p = 1, the model is as follows.

$$z_{t,\tau} = \varphi_{1,\tau} z_{t,\tau-1} + \varepsilon_{t,\tau} \qquad (3)$$

2.2   Estimation of Model's Parameters

Parameters of a first order seasonal autoregressive model are u_τ 、 σ_τ^2 、 $\varphi_{1,\tau}$ 、 and $\sigma_{\varepsilon,\tau}^2$ ($\tau = 1,2,...w$). Estimate of u_τ and σ_τ^2 can be separately reached with the average ($\bar{x}_\tau$) and variance (s_τ^2) of the sample.

$$\hat{u}_\tau = \bar{x}_\tau = \frac{1}{n}\sum_{t=1}^{n} x_{t,\tau} \qquad (4)$$

$$\hat{\sigma}_\tau^2 = s_\tau^2 = \frac{1}{n-1}\sum_{t=1}^{n}(x_{t,\tau} - \bar{x}_\tau)^2 \qquad (5)$$

$$\varphi_{1,\tau} = \hat{P}_{1,\tau} = r_{1,\tau} = \frac{\sum_{t=1}^{n}(x_{t,\tau} - \bar{x}_\tau)(x_{t,\tau-1} - \bar{x}_{\tau-1})}{(n-1)s_\tau s_{\tau-1}} \qquad (6)$$

Where $r_{1,\tau}$ is the τ season's first-order sample autocorrelation coefficient of the $x_{t,\tau}$ sequence, which indicates the linear correlation between the τ and $\tau-1$ season.

If the sequences of precipitation and evaporation are considered normal one, then the independent random sequence $\varepsilon_{t,\tau}$ is pure normal distribution. The formula is as follows.

$$\varepsilon_{t,\tau} = \sigma_{\varepsilon,\tau}\xi_{t,\tau} \qquad (7)$$

$$\sigma_{\varepsilon,\tau} = \sqrt{1 - r_{1,\tau}^2} \qquad (8)$$

$$\xi_\tau = \sqrt{-2\ln u_{1,\tau}}\cos 2\pi u_{2,\tau} \qquad (9)$$

Where $u_{1,\tau}$ and $u_{2,\tau}$ are uniformly distributed random numbers, and ξ_τ is the independent standard normal pure random sequence, namely $\xi_\tau \sim N(0,1)$. $\xi_{t,\tau}$ can be obtained by a lot of $u_{t,\tau}$.

If the sequences of precipitation and evaporation are considered skewed one, then the independent random sequence $\varepsilon_{t,\tau}$ is skewed distribution. In hydrology and water resources, the distribution of $\varepsilon_{t,\tau}$ more considers is a P-III type one. The formula is as follows.

$$\varepsilon_{t,\tau} = \sigma_{\varepsilon,\tau}\Phi_{t,\tau} \tag{10}$$

$$\Phi_{t,\tau} = \frac{2}{C_{S_{\Phi,\tau}}}(1+\frac{C_{S_{\Phi,\tau}}\xi_{t,\tau}}{6} - \frac{C_{S_{\Phi,\tau}}^2}{36})^3 - \frac{2}{C_{S_{\Phi,\tau}}} \tag{11}$$

$$C_{S_{\Phi,\tau}} = \frac{C_{S_{x,\tau}} - \varphi_{1,\tau}^3 C_{S_{x,\tau-1}}}{(1-\varphi_{1,\tau}r_{1,\tau})^{3/2}} \tag{12}$$

$$C_{S_{x,\tau}} = \frac{1}{n-3}\frac{\sum_{t=1}^{n}(x_{t,\tau} - \overline{x}_\tau)^3}{s_\tau^3} \tag{13}$$

Where $\Phi_{t,\tau}$ is standardized P-III distributed random variable, $C_{S_{\Phi,\tau}}$ is coefficient of skewness of $\Phi_{t,\tau}$, and $C_{S_{x,\tau}}$ is coefficient of skewness of $x_{t,\tau}$.

2.3   Calculation of Irrigation Water

Irrigation water is the required amount of water of irrigated land, which is drawn from the source. It depends on the factors such as the irrigated area, crop cultivation, soils, hydrogeological and meteorological conditions. Farm irrigation water is calculated as follows:

$$M = \alpha E_0 - \beta P \tag{14}$$

Where M is crop irrigation water of the crops' whole growth period (mm), α is water demand coefficient, which determined by the measured data (see table 1), E_0 is water surface evaporation of the crops' whole growth period (mm), P is precipitation of the crops' whole growth period (mm), and β is effective utilization coefficient of precipitation, which determined by relevant information, the value of β is 0.9 in the months of 11,12,1-4,and in other months is 0.8.

Table 1. The value of α

month	1	2	3	4	5	6	7	8	9	10	11	12
α	0.4	0.552	0.665	0.722	0.74	0.743	0.75	0.83	0.73	0.397	0.593	0.462

3   Results and Discussion

3.1   Applicability Analysis of the Model

Statistic was done for the precipitation and evaporation monitoring data of Yuncheng from 1971 to 2007. Considering separately for the normal and skewed distribution of rainfall and evaporation, this paper set up the first-order seasonal autoregressive models. Then three groups of one hundred length sequences were generated. And thus this paper analyzed the parameters of model and application of model. The results were shown in Table 2 and Table 3.

Table 2. Results of model suitability testing of Yuncheng district's monthly precipitation

Parameter		monthly precipitation•mm•											
		1	2	3	4	5	6	7	8	9	10	11	12
$\overline{x}$	measured	6.2	7.4	17.2	35.4	46.0	64.6	106.3	83.6	99.8	50.3	23.1	6.4
	Simulation (normal)	6.5	7.3	17.7	34.3	45.9	66.6	104.3	84.7	88.5	50.1	22.7	6.5
	Simulation (skewness)	6.1	7.6	19.2	33.4	47.0	62.9	106.2	82.3	98.1	50.1	23.1	6.9
σ_τ	Measured	7.9	7.0	15.7	26.1	32.7	50.6	61.1	55.7	97.5	35.3	30.6	7.5
	Simulation (normal)	8.2	6.9	17.0	26.2	31.3	50.4	51.7	58.0	107.3	33.8	31.2	7.2
	Simulation (skewness)	7.3	6.7	15.9	24.6	34.4	49.2	54.2	50.6	91.8	34.6	29.1	7.1
Cs	Measured	2.044	1.148	1.215	1.451	0.868	2.600	0.835	1.286	3.176	0.789	2.578	0.998
	Simulation (normal)	0.103	0.022	0.129	0.147	0.054	0.002	-0.022	-0.053	-0.247	-0.062	0.192	0.127
	Simulation (skewness)	1.525	0.810	0.932	1.074	0.754	2.144	0.680	1.004	2.269	0.773	1.841	0.871
$r_{1,\tau}$	Measured	0.061	0.117	0.113	-0.079	-0.086	0.048	-0.157	0.306	-0.403	0.053	-0.033	-0.067
	simulation (normal)	0.039	0.161	0.199	-0.066	-0.032	0.073	-0.126	0.298	-0.421	0.100	0.105	0.029
	Simulation (skewness)	0.089	0.126	0.136	-0.106	-0.081	0.156	-0.091	0.353	-0.301	0.021	-0.037	-0.121

Table 2 and Table 3 showed that the average, standard deviation, coefficient of skewness and first autocorrelation coefficient of the skewness sequence all well maintained for the measured one. However the standard deviation and the first order autocorrelation coefficient could be maintained and the other parameters varied greatly for the measured sequences, such as the average of simulated precipitation in September deviated larger from the measured one (see Figure 1), and the coefficient of normal sequence deviated largest from the measured one (see Figure 2). The results showed that the normal and the measured precipitation sequences didn't come from the same total. But the parameters of normal evaporation sequence remain good, except the coefficient of skewness.

Table 3. Results of model suitability testing of Yuncheng district's monthly Evaporation

Parameter		monthly evaporation (mm)											
		1	2	3	4	5	6	7	8	9	10	11	12
$\overline{x}$	measured	50.5	87.2	155.8	195.6	256.6	316.0	285.6	258.8	171.3	127.5	74.1	50.9
	Simulation (normal)	49.8	87.9	157.4	194.1	258.0	319.8	290.8	264.3	172.9	129.7	74.7	51.6
	Simulation (skewness)	50.2	86.2	156.3	197.0	256.6	319.5	285.6	253.1	169.8	124.7	74.8	51.7
σ_τ	measured	13.2	21.0	38.5	29.0	63.6	73.8	68.9	70.5	44.4	36.2	18.0	11.4
	Simulation (normal)	12.8	21.1	35.8	28.0	64.8	72.4	64.3	73.1	44.4	35.7	19.6	11.9
	Simulation (skewness)	13.7	19.4	37.7	30.7	64.5	76.3	65.1	73.7	42.8	36.3	19.8	10.9
	measured	0.471	0.965	1.375	-0.172	-0.134	0.240	0.233	0.776	0.126	0.306	0.509	0.633
C_s	Simulation (normal)	0.078	0.087	-0.190	-0.098	0.115	0.130	0.122	-0.142	0.016	0.369	-0.042	0.133
	Simulation (skewness)	0.681	0.696	1.590	-0.024	-0.338	0.285	0.282	1.044	0.238	0.592	0.576	0.510
	measured	0.219	0.375	0.328	0.109	0.293	0.567	0.295	0.417	0.335	0.634	0.237	0.391
$r_{1,\tau}$	Simulation (normal)	0.190	0.439	0.347	0.078	0.376	0.564	0.327	0.373	0.332	0.613	0.239	0.408
	Simulation (skewness)	0.124	0.325	0.324	0.131	0.316	0.530	0.293	0.374	0.413	0.678	0.372	0.373

In addition, negative values were found in the simulated sequence of precipitation regardless of the normal or skewed distribution. It did not match the actual situation. Thus statistic was done for the sum of the negative number and negative value in three groups of 100 length simulated sequences. The results were shown in table 4. It also could be seen from table 3 that not only the number of negative in the skewness sequence was less than in the normal one, decreased by nearly 50%, but also the negative sum in the skewness sequence was much smaller than the normal one, decreased by nearly 90%. The result showed that precipitation sequence should be simulated by the skewness model.

Table 4. Statistics of the number and the negative sum of the negative in the simulated precipitation sequence

item	Simulation (normal)				Simulation (skewness)			
	sample1	sample2	sample3	average	sample1	sample2	sample3	average
Number of the negative	150	138	168	152	67	66	93	75
Sum of the negative	-2676.5	-2800.7	-2776.4	-2751.2	-274.9	-274.6	-439.1	-329.5

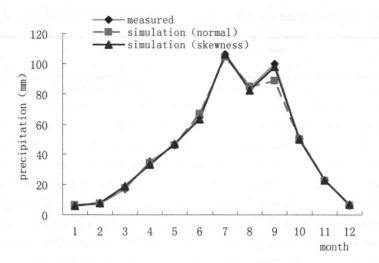

Fig. 1. Comparison of the average of precipitation simulated and measured sequences

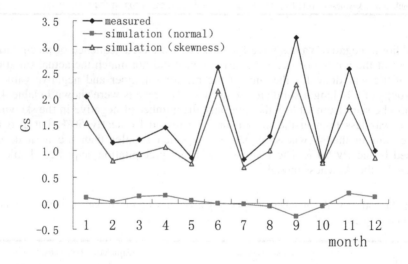

Fig. 2. Comparison of the coefficient of skewness of precipitation simulated and measured sequences

3.2   Analysis of Irrigation Water's Frequency Curve

Through the skewness model, 20 groups of equal length with the measured sequence of precipitation and evaporation sequence can be generated. The sequence of irrigation water of the winter wheat multiple corns planting type had been calculated using the formula (12). Then based on the measured sequence of precipitation and evaporation, this paper calculated the measured irrigation water sequence using

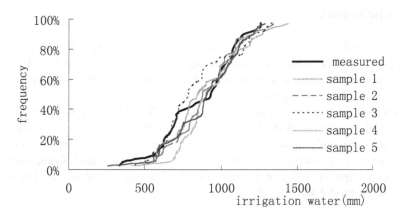

Fig. 3. The frequency curves of irrigation water

the " α " value method. Statistic was done for the data of irrigation water by frequency calculation method. Finally, five simulation groups of irrigation water's frequency curves were compared with the measured one. The results were shown in figure 3.

Figure 3 showed that simulated frequency curves fit the measured one good. 20 groups of equal length with the measured sequence of simulated irrigation water were compared with the measured one. The results were shown in table 5 and 6. As can be seen from table 5, the average, standard deviation and coefficient of skewness of simulated sequences well maintained for the measured one. Table 6 showed that the differences between simulated irrigation water sequences and measured one were little, and the difference between the maximum and minimum values was from 100 to 200 mm. If approximatively taking the average of 20 simulated groups as the average of total, the measured values of irrigation water were lager than the total's average between the 50% and 90% frequencies. Thus if the scale of irrigation projects was determined using the measured sequence, the result could be larger than the actual.

Table 5. Statistical parameters of irrigation water simulation sequences and measured values sequences

	$\bar{x}$	σ_τ	C_s
measured	863.6	242.1	-0.333
simulation average	865.9	217.8	-0.306

Table 6. Comparison of measured and simulated values at different frequencies (mm)

frequency	measured	simulation	max	min	max-min
25%	670	717	810	636	174
50%	932	883	942	824	118
75%	1069	1027	1086	944	142
90%	1141	1134	1205	1036	169
95%	1224	1256	1315	1122	193

4   Conclusions

Precipitation and evaporation of Yuncheng were simulated by the seasonal autoregressive model, which considered separately for the normal and skewed distribution of precipitation and evaporation. Long series of monthly precipitation and evaporation sequences were generated. The simulation results showed that the skewed simulation had better agreement with the measured. Then based on the data of skewed distribution, this paper established the relations formula of water evaporation and crop water requirement, and calculated the crop water requirement. The long series of crop irrigation water were obtained through the water balance calculation. Comparing with the measured irrigation water, it verified the practicability of the model. The results showed that the model was feasible in allowable error.

Acknowledgment. Funds for this research was provided by the National Science and Technology Support Program (2012BAD08B01).

References

[1] Azhar, A.H., Murty, V.V.N., Phiem, H.N.: Modeling irrigation schedules for lowland rice with stochastic rainfall. Journal of Irrigation and Drainage Engineering 118(1), 36–55 (1992)

[2] Cai, H.: Stochastic simulation of Precipitation and its application in irrigation. Irrigation and Drainage 10(2), 8–14 (1991)

[3] Zuo, X.: Irrigation process of paddy based on random simulation of precipitation and evaporation. Hohai University, 6 (2005)

[4] Wen, J., Guo, S., Guo, D.: Stochastic simulation model for winter wheat irrigation. Transactions of the CSAE 21(11), 25–28 (2005)

[5] Zai, S.-M., Wen, J., Guo, D.-D.: Stochastic Simulation for Crop Water Requirement. Journal of Irrigation and Drainage 28(5), 92–95 (2009)

[6] Wen, J., Guo, S., Lu, W.: Research of Stochastic Simulation Technology of Crop Irrigation. Yellow River 26(5), 39–41 (2004)

[7] Ding, J., Liu, Q.: Stochastic hydrology, pp. 17–27. China Water Power Press, Beijing (1997)

Elimination Method Study of Ambiguous Words in Chinese Automatic Indexing

Wang Dan[1,2], Yang Xiaorong[1,2], and Zhang Jie [1,2]

[1] Institute of Agricultural Information, Chinese Academy of Agricultural Sciences,
Beijing 100081, China
[2] Key Laboratory of Agricultural Information Service Technology (2006-2010),
Ministry of Agriculture, The People's Republic of China
{wangdan01,yangxiaorong,zhangjie02}@caas.cn

Abstract. Faced with huge amounts of information to realize the accurate retrieval under the network environment, the first step is indexing words cannot appear ambiguity word. Because Chinese's the basic unit is Chinese characters, Chinese characters form words, Word is divided into monosyllabic word and compound word, and there's no space between Chinese keywords and there are a lot of ambiguous concept. Therefore a lot of ambiguity in the indexing process will be produced. The result detected information of irrelevant or mistakenly identified. The paper focuses on a method to eliminating the crossed meanings ambiguous words in the automatic indexing. The paper puts forward a method to eliminating ambiguous words combined algorithm of exhaustive method and disambiguation rules. Experiments show that it can avoid a great lot segmenting ambiguities with better segmenting results.

Keywords: Chinese text, Automatic indexing, Keyword extraction, Ambiguous words, Elimination algorithm.

1   The Research Status

Chinese is one of the major languages in the world, and also spoken by the largest number of people in the world. It has become an important means that people get a lot of meaningful information from the voluminous network information by network technology. However, because Chinese's the basic unit is Chinese characters, Chinese characters form words, word is divided into monosyllabic word and compound word, and there's no space between Chinese keywords. Chinese words unlike western languages separated by spaces, Western processing (indexing) technology does not easily draw. This is the reason that Chinese information processing is difficult. Because of the difficulty of Chinese information processing, in order to obtain accurate information from the network information is harder. Chinese information processing, namely Chinese automatic indexing technology research began in the early 1980s, a lot of these skilled personnel on 1990s, a lot of papers published so far in the ascendant. After 30 years of research and practice, Chinese information

D. Li and Y. Chen (Eds.): CCTA 2013, Part II, IFIP AICT 420, pp. 79–88, 2014.

technology has been greatly developed, especially in the Chinese word segmentation techniques with some of the more mature approach. For example, forward sweep, reverse sweep, maximum matching algorithm for network information processing and retrieval provides a powerful tool. However, due to the complexity of Chinese information processing, the Chinese word segmentation process produces a lot of ambiguous words, resulting in retrieving information is not accurate, Especially in the vast network information, if given a search term, instead of not retrieve information, but to retrieve a large number of irrelevant information, readers need to get useful information after several rounds of screening. People prefer to go through a search operation will be able to get accurate information. In this article automatic indexing method to eliminate ambiguous words is to solve the problem of network information accurate retrieval. From the scope of information retrieval, the Chinese indexing ways have hand information indexing and automatic indexing by computer. The former is indexing staff through reading, analyzing literature, which precipitated a keyword and norms, and finally given this literature indexing terms, although indexing words good accuracy, retrieval efficiency, but the vast amounts of information need to be addressed in the current status, the artificial indexing alone is not possible, the need for computer technology to process vast amounts of information. Information processing in the network, especially when Chinese information processing automatically generate a lot of ambiguous words, for words to eliminate ambiguity generated methods of information retrieval is a hot technology, but also information retrieval technology's basic research, it is great significance to accurate network information retrieval.

2   Ask Questions

Chinese literature indexing are that indexers generally extract and record keywords or class number which have meaningful literature retrieval features from Chinese literature by analyzing the content of the literature. These keywords will be used as the basis for document retrieval. Generally to retrieve pertinent documents, First step, these keywords are subject indexing by the indexing staff. The second step, after these keywords are indexed by the computer processing, people can carry out precise searches, the prefix search operation and rear consistent retrieval operation. Chinese automatic indexing is the process of extracting keywords to achieve by computers. With the rapid development of information technology, Computer technology is increasingly used in Chinese text indexing. Keywords and class number are extracted from the Chinese text by computer technology according to some word segmentation algorithm and matching rules. For Chinese literature, these keywords for the expression of the concept of literature are contained in document titles, abstracts and text. But there are three problems which are no spaces between Chinese keywords, blurred boundaries of words and phrase, containing ambiguous words in words and phrase in Chinese literature. For example, "President Jiang Zemin" the words, has expressed a complete theme concept, which can be given as a keyword indexing. But in Chinese it also contains "Jiang Zemin", "democracy" and "chairman" three words to express the three concepts. "Jiang Zemin" and "chairman" two words can also be given as a keyword indexing, but the word "democracy" is also given as a keyword indexing is produced ambiguous word indexing. Such an ambiguous word in the case of manual indexing does not appear, but with a computer indexing, if not treated, often appear. For another example, "the People's Republic of China" in the word extraction process can also put "Chinese" is extracted, can also cause ambiguity word indexing. The paper focuses on a method to eliminating the crossed meanings ambiguous words in the automatic indexing.

3   Indexing Algorithm

There are two methods which extract the keywords or phrases from the Chinese literature in the current study. One is a method of rule-based segmentation[1], another is a method based on statistical analysis[2] of the sub-word. The former requires knowledge database for support, the latter does not need knowledge database and save part of the workload, but the search results are poor, for the accurate retrieval needs further filter the search results. I believe that the combination of the two methods used in automatic indexing will greatly improve the effectiveness of automatic indexing.

3.1   Automatic Indexing System Frame

This systematic framework for automatic indexing system and the elimination of ambiguous words is shown in Figure 1.

First step, the text is preprocessed by automatic indexing system, including some removing punctuation, extraction of feature words which are enclosed with a special symbol directly as indexing terms. The second step, the pretreatment of the text is processed and filtered by stop word list which include empty words and common words to get words or phrases, collection of short sentences. The third step, these words or phrases or short sentences are pumped word processing to obtain candidate keywords by the common vocabularies and professional vocabularies. Final step, Keywords are gave the corresponding weights based on the text of word frequency and occurrence of the word, which are sorted according to the statistics, and then according to presetting threshold value of keywords, keywords are selected by automatic indexing system.

Segmentation processing of acquired longer keywords by matching vocabulary words will produce ambiguous words. The removingmethods of ambiguous words are discussed in the next section.

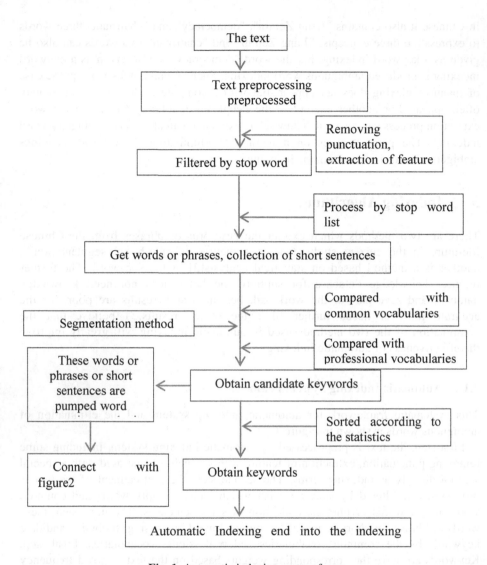

Fig. 1. Automatic indexing system frame

3.2   Text Preprocessing

First of all, information with a special symbol is extracted by automatic indexing system, including the title with quotation marks to cause, or place names and the person's name with a special symbol to cause. And then preprocessing information is preliminary segmentation processed by stop word list and removing punctuation to get words or phrases, collection of short sentences.

3.3   The Structure of Knowledge Database

Knowledge databases consists of stop word vocabularies, general and professional thesaurus, which are the basis for automatic indexing, its quality directly affects the effectiveness of automatic indexing words.

The text with removing punctuation is slicing processed by stop word vocabularies which include the commonly used function words in Chinese, for example, prepositions, conjunctions, auxiliary, and also include generic words. The use of stop words vocabularies can filter out unwanted words and rough cut of the text, in order to speed up text processing speed.

Word vocabulary is an important basis for containment of the keywords in the process of word segmentation. In order to accelerate the matching speed and accuracy of word segmentation, and general vocabulary can be divided into general vocabulary and professional vocabulary. "Classified Chinese Thesaurus" and all kinds of professional thesaurus have been expanded and modified to do common vocabularies and specialized vocabulary.

3.4   Word Segmentation Method

3.4.1   Forward Longest Matching Method
The strings obtainedby coarse segmentationhave been verbatim scanned from left to right and match with Thesaurus,and the keywords of thesauri maximum matching as the primary keywords. For example, in thesaurus in the "cadres tenure" in Chinese,and also included "cadres"、 "office"、 "age". Longest matching method is that "A short length is not taken" the word extraction rules, only extracting "cadres tenure" is used.

3.4.2   Reverse Longest Matching Method
The stringsobtained by coarse segmentation have been verbatim scanned from right to left and match with Thesaurus, and the keywords of thesauri maximum matching as the primary keywords.

First of all, according to the longest matching rule,keywords which have been matched with the thesauri have been extracted as primary keywordsin this article. Secondly,when strings length of primary keywordsare greater than or equal to four Chinese words, and then slicing process, may produce ambiguous words which this article discusses.The smallest unit of slicing process is two Chinese characters.

3.5   The Frequency and Weight of Keywords

According to the important degree of each part in the text, divided the parts identified, given the size of the contribution to content weight of the text before the text is pretreated.

3.5.1   Text Area Value

The weights of title, abstract and keywords in the text should be different, the former is big and the latter is small. Actually, the weights of the keywords from the title should be absolutely great in order to ensure that the word appears in the indexing words.

3.5.2   Important Statement

Article subtitled or weight of keyword in each of the first paragraph or the end of the paragraph statement is greater than weight of keywords in the body.

3.5.3   Word Frequency Statistics

According to the frequency and the weight of the keywords, the keywords to obtain by segmentation have been counted and sorted. According to the indexing depth which is maximum number of index words, the final text keywords of text have been gave. In accordance with the literature reports, the average depth of manual indexing are 7,usually the average depth of automatic indexing are 10 to 15.

3.6   Long Keyword Processing

After the keywords obtained by two-way scan the maximum matching have already been independent retrieval concepts, which directly access to a retrieval system for indexing processing, and providing search services in automatic indexing. But some of these words are very long term, word in the more length of the keywords not only contains the independence concept, but also has search significance, if not for slicing process, will be lost, causing leakage marked on the automatic indexing system. In general, the keywords of the more length should be re-carved process, slicing process may produce above-mentioned ambiguous words.

4   Produce Ambiguous Words

4.1   Type of Ambiguous Words

Ambiguous word is defined by different segmentation methods produce non-paper meaning of the word. Ambiguous words have two types of cross-type ambiguous words (cross ambiguity) and the combination ambiguous words (covering ambiguity). According to statistics, the cross ambiguity words accounted for 86% of the total, so to solve cross ambiguity words is the key of the segmentation of words. To exclude ambiguous words in automatic indexing algorithm is cross ambiguous words.

4.2   Methods to Disambiguate Words[4]

Currently the typical method of disambiguate words are:

4.2.1   Exhaustive Method

In general, exhaustive method is to find all possible words in the Chinese string to be split. Most matching algorithm is used in the forward or reverse matching algorithm method of exhaustion, or combination of forward or reverse matching algorithm exhaustive methods. When segmentation word is not correct, this method will produce ambiguous words.

4.2.2   Lenovo – Backtracking

Li Guochen et al [5] proposed Lenovo – Backtracking. First of all, Chinese string to be split according to feature words have been divided into several sub-strings, each sub-strings is either single word or word group, and then word group is subdivided into words by the entity thesaurus and rule base, when word Segmentation, a certain grammar knowledge is used.

4.2.3   Phrase Matching and Semantic Rules Law

Yao Jiwei, Zhao Dongfan[6] proposed a combination disambiguation method of a local single phrase matching and semantic rules based on the phrase structure grammar.

4.2.4   Part-of-speech Tagging

BaiShuanhu[7] eliminated ambiguity words by the combining method of Markov chain tagging technology and word segmentation algorithm

The paper puts forward a method to eliminating ambiguous words combined algorithm of exhaustive method and disambiguation rules.

5   Ambiguous Words Elimination Algorithm

The elimination algorithm of words ambiguous this article discusses is reprocessed the candidate words have already been cut a longer keyword or word group. Longer word is a meaningful indexing terms which appears in the dictionary as indexing words.

If no longer word segmentation processing, the leakage phenomenon of keywords may occur. In order to no-produce leakage phenomenon, we need word segmentation processing again in order to increase the literature retrieval point. Cut out of the word may appear ambiguous word, such as in the word "Jiang Zemin" the ambiguous word of "democracy".

5.1   The Disambiguation Process

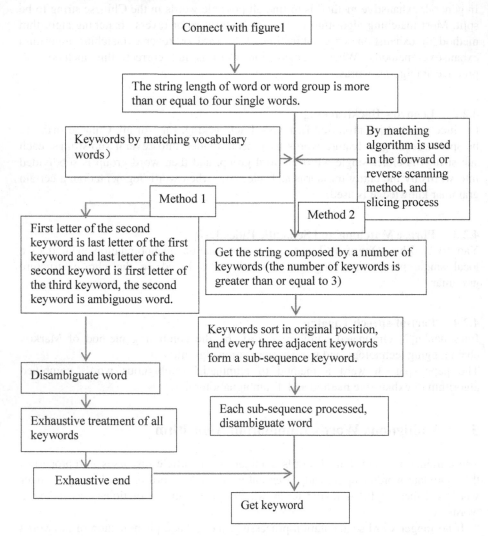

Fig. 2. The Disambiguation process

5.2   Disambiguation Algorithm

The longer Chinese words is set, it's the string length of are greater than or equal tofour single words. Without considering the single case of Chinese characters, maybe 3 or more than 3 of keywords have been cut with different methods of separation. After matching with keyword thesaurus, an N (N≥3) consisting of a sequence of keyword strings which referred to a large sequence string has been obtained. At this point, disambiguation in two ways:

Method 1: The sequence of each keyword in accordance with the Chinese characters string (longer keyword) sort order, in turn the adjacent three keyword consisting sub-sequence string, thus combined into several sub-sequence of strings, and then disambiguation process separately for each sub-sequence strings. That is, first letter of the second keyword is last letter of the first keyword and last letter of the second keyword is first letter of the third keyword, the second keyword is ambiguous word. Thus, after processing for each sub-sequence string, we eliminate all ambiguity word.

Method 2: Premise condition: Without permutations for large sequence string and do not form several sub-sequence string. That is, the first and the last letter of each keyword (referred to as A word) of large sequences check with the first and the last letter of the rest of the keywords. If the above conditions are satisfied simultaneously, namely, the first letter of A word is the last letter of one of the keywords and the last letter of A word is first letter of another keywords, then A word is ambiguous word. After the above process for each keyword, all ambiguity word have been eliminated.

5.3 Sample and Analysis

The preselected words which had been split match with the common vocabulary or professional vocabulary. Such as "President Jiang Zemin", "the People's Republic of China" ,"Major general", "East China Sea Fisheries distribution", and after segmentation produced ambiguous words "Democracy", "Chinese", "General" and "Seawater" are common vocabulary words. After the above disambiguation algorithm processing, we can eliminate these ambiguous words in turn.

In addition, specialized vocabularies such as Agricultural Thesaurus [8]often have a class of words. For example, for the following words through the segmentation processing, hyponym of "microbial fertilizer" (CLC as S114);anti bacteria fertilizer、 rhizobium fertilizer、 azotobacter fertilizer, the "fertilizer" word (hypernym) are also extracted as keywords. If "fertilizer" is used as indexing terms, the upper word indexing error occurs. According indexing rules, the upper word can't serve as index terms. Because of the massive literature retrieval operation, with the upper word as a search condition, the detection result is often a large number of irrelevant documents, the upper word indexing is the main reason. Strictly speaking, though these words are not ambiguous word, but it is ambiguous word indexing and should be eliminated.

Through the above-mentioned two types of ambiguous word indexing analysis and preliminary disambiguation experiments, this method to eliminate the ambiguous word is an effective method and provide readers reference.

6   Conclusion

Disambiguation and unknown word identification are the difficult problems in the Chinese word segmentation field. There are a lot of types and cause of ambiguous word, different word processing method produces different methods to eliminate ambiguous words. This article is related to the ambiguous word appears under certain circumstances, although conditions require to eliminate such ambiguous words, but it is still an effective way to eliminate ambiguous words. I also hope that the majority of researchers study algorithms in a wide range of methodological and propose innovative solutions to design a common method of clearing ambiguous words and improve the accuracy and speed of segmentation. In addition, the extensive literature research focuses on statistical segmentation,which also focuses on based on combination with statistical segmentation and other methods todisambiguate, they will give the Chinese word segmentation technology to bring a substantive breakthrough.

References

1. Li, D., Cao, Y., Wan, Y.: New Security Feature Extraction Method Based on Association Rules. Computer Engineering and Applications (S1), 105–107 (2006)
2. Xiao, H., Xu, S.-H.: A Method of Automatic Keyword Extraction based on Co-occurrence Model. Transactions of Shenyang Ligong University (5), 38–41 (2009)
3. Su, X., Liu, X., Shao, P.: The Word-indexand Position Retrievalforthe Document TitlesIn Chinese. Journal of Nanjing University(Natural Sciences Edition) (2), 329–333 (1990)
4. Weng, H.: Comparison Studies on Inconsistencies and Ambiguity Automatic Identification Method in Chinese Information Processing. Language Applied Research (12), 93–94 (2006)
5. Li, G., Liu, K., Zhang, Y.: Segmentating Chinese Word and Processing Different Meanings Structure. Journal of Chinese Information Processing (3), 27–32 (1988)
6. Yao, J.-W., Zhao, D.: Disambiguation Method in Chinese Word Segmentation Based on Phrase Match. Journal of Jilin University(Science Edition) 48(3), 427–432 (2010)
7. Bai, S.: Chinese word segmentation and POS integrated approach to automatic annotation. In: Advances in Computational Linguistics and Applied, Beijing, pp. 56–61. Tsinghua University Press (1995)
8. Cai, J.: "Chinese Library Classification" professional classification "Agricultural Professional Classification". Beijing. Library Press (October 1999)

Analysis and Evaluation of Soil Fertility Status Based on Weighted K-means Clustering Algorithm

Guifen Chen[1], Lixia Cai[1], Hang Chen[1,2], Liying Cao[1], and Chunan Li[1]

[1] Jilin Agricultural University, Changchun 130118, China
[2] Jilin Academy of Agricultural Sciences, Changchun 130124, China
{guifchen,lixcai,chenhang}@163.com

Abstract. Generally K-means clustering algorithm can not distinguish the imbalance between attributes, so it can only be an independent investigation situation of each attribute but can not be comprehensive analysis of the soil fertility status. To solve this problem, this paper proposes a weighted K-means clustering algorithm to evaluate the soil fertility in Nong'an County, Jilin. The algorithm uses AHP to get the weight of soil nutrient attributes. Then combined with K-means clustering algorithm. Finally through the operational efficiency and accuracy to determine the optimal classification, that can improve the clustering algorithm of intelligent. The algorithm and the traditional K-means clustering algorithm are used in the comparison, tests showed that the weighted K-means clustering algorithm has a better accuracy, operational efficiency, significantly higher than the unweighted clustering algorithm; Comprehensive evaluation of the changes in soil nutrients after precision fertilization that used algorithm. The soil fertility status has a significantly improvement after years of continuous precision fertilizing. The results show that the improved clustering algorithm is a good method to comprehensive evaluation of soil fertility.

Keywords: AHP, Weighted K-means clustering, Optimal classification, Soil fertility evaluation.

1 Introduction

3S technology (GPS, GIS and RS), networking technology and expert system (ES) technology are widely used in precision agriculture with the rapid development of information technology. That all make soil fertility data appear to rich, multidimensional, dynamic, incomplete, uncertainty and other characteristics [1].How to be more timely and accurately show the differences in temporal and spatial data, comprehensive evaluation and correct analysis of the data have an important practical significance [11]. Data mining technology [5] is the process of generating new regular, which through the massive amounts of data classification, extraction to discover the mutual contact between data.

Related data show that variable region and the traditional region of N, P, K nutrient variation coefficients in soil fertilization were compared by ZHANG. Which indicate that variable rate fertilization has a balanced effect to soil nutrient fertilization [2].

D. Li and Y. Chen (Eds.): CCTA 2013, Part II, IFIP AICT 420, pp. 89–97, 2014.
© IFIP International Federation for Information Processing 2014

The research on weighted space fuzzy dynamic clustering algorithm by CHEN Gui-fen, proved the effectiveness of soil fertility evaluation [3]. And Li Yan.etc. [4] Who used fuzzy clustering method to classify partition and introduced two kinds of partitions to compare and evaluate, such as fuzzy clustering index and normalized classification. That can offer the decision basis for soil management. Even K-means clustering algorithm based on the classification method could differentiate soil fertility according to soil nutrient, However, it can't consider the nutrient differences between each attribute. As a result, we use the improved K-means algorithm, considering the linkages between soil nutrients of the fertility in Nong'an country [8] and give a comprehensive evaluation.

2   Analysis of k-means Algorithm

K-Means algorithm is a clustering algorithm based on partitioning method [7-17], it is first suggested and one of the more classical clustering algorithms [14-15].

2.1   The Process of K-means Algorithm [6]

Algorithms: k-means. Divided k-means algorithm based on the average value of the objects in the cluster.

Input: the number of clusters (k) and the database contains n objects.
Output: k clusters, so that the minimum squared error criterion.
Method:
(1) Choose k objects as the initial cluster centers;
(2) Repeat;
(3) According to the average value of the objects in the cluster, each object is (re) assigned to the most similar clusters;
(4) Update the average value of cluster. That is to say, calculate the average value of each cluster in the object;
(5) Until no change.

2.2   Advantages and Disadvantages of k-means Algorithm

Using k-means algorithm [12] to clustering, the effect is good. While the result is a dense cluster, the differences between clusters are obvious. When we deal with large data sets, this algorithm is relatively scalable and efficient, because of its complexity is O (nkt), where, n is the number of all objects, k is the number of clusters, t is the number of iterations. Typically k«n and t«n. This algorithm often ends with a local optimum. However, k-means method [13-18] is only used in the case of the average value of clusters were defined. This attribute data is not applicable for processing symbol attribute data, it also requires user to give the value of k (the number of clusters to be generated) in advance. In addition, for the "noise" and outlier data is sensitive, a small amount of such data can have a significant impact on the average value.

3   Analysis of Weighted k-means Algorithm Based on AHP

3.1   Using AHP to Determine the Weight Coefficients

The algorithm is as follows:

Step 1: Construct paired comparison matrix;

Step 2: Take any n-dimensional normalized initial vector $\mathbf{w}^{(0)}$;

Step 3: Calculation $\tilde{\mathbf{w}}^{(k+1)} = \mathbf{A}\mathbf{w}^{(k)}, k = 1, 2, \cdots$;

Step 4: Normalization $\tilde{\mathbf{w}}^{(k+1)}$;

Step 5: For the pre-specified precision ε, when $\left| w_i^{(k+1)} - w_i^{(k)} \right| < \varepsilon, \quad i = 1, 2, \cdots, n$

Established, $\tilde{\mathbf{w}}^{(k+1)}$ shall be eigenvector; otherwise return Step 2;

Step 6: Calculate the maximum eigenvalue $\lambda = \frac{1}{n} \sum_{i=1}^{n} \frac{\tilde{w}_i^{(k+1)}}{w_i^{(k)}}$;

Step 7: Calculate consistency index $CI = \frac{\lambda - n}{n - 1}$;

Step 8: Calculate consistency ratio $CR = \frac{CI}{RI}$;

Step 9: If $CR < 0.1$ is established, through consistency test; otherwise reconstruct paired comparison matrix;

Step 10: If all the layers are calculated. And we can obtain the weight vector of total target ,

$$A = (a_1, a_2, \cdots, a_m) ;\text{ Otherwise, return back to Step 1.}$$

3.2   The Establishment of the Weighted k-means Model

In this paper, we used the weighted fuzzy dynamic clustering approach to process spatial data, which is proposed by CHEN [3].

(1) Data's standardization

Since in practical problems, different data generally have different dimensions, in order to have the amount of different dimensions can be compared, we need to standardized data, which data are compressed to the [0, 1] interval. Now we use the range transformation,

$$x_{ij}' = \frac{x_{ij} - \min\{x_{ij}\}}{\max\{xi_j\} - \min\{x_{ij}\}} \tag{1}$$

(2) Weighted calculation

$$
Y = \begin{bmatrix} x'_{11} & x'_{12} & \cdots & x'_{1m} \\ x'_{21} & x'_{22} & \cdots & x'_{2m} \\ \cdots & \cdots & \cdots & \cdots \\ x'_{n1} & x'_{n2} & \cdots & x'_{nm} \end{bmatrix} \cdot \begin{bmatrix} a_1 & 0 & \cdots & 0 \\ 0 & a_2 & \cdots & 0 \\ \cdots & \cdots & \cdots & \cdots \\ 0 & 0 & 0 & a_m \end{bmatrix}
$$
$$
= \begin{bmatrix} y_{11} & y_{12} & \cdots & y_{1m} \\ y_{21} & y_{22} & \cdots & y_{2m} \\ \cdots & \cdots & \cdots & \cdots \\ y_{n1} & y_{n2} & \cdots & y_{nm} \end{bmatrix}
$$

(2)

(3) Fuzzy similar matrix

Calculating Closeness r_{ij} of fuzzy sets i and fuzzy sets j,

$$
r_{ij} = \frac{\sum_{k=1}^{m} y_{ik} \cdot y_{jk}}{\sqrt{\sum_{k=1}^{m} y_{ik}^2} \cdot \sqrt{\sum_{k=1}^{m} y_{jk}^2}}
$$

(3)

Resulting in fuzzy similar matrix $R = (r_{ij})_{n \times n}$.

4   The Application of Weighted k-means Algorithm

4.1   Data Sources

Application and research of soil fertility data after precision fertilization for many years [9-10], which is based on the "863" plan --"research and application of corn precise operating system" project demonstration base in Nong'an County, Jilin. And we select the representative soil nutrient data to integrated analysis, such as, alkaline hydrolysis nitrogen, available potassium and available phosphorus. From Nong'an County during 2007 to 2011 years.

GIS-based sampling points shown in Figure 1:

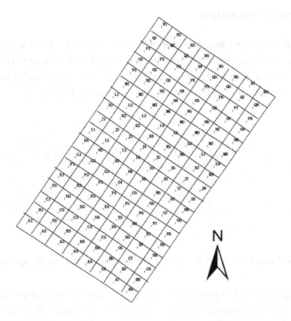

Fig. 1. GIS-based sampling points

The sampling method from figure1 is "five plum blossom sampling", namely taking soils from the four corners and the center, then blending these soils together as the sample. The samples have been taken and tested in Tab.1, from 2007 to 2011; through tests we have acquired the data of soil, such as spatial coordinates, organic matter, alkaline hydrolysis nitrogen, available potassium, available phosphorus, soil humidity and PH value. Select the main factors which affecting fertility as the sample data, part of the data shown in Table 1.

Table 1. Part of the sample data

Town name	Alkaline hydrolysis nitrogen (mg/kg)	Available phosphorus (mg/kg)	Available potassium (mg/kg)	latitude	longitude
Nong'an town	154.0	28.0	208.0	44.58417	125.2898
Nong'an town	136.0	31.3	217.0	44.49895	125.2512
Nong'an town	132.0	16.3	198.0	44.49926	125.2507
Nong'an town	125.0	36.8	140.0	44.51392	125.2540

4.2   Application of Algorithm

First, Standardization of soil nutrient data (the data come from Nong'an town during 2007 to 2011). Then we can analysis spatial patterns of soil nutrients data according to the test area of N, P and K. The results show that the test area available phosphorus in the soil spatial variability of the maximum. It is shown in Figure 2:

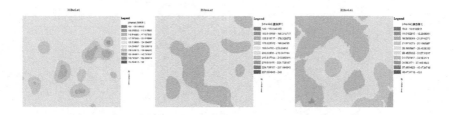

Fig. 2. Soil nutrient (N, P, K) spatial variation in Xi haolai, Nong'an town

After that, Evaluation of the results based on the status of spatial variability and local soil characteristics, then construct pair wise comparison matrix B (Equation 2).

Second, the author use AHP to get the nutrient weights of soil alkaline hydrolysis nitrogen, available potassium, available phosphorus, three nutrient weights are 0.3782, 0.2032 and 0.4185. CI = 0.026420513 ≪ 0.1, which is closer to the complete consistency. Then, compared with the classical k-mean clustering algorithm, the weighted K-means clustering algorithm has a significantly higher accuracy, and operational efficiency. We can see the results have shown in Table 2.

Table 2. The comparison result

Algorithm	Average accuracy (%)	Average running time(s)
K-means	95.03	0.08
K-Wmeans	96.91	0.06

From table 2, the weighted and unweighted k-means both are better classification methods (the accuracy are 96.91%, 95.03% respectively, and the running times are 0.06 s, 0.08 s respectively). When we use unweighted clustering, different nutrients in the soil will offset the gap between the highest and lowest and delimit in the same class, while weighting the gap will be assigned to different classes, this method can reflect the real situation of soil nutrient. Weighted k-means for the "noise" and outlier data is not very sensitive; a small amount of this kind of data does not have great influence on the average value.

Finally, using the weighted k-means algorithm, weighted clustering the soil nutrient data of Nong'an town for five consecutive years from 2007 to 2011. Experimental results as shown in Table 3 and Figure 3.

Table 3. The result of soil nutrient data weighted clustering

	2007		2008		2009		2010		2011	
	Clustered Data	Percentage (%)	Clustered Data	Percentage (%)	Clusterd Data	Percentage (%)	Clustered Data	Percentage (%)	Clustered Data	Percentage (%)
Cluster 0	310	42	247	33	161	22	198	27	99	13
Cluster 1	26	3	257	35	126	17	254	34	235	32
Cluster 2	225	30	194	26	234	31	151	20	134	18
Cluster 3	183	25	46	6	223	30	141	19	276	37

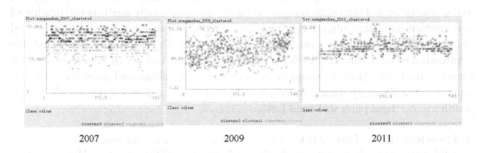

2007   2009   2011

Fig. 3. Clustering results

The Table 3 and Figure 3 show that, in the same category case, the degree of similarity between the data is gathered and the differences between clusters are decreasing year by year after a continuous precise fertilization. The value of Cluster 0 decreased from 42% in 2007 to 13% in 2011, cluster 1 from 3% in 2007 years rose to 35% in 2010.All above shows that the weighted k-means algorithm is an effective method for soil fertility evaluation. After a continuous precise fertilization, alkaline hydrolysis nitrogen,available phosphorus and available potassium, the comprehensive similarity of three kinds of nutrient data have been improved year by year. The results tally with the actual situation, so weighted k-means algorithm is an effective method of fertility evaluation.

5   Results and Discussion

Analysis and evaluation of soil nutrient data by using weighted k-means algorithm. The data of Nong'an town for five consecutive years from 2007 to 2011. We can see that significant changes in soil fertility occur after five consecutive years of precision fertilization. Experimental results show that the weighted k-means algorithm is an effective method of fertility evaluation.

(1)AHP is used to determine the initial weight values; the weighted original decentralized data can avoid the shortcomings that unweighted k-means algorithm does not distinguish between data imbalance between attributes as well as sensitive to "noise" and isolated points data .

(2)The use of weighted and unweighted k-means algorithm for comparative analysis soil nutrient data about soil alkaline hydrolysis N, available P and available K from Nong'an town in 2011, and the results showed that weighted K-means clustering algorithm has better effect than unweighted k-means algorithm, in the terms of accuracy which is increase1.88% and operating efficiency which is increase 25%.

(3)From the experimental results of the algorithm, after five consecutive years of precision fertilization on soil nutrient data in Nong'an town, Comprehensive similarity in increase year by year.This evaluation results and the actual situation is consistent, provides a new reference for analysis of soil fertility status in future.

The initialization of weighted and unweighted k-means clustering algorithm should depend on iterative method to determine the number of clusters that is relatively close to the true value and the initial center. However, this article only analysis verification of soil nutrient data that after five consecutive years of precision fertilization on soil nutrient data in Nong'an town. And the improved algorithm has not tested the application of large data sets or fully confirmed that the validity of new algorithm for massive data sets. How to simplify the clustering algorithm and combine with soil nutrient data of many years, more townships (towns) and soil types. They are all problems. So these parts still need further research.

Acknowledgment. This work was supported by the national "863" project (2006AA10A309), National Spark Plan (2008GA661003) and Shi Hang of Jilin province projects (2011- Z20).

References

1. Turner, B.L., Meyer, W.: Land use and land cover in global environmental change:considerations for study. Int. Soi. Sci. J. 130, 669–680 (1991)
2. Zhao, J., Hu, H., Zhou, X.: Application of Support Vector Machine to apple classification with near—infrared spectroscopy. Transactions of the CSAE 23(4), 149–152 (2007)
3. Chen, G.-F., Cao, L.-Y., Wang, G.-W.: Application of Weighted Spatially Fuzzy Dynamic Clustering Algorithm in Evaluation of Soil Fertility. Scientia Agricultura Sinica 42(10), 3559–3563 (2009)
4. Li, Y., Shi, Z., Wu, C.F., Li, F., Cheng, J.L.: Definition of management zones based on fuzzy clustering analysis in coastal saline land. Scientia Agricultura Sinica 40(1), 114–122 (2007)
5. Amirmahdi, N.R., Ali, N.: A Survey on Data Mining Approaches. International Journal of Computer Applications 36(6), 975–8887 (2011)
6. Shibao, S., Keyun, Q.: Improved k - average clustering algorithm research. Computer Engineering 33(13), 201–201 (2007)
7. Sun, J.-G., Liu, J., Zhao, L.-Y.: Clustering Algorithms Research. Journal of Software 19(1), 48–61 (2008)
8. Yang, C., Everitt, J.H., Bradford, J.M.: Comparisons of uniform and variable rate nitrogen and phosphorus fertilizer applications for grain sorghum. Transactions of the American Society of Agricultural Engineers 44(2), 201–209 (2001)

9. Umeda, M., Kaho, T., Michihisa, I.I.D.A., Choung, K.L.: Effect of variable rate fertilizing for paddy field. In: ASAE Annual International Meeting, Paper Number. 01(Part. II) (2001)
10. Wittry, D.J., Mallarino, A.P.: Comparison of uniform-and variable-rate phosphorus fertilization for corn-soybean rotations. Agronomy Journal 96(1), 26–33 (2004)
11. Li, L., Li, J.: Application of Clustering Analysis in Classifying Site Type and Evaluating Soil Fertility. In: 2010 Third International Conference on Education Technology and Training (ETT 2010), pp. 468–471 (2010)
12. Chawan, S.R.B., Patil, S.: Improvement of K-Means clustering Algorithm. International Journal of Engineering Research and Applications (IJERA) ISSN 2(2), 1378–1382 (2012)
13. Lai Yuxia, S.D., Yang, G.: K-means clustering analysis based on genetic algorithm. Computer Engineering 34(20), 200–202 (2008)
14. Backer, E., Jain, A.K.: A clustering performance measure based on fuzzy set decompo-sition. IEEE Trans on Pattern Analysis and Machine Intelligence 3(1), 66277 (1981)
15. Zahn, C.T.: Graph2theoreticalmethods for detecting and describing gestalt clusters. IEEE Trans. on Computers 20(1), 68286 (1971)
16. Tan, Y., Rong, Q.: Implementation of A Clustering Algorithm Based on High Density. Computer Engineering 30(13), 119–121 (2004)
17. Leung, K.W.T., Ng, W., Lee, D.L.: Personalized Concept-based Clustering of Search Engine Queries. IEEE Transactions on Knowledge and Data Engineering 20(11), 1505–1518 (2008)
18. Gao, X., Feng, Y., Feng, X.: Incremental - K Medoids clustering algorithm. Computer Engineering 7(31), 181–183 (2005)

Effect of Website Quality Factors on the Success of Agricultural Products B2C E-commerce

Ping Yu[1] and Dongmei Zhao[2]

[1] School of Economics & Management, China Agricultural University, Beijing 100083, China
yuping@cau.edu.cn
[2] China Agricultural University, Beijing 100083, China
zhaodongm@vip.163.com

Abstract. By applying analytic hierarchy process to extend the D&M IS success model, this paper builds a user preference hierarchical model for online consumers to select the most preferred website; this model is mainly constituted by the quality of information, service, system and supplier. Empirical study on agricultural products B2C field has shown that service quality is the most important factor affecting the website selections of online consumers, and e-commerce businesses should make greater efforts to improve the reliability and the rapid responsiveness of service. As supplier quality is highly correlated with user preferences, e-commerce businesses should take a balance strategy to increase product range, improve product quality with lower product price at the same time Information currency is high in the ranking, so e-commerce businesses should pay attention to update information on the website timely.

Keywords: B2C, E-commerce success, website quality factors.

1   Introduction

With the popularity of internet and people's love for shopping, consumerism has already been popular in China. In 2006, less than 10% of users shop online, while in 2012 this proportion has jumped to 39%. The intense growth of internet users promotes the rapid development of Chinese e-commerce market, data from China Electronic Commerce Research Center shows that as of June 2012, the Chinese e-commerce market transactions amounted to 3.5 trillion Yuan, increased by 18.6%.The generous profits of e-commerce market has attracted more and more businesses to enter, at the same time, the surge of e-commerce businesses makes the competition of Chinese e-commerce market fierce. In 2012, some e-commerce businesses cut down part of staffs, some e-commerce businesses went bankrupt, China Electronic Commerce Research Center analyst, Mo Dai Qing believes that the failures of e-commerce businesses will continue, superior bad discard will continue to unfold. However, many decision-makers of e-commerce businesses are continuing to make significant investment in developing e-commerce websites, in case of not being clear what factors contribute to the development of high-quality websites and how to measure their effect on e-commerce success [1-2].

D. Li and Y. Chen (Eds.): CCTA 2013, Part II, IFIP AICT 420, pp. 98–113, 2014.
© IFIP International Federation for Information Processing 2014

E-commerce is a special business industry, it is different from others, and the success of an e-commerce business depends largely on its website quality [38]. As a portal for business to interact with the current and potential consumers, website not only provides a platform for businesses to promote products and services, but also provides a way to generate more revenue by attracting more consumers. On the one hand, e-commerce businesses rely on people to visit their websites to buy their products, and more importantly, become repeated customers; On the other hand, consumers have a lot of alternative sites to choose from, so if a website's performance cannot let consumers be satisfied, they can easily switch to another alternative website. Therefore, the website quality of an e-commerce business will directly affect consumers to go or stay, and then determines e-commerce success.

Domestic and foreign researchers have made considerable efforts in determining the factors affecting e-commerce success. DeLone and McLean [3] first proposed IS success model in 1992, and subsequently based on it they proposed the updated D & M IS success model [4], which can measure the success of e-commerce systems from six dimensions: information quality, system quality, service quality, use, user satisfaction and net benefits. Molla and Licke [5] partially extended and modified the D & M IS success model, and then proposed an e-commerce success model. They regarded the consumer e-satisfaction as the dependent variable of e-commerce success, defined and discussed the relationship between the consumer e-satisfaction and e-commerce system quality, content quality, use, trust and support. Madeja and Schoder [6] conducted an empirical research on 8 website performance indicators which affect the success of e-commerce system: interactivity, immediacy, connectivity, diversity, availability, information-rich, usability, personalization and customizations. Xuan [7] studied the factors which impact the e-commerce success of network marketing companies, taking China Shanghai Brilliance Group as a case. Mainly referencing Molla & Licker (2001) e-commerce success model, he selected four indicators: system quality, content quality, trust, support and service to measure the success of e-commerce systems. Sharkey et al. [8] empirical studies the updated D & M IS success model, they found the important relationship between information quality, system quality, and three success dimensions: intention to use, user satisfaction and trading intent.

Although in order to determine the website quality factors affecting e-commerce success, researchers have conducted a number of studies and provided a deep understanding of it, but it still has exploration space being left, for example, we can add more website quality factors to extend the existing models, we can also apply the models to different e-commerce domains to compare and analyze.

This paper assumes that an e-commerce business is more likely to be success when its website has the highest quality among all the alternatives. This will result in its website being selected by online customers as the most preferred. The more customers select its website, the more likelihood the business improves its e-commerce business performance. This paper addresses this issue, limiting the study scope to an investigation of website quality of agricultural products B2C e-commerce websites.

2   Theoretical Framework

2.1   Updated D&M IS Success Model

In order to measure the success of information systems, researchers have proposed many theoretical models. Among these models, DeLone and McLean's IS success model is the most cited models.

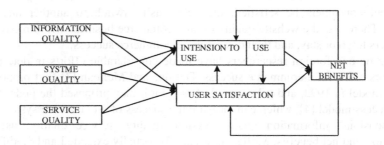

Fig. 1. Updated DeLone & McLean IS Success Model

In 2003, DeLone and McLean updated there IS success model based on other researchers' findings, as shown in Fig.1. Delone and McLean added "service quality" to the model, and merged the two dimensions: "individual impact" and "organizational impact" into a single new indicators: net benefits. DeLone and McLean believe that the IT department of an organization not only provides product information to consumers, but also provides service support to consumers, so service quality should also be one of the dimensions of information systems success; In addition, according to the different levels of analysis and research purposes, except for "individual impact", there also may have "industry impact," "social impact", "consumer impact" and so on, and "impact" may be positive or negative, so "individual impact " and "organizational impact" in the original model are not comprehensive enough and the more comprehensive and accurate concept " net benefit" should be used as the ultimate success dimension, and the situation of " net benefit "will adversely affect system use and user satisfaction.

Because e-commerce system is one kind of information system, so D&M IS success model can also be used to measure the effectiveness and the value of e-commerce system.

2.2   Analytic Hierarchy Process [9-10]

AHP is proposed by an American operational research experts Saaty in the 1970s, it is a multi-objective decision analysis method which combines qualitative and quantitative analysis.

The principle of this method is as follows:

Assuming there are n objects, denoted as $A_1, A_2,..., A_n$, their weights are $w_1, w_2,..., w_n$. If they compare their weights two by two, the ratio may constitute a $n \times n$ matrix A.

$$A = \begin{bmatrix} w_1/w_1 & w_1/w_2 & \cdots & w_1/w_n \\ w_2/w_1 & w_2/w_2 & \cdots & w_2/w_n \\ \vdots & & \ddots & \vdots \\ w_n/w_1 & w_n/w_2 & \cdots & w_n/w_n \end{bmatrix}$$

Matrix A has the following properties:

If using the weight vector $W = (w_1, w_2, \cdots, w_n)^T$ to right multiply matrices A, the results is as follows:

$$AW = \begin{bmatrix} w_1/w_1 & w_1/w_2 & \cdots & w_1/w_n \\ w_2/w_1 & w_2/w_2 & \cdots & w_2/w_n \\ \vdots & & \ddots & \vdots \\ w_n/w_1 & w_n/w_2 & \cdots & w_n/w_n \end{bmatrix} \begin{bmatrix} w_1 \\ w_2 \\ \vdots \\ w_n \end{bmatrix} = n \begin{bmatrix} w_1 \\ w_2 \\ \vdots \\ w_n \end{bmatrix} = nW$$

Namely $(A - nI)W = 0$.

Seen by the matrix theory, W is the feature vector, n is the characteristic value. If W is unknown, we can make a judgment of radio subjectively based on the relationship comparing between objects by policymakers or use the Delphi method to obtain these ratios to make the matrix A to be known, and then the judgment matrix is recorded as $\overline{A}$.

According to the theory of positive matrix, we can prove: If the matrix A has the following characteristics $((a_{ij} = w_i/w_j)$:

1) $a_{ij} = 1$

2) $a_{ij} = 1/a_{ji}$ $(i,j = 1,2,..., n)$

3) $a_{ij} = a_{ik}/a_{jk}$ $(i,j = 1,2,..., n)$

Then the matrix has a unique maximum eigenvalue nonzero λ_{max}, and $\lambda_{max} = n$.

If the given judgment matrix $\overline{A}$ having the above characteristics, then this matrix has full consistency. However, when people pairwise compares the various factors on complex things, it is impossible to achieve complete consistency of judgment, but there is estimation error, which will inevitably lead eigenvalues and eigenvectors also have bias. Then the question changes to $\overline{A}\dot{W} = \lambda_{max}\dot{W}$ from AW= nW, here λ_{max} is the largest eigenvalue of matrix $\overline{A}$, $\dot{W}$ and is the relative weight vector with deviations. This is the errors caused by incompatible judgment. To avoid the error is too large, so it need to measure the consistency of the matrix $\overline{A}$. When matrix A is fully consistent:

Because $a_{ii} = 1, \displaystyle\sum_{i=1}^{n}\lambda_i = \sum_{i=1}^{n}a_{ij} = n,$

So there exists a unique nonzero$\lambda = \lambda_{max} = n$. And when the discrimination of Matrix $\overline{A}$ exists inconsistent, in generally$\lambda_{max} \geq n$.

Here$\lambda_{max} + \displaystyle\sum_{i \neq max}\lambda_i = \sum_{i=1}^{b}a_{ii} = n$.

Due to$\lambda_{max} - n = -\displaystyle\sum_{i \neq max}\lambda_i$.

Use its mean as the index of inspection judgment matrix.

When$\lambda_{max} = n, C.I = 0$, it is fully consistent; the larger the value of C.I is, the worse the complete consistency of judgment matrix is. Usually as long as C.I ≤ 0.1, it is considered the consistency of judgment matrix is acceptable, otherwise, it needs to re-pairwise the comparison judgments. The greater the dimensions of Matrix n is, the worse the consistent of judgments is. Thus the correction value R.I is introduced. As shown in table 1. And taking a more reasonable value of C.R as the indicators to measure the consistency of judgment matrix.

$$C.R = \frac{C.I}{R.I}$$

Table 1. Correction value R.I

Dimension	1	2	3	4	5	6	7	8	9
R.I	0.00	0.00	0.58	0.96	1.12	1.24	1.32	1.41	1.45

3   Research Model

Base on D & M updated IS success model (2004) and AHP, this paper proposes a research model (Fig.2) for online consumers to select the most preferred website. The model consists of four main website quality factors including information quality, system quality, service quality, and supplier quality. We believe that these four website quality factors significantly affect the online consumers to choose the most preferred website.

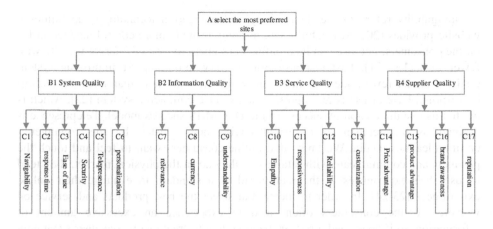

Fig. 2. User Preference Hierarchy Mode

3.1   Information Quality

Because of the characteristics of virtual environment, online consumers can not physically touch the products, therefore, product information on the website is the only way by which online consumers can understand products. In some studies information quality is also known as content quality. Some researchers define information quality as the characteristics and expression of information in the e-commerce system [11-12]. Alba and other studies suggest that information quality refers to to what extent online consumers can use information to predict their subsequent consumer satisfaction prior to purchase [15]. Previous studies have demonstrated that information quality and its sub-properties can improve customer satisfaction and maintain a website's attraction to consumers [13-14].

Information quality can be measured by information relevance, currency and understandability. Information relevance includes the relative depth, range and integrity of information. Currency refers that information should be updated timely, studies have found that frequent updates can improve the access rate [15], while outdated information is a primary factor that causes website closure and business failures [16]. Understandability means information content should be clear and easy to be understood.

3.2   System Quality

System quality used to measure the characteristics e-commerce system requires [17]. System quality is believed to be a critical success factor affects technology use and user satisfaction. Research shows that websites with poor system quality will let consumers to have a negative impact on online experience and consumer satisfaction [18-19]. System quality can be measured by navigability, response time, ease of use, security, telepresence and personalization.

Navigability refers to the interactivity and navigation technology capabilities a website provides [20]. Navigability can not only provide more control navigation for online consumers, but also can help online consumers to reach the target website with fewer obstacles [21]. Fast response time is very important for improving system quality because few online consumers are willing to wait more than a few seconds to be respond. Ease of use is an important factor for e-commerce system [22], which is treated as one dimension of success in the D & M IS success model. Telepresence is defined as the realism in a virtual environment created by the computer / communications media. We know that online consumers want to feel and touch the products and communicate with retailers as they do in the physical market, they tend to use their experiences in the real world as a standard to evaluate their online experiences [23-24]. In order to locate and select the best products and service on websites, online consumers often need to experience an excessive amount of information, so it does need a personalized system to treat every consumer separately [25-26], a personalized system can provide online consumers with personalized interfaces, effective one to one information and personalized service [27-28].Finally, security is one of the biggest obstacles for e-commerce. Online consumers will not disclose their personal information and financial information until they are sure a website is secure, so the website should achieve a variety of functions (for example: encryption, third-party certification, Security Statement) to ensure the security of online shopping [29].

3.3   Service Quality

Service Quality is the comprehensive support offered by the online retailer. Because online consumers are trading with invisible retailers, so service quality has become particularly important in e-commerce [20]. Service quality can be measured by empathy, responsiveness, reliability and customization.

Empathy refers to the care and attention provided by e-commerce businesses to online consumers [30].Responsiveness is the willingness e-commerce businesses provide timely services to help online shoppers [30]. Reliability is the capabilities e-commerce businesses fulfill service warranty reliably and accurately [30]. Customization is another important service activity on the website [31], it is the customized information, products and prices provided by e-commerce businesses to online consumers based on the track of the interaction between online consumers and their websites. [32]; through customization e-commerce businesses can further differentiate their competitors, and encourage online consumers to return to their websites for the next shopping event [33].

3.4   Supplier Quality

Supplier quality is also considered an important factor of e-commerce success. Supplier quality can be measured by price advantage, product advantage, brand awareness and reputation.

Price advantage is a measure of store efficiency, because a valid store can reduce transaction costs and thus provide consumers with better prices [34]. It is discovered that price advantage has important implications for online purchase, for example, Devaraj et al found that price advantage significantly affects e-commerce channel satisfaction of books or CDs [35]; Chen and Dubinsky's studies have shown that high price has a negative impact on online consumers [36]. Product categories is one of the most important factors that affect e-commerce success [37], providing a range of products can improve the efficiency of e-commerce transactions, because online consumers will not have to search products on other websites and thus saving time for online consumers, as well as increasing trading revenues for e-commerce businesses; while quality problems will affect consumers' confidence in e-commerce suppliers [38]. Websites with good brand positioning will attract high click-through rate [39], a customer's brand loyalty is positively correlated with his website loyalty [40], and users are more willing to choose the technology choose by a large number of other users [41]. E-commerce business spends millions of dollars to advertise its website to improve its brand awareness, when a lot of people know and want to experience a website, the website's brand awareness has been improved. Economists found reputation is positively correlated with prices, the higher an e-commerce business's reputation is, and the more consumers are willing to pay for the e-commerce business.

4   Research Method

4.1   Data Collection

A questionnaire-based online survey was conducted to investigate the relative importance of website quality factors for online customers to select the most preferred website. The factor and their measurement items were initially developed based on a literature review. Then, we invited three business doctoral tutor and six doctoral students to check the wording, content, and format of the questionnaire, and according to their views the questionnaire was modified.

The questionnaire is divided into three parts. The first part investigates the demographic characteristics of the survey respondents, including: gender, age, educational background and career; the second part investigates the network familiarity and shopping habits of the survey respondents, including: contact time of online shopping, number of online shopping average weekly; The third part investigate the relative importance of website quality factors for the survey respondents by pairwise comparing.

First, each survey respondent was required to navigate an agricultural product B2C websites to conduct tasks based on a given online purchasing scenario. Then each survey respondent was asked to fill out the questions about the relative weight of one factor over another. Finally, each survey respondent was demand to visit all the target websites and answer the questions about relative strength of each alternative website based on each factor.

We adopted online survey questionnaire to collect data. The web-based survey began from February 15 and ended at March 15 in 2012. Finally we received 338 questionnaires and 310 were valid.

4.2 Descriptive Analysis

In the effective research samples, male and female ratio is about 1:1, men slightly more than women. Samples under 35 years account for 82.91 percent, which have an absolute advantage. Undergraduate and equivalent account for 59.03% in the total sample. AS to thire average monthly expenditure for online shopping in the last three months, 3000-5000 Yuan accounts for 20.65%, which ranked first. And among their average monthly expenditure for online shopping, 500-1000 Yuan shares the largest proportion, accounts for 43.87%. Company or enterprise managers share the highest percentage 33.35%, followed by the general staff, accounting for 28.39%, and then followed by professional and technical personnel, accounting for 18.71 percent. Analysis of the basic demographic characteristics of samples illustrates they have the very similar online shopping user characteristics published by CNNIC.

In addition, we can also see that 37.74% of samples have more than 4 years online shopping experiences; 79.35% of samples shopped online more than twice an week in the recent three months. More than 75% of samples spent more than to 2 hours a week to browse online shopping information. Analysis of shopping habits of samples illustrates they have more frequent and stable shopping habits.

5   Results and Discussion

The data was analyzed by Expert Choice. Expert Choice is an AHP software, which can provide analysis results including local weights, overall weight, the priority of alternatives and sensitivity.

5.1 Comparison of Website Quality Factors

First, we analyze the relative importance of website quality factors, check the importance of each factor contributing for consumers to choose the most preferred online website. In order to obtain the relative importance of each factor, we conducted a pairwise comparison of factors. The weight of each website quality factor is shown in Fig.3.

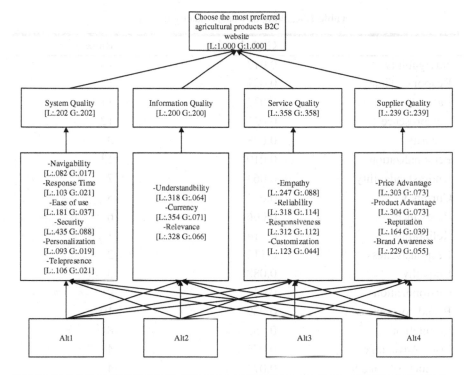

Fig. 3. Relative importance of website quality factors

From Fig.3, we can see that when online consumers choose agricultural products e-commerce websites, service quality has the highest global weight 0.358, followed by supplier quality (0.239) and system quality (0.202). This fully verifies that service quality is increasingly becoming an important aspect of e-commerce [42]. For system quality, safety has the highest local weights 0.435, followed by ease of use (0.181), telepresence (0.106), personalization (0.093), and navigability (0.082). For information quality, currency has the highest local weights 0.354, followed by relevance (0.328), understandability (0.318). For service quality, reliability has the highest local weights 0.318, followed by responsiveness (0.312), empathy (0.247) and customization (0.123). For supplier quality, price advantage tied with product advantages have the highest local weights 0.73, followed by brand awareness (0.229) and reputation (0.164).

The ranking of website quality factors is shown in Table 2.

From Table 2 we can see that when online consumers choose the most preferred website, the reliability, responsiveness, empathy of service quality, the security of system quality, the price advantage and product advantages of supplier quality, and the currency of information quality are among the top five website quality factors.

Table 2. Ranking of website quality factors

	Global weights	Rank
Navigability	0.017	14
Response Time	0.021	12
Ease of use	0.037	11
Telepresence	0.021	12
Security	0.088	3
Personalization	0.019	13
Understandability	0.064	7
Currency	0.071	5
Relevance	0.066	6
Reliability	0.114	1
Responsiveness	0.112	2
Empathy	0.088	3
Customization	0.044	9
Brand Awareness	0.039	10
Reputation	0.055	8
Price Advantage	0.073	4
Product Advantage	0.073	4

5.2   Comparisons of Alternative Websites

Alternative websites are websites which can substitute for each other to meet a certain kind of consumers' desire. As more and more e-commerce websites being created, the homogenization of e-commerce websites becomes increasingly serious, how to create a differentiated competitive advantage, is the problem which e-commerce businesses have to face if they want to successfully survive.

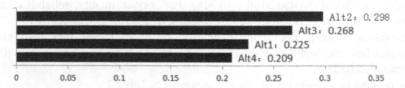

Fig. 4. Ranking of Alternative websites

Fig.4 shows the result of the most preferred website choose by online consumers. We can see Alt2 (0.298) are the most preferred site of online consumers, followed by Alt3 (0.268), Alt1 (0.225), and Alt4 (0.209).

Table 3. Ranking of website quality factors

	Alt1	Alt2	Alt3	Alt4
Navigability	0.218	0.295	0.256	0.181
Response Time	0.337	0.245	0.263	0.115
Ease of use	0.283	0.304	0.243	0.170
Telepresence	0.267	0.264	0.252	0.237
Security	0.279	0.309	0.225	0.187
Personalization	0.278	0.283	0.268	0.172
Understandability	0.262	0.236	0.273	0.229
Currency	0.239	0.328	0.226	0.208
Relevance	0.156	0.335	0.307	0.203
Reliability	0.194	0.308	0.250	0.248
Responsiveness	0.212	0.234	0.343	0.211
Empathy	0.224	0.355	0.233	0.189
Customization	0.180	0.308	0.268	0.244
Brand Awareness	0.231	0.323	0.263	0.183
Reputation	0.172	0.295	0.302	0.231
Price Advantage	0.271	0.305	0.211	0.212
Product Advantage	0.182	0.302	0.320	0.196

Table 3 shows the normalized priority weights of website quality factors. Through this table we can see that Alt2 which is the most preferred alternative website of online consumers has the highest navigability , ease of use, security, personalization, currency, relevance, reliability, customization, brand awareness and price advantage. Alt3 which ranks second has the highest service responsiveness, information understandability, and reputation and product advantages. Alt1 ranks third has the highest system response time and telepresence, while alt4 ranks last does not have outstanding website quality factors. Through this table, an e-commerce business can identify the gap of website quality factors between its website and their competitors', and thus develops the relevant strategies to improve the quality level of its website.

5.3   Sensitivity Analysis

Sensitivity analysis is to investigate when the weight of a factor or an indicator changes, how other factors or indicators change relatively. Expert Choice software provides three kinds of sensitivity analysis [43]: Dynamic Sensitivity Analysis, Performance Sensitivity Analysis, and Gradient Sensitivity Analysis.

Through sensitivity analysis, we can provide more information about how online retailers improve their website quality to catch up with competitors or how to maintain its position as the most preferred website.

The Gradient Sensitivity Analysis chart of System quality is shown in Figure 5. From Fig.5, we can see when the weight of system quality is greater than 0.646, it may lead to Alt3 whose user preference ranking is second interchanges its ranking with Alt1 whose user preference ranking is third. This shows that if Alt1 greatly improves its system quality, it can improve its ranking among online consumers. When the weight of system's response time is greater than 0.08, online consumers may replace the ranking of Alt2 and Alt3. This shows that if Alt 3 speed up its response time, then its user preference ranking may exceed Alt2.

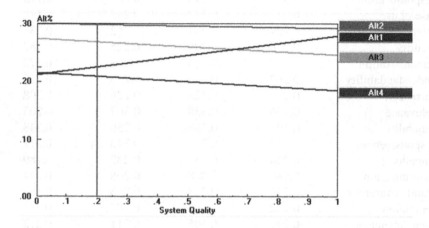

Fig. 5. Gradient Sensitivity Analysis chart of System quality

The Gradient Sensitivity Analysis chart of Service quality is shown in Fig.6, which indicates that when the weight of service quality is greater than 0.657, online consumers may replace the ranking of Alt4 and Alt1. Therefore, if Alt4 wants to improve its user preference ranking, it needs to improve its service quality.

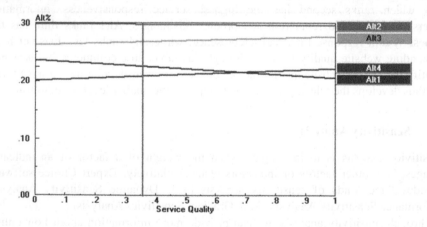

Fig. 6. Gradient Sensitivity Analysis chart of Service quality

6   Conclusions

Through modifying updated D & M IS success model and applying AHP, this paper explores the relative importance of website quality factors which affect online consumers to choose the most preferred websites and the priority of alternative websites. We found that when online consumers choose the most preferred website, service quality is considered to be the most important factor, among which reliability and responsiveness rank the highest, this suggests that e-commerce businesses should pay more effort to improve service reliability and rapid response. Supplier quality is highly relevant, product advantage and price advantage causes online consumers' concerns which shows e-commerce businesses should take the balancing strategy to increase product variety, improve product quality and at the same time lower product price. The currency of information quality is also top-ranking, indicating that e-commerce businesses should pay attention to update the information including product information and contact information on their website timely.

This paper has limitations, which needs to revisit in future studies. First, the results of this paper confines to the field of agricultural products B2C e-commerce, not represents all e-commerce. Secondly, the evaluation factor is based on the updated D & M IS success model, which may rule out some factors strongly affect the website quality. Third, because the target website is e-commerce website, which makes the validity study of each factor not in the controllable environment. Fourth, the sample used in this paper are not enough, it may lead to sample selection bias problem. Fifth, this paper is based on online survey, the sample's behavior is uncontrollable, which may result in deviation of sample data.

This paper empirically validates that the updated D&M IS success model can successfully reveal e-commerce success; by adding ease of use factor in system quality, adding customization factor in service quality, adding product advantages factor in supplier quality, the initial model is extended. The results showed a significant effect. The extended model can be used to guide managers / designers to measure the level of their websites quality. With comparing with competitor's website quality, E-commerce businesses can develop strategies and resource allocation decisions to improve the current websites to obtain e-commerce success; Research models, standards and their relative impact not only provide useful information for decision-makers of e-commerce businesses to develop decision support systems to monitor the performance of the current websites, but also provides strategic advice for businesses to develop better e-commerce websites.

Acknowledgment. Funds for this research was provided by the National Natural Science Foundation of China (No.71271207).

References

1. Irani, Z.: Information systems evaluation: navigating through the problem domain. Information & Management 40(1), 11–24 (2002)
2. Thornton, J., Marche, S.: Sorting through the dot bomb rubble: how did the high-profile e-tailers fail? International Journal of Information Management 23(2), 121–138 (2003)

3. DeLone, W.H., McLean, E.R.: Information systems success: the quest for the dependent variable. Information Systems Research 3(1), 60–95 (1992)
4. Delone, W.H.: The DeLone and McLean model of information systems success: a ten-year update. Journal of Management Information Systems 19(4), 9–30 (2003)
5. Molla, A., Licker, P.S.: E-commerce systems success: An attempt to extend and respecify the DeLone and McLean model of IS success. Journal of Electronic Commerce Research 2(4), 131–141 (2001)
6. Madeja, N., Schoder, D.: Designed for success-empirical evidence on features of corporate web pages. In: Proceedings of the 36th Annual Hawaii International Conference on System Sciences, 2003, p. 10. IEEE (2003)
7. Xuan, W.: Factors affecting the achievement of success in e-tailing in China's retail industry: a case study of the Shanghai Brilliance Group. College of Management, Southern Cross University, Australia (2007)
8. Sharkey, U., Scott, M., Acton, T.: The influence of quality on e-commerce success: an empirical application of the Delone and Mclean IS success model. International Journal of E-Business Research (IJEBR) 6(1), 68–84 (2010)
9. Bo, Z.: AHP Fundamentals Introduction. Journal of Northwestern University (Natural Science Edition) 28(2), 109–113 (1998)
10. Wang, Y.: Systems Engineering, 2nd edn. Mechanical Industry Press (1999)
11. Von Dran, G.M., Zhang, P., Small, R.: Quality websites: An application of the Kano model to website design. In: Proceedings of the Fifth Americas Conference on Information Systems, pp. 898–900 (1999)
12. Zhang, X., Keeling, K.B., Pavur, R.J.: Information quality of commericial web site home pages: an explorative analysis. In: Proceedings of the Twenty First International Conference on Information Systems, pp. 164–175. Association for Information Systems (2000)
13. Alba, J., Lynch, J., Weitz, B., et al.: Interactive home shopping: consumer, retailer, and manufacturer incentives to participate in electronic marketplaces. The Journal of Marketing, 38–53 (1997)
14. Calkins, J., Farello, M., Shi, C.: From retailing to e-tailing. Strategic Direction 16(6), 17 (2000)
15. Dholakia, U.M., Rego, L.L.: What makes commercial Web pages popular? An empirical investigation of Web page effectiveness. European Journal of Marketing 32(7/8), 724–736 (1998)
16. Cogitative Building maintaining & repairing web brand loyalty (EB/OL) (1999), http://www.cognitiative.com
17. Delone, W.H., Mclean, E.R.: Measuring e-commerce success: applying the DeLone & McLean information systems success model. International Journal of Electronic Commerce 9(1), 31–47 (2004)
18. Chiger, S.: List shopping online. Catalog Age 14(7), 95–97 (1997)
19. Nielsen, J.: Seven deadly sins for web design. Technology Review 73(1), 9A–10A (1998)
20. Marsico, M.D., Levialdi, S.: Evaluating web sites: exploiting user's expectations. International Journal of Human-Computer Studies 60(3), 381–416 (2004)
21. Carver, D.L.: Designing the user interface, strategies for effective human-computer interaction. Journal of the American Society for Information Science 39(1), 22–22 (1988)
22. Horsti, A., Tuunainen, V.K., Tolonen, J.: Evaluation of electronic business model success: Survey among leading Finnish companies. In: Proceedings of the 38th Annual Hawaii International Conference on System Sciences, HICSS 2005, p. 189c. IEEE (2005)

23. Steuer, J.: Defining virtual reality: Dimensions determining telepresence. Journal of Communication 42(4), 73–93 (1992)
24. Klein, L.R.: Creating virtual product experiences: the role of telepresence. Journal of Interactive Marketing 17(1), 41–55 (2003)
25. Lombard, M., Snyder-Duch, J.: Interactive advertising and presence: a framework. Journal of Interactive Advertising 1(2), 56–65 (2001)
26. Riecken, D.: Personalized views of personalization. Communications of the ACM 43(8), 27–28 (2000)
27. Zhang, Y., Im, I.: Recommendation systems: a framework and research issues. In: The Proceedings of Americas Conference on Information Systems, pp. 468–473 (2002)
28. Murthi, B.P.S., Sarkar, S.: The role of the management sciences in research on personalization. Management Science 49(10), 1344–1362 (2003)
29. Schonberg, E., Cofino, T., Hoch, R., et al.: Measuring success. Communications of the ACM 43(8), 53–57 (2000)
30. Singh, M.: A primer on Developing an E-business Strategy. Western Illinois University (2002)
31. Morelli, W., Clark, G., Tesler, S.: Up close and personal. Best's Review 101(9), 99–100 (2001)
32. Roman, E.: Ready, fire, aim (EB/OL) (2005), http://www.Direct.com
33. Epstein, M.J.: Implementing successful e-commerce Initiatives. Strategic Finance 86(9), 22 (2005)
34. Bakos, J.Y.: Reducing buyer search costs: implications for electronic marketplaces. Management Science 43, 1676 (1997)
35. Devaraj, S., Fan, M., Kohli, R.: Antecedents of B2C channel satisfaction and preference: validating e-commerce metrics. Information Systems Research 13, 316–333 (2002)
36. Chen, Z., Dubinsky, A.J.: A conceptual model of perceived customer value in e-commerce: a preliminary investigation. Psychology and Marketing 20, 323–347 (2003)
37. Smith, P.: Hi tail to E-tail. New Zealand Management 47(6), 22 (2000)
38. Ba, S., Pavlou, P.A.: Evidence of the effect of Trust building technology in electronic markets: price premiums and buyer behavior. MIS Quarterly 26(3), 243–268 (2002)
39. Epstein, M.J.: Implementing successful e-commerce Initiatives. Strategic Finance 86(9), 22 (2005)
40. Teerling, M.L., Huizingh, E.K.R.E.: How about integration: the impact of online activities on store satisfaction and loyalty. University of Groningen (2003)
41. Markus, M.L.: Electronic mail as the medium of managerial choice. Organization Science 5, 502–527 (1994)
42. Santos, J.: E-service quality: a model of virtual service quality dimensions. Managing Service Quality 13(3), 233–246 (2003)
43. T.C. & Forman. Pennsylvania: Expert Choice, Pittsburgh (1996)

Importance of Information Systems in the Evaluation and Research of Nutrition and Health of Key Groups in China's Rural Areas

Liqun Guo, Bo Peng[*], and Zhenxiang Huang

College of Information and Electrical Engineering, China Agricultural University
No. 17 Qinghua East Road, Haidian District, Beijing 100083, China
{pengbo,guoliqun,huangzx}@cau.edu.cn

Abstract. The improvement of nutrition and health evaluation research will better help people to strengthen food safety, prevent and control the major diseases, improve medical and health services. This is especially important with key groups in the vast rural areas in China. Under the background of information era, if the nutrition and health assessment methods research can be combined with the actual research findings of statistical analysis using iinformation systems and big data, the research level will be achieved and be good for the reducing of regional differences in levels of development.This paper summarizes and analyzes the important indicators and elements that have great impacts on the level of nutrition and health for the population through the latest development about nutrition and health research at home and abroad first, then discribes the parametric modeling research ideas and methods under the conditions of modern information technology,at last revealed widely practical prospect of information technology in the field of nutrition and health. This research has practical significance to the promotion of economic and social development in rural areas and the enhancement of the overall nutritional and health level of the population.

Keywords: information system, rural population, nutritive index, health conditions, evaluation model.

1 Introduction

Chinese nutrition health and food nutrition safety are gradually becoming one of the major strategic issues in the construction of national quality in China; thus the research on national nutrition health level becomes imperative. Since 2010, aiming at the sustainable development of national fitness and health, the Chinese government has launched and implemented "The 12th Five Year Plan of Food Industry" and "The 12th Five Year Plan of Health Protection" based on the domestic food nutrition safety situation and the development process of nutrition and population health at home and abroad. This is the top-level design and systematic plan for the mid and long term

---

[*] Corresponding author.

D. Li and Y. Chen (Eds.): CCTA 2013, Part II, IFIP AICT 420, pp. 114–128, 2014.
© IFIP International Federation for Information Processing 2014

development of national food and nutrition safety and health of the population. Due to the large population of China, health protection development levels varied in different places; health evaluation indicators varied for different key groups, such as for the elderly, women and children; chronic, endemic, occupational and other kinds of diseases are crossed and frequently occurred; all these situations caused the lagging of the investigation and research of nutrition and health of key groups outside the urban areas. However, with the implementation of new health insurance and new rural cooperative medical system (NCMS) and the emergence of potential technologies like the Internet of Things and Big Data, it is predicted that the modern information will become the most effective means in the investigation and research of nutrition and health of the population in China.

Research of nutritional and health level mainly involves two major fields: human nutrition and food nutrition. From the view of academic development, nutriology is closely linked with biochemistry, physiology, pathology, clinical medicine, food science, agricultural science and other disciplines. From the micro perspective, it can provide guidance for the reasonable arrangement of the diet of an individual, a family or a group; it is closely connected with the life processes of human beings, such as the growth and development, physiological function, operational capacity, disease prevention, health protection and longevity; from a macro perspective, it is related to the agricultural production, food processing, the populace's cultivation and economic levels of a country. Therefore, nutriology is a natural science discipline with high potential for scientific applications. For example, the incidence of chronic diseases and lifestyle are closely related, which accounting for about 60% of health issues. The two major factors affecting chronic diseases are nutrition and exercise. People with obesity and overweight have a high risk of chronic diseases. Thus nutrition is the most important factor which affects public health. This paper provides an overview of current studies on the evaluation of nutrition health level and puts forward that the information system can play an important role in this evaluation, particularly for the key groups in Chinese rural areas.

2   Analysis of Research Progress

2.1   Research Status Worldwide

Trophic level evaluation is an important reference for the measurement of population's health level. Currently, a large number of studies have been conducted to evaluate population's nutrition health level based on the indicators, such as heart health, digestive system health, bone health, weight control, mental supplement and blood sugar control, etc. Also, the evaluation methods for nutrition health level have been well documented.

In the field of evaluation for nutrition health level, worldwide, according to the Report of United Nations Standing Committee on Nutrition (2009), the Sixth World Nutrition Situation Report showed obvious progress of nutrition improvement have been made in some countries with nutritional deficiencies, in which the control of iodine deficiency can be taken as one of the very successful cases. In addition, the tendency of dysplasia and underweight in different populations and areas has been effectively monitored. In the end, this report indicated the need of a sustainable solution and a more lasting change to enhance the connection between food safety and nutrition and make it closer. Gu et al. (2007) compared the bone mineral content (BMC) and bone mineral density (BMD) of elderly population. The study included 490 men aged 50-70 and 689 women. The results showed that the BMC and BMD samples of urban population were significantly higher than the samples of rural population. Differences of BMC and BMD of women from urban and rural areas were not limited to the lifestyle, but also included other activities, such as income, milk intake, vitamin D and calcium, the general level of physical activity, walking, and social activities. Aubel (2011) indicated that improvement of nutritional status of infants and young children in developing countries was largely depended on whether the family adopted the optimum nutrient supply, especially, the impacts and functions of women (e.g. grandmother) could not be ignored. Three different cultural backgrounds, namely Africa, Asia and Latin America and social dynamics factors influencing the commonly seen models in child nutrition were involved in this research, including: grandmother's influence on the child nutrition and health issues as the central role; the impact of grandmother on the practices of pregnant women and child nutrition, especially other nutrient levels for children with regard to pregnancy, feeding and care of infants, toddlers and sick children. It also pointed out that the impacts of male members of the families on child nutrition were relatively limited. Brauw et al.(2011) studied children's weight status in rural areas and indicated that weight of children aged 7-12 inclined to be lighter than the standard weight, and if not taken care of grandparents, children aged 2-6 were less likely to excess the standard weight. Migration of the parents was positively correlated with the underweight of children aged 7-12. Because members in families that had migrant workers tended to spend less time in cooking. And children aged 7-12 in such families would have to bear part of the house works, mainly cooking. Nutritional status of children aged 2-6 was largely unaffected in families with migrant workers, especially when they were living with grandparents. Fox et al. (2012) reported a survey focused on the health status of children from 35 developing countries and the multi-level regression results of the data showed that the problem of child malnutrition in rural areas was very serious, which was affected and restricted by the socio-economic level, the medical care level as well as the supply of nutrients and other conditions. Meanwhile, the urban-rural gap was shrinking with the countries' gradual development. Delpier et al. (2012) demonstrated that rapid growth of young people's consumption of sugared beverages would cause weight gain, bone damage and tooth decay, as well as the situation of type II diabetes. Also, the results showed that reduction of the consumption of highly sugared beverages had high statistical significance. It also reflected the significance of applications of Internet and smart phone technology in the analysis and nutritional advices. In America, studies showed (Pan American Health Organization, 1998) that in the last decade, health conditions of Americans was steadily improved, which should give the credit to social, environmental, cultural, and technological development and improvement of medical and health conditions. The characteristics of the progress and speed were unique and could not be reproduced in other countries, or other populations. Some countries, like Latin America and the Caribbean, are still facing the distresses brought by traditional health problems such as famine, environmental degradation and the deterioration of living conditions. In many areas of America, nutrient level has been greatly improved, which can be proved by conventional measurements of body weight and height, but lower weight to age ratio and weight to height ratio are still serious, especially in children, the number of children with lower weight and age ratio caused by malnutrition accounted for 50% of the preschool and school-age children. This problem is not only related to the height, but also reflects the physical and mental development. And obesity mainly refers to excessive weight to height ratio, which mainly present in areas with lower level of development. Among urban population and women, this obesity would normally be misunderstood as excess nutrients, but it is actually accompanied by deficiency of some microelements, such as iron, folic acid and zinc. And its prevention and eradication is very complex. Among the deficiency of these microelements, the deficiencies of iodine and vitamin A had been controlled, the deficiency of iron remains as the most commonly seen nutritional problems, especially among pre-school children and pregnant women.

In brazil, Vieria et al. (2011) studied the socio-economic, diets and anthropological characteristics of school-age population. The research of nutritional conditions mainly focused on the BMI/Age ratio (Body Mass Index) and Height/Age ratio (Height Indicator). The research of food intakes used the 24 hours recall method and the results were analyzed through the comparison with Dietary Reference Intake indicators. The population studied included 145 school-age children and adolescents, of which 79% of their legal guardians had no formal occupation (referring to agriculture and handicrafts industries); average monthly income of 82% of the legal guardians were less than the standard level, 35% mothers of the investigated population received less than 3 years of school education. Analysis of the Body Mass Index and Height Indicator results of the population studied, demographic indicators of results indicated that 7.1% of children and 14.8% of adolescents showed weight and height defects. If the results were separated by gender, female children showed more height defect ratio. Compared with the recommended value of daily energy intake, approximately 72.6% of children and 63.9% of adolescents have energy intake deficiency. Analysis on these two groups showed that the intake of microelements such as iron, zinc, vitamin A and calcium are seriously in shortage. These findings suggest that the level of social and economic development as well as nutritional status is important factors that determine the level of children's nutritional health.

In the field evaluation methods for nutrition health level, for instance, Klein et al. (1997) reported the reality of urgently needed health care resources and intensive medical models for elder population in rural areas. A nutritional risk screening model suitable for the managed health care model targeting elder population in rural areas was also proposed in the research. The nutritional risk screening was achieved through the Geisinger Health Care System; the managed health care model was embedded in its individual remote diagnostics site, which made further screening and case management of malnourished populations possible. And the screening and intervention would be conducted at the clinic sties, which were selected based on the integrated professional knowledge and resources provided by this research. A rational clinic case manager would be developed based on the personalized assessment and intervention programs. Research subjects completed the screening at the remote sites, and the medical records management and selection of nutritional status would be developed according to risk criteria. Tham et al. (2010) reported the evaluation method of integrated health service level in rural areas of Victoria region, Australia. The optimized combination of important parts of the successful basic medical insurance and the target health services and health status indicators was used to establish a conceptual evaluation system. The promotion of this kind of service in Victoria region indicated that there was not enough evidence to prove that this system ran better in rural areas. Although the health service model might have minor differences due to geographical, environmental and other factors, there was evidence to suggest that the health service model could be sustainable be able to feedback and could meet the local medical standards. This evaluation system could provide guidance for evaluation of future health services and provide new ideas for research of health services' influences on the community and residents.

2.2   Research Status in China

Nutrition and health condition of the population are indicators that could reflect the economic and social development, health care level and population quality of a country or region. Good nutrition and health status are both the foundation of social and economic development and the important social and economic development goals. However, many regularity things still needed to be explore in the academic research of nutrition and health levels of key groups in rural areas, in practice, there are still many problems to be solved. To study the development law of nutrition and health levels of key groups in rural areas, its system and structure shall be firstly clarified, including the composition of the various elements within the system and their corresponding functions. The introduction and presentation of related concepts concerning nutrition and health levels of key groups in rural areas are the basis and premise for the research of its evaluation and classification.

In China, according to the level of social development, nutrition and health research conducted by domestic scholars mainly focused on: analysis and assessment of health conditions of different population groups, analysis and assessment of nutritional conditions of different groups, economical analysis and assessment of trophic levels of different groups, economical analysis and assessment of health levels of different groups, as well as research on the specific nutrition evaluation of people in the hospital that already had health problems, application of information technology in health management of different population groups and application of information technology in the field of food safety.

2.2.1   Analysis and Assessment of Health Conditions of Different Population Groups

Based on the analysis and assessment of health conditions of different population groups, an indicator system on nutrition and health level of key rural population can be proposed to build its comprehensive evaluation model, which is the Grey Clustering Evaluation Model for nutrition and health level of key rural population.

Based on the WHO definition of health, conformed to the shift from biomedical model to organisms-psychological paradigm-social medicine model and the shift of health measurement from one-dimensional to multi-dimensional and with the introduction of "three elements" in health care, Yao Xuyi (2005) quantified the above elements to establish a three-dimensional mathematical model of integrated health evaluation, which could provide more intuitive, comprehensive, and accurate reflection of the true meaning of individuals or group health from the quantitative point of view, which made the realization of people's desire in practice became possible.

He Liping (2010) evaluated the health fairness of farmers in three counties in Yunnan Province with the application of range method, Gini coefficient, concentration index and Logistic regression; the results showed that if the impacts of only one factor was taken into consideration, the concentration index would applicable; but multivariate analysis should be used if impacts of more factors were to be considered.

Based on disease surveillance and the NCMS information system, Wang Hongjuan (2012) established the evaluation index system and evaluation methods for health status of rural residents. The researcher conducted comprehensive evaluation of the health status of rural residents in Miji District with the application of the established evaluation methods for health status of rural residents, which filtered out the major public health problems influenced the health status of rural residents to provide scientific basis for decision-making of local government on the development of health services. Issues of exploration of how to share existing diseases and health monitoring system data and how to effectively apply the evaluation methods of health status of rural residents were also studied.

Zhao Huashuo (2011) studied the grading evaluation of Quality of Live (QOL). Three kinds of multivariate statistical analysis methods (principal component analysis (PCA), cluster analysis, discriminant analysis) were used to conduct grading evaluation on QOL data. The results classified the QOL of 209 elderly people into 3 grades of good, medium and poor; the percentage of each grade was 45.93%, 33.02% and 21.05% respectively. The conclusion was that the comprehensive application of a variety of multivariate statistical methods can successfully solve the problem of grading evaluation of Quality of Live (QOL).

Liu Tanghong (2010) found the best way to evaluate the comprehensive health status through the assessment of the health status of rural residents in Dongying. The understanding of the multi-dimensional health status and comprehensive health status of rural residents in Dongying provided basis for the formulation of various policies concerning the improvement of health level of rural residents and thus could effectively improve the comprehensive health status of rural residents. A combination of quantitative and qualitative research methods was used in the research: in the quantitative study, survey respondents were randomly selected through stratified cluster; household survey were completed by face to face interviews; in qualitative research, health assessment and recommendations from experts were obtained through group discussion and the statistical analysis of the collected data was conducted with the application of SPSS16.0 software.

Hu Yong (2007) described the current status of the overall health of farmers in China from three dimensions, namely: health outcomes, health care utilization and availability, health financing and health insurance and analyzed the major influence factors from the perspective of sociology of health and illnesses, including the public health environmental degradation in rural areas, laggard health concepts of farmers, weak theoretical basis for health promotion of farmers, constraints of dualistic urban-rural social and economic structure, the coexist "absence" and "overdone" of the government functions.

Ma Xiaorong (2010) believed that the empirical research results of the health needs of rural residents showed that: age and health service prices and had significant negative effects on health; but education level and household income per capita had significant positive effects on health; self-rated health evaluation of employed rural residents was better than that of unemployed urban residents; QOL indicators of rural residents in marriage were higher than that of the unmarried, divorced or widowed residents. In general, as a health measurement method, regression results of self-rated health evaluation fitted the predictions of Grossman Model better than the regression results of the QOL indicators.

Nie CuiFang (2007) used the random cluster sampling method to select 1661 elderly person over 45 years old from Lacey City as the research object to conduct physical examination, including measurement of height, weight, waist circumference (WC), hip circumference (HC), blood pressure, blood glucose and hemoglobin, etc. and calculated the body mass index (BMI), waist-hip ratio (WHR), waist / height ratio (WHtR) and other indicators. Correlation analysis was conducted to understand the nutrition-related diseases situation among middle age and elderly population in rural areas of Lacey City. The studies showed that the prevalence of nutrition-related diseases among the elderly population in rural areas remains high. Nutrition and health education could reduce the prevalence of some nutrition-related diseases of the elderly population in rural areas, but chronic disease control was a long-term process. It was recommended that regular and continuous health education for the elderly in rural areas should be conducted.

Gao Hong (2011) reviewed that analysis method had been established on personal health evaluation swarms for Chinese people. Suggestions and opinions of experts on the content arrangement and index selection of the personal health evaluation index for Chinese were collected through Delphi method using questionnaires. A personal health evaluation index system for Chinese was established after analysis and summarization. Then the normal distribution method and percentile method were used to define the medical reference range of physical, psychological, social, medical and behavioral health dimensions of Chinese people. In the end, the reliability and validity of the index system were tested by the application of questionnaire.

2.2.2   Analysis and Assessment of Nutritional Conditions of Different Groups

To establish an assessment method for the evaluation of nutritional and health conditions of key groups in rural areas, a set of food accessibility –based nutritional level indicators and a set of health condition indicators under the environmental stress conditions shall be defined. Through questionnaire design, representative sample selection; with the application of combination of field research and literature analysis, this research was able to acquire measured data of nutritional and health conditions of key groups in rural areas. Gray cluster theory was used in the data analysis and the interaction between the two sets of indicators mechanisms was discussed in the research. This research also constructed the comprehensive evaluation model for the nutritional and health conditions of key groups in rural areas, which was also modified according to the rural development in China to establish the evaluation methods for the nutritional and health conditions of key groups in rural areas.

Through the investigation of health literacy of residents in Hubei Province, Hu Xiaoyun (2009) analyzed the influencing factors on health literacy of residents and how these factors in turn impacted the health conditions of the residents. She assessed health literacy of individuals and groups by selecting individuals and groups with lower health literacy as the subjects. Her research provided tools for the evaluation and assessment of the effects and achievements of health promotion/education works or projects and further provided reference for the promotion of health literacy monitoring; for the setting of health promotion and health education related strategies and standards. This research was the first relatively comprehensive description and analysis of the health literacy of residents in Hubei Province. It used the SEM model to build up the relations between the basic information, health literacy and health conditions of the residents and conducted quantitative analysis of the relations to provide reference for the decision-making in residents' health literacy improvement.

Li Jing (2009) carried out a nutrition intervention study about the rural communities in Tianjin from November 2007 to January 2009. The result indicated that rural residents' knowledge in terms of nutrition and other health knowledge had been significantly improved and their attitudes also greatly changed after the intervention, but the changes concerning diet and living habits were not significant. Intake of vegetables and fruits was increased in rural residents since they begun to establish a sense of eating fruits and vegetables.

Xiong Guohong (2009) developed and designed a professional health advice website based on nutrition counseling called "My health, My say". With its wide application in community service, people's nutritional awareness had been improved and nutrition knowledge was universally popularized.

Research result of Rao Jianjun (2009) suggested that a significant number of the rural elderly people were in a dangerous state of malnutrition with the coexisting of weight loss and overweight problems. Therefore, nutrition and health issue was a problem existing and needing to be handled among rural elderly people. The overall nutritional status of the elderly people in rural areas of Tongzhou District was higher than that of two other areas in the country (Anhui and Jiangsu Nantong) but lower than the level of elderly people living in urban communities in Wuhan and Shanghai. Among the surveyed elderly people, population with malnutrition accounted for a small proportion (7.6%), but population at risk of malnutrition is relatively high, accounting for 45.7%. Relatively large differences existed between individuals in terms of nutritional status. Appropriate care and interventions should be taken by relevant agencies according to the major influencing factors of the nutritional status and the specific circumstances of different groups towards the elderly.

Li Jing (2011) conducted primary exploration on nutritional intervention and provided reasonable proposals through her investigation of collective meal quality of children in urban areas of Lanzhou City; her physical testing and determination of mineral elements and her assessment of their nutrition and growth conditions, including: 1. dietary and nutritional status: method of continuous 5 days weighing was used in this survey and the evaluation of the results referred to the Diet Guidance for Chinese People and Chinese DRIS; 2. the growth and development status: the evaluation of physical development was done through the tests of height and weight; the recommended height and weight provided by World Health Organization (WI10, 2006) were used as the reference standard, weight for age Z score (WAZ), height for age Z score (HAZ) and weight for height Z score (WHZ) of each child were calculated for the evaluation of their growth and development; integrated growth retardation, low birth weight and weight loss were the three indicators used in malnutrition evaluation; three, minerals determination: fingertip peripheral blood of children were acquired and analyzed with the use of atomic absorption spectrometry to determine the levels of calcium, iron and zinc.

2.2.3   Research on the Specific Nutrition Evaluation of People in the Hospital That Already Had Health Problems

Wu Kun's (2005) definition of the original meaning of the nutrition was "to seek health", which referred to the process of intake, digestion, absorption and utilization of nutrients in food to meet the body's physiological needs of human body. That proper nutrition meant through a scientific cooking process, reasonable diet could provide sufficient energy a variety of nutrients to the body and maintain a balance between the various nutrients to meet the body's normal physiological needs and to maintain the healthy nutrition in human body.

Xu Shiwei (2008) believed that development of modern agricultural aiming at nutrition and health improvement involved a wide range of research work. Research of agriculture as the foundation of scientific research should be strengthened, such as research on the "high-yield, high-quality, high-efficiency, ecological and safe" agricultural production theory and technology; research on the collaborative development of production and environment approach and study of the impacts of agricultural investment on food quality and safety and its reduction ways. Currently, research on the food safety risk assessment theory and method should be strengthened; the stimulation system of plant food quality and safety risks assessment should be established; the ancient analog systems of animal food quality and safety risks assessment should be established; processed food quality and safety risk assessment system taken animal and plant agricultural products as raw materials should be established; quality and safety of analog systems of food from farm to fork shall be established; interdisciplinary integrated research should be deepened to provide a scientific basis for the food quality and safety risk assessment and early warning. Research on agricultural early warning should be strengthened at the same time reduce agricultural production risks and promote the healthy development of modern agriculture.

2.2.4   Applications of Information Technology in the Field of Nutrition and Health Education

Currently, in many rural areas, nutrition and health information took on the one-way propagation, which was confined to merely pass information to the farmers and failed to timely feedback the needs of farmers and various customer-tailored information services were relatively weak. Meanwhile, the farmers' needs for nutrition and health information became more widely, but the existing services lacked of pertinence and thus failed to meet the huge needs of the farmer groups for finer and real information more suitable to local needs, thus the problem of dispersed nutrition and health information failed to match with the needs of farmers occurred. In addition, nutrition and health information provided lack of time-validity, mainly showed by the weak ability in the collection, analysis, processing and dissemination nutrition and health information and the updates of website content were slow, including too much outdated information and inadequate up-to-date information.

Zhu Xiumin (2009) believed that modern information technology provided a framework for modernization model of the nutrition and health education. Its conception was based on the thought of taking computer as the basis and carrier of the transmission of digital contents of nutrition and health education; the "networked" and "intelligent" information technology was taken as the driving power for the efficient   storage   and   transmission   of   education   contents   in   the   carriers. "Digitalization" was the opportunity that triggered IT revolution for its realization. Text, graphics, images, sounds, videos, animations and other teaching content elements can be input into the computer in a certain number format, so as to achieve the purposes of using computer for storage and transmission. Popularization of internet facilitated the IT take-off and broadened the spreading time of information technology. "Intelligentized" multimedia, hypermedia and artificial intelligence, etc. could improve the performances of nutrition and health education software.

Information resources construction is the foundation and guarantee for the nutrition and health status of key groups in rural areas. Information resources and energy resources, material resources together constitute the three pillars of resources of the modern socio-economic and technological development, which plays an important role in all aspects of social development. Database construction of nutrition and health information resource is the major form for the large-scale, high-efficiency development and utilization of information resources in rural areas. After processing, handling and ordering, these information resources could be massive accumulated to form the formatting information resource database, which can easily store, retrieve, transmit, publish and share information with the application of modern information technology.

2.2.5   Applications of Information Technology in the Food Industry

Fang Hai (2006) believed that in order to maximize the control of China's food safety incidents, the using modern information in food safety management was imperative. In terms of construction of the food safety expert advisory system and the perfection of the database system, the expert consultation work should be normalized and institutionalized. Development of expert system, database, knowledge base and rule base, etc. that would be used in a variety of food safety-related aspects should be included to substitute for the relevant experts in providing technical guidance.

Liu Zhen (2008) introduced food safety systems in the United States, European Union, Japan and other developed countries; summed up food safety problems currently existing in China; discussed and analyzed these various issues. Combined with foreign management experiences and the actual conditions in China, based on the ISO9001, HACCP system and GAP, SSOP norms, he focused on the research of food traceability system; analyzed the difficulties existed and put forward prospects based on the principles of this system.

Based on the introduction of relevant theories of database systems, food consumption and dietary balance, Su Yanyan (2007) explained the transformation basis and methods of food consumption and nutrients and described the sources of essential data in detail. The research introduced the requirements analysis, data table structure design choice of development environment and data processing instructions and other system design related works of the system and made a detailed description of how to create a database and how to achieve the functions of the system through code design. The research also designed the application system of the database for the conversion between food consumption and nutrient of Chinese residents, which solved the problem and difficulty lied in the conversion between food consumption and nutrient over the years and also provided a possible way to the offer comprehensive, continuous dietary intake data.

With the Hospital Information System (HIS) as the basis and the smart card reservation system as the backbone, Zhao Hesong (2007) achieved the comprehensive information management of nutritious diet center. This integrated hospital nutritional and dietary management information system integrated the functions of reservation management, nutrient management and inventory management and it was able to share information with the HIS system. The introduction of information technology into the present nutritional and dietary safeguard works greatly reduced the manual errors and improved work efficiency through the application of automatically customized recipes and enhanced ordering data examination and verification, which was also greatly improved implementation rate of diet therapy and thus made the dietary treatment security of the PLA General Hospital reached domestic advanced level.

2.2.6 Application of Information Technology in Health Management of Different Population Groups

The information era brought new opportunities for the nutrition and health research of the key groups in rural areas since the information science provided modern theory and methods for the development of nutrition and health level of the key groups in rural areas while the information technology provided more advanced and powerful research tools for the decision-making concerning the nutrition and health level of the key groups in rural areas. The establishment of the online evaluation system of the nutrition and health level of the key groups in rural areas would provide strong support for the improvement of the nutrition and health level of the key groups in rural areas.

Combined with the health needs of the elderly population, Xu Xiangyi (2009) designed and implemented an all round elderly health management system that covered four aspects, including: management of personal health information, health assessment, preventive health checks and health knowledge consultation.

Taken the traditional Chinese medicine theory as the core of the clinical actual needs center of Guangzhou Hospital of Traditional Chinese medicines, Chen Xiao (2010) made specific and detailed design plan of the construction of the system based on the analysis of the detailed needs. Through the jointly work with software development company, the advanced three-tier B / S architecture Microsoft.net development tools were combined with large databases to achieve the research and development of computerized health management system software with Chinese characteristics and the successfully trial application of the system in clinical practices.

Through the usage of software engineering methods, Ji Yu (2008) developed "health management system for cadre population in armed forces" to provide an information platform for the good health management of the cadre population in armed forces. Used the software as the electronic platform, 6 months of health management services were provided for 209 armed police cadres at their posts. The effects evaluation of the services was conducted from the aspects of relevance, feasibility, appropriateness, effectiveness, efficiency, impacts and sustainability.

The theory of system interpretative structure model (ISM) was used to analyze the elements that influenced the key groups in rural areas and the ISM method was used to build its multi-layered structure. The results would show that the factors that had profound impacts on the entire health and nutrition level of key groups in rural areas were a number of factors; the factors that had medium impacts were also a number of factors and the factors that had surface impacts were a number of factors, which indicated that the current development of information technology in rural areas was closely related to the factors that had profound impacts. Those factors were the key factors in the improvement of health and nutrition level of key groups in rural areas, which were also the most crucial and nuclear factors that would have far-reaching influences.

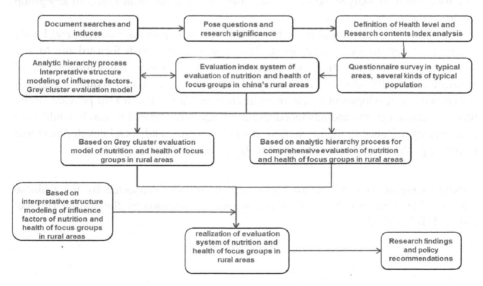

We should search and induce relavant document of the evaluation of nutrition and health of focus groups in china's rural areas. Pose out the relavant questions and research significance of the research. Definition of health level and index analysis of research contents should be clarified. Then do the questionnaire survey in typical areas and several kinds of typical population. Evaluate nutrition and health of focus groups in china's rural areas based on the different evaluation model such as analytic hierarchy process, interpretative structure modeling of influence factors, grey cluster evaluation model. Especially study on the optimization of grey cluster evaluation model and analytic hierarchy process. Combined to the Interpretative structure model, the evaluation system of nutrition and health of focus groups in rural areas will be realized. At last, summarize the research findings and propose policy recommendations of the evaluation of nutrition and health of focus groups in china's rural areas.

3   Conclusions

Based on current studies worldwide, it was obvious that nutrition and health are taken as an holistic object to conduct systematic researched from the disciplinary perspective, while the domestic researchers still take nutrition and health as two separate objects and the scientific and rational link between the nutrition and health for the key groups in rural areas had not been established yet. Thus, by means of modern information technology, the investigation on the health and nutritional needs of the key groups included women, children, the elderly, the disabled and the mentally ill in China's rural areas should be carried out as soon as possible. It will be beneficial to obtain basic data for the construction of evaluation models and methods. Then combining the the evaluation methods of nutrition and health and the practical research results to analyze the factors affecting nutrition and health level of key group in vast rural area.

Information system can play an important role in monitoring nutrition and health situation of key group in rural areas. This would greatly push forward the further strengthen of the prevention and control of major diseases, the improvement of medical and health services, the strengthen of food security and reducing regional differences in development levels in rural areas in China. It can also provide strong theoretical foundations and advanced technical supports. Related research would have practical significance in the promotion of rural economic and social development and improving   the nutritional and health levels of whole chinese population.

Acknowledgements. This investigation was financially supported by the Chinese National "Twelfth Five-Year" Plan for Science & Technology Supporting (Project No. 2012BAJ18B07).

References

1. Aubel, J.: The role and influence of grandmothers on child nutrition: culturally designated advisors and caregivers. Maternal & Child Nutrition 8(1), 19–35 (2012)
2. Brauw, A., Mu, R.: Migration and the overweight and underweight status of children in rural China. Assessing the Impact of Migration on Food and Nutrition Security 36(1), 88–100 (2011)
3. Xiao, C.: Construction and application of the health management system of traditional Chinese medicine. Guangzhou University of Chinese Medicine, Guangzhou (2010) (in Chinese)
4. Delpier, T., Giordana, S., Wedin, B.M.: Decreasing Sugar-Sweetened Beverage Consumption in the Rural Adolescent Population. Journal of Pediatric Health Care (in press) Available online August 26, 2012
5. Hai, F.: Studies on information-based management structure of food safety in developed countries and suggestions on establishing China food safety management structure. East China Normal University, Shanghai (2006) (in Chinese)
6. Fox, K., Heaton, T.B.: Child Nutritional Status by Rural/Urban Residence: A Cross-National Analysis. The Journal of Rural Health 28(4), 380–391 (2012)
7. Gao, H.: Study on Chinese Health Status Assessment Indicators System. Huazhong University of Science and Technology, Wuhan (2011) (in Chinese)
8. Gu, W., Rennie, K.L., Lin, X., Wang, Y.F., Yu, Z.J.: Differences in bone mineral status between urban and rural Chinese men and women. Bone 41(3), 393–399 (2007)
9. He, L., et al.: Study on the evaluation methods of health equity. Soft Science of Health 24(2) (2010) (in Chinese)
10. Hu, X.: The characteristics of health Literacy of residents in Hubei Province and its impact on health status. Huazhong University of Science and Technology, Wuhan (2009) (in Chinese)
11. Hu, Y., et al.: Analysis on actuality of peasant's health in China and it 's main influencing factors. Chinese Primary Health Care 21(3), 1–2 (2007) (in Chinese)
12. Yu, J.: The development and application of armed cadre health management software system. Hebei Medical University, Shi jiazhuang (2008) (in Chinese)
13. Klein, G., Kimberly, K., Judith, F., Sinkus, B., Jensen, G.L.: Nutrition and health for older persons in rural American. J. Am. Diet Assoc. 97, 885–888 (1997)
14. Li, J., et al.: The primary effect of nutritional intervention in farmers. China Academic Journal Electronic Publishing House 17(2), 161–163 (2009) (in Chinese)
15. Li, J.: Studies on nutrition and growth of children aged 3-6 years on Lanzhou city urban area in 2010. Lanzhou University, Lanzhou (2011) (in Chinese)
16. Liao, T.: Comprehensive health evaluation research on rural residents living in dongying region. Shandong University, Jinan (2010) (in Chinese)
17. Liu, Z.: Research and application of the traceable system of food. Xiamen University, Xiamen (2008) (in Chinese)
18. Ma, X.: An empirical research on residents' health demand in rural china. Nanjing Agricultural University, Nanjing (2010) (in Chinese)
19. Nie, C.: Health status of middle-aged and elderly People in rural area of Laixi and effects of health education. Qingdao University, Qingdao (2007) (in Chinese)
20. Pan American Health Organization. Health in the Americas, edition. PAHO, Washington, D.C. (1998)

21. Rao, J.: A survey of the rural elderly nutrition state in Tongzhou district of Beijing and analysis of the related factors. Hubei University of Traditional Chinese Medicine, Wuhan (2009) (in Chinese)

22. Su, Y.: The design of database application on transform from food consumption to nutrients of Chinese residents. Chinese Academy of Agricultural Sciences, Beijing (2007) (in Chinese)

23. Tham, R., Humphreys, J., Kinsman, L., Buykx, P., Asaid, A., Tuohey, K., Riley, K.: Evaluating the impact of sustainable comprehensive primary health care on rural health. Australian Journal of Rural Health 18, 166–172 (2010), doi:10.1111/j.1440-1584.2010.01145.x

24. United Nations Standing Committee on Nutrition. 6th report on the world nutrition situation. Switzerland: Geneva (2009)

25. Vieira, D.A., dos, S., Costa, D., da Costa, J.O., Curado, F.F., Mendes-Netto, R.S.: Socio-economical characteristics and nutritional status of children and adolescents in rural settlements in Pacatuba, Sergipe (Caracteristicas socioeconomicas e estado nutricional de criancas e adolescentes de assentamentos rurais de Pacatuba, Sergipe.). Nutrire - Revista da Sociedade Brasileira de Alimentacao e Nutricao 36(1), 49–69 (2011)

26. Wang, H.: Study of rural residents health evaluation based on the disease monitoring and new-rural cooperative medical system-to take Maiji district of Tianshui for example. Lanzhou University, Lanzhou (2012) (in Chinese)

27. Wu, K., et al.: Nutrition and food hygiene. People's Medical Publishing House, Beijing (2005) (in Chinese)

28. Xiong, G.: A preliminary design of nutrition information sharing system. Central South University of Forestry and Technology, changsha (2009) (in Chinese)

29. Xu, S., et al.: The modern agricultural development for Nutrition and health goals. Food and Nutrition in China, 1 (2008) (in Chinese)

30. Xu, X.: The researeh and development of old People's Health management and analysis system. Xidian University, Xi'an (2009) (in Chinese)

31. Yao, X., et al.: Human health state comprehensive evaluation method research. Chinese Journal of Preventive Medicine 6(5) (2005) (in Chinese)

32. Zhao, H.: Hospital nutrition and meal management system development based on the HL7 agreement. Nankai University, Tianjin (2007) (in Chinese)

33. Zhao, H., et al.: Study on grading assessm in the quality of life of family with children absent elderly in Xuzhou countryside. China Health Statistics 05, 492–494 (2011) (in Chinese)

34. Zhu, X., et al.: Modern information technology and the nutrition health education. Examination Weekly 51, 166 (2009) (in Chinese)

The Classification of Pavement Crack Image Based on Beamlet Algorithm

Aiguo Ouyang[*], Qin Dong, Yaping Wang, and Yande Liu

Institute of Optics-Mechanics-Electronics Technology and Application (OMETA),
School of Mechanical and Electronic Engineering, East China Jiaotong University,
Nanchang 330013, P.R. China
ouyang1968711@163.com

Abstract. Pavement distress, the various defects such as holes and cracks, represent a significant engineering and economic concern. This paper based on Beamlet algorithm using MATLAB software to process the pavement crack images and classify the different cracks into four types: horizontal, vertical, alligator, and block types. Experiment results show that the proposed method can effectively detect and classify of the pavement cracks with a high success rate, in which transverse crack and longitudinal crack detection rate reach to 100%, and alligator crack and block crack reach more than 85%.

Keywords: Pavement Crack, Beamlet Algorithm, Classification, Transform.

1   Introduction

Pavement distress, the various defects such as cracks illustrated in Fig.1, represent a significant engineering and economic concern. Pavement crack image classification is important in an automated pavement inspection system, because it can provide critical information for pavement maintenance.

There has been a significant amount of research during the last two decades in developing image processing algorithm for pavement crack inspection. Chou et al [1]. approached the problem of pavement crack classification by using moment invariant and neural networks. After preprocessing and thresholding into binary images, they calculated Hu, Bamieh, and Zemike moments. Teomete et al [2]. proposed histogram projection to identify cracks within a cropped image. While focused on the severity of cracks, crack classification, was not performed. Moreover, the system cannot detect multiple cracks within an image.In a paper by Bray [3], cracks is performed using a neural network while classification is performed by another neural network. The proposed algorithm has not been tested on real images. Cheng et al [4]. described a neural network based thresholding method to segment and classify pavement images that can be implemented in real time. Tsai et al [5]. presented a critical assessment of various segmentation algorithms for pavement distress detection and classification.

---

[*] Corresponding author.

D. Li and Y. Chen (Eds.): CCTA 2013, Part II, IFIP AICT 420, pp. 129–137, 2014.

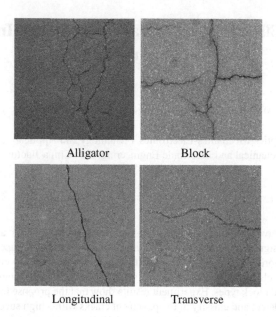

Alligator   Block

Longitudinal   Transverse

Fig. 1. Types of pavement cracks

2   Beamlet Alogrithm

2.1   Beamlet Dictionary

The beamlet transform is performed in the dynamically partitioned squares of an image. Images are viewed as the continuum square $[0, 1]^2$ and the pixels as an array of 1/n by 1/n squares arranged in a grid in $[0, 1]^2$. The collection of beamlets is a multiscale collection of line segments occurring at a full range of orientations, positions, and scales[6], as illustrated in Fig.2.

2.2   Beamlet Transform

The beamlet transform is defined as the collection of line integrals along the set of all beamlets. Let $f(x_1, x_2)$ be a continuous function on 2-D space, where x_1 and x_2 are coordinates. The beamlet transform T_f of function f is defined as follows:

$$T_f(b) = \int_b f(x(l))dl, \quad b \in B_E \tag{1}$$

Where B_E is the collection of all beamlets.

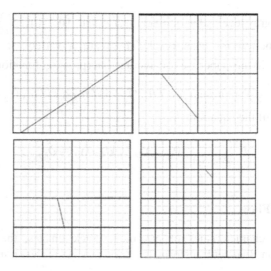

Fig. 2. Four beamlets, at various scales, locations, and orientations

For a digital image, the beamlet transform is a measure of the line integral in the discrete domain. As Fig.3 shows, the beamlet transform for all the points along the beamlet b is defined as,

$$f(x_1, x_2) = \sum_{i_1, i_2} f_{i_1, i_2} \Phi_{i_1, i_2}(x_1, x_2) \qquad (2)$$

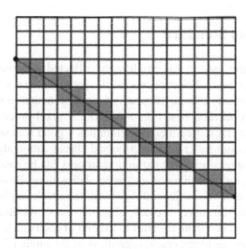

Fig. 3. Beamlet transform as a weighted sum of pixel values along the s headed line

Where f_{i_1,i_2} is the gray level value of pixel (i_1,i_2) and $\Phi_{i_1,i_2}(x_1,x_2)$ is considered to be the weight function for each pixel. There are a variety of ways to choose $\Phi_{i_1,i_2}(x_1,x_2)$, and in this paper we use average interpolation function.

If $p(x_1,x_2)$ represent $[i_1/n, (i_1+1)/n] \times [i_2/n, (i_2+1)/n]$, choose function Φ_{i_1,i_2} fulfill the equation:

$$n^2 \int_{P(x_1,x_2)} \Phi_{i_1,i_2}(x_1,x_2)dx_1dx_2 = \delta_{i_1,i_2} \tag{3}$$

2.3   Algorithm Flow

According to the characteristics of pavement crack images, the images can be identifying based on Beamlet transform, the basic process is shown in Fig.4.

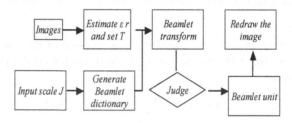

Fig. 4. Algorithm Flowchart

3   Classification Standard

The experimental images used in this article were collected on cement road and asphalt road of campus. Under normal circumstances, the pavement crack images have linear characteristic and contain a large number of environmental noise which are not continuous, so that the method based on conventional pixel processing is difficult to detect and classify the crack. Beamlet algorithm has a good robustness because of its line detection, and was suit for crack detection and classification algorithm.

Pavement crack Images during the extraction process can be easily measured by the projection of the crack on the horizontal and vertical directions. Based on the number of branches and the angle of the horizontal direction, the crack can be divided into four types: longitudinal crack, transverse crack, alligator crack and block crack (Table 1). The crack angle is calculated from the start to the end of each crack. If there is a branch exists, the crack deemed as block crack, regardless of the angle of the crack. For each cell block of a crack image, the largest value of Beamlet transform is defined as block cracks length, and the total length of the cracks is the sum of all the blocks along the crack.

Table 1. The Characteristics of Different Types of Cracks

Crack types	Crack angle Ω	Branch
Longitudinal	$\Omega \geq 60°$	no
Transverse	$\Omega \leq 30°$	no
Alligator	$60° > \Omega > 30°$	no
Block	-	yes

4   Experiment and Discuss

According to the Algorithm flow (fig.4), pavement crack image was processed through image improvement, image enhancement, thresholding, binarization denoising processing and beamlet transform, and the results obtained as shown below. The following figures show the results of the different types of pavement crack image after Beamlet algorithm processing.

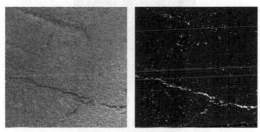

Original image   Binary image enhanced and threshold

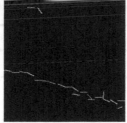

Beamlet transform processing image

Fig. 5. Transverse Crack

Table 2. Transverse Cracks Category

Number	Crack angle Ω	Branch	Crack length	Crack types
1	21°	yes	18.25	Transv-erse
2	6°	yes	179.31	Transv-erse

Note that Fig.5 comprises with two transverse cracks, one long and one short. Original crack image was processed through image enhancement, thresholding and beamlet transform, and obtained a good denoising result. The crack information in the original image retained very well, and can be effectively distinction between two types of cracks.

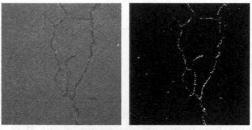

Original image   Binary image enhanced and threshold

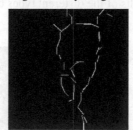

Beamlet transform processing image

Fig. 6. Alligator

Table 3. Alligator Classification

Number	Crack angle Ω	Branch	Crack length	Crack types
1	-	4	448	Alligator

Fig.6 is alligator crack, in which have four branches. According to the presence or absence of the branch, original image can be easily determined as alligator crack. And it can be seen that the method is not susceptible to noise interference, and able to detect the real weak edge.

There was a horizontal crack and a vertical crack in Fig.7. The crack angle of Number 1 is 9°, which is less than 30°, is judged to be transverse cracks. And the crack angle of Number 2 is 67°, which is more than 60°, is judged to be longitudinal cracks.

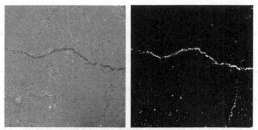

Original image   Binary image enhanced and threshold

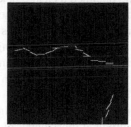

Beamlet transform processing image

Fig. 7. Transverse Crack and Longitudinal Crack

Table 4. Transverse Crack and Longitudinal Crack Classification

Number	Crack angle Ω	Branch	Crack length	Crack types
1	9°	No	154.26	Transve-rse
2	67°	No	53.09	Longitu-dinal

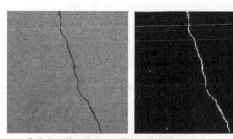

Original image   Binary image enhanced and threshold

Beamlet transform processing image

Fig. 8. Longitudinal Crack

Table 5. Longitudinal Crack Classification

Number	Crack angle Ω	Branch	Crack length	Crack types
1	71°	No	126.39	Longitu-dinal

Fig. 8 only has one longitudinal crack after image processing. The noise in Original image has removed very well, and obtained a clear crack edge. To test the success rate of the proposed beamlet algorithm, a group of 80 pavement images with different types of cracks is chosen for the experiment. Table 6 presents the statistical results of crack classification. The success rate is calculated by dividing the number of the correctly classified cracks over the total number of cracks chosen. Thus the success rate of the block cracks classification is 90%. From the table it can be seen that the single cracks are easier to classify than the alligator or block cracks.

The results of the experiment show that: Beamlet algorithm can smoothly classify pavement crack images into longitudinal, horizontal, block and Alligator with a good classification results.

Table 6. Success Rate of Crack Classification

Cracks	Number	Longitudinal	Transverse	Alligator	Block
Longitudinal	20	20			
Transverse	20		20		
Alligator	20	1	2	17	
Block	20			1	19
Success rate		100%	100%	85%	95%

5   Conclusion

This paper presents a Beamlet transform-based technique to classify the pavement crack images into four types: horizontal, vertical, alligator, and block types. Experiment results demonstrated that the proposed method is very effective with the presence of noise in pavement images. It can be applied on noisy pavement images and classify different types of cracks with a high rate of detection and very low rate of false detection. However, since the Beamlet transform is used to extract linear features, it cannot be used to detect the defects with large area, such as pot holes. In addition more extensive testing is needed to make the algorithm more practical to detect other types of cracks. And the proposed method is capable of finding the length of the cracks but not its width.

References

1. Chou, O'Neill, Cheng: Pavement Distress Classification Using Neural Networks. IEEE-94, 0-7803-2129-4/94 (1994)
2. Teomete, E., Amin, V.R., Ceylan, H., Smadi, O.: Digital image processing for pavement distress analyses. In: Proc. of the 2005 Mid-Continent Transportation Research Symposium, Ames, Iowa (August 2005)
3. Bray, J., Verma, B., Li, X., He, W.: A neural network based technique for automatic classification of road
4. Cheng, H.D., Shi, X.J., Glazier, C.: Real-time image thresholding based on sample space reduction and interpolation approach. Journal of Computing and Civil Engineering 17(4), 264–272 (2003)
5. Tsai, Y., Kaul, V., Mersereau, R.M.: Critical assessment of pavement distress segmentation methods. Journal of Transportation Engineering, ASCE 136(1), 11–19 (2010)
6. Ming, Y., Yon, Y., Yuhua, P., et al.: Beamlet transfrom and multiscale linear feature extraction. Acta Electronica Sinica 35(1) (2007)

Research on the Construction and Implementation of Soil Fertility Knowledge Based on Ontology

Li Ma, Helong Yu, Guifen Chen, Liying Cao, and Yue Wang

College of Information and Technology Science, Jilin Agricultural University,
Chang Chun, Jilin, China
mary19801976@sohu.com

Abstract. Soil fertility is the comprehensive reflection of related factors and the related factors. Soil fertility evaluation knowledge is stored by relational database as usually, and it is difficult to show the correlation and constraints among attributes .In this paper, Nongan county farmland productivity data is as the research object, Using rough set approach to do attribute reduction, using ontology method to establish the soil fertility level knowledge base, using multi Agent technology to implement the prototype system, and complete the reuse and sharing of knowledge, lay the foundation for semantic level reasoning.

Keywords: data mining, ontology, soil fertility, multi-Agent.

1   Introduction

The cultivated land fertility is integrated by soil characteristics, natural conditions and farm management and other elements of the productive capacity of cultivated land, is the comprehensive reflection of related factors and the influence of the nature of the soil properties [1-3]. In soil fertility evaluation, the description of the knowledge and rules is mostly used relational database. And is less considering the relationship among properties of the soil, and there are no sound constraints on attributes, Can't describe the implicit relationships between concepts, Cannot carry out the sharing and reuse of the data, cause a large amount of data redundancy. Rough set can reduce the redundant attribute in database under the premise of guarantee the same classification and decision, simplify the knowledge representation, and improve the efficiency of system processing [4]. Ontology is a new method of data description, Can be a clear description of concept in concept level, and expression association and constraints between concepts [5].

In this paper, on the basis of the research of data mining and ontology technology, the cultivated land of NongAn of Jilin province is as the research object, make extraction and classification of the knowledge in the field, use of ontology technology for knowledge representation, Construct of soil fertility of ontology library, and realize knowledge sharing using Multi-agent Technology, present a new method for data storage.

D. Li and Y. Chen (Eds.): CCTA 2013, Part II, IFIP AICT 420, pp. 138–144, 2014.

2   The Research Data and Related Technologies

2.1   The Data Source

The data this paper used is from the cultivated land fertility survey data, the NongAn (2006), offers by agricultural technology extension center of NongAn. The data includes 25 attributes, such as soil humidity, groundwater depth, light radiation intensity, soil irrigation capacity, annual rainfall, soil drought resistance and soil erosion degree, soil texture, crop rotation suitability, topography, soil parent material, part into layer thickness, salt concentration, humus soil pH value, effective copper, iron, effective slowly available k, effective k, effective fierce, total nitrogen, phosphorus, organic matter and cationic content, effective zinc and productivity grade. Some data are shown in table 1.

Table 1. Some fertility data

ground water depth	soil irrigation capacity	annual rainfall	soil drought resistance	soil texture	soil parent materia	part into layer thickness	salt concen-tration	pH value	effective copper	effective iron
				Light	alluvial					
3-5m	no	400-450mm	strong	clay		10-20cm	<0.1	6.6	1.25	8.72
	strong		strong	Light	alluvial	10-20cm				
<3m		400-450mm		clay			>0.1	6.7	1.26	8.47
3-5m	strong	>450mm	strong	loam	alluvial	10-20cm	<0.1	6.7	1.39	4.70
	strong		strong	sandy	alluvial	10-20cm	<0.1			
3-5m		400-450mm		loam				6.5	1.21	9.45
	strong		strong	sandy	alluvial	10-20cm	<0.1			
<3m		400-450mm		loam				6.7	1.27	9.71
				sandy			<0.1			
5-8m	no	400-450mm	weak	loam	loess	0-10cm		6.6	1.20	8.30
				sandy			<0.1			
3-5m	no	400-450mm	weak	loam	diluvial	20-30cm		6.6	1.30	8.50
	strong		strong	sandy	alluvial	10-20cm				
<3m		400-450mm		loam			>0.1	6.7	1.28	8.48
	strong		strong	sandy	alluvial	10-20cm				
<3m		400-450mm		loam			>0.1	6.7	1.30	0.10
			strong	sandy	alluvial	10-20cm	<0.1			
3-5m	no	400-450mm		loam				6.6	1.27	8.28

2.2   The Rough Set Theory

The rough set (RS) theory is a new mathematical tool which can process fuzzy and uncertainty knowledge, its characteristic is that it does not need to assign quantity descriptions of some characteristics and attributes in advance, starts from the description of given problems, discovers the inherent laws, its basic philosophy is closer to the realistic situation[6].

The basic philosophy of is rough set is that $S=(U, A, \{Va\}, a)$ is called the knowledge expression system, and U is a non-spatial finite set, called the universe ; A is a non-spatial finite set, called attributes set; Va is the range of $a \in A$, $a:U \rightarrow Va$ is a injective maps. If A composes of condition attribute set C and conclusion attribute set D, C and D satisfy $C \cup D = A$, $C \cap D = \Phi$, then S is called the decision system.

In a decision system, the dependence or the connection are existence in certain degrees between each condition attribute, the reduce may be considered that under the premise of losing no information ,using simple description to express the dependence and connection between conclusion attributes to condition attributes in a decision system(Z.Pawlak , 1995). Indiscernibility relation ind(C) divide U into t indiscernibility classes $X1$, $X2$, …, Xt, makes $D(Xi)$ is a set of All values of conclusion attribute d of Xi, that is $D(Xi)=\{v=d(x):x \in Xi\}$, if $D(\ [\ Xi\ ]\ ind(C-\{a\}))=D(Xi)$, then said that the condition attribute $a \in C$ can be removed compared with indiscernibility classes Xi. $C' \subseteq C$ is called the reduce of C relative to indiscernibility classes Xi , if $a \in C$ ', then a cannot be removed relative to Xi.All reduce sets compared with Xi is recorded $SRED(C, Xi)$, $Score(C, Xi)= \cap SRED(C, Xi)$ is called core of Xi.

2.3   The Ontology

The concept of ontology is from the field of philosophy. In the 1960 s, computers started to use ontology. At present, the definition of ontology is a shared standard conceptual model explicit formal specification[7].

As to how to use ontology to organize knowledge, Perez et al. summed up the 5 basic modeling primitives using classification: class, relation, function, the axiom and examples.

(1) classes or Concepts. Classes can be any transaction, such as job, function, behavior, strategies, and reasoning process, and so on. Semantically, the meaning of class is a collection of objects, Its definition generally adopts frame structure, including the name of a concept, with the rest of the concept of the relationship between the collections, as well as in natural language description of the concept, etc.
(2) relations. Formal definitions for the n dimension of a subset of the Cartesian product: $R : C1 \times C2 \times … \times Cn$. For example, the subclass relationship (subclass-of), which represents the interaction between domain concepts.
(3) functions. Function is a kind of special relationship. The formal definition of F is $C1 \times C2 \times … \times Cn-1 \rightarrow Cn$ .The n element can be uniquely determined by n-1 elements before of the relationship, such as is-a is a function, is-a (m, n) means that n is an instance of m.

(4) axioms. Axiom represents the eternal truth assert, is defined in the "concept" and "property" on the limit and rules, such as the concept of A belongs to the scope of the concept of B.

(5) the instance (instances). Instances represent elements. Semantically representation is the object, is a concrete entity refers to a concept class.

3   Construction of Soil Fertility Ontology

3.1   Attribute Reduction in Rough Set

This paper uses the genetic algorithm for attribute reduction, the process is: (1) A randomly generated population. (2) Evaluation of the merits of each chromosome, then choose the excellent chromosome, format a new species. (3) Operators (crossover and mutation) to the new population genetic, then get new specie. (4) Repeat the genetic operation chose the best chromosomes as a solution [8].

According to rough set theory, data set is divided into attribute set and decision set. Attribute set includes 24 attributes, those are soil humidity, groundwater depth, light radiation intensity, soil irrigation capacity, annual rainfall, soil drought resistance and soil erosion degree, soil texture, crop rotation suitability, topography, soil parent material, part into layer thickness, salt concentration, humus soil pH value, effective copper, iron, effective slowly available k, effective k, effective fierce, total nitrogen, phosphorus, organic matter and cationic content, effective zinc. Decision set contains 1 attributes that is fertility level.

Using genetic algorithms to do the reduction, attribute set contains soil humidity, groundwater depth, light radiation intensity, soil irrigation capacity, annual rainfall, soil drought resistance and soil erosion degree, soil texture, crop rotation suitability, topography, soil parent material, part into layer thickness, salt concentration, humus soil pH value, iron, effective slowly available k, effective k, total nitrogen, phosphorus, organic matter and cationic content, A total of 21 properties. Reduces 3 attributes including effective copper, effective fierce, effective zinc The decision set contains an attribute {soil fertility level}.

3.2   Construct the Soil Ontology

Using the seven step of ontology construction as the guidance, taking fertility evaluation based on the series of elements, Soil data based on the actual situation, according to the actual situation of the soil data, establishing the ontology classes and attributes. Soil ontology construction process is as follows:

Define classes and class rating system: (1)to establish the section and physical and chemical properties of physical attributes of, ph value, soil moisture, salt concentration, humus layer thickness, soil texture, cationic content and groundwater depth; (2)The site conditions, including attribute of site topography, soil erosion degree and parts into soil organic; (3)Meteorological conditions of weather, contain attributes rainfall and light radiation intensity; (4)Establish the nutrient content of nutrient, contain attributes effective iron, slowly available k, effective, total nitrogen, phosphorus, effective organic matter; (5) Establish the soil management class management, including attribute soil drought resistance, irrigation and crop rotation suitability; (6)Set up the soil, the soil as a kind of all kinds of the father; (7)Established the kind of equipment, including productivity grade attributes.

Define a relationship class_is, used to determine a specific example belonging to which productivity grade, its domain of definition is soil, its range is class.

The exact makeup of soil ontology is :

Class soilclass {
    Class physical (ph , moisture , salinity , humus , textrue , cation , groundwater)
    Class site (position, erosion, material)
    Class weather (rainfall, light)
    Class nutrient (fe, slow k, k, n, p, om)
    Class management (irrigation, drought, crop)
    }
Class class (attribute is soil_class)
Class relations attribute : class_is

The domain of definition of these attributes including pH, fe, slow k, k, n, p, om, cation is soil, and the range is int. The domain of definition of these attributes including moisture, groundwater, light, irrigation, rainfall, drought, erosion, textrue, crop, position, material, humus, salinity, is soil, and the range is float.   The domain of definition of soil_class is class, and the range is int. Ontology construction result is shown in figure 1.

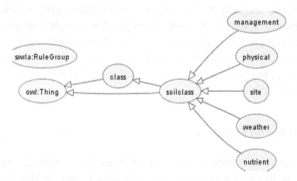

Fig. 1. Ontology Construction

4   Realization System Model

Multi - Agent technology with the development of distributed artificial intelligence can effectively solve the problem of distributed data integration. This paper uses the Agent oriented software development method,   designs and develops data integration and exchange system for the soil fertility data. System development tools is VC++, it is used to achieve the Agent procedures for the preparation and encapsulation. The main function of Agent consists of information exploration of Agent, Agent, intelligent evaluation task decomposition and scheduling Agent, ontology evaluation Agent and ontology information processing Agent and Ontology learning Agent [9]. Using KQML as the message protocol Agent interaction, Enables Agent to exchange knowledge and information to other Agent and Agent running environment. System model is shown as figure 2.

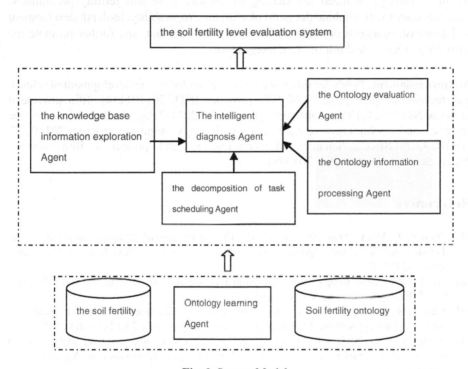

Fig. 2. System Model

Soil fertility level evaluation process can be described as: User submitted data remotely through the client, after the system receives the data, the system preliminary analysis, using the evaluation model for data that is stored in advance, if no match is found in the model, then uses of ontology learning Agent, to study the corresponding evaluation model. Task scheduling of Agent will evaluate the task decomposition, Uses intelligent evaluation Agent call the evaluation module algorithm, then ontology evaluation evaluates the algorithm Agent, Or the ontology information processing Agent deals with the corresponding evaluation results. The final conclusion is reached, and the evaluation results will be output to the user.

5   Summary

In this paper, we study the rough set theory and ontology technology, propose a new method to describe the soil fertility data, realize the sharing and reuse of heterogeneous system knowledge. Using rough set approach to attribute reduction of data, can remove redundant attribute data, reduced data set; Using ontology can make the concept in the field of pattern be understood more accurately, can more clearly express the relationships and rules among the concepts. In this paper, the application of multi Agent technology, realized the sharing of knowledge of soil fertility preliminary. Future research will combine the multi platform interoperability, in-depth development and establish standards ontology of the application domain, and further promote the knowledge sharing and information integration.

Acknowledgment. Funds for this research was provided by the development of science and technology plan projects of Jilin province (NO. 201101114), Jilin provincial projects(NO.201248),The word bank project(No. 2011-Z20), Hall of Jilin province science and technology science and technology support program(20110237, 20110237,20120802), Science and technology research project of Jilin province department of education(No. 2013-68)

References

[1] Chen, G.-F., Ma, L., Dong, W., Xin, M.-G.: The combination of clustering and rough set and decision tree algorithm applied in the evaluation of soil fertility. Scientia Agricultura Sinica 12 (2011)

[2] Li, M.: Research on Rough Set and Decision Tree Application in Evaluation of Soil Fertility Level. Jilin agricultural university (2010)

[3] Tian, J., Hu, Y., Wang, C., et al.: Clustering support decision tree model in the evaluation of cultivated land application. Journal of Agricultural Engineering 23(12), 58–63 (2007)

[4] Huang, Y., Lan, Y., Thomson, S.J.: Development of soft computing and applications in agricultural and biological engineering. Computers and Electronics in Agriculture 71, 107–127 (2010)

[5] Ma, L., Chen, G.: The Knowledge Representation and Semantic Reasoning Realization of Productivity rade Based on Ontology and SWRL. Computer and Computing Technologies in Agriculture 11 (2011)

[6] Chen, G.: For the spatial data mining technology research and application of precision agriculture. Jilin university (2009)

[7] Du, X.: lee, king of bashan. Ontology learning research review. Journal of Software 12, 1837–1847 (2006)

[8] Cui, G., Yin, Q.T.: A kind of attribute reduction algorithm based on genetic algorithm. Journal of Changchun University of Science and Technology 9, 4–7 (2003)

[9] Li, C., Chen, Q.: Multi Agent technology applications in disease diagnosis system. Computer Engineering 33, 182–184 (2008)

Virtual Prototype Design of Double Disc Mower Drive Bracket Based on ANSYS Workbench

Ning Zhang[1], Manquan Zhao[1], Yanhua Shi[2], and Yueqin Liu[3]

[1] Mechanical and Electrical Engineering College of Inner Mongolia Agricultural University,
Hohhot, 010018, China
{lyuzhangning,nmgzmq}@163.com
[2] College of Mechanical and Electrical Engineering of Wuhan Polytechnic University,
Wuhan, 430023, China
shiyanhua122@163.com
[3] Mechanical and Electrical Engineering of Inner Mongolia Technical
College of Mechanics and Electrics, Hohhot, 010070, China
853945990@qq.com

Abstract. During the double disc mower structural design process, static analysis is an extremely important field, not only decide to what the structure size is, but also for the subsequent fatigue analysis, providing the basis for the overall stability analysis. Now the traditional domestic design mainly based on the empirical design, which is resulted in Performance indicators lag behind of the mower, focusing on the performance of vibration is too large and the structure is irrational. Therefore, optimizing the structure of the mower to improve the vibration situation has important significance. Using SolidWorks software to build the three-dimensional solid model of the first mower driveline, and then import the model into ANSYS Workbench to establish the finite element analysis model, through the finite element and modal analysis of the results obtained the stress and strain distribution of the natural frequencies and mode shapes characteristic, providing an important basis for further design optimization of mower structure.

Keywords: Double disc mower, Virtual Prototyping, Optimization design, Finite element analysis, ANSYS workbench.

1 Introduction

For a long time, due to the complexity of agricultural machinery work objects, there is a big difficulty during the process of the agricultural machinery design development, some theoretical analysis are complex, in addition, computation is intensive[1-3]. Traditional agricultural machinery product development process usually go through the prototype design, test, field test, improve the design, retrial and other steps, during the design development process, there are some fatal flaws, namely the high cost, long cycle, poorly designed, low precision relatively complex institutional analysis and synthesis problems, many mechanism parameters are depended on the designers' experience, which is seriously hampered the improvement of the products quality.

D. Li and Y. Chen (Eds.): CCTA 2013, Part II, IFIP AICT 420, pp. 145–151, 2014.
© IFIP International Federation for Information Processing 2014

During the modern machine design, optimal design of the mechanical structure is an indispensable part, it has a significant effect on the performance of the whole mechanical structure [4]. Traditional optimization design method is cumbersome and solving complex, accuracy is also difficult to meet the requirements. With the rapid development of the computer technology, a variety of simulation software capabilities matured and improved, and then the process of structural optimization is feasible, analysis results is visualized, to infuse a new strength into the mechanical optimizing design of the structural design. Using ANSYS workbench ANSYS Workbench to establish the finite element analysis model, which is an inevitable trend for the future of mechanical design optimization design to improve the structural optimization of efficiency and accuracy and shorten the production development cycles.

2   Overall Program to Determine

9YG-130 front mounted dual disc mower, is mainly used in cutting and laying grass strip operation, the machine design is reasonable, easy to operate, reliable, and stable performance et al, which is an ideal promotion models for farmers to use.

9YG-130 dual disc mower is connected by tractor yoke1, the first stage drive system 2, transmission and manual clutch 3, mower frame 4, the cutting member 5, four-bar linkage 6, the lift cylinder 7 and the other components, as is shown in Fig.1.

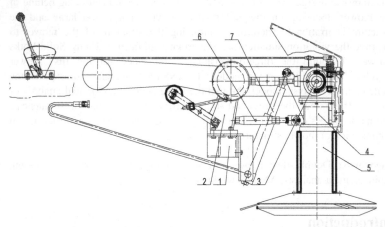

1. Tractor yoke 2.The first stage drive system 3. Transmission and manual clutch 4. Mower frame 5. The cutting member 6. Four bar linkage 7. Lift cylinder

Fig. 1. 9YG-130 model double disc mower

2.1   Create the Double Disc Mower Three-dimensional Model

Founding accurate, and reliable compute model is one of the important steps for dual disc mower finite element analysis to apply the finite element method [5]. Because the modeling capabilities of three-dimensional ANSYS Workbench is weaken, Using

SolidWorks software to build the first stage drive system 3D solid model. To facilitate the analysis, under the premise of the results without prejudice aim at for the model simplify moderately. SolidWorks is applied more widely and used as a feature-based parametric 3D solid design software. Solid designers by stretching, rotating, shelling, wall thinning, arraying and punching achieve product dimensions and structural design. As is shown in Fig.2.

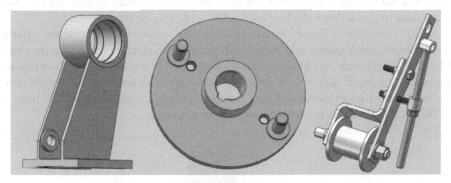

Fig. 2. Double disc mower three-dimensional model

According to the completed three dimensional parts model, using the design method of from bottom to top, the assembly of various components through constraints, cooperate relationship and order together, and finally the entire drive train components assemble the parts made use of 3D entity model. The final double disc mower the first level of the drive train assembly as is shown in Fig.3.

Fig. 3. Double disc mower assembly model

2.2   Importing the Three-dimensional Model

ANSYS for CAD/CAE collaborative environment can be read directly into a variety of CAD software part model, and in its unified environment to achieve arbitrary model assembly and CAE analysis, by connecting technology and sharing with CAD software. In this study, using the current professional graphics software SolidWorks to build the first level and the tension wheel drive system of the dual disc mower three-dimensional model, and then assembled. Through the seamless connection of the data between SolidWorks and workbench. Imported it into ANSYS Workbench, improving the efficiency of modeling, to prepare for finite element analysis.

Firstly, gray iron material was determined as the model material [6], Elastic Modulus is E=113MPa, Tensile Strength is σ=270MPa, Density is ρ=7.25e3kg/m3, Poisson ratio is 0.25, load is applied to the model, after analysis, the load of drive bracket mainly from the ends of the two belt drive transmitted to the pressure bearing, drive bracket to establish the center of the Cartesian coordinate system, the calculated equivalent in contact with the bearing load is applied to the inner surface.

Mesh Generation. Mesh Generation is the main work of finite element pretreatment, mesh quality and merits will have a considerable impact on the results, ANSYS Workbench meshing is more intelligent, there are a variety of control methods, this paper uses size control method to mesh Generation. In the division of the value set of smart sizing is 20mm for free meshing, divided into 45,249 units and 75,462 nodes. Analysis mesh model as is shown in Fig.4.

Fig. 4. Mesh generation model

Impose Constraints. Then set at a fixed constraint on stand base four holes, bearing contact with the inner surface of is given a vertical downward pressure, and a belt drive pressure, and contrary to the overall stent carry on drive and the local stress and strain analysis.

3   Results and Discussion

By the computer automatic calculation of the optimal final result, meanwhile Design Explorer offers, included response surface contour, response curve, sensitivity, and histogram display, including a variety of results [7, 8]. As is shown in Fig.5, materials were made of gray iron castings, and while the production is small batch, the production cost is too high, the safety factor is too large, we can take use ordinary carbon steel of welded parts as an alternative, so as to reduce production costs. As is shown Fig.6, Fig.7, Fig.8, We can see from the figure, the stress and strain are less than the allowable stress and strain, the maximum stress is about 174.03MPa, these results reflect the relationship between different angles input parameters and output parameters. Laid the foundation for the subsequent topology optimization.

Fig. 5. Safety factor

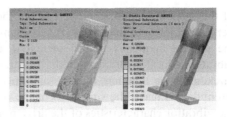

Fig. 6. Total strain and strain direction

Fig. 7. Strain and stress Cloud images

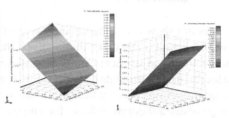

Fig. 8. Response surface Cloud images

Modal analysis technology is used to determine the design organization or vibration properties of machine parts, they are the important parameters in the design of structure under dynamic load.

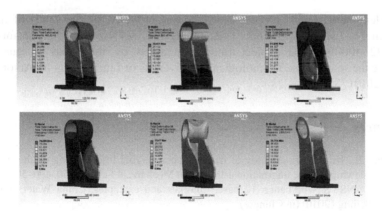

Fig. 9. The first to sixth order natural frequency displacement contours

Lawn mower drive bracket as an important component of dual disc mower, the strong vibration will cause the resonance of structures or fatigue, using ANSYS Workbench software for modal analysis, obtained the double disc mower characteristic of support the natural frequency and vibration mode of transmission, analyses the inherent frequency and vibration frequency displacement nephogram, As is shown Fig.9.

Under the action of the external excitation of bending, torsional vibration, not only caused the fatigue damage of primary vertical shaft, but also affected the smoothness of the overall transmission, we can see the natural frequency of the drive bracket is between 2369.5HZ and 463.83HZ to change, with the modes increase of the order of the natural frequency is also increased, the most of vibration is belongs to local vibration, local vibration mode illustrated its structure is low local stiffness, thickening the ligament may be appropriate to increase the structural stiffness and strength and make the structure change into the overall local vibration and improve the vibration characteristics of the drive bracket.

4   Conclusions

Results are shown that the development of virtual prototype technology creative the new theoretical analysis method, solved the multi-objective optimization request for multi-variable adjustment cannot be replaced by the physical prototype design means perfectly, using virtual design can significantly speed up the design progress, improve the quality of our product design for the subsequent design of agricultural machinery and provides a theoretical basis for vibration analysis. China's agricultural machinery design, due to the application of the finite element method and the development of experimental techniques have been gradually emerging from the rule of thumb approach. Structural optimization work in depth and more widespread, will further improve the design level of farm machinery, improved reliability and shorten the design cycle, so that guide the agricultural machinery design into a new phase.

Acknowledgment. Funds for this research was provided by the funds of national Agricultural science and technology achievement transformation (2009GB2A400054), the project financially aided by the funds to guide and encourage the science and technological creativeness of Inner Mongolia Autonomous Region (2010173).

References

1. Jerman, B., Kramar, J.: A study of the horizontal inertial forces acting on the suspended load of slewing cranes. International Journal of Mechanical Sciences 50, 490–500 (2008)
2. Pu, G.: ANSYS Workbench 12 based tutorials and examples. China Water and Power Press, Beijing (2010) (in Chinese)
3. Zhang, G., Tian, Y.: Variable speed bicycle frame finite element analysis based on ANSYS Workbench. ITS Applications 28(6), 63–65 (2009) (in Chinese)
4. Shang, Y.: Principles and ANSYS FEM Application Guide, pp. 280–281. Tsinghua University Press, Beijing (2005) (in Chinese)
5. Qian, J.M., Jiang, X., Zhong, X., Fan, J.: Discussion on contact analysis based on ANSYS software problems. Mining Machinery 7, 69–72 (2006) (in Chinese)
6. Li, B., He, Z., Chen, X.: ANSYS Workbench design, simulation and optimization, pp. 8–50. Tsinghua University Press, Beijing (2008) (in Chinese)
7. Wu, Z., Wang, Y.: The Whole Hydraulic Support finite element analysis based on ANSYS Workbench. Mining Machinery 30(9), 106–107 (2009) (in Chinese)

8. Wang, Y., Wang, P., Li, L.: The development of parametric modeling technology for truck frame. Agricultural Equipment Vehicle Engineering (4), 13–15 (2010) (in Chinese)
9. Xie, Z., Yu, S., Li, C., et al.: The handling robot structural optimization design based on ANSYS Workbench. Machinery & Electronics (1), 65–67 (2010) (in Chinese)
10. Mao, X., Wen, T.: Based on finite element analysis, optimization design motorcycle frame. Motorcycle Technology 37(5), 35–37 (2007) (in Chinese)
11. Lu, L., Li, Y.: Virtual Prototyping Technology in Agricultural Machinery Design. Chinese Agricultural Mechanization (4), 59–61 (2004) (in Chinese)
12. Zu, X., Huang, H., Zhang, X.: Virtual Prototype Technology and Its Development. Agricultural Machinery 35(2), 168–171 (2004) (in Chinese)

Research on 3G Terminal-Based Agricultural Information Service

Neng-fu Xie and Xuefu Zhang

Agricultural Information Institute, The Chinese Academy of Agricultural Sciences Key Laboratory of Digital Agricultural Early-warning Technology, Ministry of Agriculture, Beijing 100081, P.R. China
{xienengfu,zhangxuefu}@caas.cn

Abstract. In order to solve the farmer's agricultural production information' acquisition problems with a 3G terminal, a 3G-terminial agricultural information service system method is proposed in the paper. With the advent of 3G and WLAN technologies, the rural and urban areas in which 3G coverage is complemented by WLAN deployments is becoming available. The agricultural information services limited by network bandwidth and geography will be changed completely. In the paper, we will propose an architecture of the 3G-terminial agricultural information service. First, we present the design and implement of the GAIS architecture for mobile multimedia information service. Then the agricultural content-based information services are described in detail. Finally, we present a prototype agricultural information service to show our experiment for the agricultural information service in a mobile information context.

Keywords: 3G terminal, service design, agricultural information, push technology.

1   Introduction

With the urgent demand of agricultural information for farmers and quick development of telecommunication industry in china, the convenience, quickness and validity of agricultural information service are becoming more important [1]. The mobile phone has been described as the most likely modern digital device to support economic development in developing nations [5, 7]. Tapan presented CAM - a framework for developing and deploying mobile applications in the rural developing world. Supporting minimal, paper-based navigation, a simple scripted programming model and offline multimedia interaction, CAM is uniquely adapted to rural user, application and infrastructure constraints [3, 4]. The shortage of agri cultural information, poor telecommunication networks and uncoordinated information resources are crucial problems facing agricultural information sources in china. The 3G technology will enable a wide range of services and applications in rural areas with support of multimedia and positioning, which will bring novel applications with high impact for the farmers. Combined with modern web and data management

D. Li and Y. Chen (Eds.): CCTA 2013, Part II, IFIP AICT 420, pp. 152–157, 2014.

technologies application in agricultural domain can bring new agricultural applications and services for agricultural production. The mobile phone is no longer just an audio communication tool but capable of providing additional integrated functions. The various features of mobile phone, which make it versatile, are SMS, MMS, GPRS [2].

The advent of 3G and WLAN technologies in the recent years open a way for sign language based services in mobile and wireless settings, which bring the applications and services come on stream. Third generation mobile network (3G) is the latest advancement in the field of mobile technology. Providing high bandwidth communication of 8kbit/s-2Mbit/s and a revolutionary introduction of multimedia services over mobile communication, it aims to make mobile devices into versatile mobile user terminals[8]. TD-CDMA (Time Division CDMA) wireless technology in china can meet the rapidly growing demand for mobile broadband services in agricultural information systems. The mobile information services will make it easier and quicker to provide agricultural information services in today's more complex web environment.

The agricultural information services limited by network bandwidth and geography will be changed completely with the 3G network coverage to rural area, which will help solve the problem of "last kilometer" in agricultural informationization. In the paper, we will develop an agricultural information system, which can gather, manage agricultural multimedia information resources, and push certain agricultural production technology information to the customized terminals.

The paper is structured into four main sections. Section 2 presents the terminal - based agricultural information service architecture, which provides a software framework that aims to describe our need and the main components. In section 3, the GAIS-based prototype system is presented and describes the agricultural content-based information services in detail. Finally, Section 5 provides some conclusions and summary remarks.

2   The GAIS Architecture Design

The terminal -based agricultural information service (GAIS) architecture is the framework and soul of the agriculture information system deciding the system functions and information service. As showed in Fig. 1, the GAIS architecture is composed of our logic tires from the terminal client side to the server side: 3G terminal tier, portal service tier, 3G network tier and service center tier. At the fat server side, we provide an environment that integrates related different agricultural information sources into the agricultural information center. The source can be deleted, added and indexed diametrically. The center will also provide Update Tool guarantees that all data are always up to- date, manage the 3G terminal configure, and present enriched content on 3G terminals with limited user interface i.e. small screens and simple keyboards.

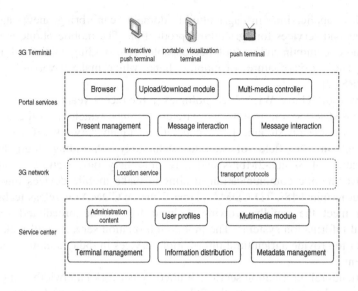

Fig. 1. The GAIS Architecture

2.1   Agricultural Information Service Center Tier

The service center tier manages and stores the agricultural information of the whole system, and decides the information exchange between the 3G terminal and service center according the setting of the terminal in the system.

The administration content module is designed for the data adding deleting and querying in agricultural information center. It will control the data quality of different databases, data service, and data clustering. The metadata management will help users find the information they required easily. The information meta-data is building an object-oriented repository technology that is integrated with agricultural meta data management that process metadata, which will provide a united category that organize the information in information center, and a standard of the information' store, and support exchange of model data. The metadata-based information service will provide a united interface for users to access the information by metadata-based information category. The information distribution is about assuring that the right agricultural information is available to the right agricultural users. In order to distribute the information to the users timely, we will build 3G-based information distribution service, by which the information will be accessed with the 3G terminals. In the means, the information distribution will help the communication between the users and the information center.

2.2   3 3G Network Tier

In the near future, wireless networks will appear with greater ubiquity, from public access networks offering connectivity in agricultural information services, to telecoms operated 3G and 4G networks, especially TD-SCDMA network into rural areas in china, which allows 9 hundred million of farmers access to services as important and varied as information acquisition, health care, education, and financial and governmental services. The 3G network tier is mainly implementing the information exchange the 3G terminal and data server in TD-SCDMA network. Indeed our simulations are based on the following two assumptions: 1) full TD-SCDMA network coverage to rural area; 2) TD-SCDMA link always on, which we argue that are realistic assumptions in typical scenarios.

2.3   Portal Service Tier

The portal service tire contains a group service functions for the terminal, which can customized its necessary function, such as multimedia browsing, upload/download, multimedia controller, present management, message interaction and message interaction. The portal service tire can optimize the communication process and may offer mobile service enhancements, such as location, privacy, and presence based services. It communicates with the Web Server using the standard Internet protocols such as HTTP/HTTPS.

2.4   3 3G Terminal Tier

The mobile terminals must be seen as complex microelectronics assemblies that are attached to sophisticated network-based system. The 3G terminal tier is a carrier of the client of 3G-based Mobile agricultural information service. The client can store the information coming from the service center for offline browsing, and contain a microbrowser that the GUI and is analogous to a standard Web browser, which can accept the request agricultural information and send message to the service center that makes a plan, and pushes the information to the users' 3G mobile terminal by 3G network.

3   GAIS-Based Prototype System

Based on GAIS hierarchy and the actual needs of the project, the entire system is divided into two parts: service centers subsystem and mobile terminal subsystem, which constitute a service whole. The service center' functions contain: 1) It manages all the registered 3G terminals and gives orders for pushing information and emergence notification to the terminals; 2) It can put the 3G terminals into different groups according to the region to ensure that the pushed information and emergence notification is closely related with the farmers' requirements; 3) The service center manages the agricultural information and provides real-time dynamic information to the client-side. And it receives the request and processes it then return the result. The service center operation interface is showed in the Fig. 2. The mobile terminal subsystem is mainly used to store the information pushed by the service center, and browses video on demand. The mobile terminal subsystem can also send technology questions using text and picture to the server, and accept the result. The service center operation interface is showed in the Fig. 3.

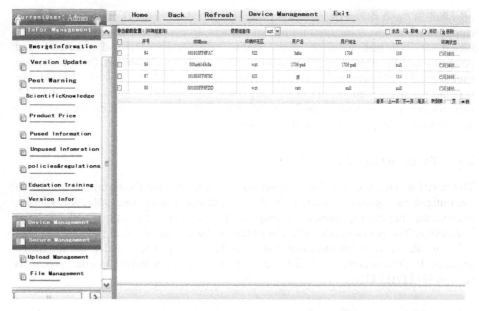

(a) The service center portal interface

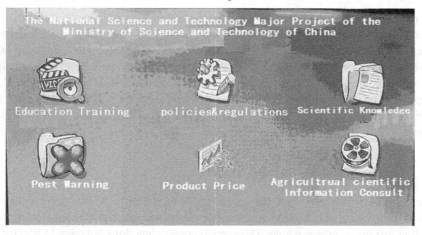

(b) The mobile terminal operation interface

Fig. 2. The system inferace based on GAIS hierarchy

The first applications testing wireless accesses was implemented for agricultural information provided by central information center, such as agricultural production technology, pest warning, Agricultural science and technology popularization, and farm product price, etc. The application have been supported by Special Project on of The National Department of Science and Technology "TD-SCDMA based application development and demonstration validation in agriculture informationization", which got very good effect to solve the problem of "last kilometer" in Chinese rural.

4   Conclusion

In the paper, we have proposed an architecture of the 3G-terminial agricultural information service, which is the key and core of 3G-based information service system. In the paper, we also discussed how to push agricultural the information to the farmers and how the farmers get the agricultural production needs' technologies by communication with the system. And a prototype system is designed and developed based on the GAIS architecture which is composed of four logic tiers. The prototype system can push the information to the 3G terminal for the farmer' demands and the 3G terminal can also communicate with the system for getting the agricultural expert answer by asking questions, which proves that the architecture is feasible. In the future, we will make further study on the terminal information service development and the system demonstration.

Acknowledgments. The work is supported by the National Science and Technology Support Program (No.2011BAH10B01) and the special fund project for The CAAS scientific and technological fund project "Research on 3G information terminal-based rural multimedia information service" (No.201219).

References

1. Xie, N.F., Wang, W.S.H.: Research on 3G Technologies-Based Agricultural Information Resource Integration and Service. In: Li, D., Zhao, C. (eds.) CCTA 2009. IFIP AICT, vol. 317, pp. 114–120. Springer, Heidelberg (2010)
2. Dhaliwal, R.K., Joshi, V.: Mobile Phones - Boon to Rural Social System. Literacy Information and Computer Education Journal (LICEJ) 1(4), 261–265 (2010)
3. Parikh, T.S.: Using Mobile Phones for Secure, Distributed Document Processing in the Developing World. IEEE Pervasive Computing 4(2), 74–81 (2005)
4. Parikh, T.S., Lazowska, E.D.: Designing an architecture for delivering mobile information services to the rural developing world. In: Proc. WWW 2006, pp. 791–800 (2006)
5. The real digital divide. The Economist (March 2005)
6. Bi, J.F., Wang, L., Wang, Y., Liang, D.S.: Research of Mobile Spatial Information Service Based on Open Framework. In: GSEM 2009, pp. 106–108 (2009)
7. Guan, F.-Y., Long, S.-T., Huang, J.: Research on Rural Mobile Information Service in 3G Era. Sci-Tech Information Development & Economy (2011)
8. Yilin, Z.: Standardization of mobile phone positioning for 3G systems. IEEE Communications Magazine 40(7), 108–116 (2012)

Study on Semantic Heterogeneity Elimination of Agricultural Product Price Information in Multi-source Network

Jing Zhang, Guo-min Zhou, Jian Wang, Jie Zhang, and Fangli Xie

Agricultural Information Institute of CAAS, Beijing 100081, China
{zuer0101,tulaapple,fangli_x}@163.com,
{zhouguomin,wangjian01}@caas.cn,

Abstract. With development of web information technologies, internet applications have come into burst growth, which brings different kinds of standard, database and description methods. This condition causes the "information islands" that make a lot of trouble in data sharing. The root cause of these problems is information source heterogeneity, which consists of four levels, and among them the semantic heterogeneity is the most difficult issue. The reasons of semantic heterogeneity existing are pointed out in this article and the research status of heterogeneity elimination is introduced. At last the solutions of agricultural product price information in multi-source network are summarized and the research emphasis in future is prospected.

Keywords: network information technology, semantic heterogeneity, agricultural product price information.

1 Introduction

Network information technology has penetrated into human production and other aspects of our life, and so as agricultural sector as a basis for human survival and development. At present, agriculture website is developing rapidly, but the content is very complex. It's a great difficult for the peasantry and agriculturists to get the information conveniently and effectively. Therefore, the development of agriculture search engine has become their urgent needs. Agricultural product price information is the most important part of the rural information service system. However, the existence of heterogeneity led to disconnect and asymmetries appearance in prices of agriculture search engine [1]. In order to make agriculture search engine more accurate and effective, it is imperatively to eliminate semantic heterogeneity.

This article explained the reason of semantic heterogeneity, discussed the treating mode of semantic heterogeneity which have three levels (schematic heterogeneous, context heterogeneous, individual heterogeneous), and then discussed semantic heterogeneity of agricultural product price information in multi-source network, at last summarized the research progress of the price information semantic heterogeneity's eliminating.

D. Li and Y. Chen (Eds.): CCTA 2013, Part II, IFIP AICT 420, pp. 158–164, 2014.

2   Generation and Research Progress of Semantic Heterogeneity

Information sources heterogeneity can be divided into four levels: system, syntax, structural and semantic, during these the most difficult to eliminate is semantic heterogeneity [2]. To some extent, systems, syntax and structure heterogeneity has been resolved through the traditional information integration. Therefore, the identification and elimination of semantic heterogeneity has become a difficulty and then the research hotspot.

The generation of semantic heterogeneity has been caused by different types of semantic conflict [3]. Park and Ram (2004) proposed classification of semantic heterogeneity conflicts: scheme-lever conflict and data-lever conflict [4]. Scheme-lever conflict is the conflicts caused by the logical structure that used by the same concepts in different information sources. Data-lever conflict is reflected in same aspects, such as naming identifiers, precision, representation, reliability, etc., which is due to the different perception of the same concept [5]. Zhou Jianfang et al (2008), from Huazhong University of Science and Technology, proposed semantic heterogeneity between the data sources that are not independent have interrelationship with each other. Semantic heterogeneity is manifested as schematic heterogeneity, context heterogeneity, and individual heterogeneity [6, 11]. In order to enable users to obtain efficient and accurate data, all of the three levels of conflicts need eliminating.

(1) Schematic heterogeneous elimination
Schematic heterogeneity is characterized by differences in logical structures and/or inconsistencies in metadata (i.e., schemas) of the same application domain [4]. At present, schematic heterogeneous data sources have been widely studied. The main eliminating method is to use the existing metadata and a small amount of users' intervention to achieve the automatic schema mapping. Mediator Environment for Multiple Information Sources (MOMIS) [3], Developed by Bergamaschi et al in 1998, is one of the typical representatives.

(2) Context heterogeneous elimination
Context heterogeneity is the different data sources that have the same semantic information (including entities and attributes). Data interpretation deals with the existence of heterogeneous contexts, whereby each source of information and potential receiver of that information may operate with a different context which leads to large-scale semantic heterogeneity [7]. At present, among all the methods of context heterogeneous elimination based on the context mediation, the typical representative is COIN project team in the MIT [8]. The project solves four kinds of context heterogeneities.

(3) Individual heterogeneous elimination
Individual heterogeneity mainly refers to individual identification in different data sources [6]. After schematic and context heterogeneity elimination had been implemented, the same individual in different data sources would still use different forms of representation and description. Thus, the traditional method to exactly match the property value comparison can't eliminate individual heterogeneity. Active Atlas system is a typical representative, which has been developed by Tejada et al. on

University of Southern California [9-10]. However, the solutions in the system can only applied to string matching without other types of data matching. What is more, this solution does not apply to Chinese.

Zhou Jianfang proposed a general framework for semantic information integration, which is used to solve the three levels semantic heterogeneity [11]. In the framework, query engine, context mediation and results processing components are the core processing components. They work together in order to solve schematic, context and individual heterogeneity. However, each core processing component needs to reference the relevant metadata. The acquisition of this metadata requires the domain experts and system designers to work together.

3   Semantic Heterogeneity of the Agricultural Product Price Information

3.1   Influence and Relative Research Status

Nowadays, with the rapid development of Chinese agricultural internet, there have been more than 30,000 agricultural website. Many problems would occur if researchers just simply collect price information in multi-source network without eliminating its semantic heterogeneity:

(1) Information structure is inconsistent. The different sources release price information in different formats. Some only publish average price and specifications while some else provide wholesale market, quality and other related content as well. Therefore, agricultural search cannot be used if we do not solve this kind of semantic heterogeneity.

(2) With information's complexity and redundancy, it is difficult to meet all the needs of users. Due to geographical and cultural differences, the same kind of agricultural commodities have different titles, such as 'huanggua', also known as 'qinggua' in Guangzhou. The release of this kind of price information will bring about unnecessarily redundant information if researchers do not eliminate such a semantic heterogeneity.

(3) Actually practical value is reduced. The agricultural product price information without eliminating its semantic heterogeneity lacks unified comparison and overview. The price information cannot either help majority famers to understand the market or be able to give a supporting guidance to agricultural-related policy.

Through above analysis, the identification and elimination technology of semantic heterogeneity of agricultural product price information in multi-source network is of vital importance to improve agricultural information services' quality and provide information that is more consistent with the needs of farmers and related agricultural workers. Related research about the technology to eliminate the semantic heterogeneity of agricultural product price information in the multi-source network is limited currently; it is University of Science and Technology of China that mainly conducted some research. According to the characteristics of agricultural product price information given by the Internet, Lei Ying (2010) uses a seven-tuple (price, unit, name of agricultural products, source, products trading market, province, transaction date) to represent the properties of agricultural product price information, so as to solve semantic heterogeneity [12]. But not all sources are able to meet the seven-tuple. As many sources just provide products trading market, it is necessary to get specific province information. He Huang (2010) treated the market name of agricultural product price information by dividing them into two cases, and establish a uniform location data table to solve spatial information heterogeneous through a combination of automatic annotation of spatial properties using administrative divisions indexing library and extracting location information for labeling using the general search engines [13]. What's more, he eliminating the data reliability conflicts through the establishment of abnormal data detection system and redundant data processing methods based on the semantics which could develop the quality and availability of data. With further research, the elimination of semantic heterogeneity in order to clean a lot of redundant data comes true with the system's combination with agricultural classification ontology, geographic name resolution system, unit conversion rules, and several other modules [14].

The current study of the semantic heterogeneity of agricultural product price information in multi-source network does not have a comprehensive summary and classification, what the elimination technology could solve is only parts of semantic phenomenon. What the problem that needs to be focused on next step is that to explicit the classification of semantic heterogeneity of agricultural product price information through a lot of research and the participation of experts so as to gradually improve the semantic heterogeneity elimination.

3.2   Study on the Semantic Heterogeneity in the Agricultural Product Price Information

3.2.1   Classification of Semantic Heterogeneity of Agricultural Product Price Information

In order to get the semantic heterogeneity of agricultural product price information existing in the current agricultural sources, the survey from a large of agricultural network have many specification and standard information about the price information. The multi-source network mainly includes China Agricultural Information Network (http://www.agri.gov.cn) which provides regional agricultural websites of wholesale market, and provincial and municipal government networks which provides agricultural product price information. Through sampling survey and combination with the basic principles of semantic heterogeneity, the semantic heterogeneity of the agricultural product price information in multi-source network has been divided into three categories: semantic heterogeneity based on schematic, context data, and abnormal individual data.

(1) Semantic heterogeneity based on schematic

① The semantic homogeneous conflict is a same concept in different sources has different set of attributes. For instance, with regard to price of a certain agricultural product, some sources only provide basic information of agricultural products about varieties, prices, units and wholesale. However, other sources include origin, highest price, lowest price, specifications and other information.

② The aggregate conflict is some sources use a concept, which the others of the same kind use a set of concepts. For instance, in the description of a kind of agricultural products, product name and specifications would be compressed into a single attribute in some sources.

③ Ancestor or descendant conflict is that the same concept is a superclass in one source and a subclass in another source. For instance, 'suan' is a subclass in some sources while it is a superclass for 'qingsuan', 'dasuan' in other source webs.

④ Classification conflict is that a same concept in different sources belongs to different superclass. For instance, some sources would be divided into two categories of fresh meat and eggs, and in the other sources meat and egg would be placed in the one category.

⑤ Naming conflict is that a conflict resulting from a same concept subjectively determined by designers of different fields or from different sources. This is more common in semantic heterogeneity, which in the price information is often highlighted in the two properties: the name of agricultural products and the wholesale market.

(2) Semantic heterogeneity based on context data

① Data value conflict is a same data in different sources to explain inconsistencies because of the different data representation, scales or coding. In different sources, the same price data have a different meaning for wholesale prices or retail prices.

② Data presentation conflict is a same concept with different data types or different formats. For instance, representation of date may be conflict on different systems, such as 'yyyy-MM-dd' or 'MM/dd/yyyy'.

③ Data unit field conflict is to use of different units of measurement resulting from the conflict. The conflict is particularly prominent in the agricultural product price information. For instance, some sources use 'kg' as the unit of measure and some use 'jin'.

④ Data Accuracy conflict is the concept resulting from a same concept described in different precisions or the different domain and range. For instance, price information from different sources has different data accuracy, such as 'yuan' or 'fen'.

(3) Semantic heterogeneity based on abnormal individual data

① Individual description conflict is that different systems use different methods to describe the same individuality. As price information of agricultural product consists of many kinds of characterized properties, different describing methods for them are involved and so conflicts are inevitable.

② Data reliability conflict is that different reliable levels are assigned to a certain conception. This comes from possible incorrect data filling of information collection sources and it is necessary to correct the data without semantic heterogeneity.

The sorting of semantic heterogeneity on agricultural product price in multi-source network has been listed above. The conflicts need cleaning in sequence by three levels. At first, semantic heterogeneity based on schematic is the most important and influencing most among all the conflicts. And its elimination is the prerequisite of solving semantic heterogeneity of context data and abnormal individual data. Secondly, the context data heterogeneity brings incorrect calculations and comparison, which make it necessary to be dealt with based on the cleaning of schematic heterogeneity before solving abnormal individual data heterogeneity. In the third level of priority, schematic and context data heterogeneity has been solved while individual abnormal data still remains in information which will cause semantic heterogeneity. Therefore to obtain the final valuable price information of agricultural products, a certain individuality data from different sources need to be intelligent calculated including combining and duplicates eliminating.

3.2.2   Expectations for Sematic Heterogeneity Research of Agricultural Product Price Information

(1) Improve the sematic heterogeneity correcting system for agricultural product price information

Remaining problems will be resolved in sequence by levels according to sorting results of semantic heterogeneity research.

① Semantic heterogeneity based on schematic will be resolved through completing domain ontology and reducing naming conflicts of product types and markets.

② Context mediation technology will be involved to remove semantic heterogeneity based on context data and accurate, unified and precise data is expectable.

③ Individual conflicting information detecting and removing system will be actualized based on data conflict detecting and duplicates eliminating.

④ System efficiency will be considered in completing the whole system to achieve its fast running.

(2) Achieving personal service

The building up of semantic heterogeneity eliminating system for agricultural product price is connected with improving price showing system of agricultural product and different requests of users for data access and query are expectable. Personal service is important along with common usage. Therefore it comes into a promising research filed in building personalized data space and pushing characterized information.

4   Conclusions

Based on semantic heterogeneity research and analysis of agricultural product price information, the semantic heterogeneity is sorted into three levels as schematic, context data and abnormal individual data. Accordingly it is proposed that the heterogeneity will be eliminated in sequence of the three priorities. As a result, it is necessary to build up a complete semantic heterogeneity eliminating system to obtain the ideal information with high quality for users. Furthermore, efficient combination of the heterogeneity eliminating system and professional agricultural search engine will dramatically improve the information provided for peasantry and agriculturists.

Acknowledgment. The research is funded by the National High Technology Research and Development Program of China (2013AA102405) and the basic scientific research special fund of nonprofit research institutions at the central level (2012-J-07).

References

1. Yu, J.-X.: The Study on the Outline Model of the Agriculture Product Price Information Service. Agriculture Network Information (02) (2004)
2. Sheth, A.P.: Changing Focus on Interoperability in Information Systems: From System, Syntax, Structure to Semantics. Kluwer Academic Publishers, Boston (1999)
3. Bergamashi, S., Castano, S., De Capitani di Vimercati, S., et al.: An intelligent approach to information integration. In: Formal Ontology in Information Systems. IOS Press, Italy (1998)
4. Park, J., Ram, S.: Information Systems Interoperability: What Lies Beneath? ACM Trans. on Information Systems 22(4), 595–632 (2004)
5. Li, Y.: Research on Semantic Heterogeneity Elimination for information integration. Northwest Normal University (2009)
6. Zhou, J.-F., Xu, H.-Y., Lu, Z.-D.: Research of semantic heterogeneity in information integration. Application Research of Computers (08), 2349–2351 (2008)
7. Aykut, F.: Information Integration Using Contextual Knowledge and Ontology Merging. Massachusetts Institute of Technology, Cambridge (2003)
8. Zhu, H., Madnick, S.E.: Context Interchange as a Scalable Solution to Interoperating Amongst Heterogeneous Dynamic Services. In: Proc. of the 3rdWorkshop on eBusiness, Washington DC, pp. 150–161 (2004)
9. Tejada, S., Knoblock, C.A., Minton, S.: Learning Object Identification Rules for Information Integration. Special Issue on Data Extraction, Cleaning, and Reconciliation, Information Systems Journal 26(8) (2001)
10. Tejada, S., Knoblock, C.A., Minton, S.: Learning Domain-Independent String Transformation Weights for High Accuracy Object Identification. In: Proc. of the 8th ACM SIGKDD International Conference on Knowledge Discovering and DataMining, pp. 350–359. ACM Press, New York (2002)
11. Zhou, J.-F.: Research on Context Mediation Based Semantic Information Integration Method. Huazhong University of Science & Technology (2009)
12. Lei, Y.: Abnormal data detection in agriculture search engine. University of Science and Technology of China (2010)
13. He, H.: Complex Adaptive Agriculture Vertical Search Model and its Implementation. University of Science and Technology of China (2010)
14. Hu, Y.: Semantic Research and Implementation on Agricultural Vertical Search Engine. University of Science and Technology of China (2012)

Study on the Application of Information Technologies on Suitability Evaluation Analysis in Agriculture

Ying Yu, Leigang Shi, Heju Huai, and Cunjun Li[*]

Beijing Research Center for Information Technology in Agriculture, Beijng 100097, China
{yuy,licj,shilg,huaihj}@nercita.org.cn

Abstract. It is expounded the suitability evaluation research in agriculture in three aspects: land suitability, climatic suitability and crop ecological suitability in this paper. The suitability evaluation methods in agriculture are summarized systematically, which including of traditional mathematic model methods (fuzzy comprehensive assessment analysis, AHP etc.) and modern information technologies (GIS, RS and ES etc.). The future development trends of suitability evaluation in agriculture is pointed out that GIS etc. other information technologies and systems can be used and developed for impact assessments of agricultural practices and for studying the effects of land, climate and ecology etc. change.

Keywords: Suitability, evaluation, agriculture, information technology.

1   Introduction

The original study of suitability evaluation in agriculture was mainly focused on land suitability for land use planning and city planning et al. Along with the development of suitability evaluation theory and method, the meteorologists and ecologists were also referring to the suitability evaluation analysis in their research. Most suitability evaluation was discussed in three aspects in agriculture: farming land suitability, climatic suitability and ecological suitability.

In the past, many mathematic-based methods have been used for the evaluation of the suitability in agriculture, which mainly concluded mathematic model, fuzzy comprehensive assessment analysis [1, 2], AHP [3, 4], grey correlation degree analysis and cluster analysis et al. Then artificial neural network (ANN)[5] and information technologies, such as GIS[6], RS[7,8], ES[9], SOTER database[10], AEZ (Agro ecological Zone) and ARC/INFO software[11] et al. were also wildly application. The combination of qualitative and quantitative methods, information technology and mathematical methods combining the integrated use of research methods, making suitability evaluation results more scientific and precision[12,13,14,15].

The research of suitability is not only limited on land, climatic and ecology et al. science area but also combined with economy and society. LiJing in her master dissertation evaluated the eco-economic adaptability and development potential of crops production by using the theory of ecology and economic. They put forward the

D. Li and Y. Chen (Eds.): CCTA 2013, Part II, IFIP AICT 420, pp. 165–176, 2014.
© IFIP International Federation for Information Processing 2014

concept of the apparent eco-economic adaptability of crop-region. They calculated 6 crops' apparent eco-economic adaptability indexes (AEEA) in31 regions by using synthesis scale level and relative output level and scale stability and yield stability. They compartmentalized the adaptability grades of 6 crops in 31 regions, comparatively analyzed the AEEA of crops in each region, and got 5 adaptability grade areas of each crop and dominant crops in each region. Based on the theory of three critical points in ecology and comparative advantage in economics, they put forward the evaluation of eco-economic adaptability of crop production along with the mind of combining the ecology, economics and social background [16]. That extended the ecological suitability evaluation to economy and society which making agricultural suitability evaluation more comprehensive and integrated.

1.1   Traditional Mathematic Methods

As we all know, some scientists have done a lot of research work on land, climatic, ecological suitability evaluation, but we can also find out that their studies were often limited in one or several factors, which were almost the climatic factors. On the other hand, we know that many other factors can influence the crop growth and develop. Taking into account these factors, they are very complex and numerous even impossible if we continue to use the previous ways and the routine means. With development of the computer, especially rapid progress of GIS, it is possible to consider so many factors to evaluate the suitability [17].

On the one hand, the analysis of great deal of spatial data is needed for the eco-suitability evaluation and planting regionalization of the crop, on the other hand, the comparison of all kind of the evaluation project is also needed in order to support the establishment of reasonable project. The work is heavy and complicated if using the routine appraisement method. Therefore, by using the important factors to the evaluation objects, the comprehensive evaluation model of suitability was set up. The method of comprehensive assessment analysis was wildly used in the evaluation of the suitability of agriculture. The evaluation index system was established according to the factors selection principle and the actual condition of agricultural region [18]. Actually most of the factors selection principle was subjective and empirical, even by using the Delphi methods to quantify in some degree the experts' subjective conclusion. They selected the factors from professional books, literatures and some experts' experience, such as the light intensity, temperature and precipitation et al. some crop living factors' data. At the same time they will combine with the practical production situation data in the research region, such as more than 20 years climatic data from meteorological stations and crop production data from statistical yearbook et al., then finally the important factors which using in the index system will be determined.

The evaluation factors were selected empirically for each evaluation target. Thus, each evaluation target has its own evaluation factors, and then each evaluation factors should be given its rating values. Weight determination is also a difficult points and key problem in the evaluation system. The grey relational analysis (GRA) was combined with the analytic hierarchy process (AHP) to address the uncertainties during the process of evaluation, especially of the fuzzy comprehensive suitability evaluation

[15]. The growing areas of natural sweet wine material were studied by Song Yuyang et al., using of the method of AHP and multi- factors evaluation, and the mathematics model was established [4]. Li Baoguo et al. chose different red Fuji apple growing areas in Hebei , Shandong , Shanxi , Henan in their study , the index which determined fruit qualities in these areas were collected, ecology environment data were gathered at the same time, and then the evaluation equation of red Fuji apple suitable cultivation area were established [19]. Zhan Xiangwen et al. based on black-box theory, a lot of work on data processing and regression analysis had been done, then the parameters required in the evaluation model were got [20]. ZhangJing et al. on the basis of limitation law of ecological factors, variable weight principle and method were introduced into establishing a systemic approach of crop ecological adaptability evaluation to avoid the disadvantages induced by subjective weighting method [21].

In the fuzzy synthetically judgment for suitability evaluation, different type of factors organizational state, each factors importance and main limited factors are different, therefore weight vector to fuzzy weight matrix were extended and rebuilt by Lu Enshuang et al. According to the theory, they found the positive method of fuzzy weight matrix and applied it to Kiwifruit suitability synthetic judgment in mountain area of South Shaanxi, the consequence showed that the method can correctly reflect short-factor's constraint function and the conclusion is in accordance with practice [1]. Wu Kening et.al due to fuzziness in adaptability assessment, established tobacco eco-adaptability assessment models by applying fuzzy comprehensive evaluation [2]. Luolin et al. used the annual mean temperature, annual precipitation, and soil pH which are vital to the fruit tree growth and development to set up the fuzzy evaluation model of ecological suitability. A comprehensive evaluation was carried out for the chestnut grown in Bijie in western Guizhou Province, and a designation of the most favorable region, favorable region, suitable region, and undesirable region was generated [3].

1.2   Information Technologies

Since 1900s, modern new and high technology which concludes space, remote sensing(RS), geographical information system(GIS), computer, et al. and modern scientific method which includes systematology, informationism, cybernetics et al. are got extensive and deep application. In developed countries, the application of GIS technology in ecological suitability evaluation could be ascended to 1960s. At that time, overlay aerial image had been applied on urban land use suitability evaluation by the planning designers in American and West European, and fast developed to agriculture planning and urban construction. With the improvement of image resolution of GIS and spatial analysis technology, the land and ecological suitability evaluation methods have been greatly expanded in agriculture [22].

In China, the deep and systemic research in this respect started in the late 1970s, and the rapid development focused on evaluation index system, evaluation methods and evaluation technology. In the beginning of 1990s, Huang Xingyuan et al. [23] first used the theory and method of GIS to land evaluation. Until the late of 1990s, most land evaluation was all had GIS technology application [24].

On the basis of mathematic method, many researchers combined GIS et al. modern information technologies in the suitability evaluation in agriculture. Based on GIS-fuzzy comprehensive evaluation method, the land, climatic or ecological suitability of the crop in some region was evaluated. The comprehensive consideration was taken on the climatic, soil, and topographic factors related to crop growth. The spatial data of the factors were organized and computed with GIS method; the weights of the factors were derived by using AHP method; and the proper membership function and fuzzy arithmetic operators were selected to conduct the comprehensive ecological suitability evaluation. Compared with traditional evaluation methods, the GIS-fuzzy comprehensive evaluation method had the advantages of short-term,more fine and detailed, and more suitable for small spatial scale areas [24]. GIS technology and fuzzy mathematics methods were used on studying the relationship between eco-physiological features and environment, selecting evaluating indexes of ecological adaptability and constructing evaluation model [25]. Li Qifeng et al. selected single pollution index and comprehensive pollution index to analysis the three bases of non-pollution food, green food and organic food, and then the evaluation unit with GIS technology was built. On this basis, considering of the surrounding environmental conditions, road conditions, farmland soil quality, industrial development conditions, economy and society level et al. factors, the evaluation system of three bases was constructed [26].

In order to establish a framework of quantitative evaluation method and implement a universal tool (software) for evaluating crop ecological suitability based on the framework. Three improvements were made by Lu Zhou et al. [27] in quantitative evaluation method. With the improved method, they had developed a universal evaluation tool implementing a quantitative evaluation, which can help agriculture experts without IT engineering knowledge. Song Ruhua et al. built land source management information system for land suitability evaluation and land use planning[28], Cheng Jianquan summarized the space analysis methods by using GIS to quantified spatial indicators, and applied to optimize the urban dimensional development[29]; otherwise, GIS also be widely applied in tourism land evaluation[30], Land reclamation[31] et al. aspects.

Not only information technology could combine with mathematic method, but also different information technologies could combined. Expert system and Geographic information system are both new high-technology. Xie Yu et al. probed into the study on the application of the integration of ES and GIS in the rice cultivated adaptability analysis. They applied system science, mathematics, crop planting science and rice weather ecology, established an expert system of rice cultivated adaptability analysis, accordingly added space database and space knowledge base which realized by GIS to its database and knowledge base systems, integrated the managing function of space information of GIS into the ES, which made the expert system have the deductive and ratiocinative ability in space. That realized the integration of ES and GIS, successfully used in the rice cultivated adaptability analysis [9]. Remote sensing(RS) also be applied in land evaluation, Fang Linna et al. based on SPOT multispectral remote sensing image and data using in cultivated land fertility survey, the cultivated land quality assessment study was carried out. Cultivated land quality assessment indicators were abstracted from SPOT multispectral image, e.g. NDVI, DVI and RVI, which represented soil fertility, water availability and soil degradation respectively. The assessment indicator system was constructed using the indicators mentioned above. By virtue of PSR framework, the assessment model was developed in order to explore the feasibility of RS technology in cultivated land quality assessment [33].

2   Suitability Evaluations in Agriculture

2.1   Farming Land Suitability Evaluation

Land suitability evaluation is the appraisal of the suitability and its extent of land for a purpose, it is the basis for land-use decision-making and using direction, it is also the content of the land resource research in the near past 20 years [34]. The Food and Agricultural Organization (FAO) [35] recommended an approach for land suitability evaluation for crops in terms of suitability ratings from highly suitable to not suitable based on climatic and terrain data and soil properties crop-wise, in which the procedures and methods of land suitability evaluation were explained. However, the framework is basically a qualitative one and it is difficult to make a direct connection of evaluation results with decision-making on land use planning [6].

Recently years, the integration of GIS and assessment model have been a new trend, Xie Shuchun et al. based on VB and MapX to expounded how to make use of the mighty special analysis function of GIS to realize the suitability evaluation of testing land[36]. The whole evaluation course regards GIS and RS as the technologies platform of the work, has realized greatly, formed a set of intact technological routes of evaluation, at the same time automation basically from beginning to the output of the achievement pictures. Relatively traditional method, this method, which improved the working efficiency and evaluating precision, has certain reference value to the suitability evaluation of other land [37]. Wang Dacheng etc. employed artificial neural network (ANN) analysis to select factors and evaluate the relative importance of selected environment factors, and the spatial models were developed and demonstrated their use in selecting the most suitable areas for the winter wheat cultivation. Satellite images, top sheet, and ancillary data of the study area were used to find tillable land. These categories were formed by integrating the various layers with corresponding weights in GIS. An integrated land suitability potential (LSP) index was computed considering the contribution of various parameters of land suitability [38]. Xia Min et al. studied on the components and their realization of farmland suitability evaluation spatial support system (FSESDSS), probed into all methods and their suitable area used in land suitability evaluation [39].

The suitability evaluation for farming land is one of the most major content of land suitability evaluation. By grading the appropriate level of the farming land into several levels and opening out the suitability for faming land, it can provide basis for adjusting and optimizing the land-use structure and making rational land exploitation layout. The research is focused on two aspects of farming land in suitability evaluation: cultivated land quality evaluation and cultivated land fertility.

2.1.1   Cultivated Land Quality Evaluation

Cultivation land is an important base of grain production, in order to protect the farmland, the overall investigation and analysis about the present status of it is required.

Nong Xiao-xiao et al. [40] based on the spatial analysis model of ArcGIS, evaluated the cultivation land quality with the method of multi-factor comprehensive judgment, which basis data were the topographic map, land use map, soil type map. Shi Changyun et al. pointed out that a quantitative and scientific evaluation method of land quality based on GIS was made. Mathematical models, such as correlation analysis, hierarchical analysis, fuzzy evaluation, were applied in their study [41]. Kou Jinmei had set up the information system for comprehensive evaluation of farmland quality which takes Microsoft SQL Server 2000 as the backstage database, edited under the condition of Delphi which was new visual editorial environment and supported by GIS software (Map/Info) [42]. Nie Yan et al. by using the new and high technology as computer, ComGIS, UML, workflow, expert system, combining such multidisciplinary theory as soil, land, landscape, ecology, information and modern mathematics, they carried out the information system of classification, graduation and evaluation on cultivated land (ALEIS). Under the support of AELIS, the quantitative appraisal models and methods were developed [43].

All the cultivated land quality evaluation result is almost rational and provided the reference for the farmland protection.

2.1.2   Cultivated Land Fertility Evaluation

Recent years, cultivated land fertility evaluation is developing toward quantitative and practical direction, especially the wildly application of GIS, including of complicated mathematic model combined with GIS, RS technology combined with GIS and expert system (ES) combined with GIS etc.

Wang Ruiyan etc. took Qingzhou City as their study area, intended to research for quantitative methods for cultivated land fertility evaluation. Based on the plentiful information that obtained by remote sensing technique, field-survey and lab analysis, the automatic and quantitative evaluation procedure was realized by adopting various mathematical models and methods such as system-cluster, analytical hierarchy program, fuzzy math, etc. and supported by GIS techniques [44]. In Niu Yanbin's study, farmland evaluation in Quzhou county of Hebei Province based on GIS was made, mathematical models, such as analytical hierarchy process (AHP), fuzzy evaluation, were applied in this study. The productivity of farmland is evaluated rapidly and exactly, with the powerful functions of GIS software [45]. Chen Haisheng etc. based on the analysis of the physical and chemical properties of soil samples collected from Henan tobacco plantation area, established the index system of soil fertility adaptability of tobacco. The fertility level was evaluated and classified by fuzzy and analysis. And the fertility map of Henan tobacco plantation area was drawn with GIS software MapGIS [46].

The approach of combination of GIS, RS, and ES etc. was feasible and effective in cultivated land fertility assessment, and the results of evaluation were almost in accorded with the real circumstances.

2.2   Agriculture Climatic Suitability Evaluation

In climatic suitability evaluation, the maximum entropy (MaxEnt) model was introduced in recently two or three years. Many researchers combined it with GIS to establish the relationship between climate and climatic suitability regionalization and potential cultivation distribution.

He Qijin etc. [47] based on the potential climate indices at national and annual scales influencing maize cultivation distribution from the references, together with the maximum entropy (MaxEnt) model as well as ArcGIS spatial analysis technique, the relationship between potential spring maize cultivation distribution and climate and the climatic suitability regionalization of potential spring maize cultivation in China were studied. Based on published data, geographical information, national climate data, and the MaxEnt model, the relationship between the distribution of the winter wheat cultivation zone and climate was established by Sun Jing-Song etc [48].

For the crop suitability zoning, the concept of growing period was introduced into the traditional approach by Araya etc. [49], to produce agro-climatic zones. This method could be used to develop agronomic strategies to cope with the anticipated increase in drought in the semi-arid tropics under climate change. Accordingly, quick maturing and drought-resistant varieties of teff and barley can be grown in the center and in the east, while medium-maturing cultivars should do well in the south-west. The method requires limited input data and is simple in its use.

An agro-climatic suitability library for crop production was generated by using climatic data sets from 20 to 33 years for 41meteorological stations in the Bolivian Altiplano by Sam Geerts [50]. Four agro-climatic indicators for the region were obtained by validated calculation procedures. The reference evapotranspiration, the length of the rainy season, the severity of intra-seasonal dry spells and the monthly frost risks were determined for each of the stations. To get a geographical coverage, the point data were subsequently entered in a GIS environment and interpolated using ordinary Kriging, with or without incorporating anisotropy.

The actual distribution of crop cultivation depends not only on climate, socio-economic conditions, and local production technologies, but also on soil type, geographic characteristics, crop varieties,human activity and so on, especially in relation to its yield and economic value. All the above research provided scientific support for planning crop production and designing the countermeasures against the effects of climate change on crop.

2.3   Crop Ecological Suitability Evaluation

In the crop ecological suitability evaluation, there were many researchers using the Analytical Hierarchical Process (AHP) technique and combining with GIS etc. information technologies.

Chen Haisheng et al. [51] based on the principal of hierarchy analysis and fuzzy mathematics and the technique of GIS, the comprehensive evaluation of tobacco ecology suitability were studied according to the actual circumstances of the whole Henan tobacco planting regions. The evaluation index system of tobacco ecology suitability of Henan tobacco planting regions was established by choosing 17 evaluation indexes from 3 respects of climate, soil and landform with Delphi method. Furthermore, the membership function was set up according to the effects of each ecology factors on the growth and quality of tobacco suitability. And the AHP was used to determine the weight of indexes by using quantitative analysis. Then the tobacco ecology suitability map of Henan tobacco plantation was drawn with GIS software MapGIS. Li Bo etc. [19] using a geographic information system (GIS), there nine factors were quantitatively analyzed. The grey relational analysis (GRA) was combined with the analytic hierarchy process (AHP) to address the uncertainties during the process of evaluating the traditional land ecological suitability, and a modified land ecological suitability evaluation (LESE) model was built. Mo Jingjing et al. [52] by using Delphi method and based on the ecological conditions of eight tobacco planting counties of Nanyang, 6 ecological factors related to the tobacco ecological suitability evaluation were established. According to the 6 various factors which affect the tobacco growth and quality, the corresponding membership degree function was established, and the original values of the factors were transformed into degree of membership, the weight of the factors were confirmed through Analytic Hierarchy Process (AHP), and overlaid into a raster map through using the spatial overlay analysis of GIS. Finally, the tobacco cultivation ecological suitability classification map of Nanyang was gained.

The study objective of Chavez, MD et al. [53] is to develop useful criteria for assessing diversification activities and to provide a ranking of different diversification activities on these criteria. The Analytical Hierarchical Process (AHP) technique is applied to get consistent assessments of criteria and activities from experts and stakeholders. Next, goal programming methods are used to aggregate individual assessments in order to arrive at the final ranking of farming activities for diversification. The results of this research can be used in optimization models for determining the optimal mix of farming activities in combination with tobacco production. Such models can provide further insights into factors determining diversification.

The software component suitability presented herein implements several published approaches for computing crop suitability, based on available climate, soil and crop information. Users can access the suitability software component via two application programming interfaces for single- and multi-cell estimations, the latter based on multiple regression methods. The component, extensible by third parties, is released as .NET 3.5 DLL, thus targeting the development of .NET clients [54].

An integrated indicator-based system was established to map the suitability of spring soybean cultivation in northeast China by He Yingbin [55]. The indicator system incorporated both biophysical and socioeconomic factors, including the effects of temperature, precipitation, and sunshine on the individual development stages of the spring soybean life cycle. Spatial estimates of crop suitability derived using this indicator system were also compared with spring soybean planting areas to identify locations where there was scope for structural adjustment in soybean farming. It is anticipated that this study will provide a basis for follow-up studies on crop cultivation suitability.

3   Conclusions and Future Perspectives

The theoretical foundation of suitability evaluation is building suitable analysis model by the statistical relationship between research targets and each variables. The support of technology is combining GIS etc. technologies with mathematic model effectively, which based on the multi-criteria analysis function of GIS. By the software of GIS to deal with the original data and derived data to send command which is ordered and interactional, to simulated the spatial decision-making process, for achieving the aim of analysis and evaluation on the research object.

The scientific and reliable of the evaluation result depends on the integrality of the basic data and the rationality of the evaluation method. It can be conducted rapidly and correctly by combining AHP and multi-factor fuzzy comprehensive evaluation method which also supported by GIS. That could overcome and avoid the drawbacks of empirically determined classification and reflect crop areas climate, ecological suitability level difference accurately.

In the future, by redefining query limits and incorporating other data, the GIS etc. other information technologies and systems can be used and developed for impact assessments of agricultural practices and for studying the effects of land, climate and ecology etc. change.

Acknowledgment. Funds for this research was provided by the construction of integrated information "three dimensional rural" service platform in National Modern Agricultural City for Science and Technology Projects (D121100003212003).

References

1. Lu, E., Sun, Q., Liang, X.: Quantification method of weight matrix in fuzzy synthetically judgment for corps ecology suitability and its application. Mathematics in Practice and Theory 34(6), 70–76 (2004)
2. Wu, K., Yang, Y., Lv, Q.: Application of Fuzzy Comprehensive Evaluation to Tobacco Eco-adaptability Assessment. Chinese Journal of Soil Science 38(4), 631–634 (2007)
3. Luo, L., Zhou, Y., Wang, M., Liu, C.: Fuzzy Evaluation Model of Ecological Suitability in Chinese chestnut. Economic Forest Researches 23(1), 27–29 (2005)
4. Song, Y., Ta, Y.: Evaluation on the Growing Areas of Natural Sweet Wine Material Based on AHP. Hubei Agricultural Sciences 46(1), 94–96 (2007)
5. Wang, X., Xu, X., Lv, J., Wei, C., Xie, D.: GIS-fuzzy neural network-based evaluation of tobacco ecological suitability in southwest mountains of China. Chinese Journal of Eco-Agriculture 20(10), 1366–1374 (2012)
6. Ni, S., Huang, X., Hu, Y., Xu, S., Gao, W.: GIS application in land suitability evaluation. Chinese Science Bulletin 37(22), 1911–1914 (1992)
7. Nie, Q., Yan, L., Cai, Y.: Evaluation of Land Suitability Based on Remote Sensing and GIS. Geospatial information 7(2), 28–30 (2009)
8. Wu, W.: Study on Land Suitability Evaluation Based on Remote Sensing and GIS——A Case Study of Zaghouan Province in Tunisia. Chinese Academy of Agricultural Sciences Master Dissertation (2005)
9. Xie, Y.: Study on the Application of the Integration of ES and GIS in the Rice Cultivated Adaptability Analysis. Guangxi University Master Dissertation (2001)
10. Wang, Z., Zhang, H., Zhou, Y.: Application of Matter Element Model in Farm Land Suitability Evaluation Based on SOTER Database——A Case Study from Hubei Province. Journal of Henan Agricultural Science 1, 41–45 (2005)
11. Zhang, M., Sun, L., Li, M., Wang, Y.: Appraising the Crops Soil Suitability Property by Utilizing the AEZ and ARC/INFO Softwares. Chinese Agricultural Resoures and Regional Planning 5, 19–23 (1998)
12. Chavez, M.D., Berentsen, P.B.M., Lansink, A.G.J.M., Oude: Assessment of criteria and farming activities for tobacco diversification using the Analytical Hierarchical Process (AHP) technique. Agricultural Systems 111, 53–62 (2012)

13. Reshmidevi, T.V., Eldho, T.I., Jana, R.: A GIS-integrated fuzzy rule-based inference system for land suitability evaluation in agricultural watersheds. Agricultural Systems 101, 101–109 (2009)
14. Chen, H., Liu, G., Yang, Y., Ye, X., Shi, Z.: Comprehensive Evaluation of Tobacco Ecological Suitability of Henan Province Based on GIS. Agricultural Science in China 9(4), 583–592 (2010)
15. Li, B., Zhang, F., Zang, L., Huang, J., Jin, Z., Gupta, D.K.: Comprehensive Suitability Evaluation of Tea Crops Using GIS and a Modified Land Ecological Suitability Evaluation Model. Pedosphere 22(1), 122–130 (2012)
16. Li, J.: The evaluation of eco-economic adaptability and development potential of main crops productive regions in China. Nanjing Agricultural University Master Dissertation (2007)
17. Yang, Y.: Research on Eco-suitability Evaluation and Planting Regionalization of Tobacco in Henan Province. Henan Agricultural University Master Dissertation (2006)
18. Feng, X., He, W., Jiang, G., Pan Hongyi, F.: Fuzzy Comprehensive Assessment Analysis of Agricultural Land Suitability Evaluation in Shuangliu County. Southwest China Journal of Agricultural Sciences 25(3), 982–988 (2012)
19. Li, B., Guo, S., Qi, G., Yang, B., Gu, Y., Cui, H.: Study on Evaluation Method of Ecological Optimum Growing Area of Red Fuji Apple. Journal of North West Forestry University 21(5), 78–80 (2006)
20. Zhan, X., Yang, Y., Li, S.: Study on suitability evaluation method of new maize varieties. Journal of Anhui Agricultural Sciences 37(1), 303–304 (2009)
21. Zhang, J., Feng, J., Bian, X.: Variable weight approach in evaluation of crops ecological adaptability. Journal of Nanjing Agricultural University 29(1), 13–17 (2006)
22. Yu, H., Wu, J., Xiao, M., Ge, C.: Utilization of GIS in the Evaluation on Crop Ecological Suitability and Rubber Planting. Journal of Tropical Organisms 2(3), 277–281 (2011)
23. Huang, X., Ni, S., Xu, S., et al.: Study on regional land use decision making supported by GIS. Acta Geographica Sinica 48(3), 114–121 (1993)
24. Shi, T., Zhang, G., Wang, Z., Wang, L.: Progress in Research on Land Suitability Evaluation in China. Progress in Geography 26(2), 106–115 (2007)
25. Guo, X., Fan, J., Zhu, W., et al.: Ecological suitability of olive in Sichuan Province: Fuzzy comprehensive evaluation based on GIS. Chinese Journal of Ecology 29(3), 586–591 (2010)
26. Liu, D., Du, C., Yu, C.: Adaptability evaluation and planting division of maize in Heilongjiang Province. Journal of Maize Sciences 17(5), 160–163 (2009)
27. Li, Q., Liu, X., Kong, Q., et al.: The research of the base of "three grades" suitability evaluation based on GIS technology. Chinese Agricultural Science Bulletin 27(14), 192–194 (2011)
28. Lu, Z., Qin, X., Li, Q., Yu, Y., Zang, C., Huai, H.: Quantitative evaluation method and universal tool for crop ecological suitability. Transactions of the Chinese Society of Agricultural Engineering 208(20), 195–201 (2012)
29. Song, R., Qi, S., Sun, B.: Suitability Assessment of Regional Land Resources and Its Spatial Distribution. Journal of Soil Erosion and Soil and Water Conservation 3(3), 23–30 (1997)
30. Cheng, J.: GIS Support Multi-Criteria Evaluation. Systems Engineering 15(1), 50–56 (1997)
31. Zhong, L., Xiao, D., Zhao, S.: Ecotourism suitability evaluation: the case of Wusuli River National Forest Park. Journal of Natural Resources 17(1), 71–77 (2002)
32. Liu, C., Lu, W., Jin, X.: Assessment on the suitability of unused land resources based on geographic information system in the course of land exploitation and arrangement-—taking Liucheng County in Guangxi province as an example. Resources and Environment in the Yangtze Basin 13(4), 333–337 (2004)

33. Fang, L., Song, J.: Cultivated Land Quality Assessment Based on SPOT Multispectral Remote Sensing Image: A Case Study in Jimo City of Shandong Province. Progress in Geography 27(5), 71–78 (2008)
34. Deng, Q.: The Application of GIS in the Evaluation of Farming Land Suitability—Taking LongQuanyi district in ChengDu for Example. Sichuan normal university Master Dissertation (2008)
35. FAO: A framework for land evaluation. Soils Bulletin 32, Rome (1976)
36. Xie, S., Zhao, L.: Land suitability evaluation based on GIS for the purple soil upland region in the middle part of Hunan province. Economic Geography 25(1), 101–105 (2005)
37. Zhang, C.: The Suitability Evaluation Based on GIS/RS of Farming Land in North Hebei Province Areas. Hebei Normal University Master Dissertation (2005)
38. Wang, D., Li, C., Song, X., et al.: Assessment of land suitability potentials for selecting winter wheat cultivation areas in Beijing, China, using RS and GIS. Agricultural Sciences in China 10(9), 1419–1430 (2011)
39. Xia, M.: On Spatial Decision Support System of Farmland Suitability Evaluation Nanjing Agricultural University Doctor Dissertation (2007)
40. Nong, X., He, Z., Wu, B.: Application of ARCGIS Spatial Analysis Model in Evaluating Cultivated Land Quality. Research of Soil and Water Conservation 16(1), 234–236 (2009)
41. Shi, C., Zhou, H.: Evaluation of land quality based on GIS—A case study on paddy field in Suzhou. Acta Pedologica Sinica 38(3), 248–255 (2001)
42. Kou, J., Hao, Y., Pan, D., Lv, X.: Research on Information System for Comprehensive Evaluation of Farmland Quality Supported by GIS. Chinese Agricultural Science Bulletin 22(7), 535–538 (2006)
43. Nie, Y.: Research models, methods and information system integration and application of crop land quality evaluation. Huazhong Agricultural University Doctor Dissertation (2005)
44. Wang, R., Zhao, G., Li, T., Yue, Y.: GIS supported quantitative evaluation of cultivated land fertility. Transactions of the Chinese Society of Agricultural Engineering 20(1), 307–310 (2004)
45. Niu, Y., Xu, H., Qin, S., Zhou, Y.: Research on the evaluation method of farmland supported by GIS. Journal of Agricultural University of Hebei 27(3), 84–88 (2004)
46. Chen, H., Ye, X., Liu, G., Li, Y.: Comprehensive Fertility Evaluation of Soil for Tobacco Plantation in Henan Province Based on GIS. Chinese Journal of Soil Science 38(6), 1081–1085 (2007)
47. He, Q., Zhou, G.: Climatic suitability of potential spring maize cultivation distribution in China. Acta Ecological Sinica 32(12), 3931–3939 (2012)
48. Sun, J., Zhou, G., Sui, X.: Climatic suitability of the distribution of the winter wheat cultivation zone in China. European Journal of Agronomy 43, 77–86 (2012)
49. Arayaa, A., Keesstrab, S.D., Stroosnijder, L.: A new agro-climatic classification for crop suitability zoning in northern semi-arid Ethiopia. Agricultural and Forest Meteorology 150, 1057–1064 (2010)
50. Geerts, S., Raes, D., Garcia, M., Castillo, C.D., Buytaert, W.: Agro-climatic suitability mapping for crop production in the Bolivian Altiplano: A case study for quinoa. Agricultural and Forest Meteorology 139, 399–412 (2006)
51. Chen, H., Liu, G., Liu, D.: Studies on comprehensive evaluation of tobacco ecological suitability of Henan Province supported by GIS. Scientia Agricultura Sinica 42(7), 2425–2433 (2009)

52. Mo, J., Liu, G., Ye, X., Shi, H.: Ecological suitability evaluation of Nanyang tobacco planting areas based on AHP and GIS. Journal of Henan Agricultural University 43(3), 331–334 (2009)
53. Chavez, M.D., Berentsen, P.B.M., Lansink, A.G.J.M., Oude: Assessment of criteria and farming activities for tobacco diversification using the Analytical Hierarchical Process (AHP) technique. Agricultural Systems 111, 53–62 (2012)
54. Confalonieri, R., Francone, C., Cappelli, G., Stella, T., Frasso, N., Carpani, M., Bregaglio, S., Acutis, M., Tubiello, F.N., Fernandes, E.: A multi-approach software library for estimating crop suitability to environment. Computers and Electronics in Agriculture 90, 170–175 (2013)
55. He, Y., Liu, D., Yao, Y., Huang, Q., Li, J., Chen, Y., Shi, S., Wan, L., Yu, S., Wang, D.: Specializing Growth Suitability for Spring Soybean Cultivation in Northeast China. Journal of Applied Meteorology and Climatology 52(4), 773–783 (2013)

Research of the Early Warning Analysis
of Crop Diseases and Insect Pests

Dengwei Wang[1,2], Tian'en Chen[1,*], and Jing Dong[1]

[1] National Engineering Research Center for Information Technology
in Agriculture(NERCITA), Beijing 100097, China
[2] School of Mathematics and Computer Science Ningxia University,Yinchuan, 750021, China
wangdws@126.com, {chente,dongj}@nercita.org.cn

Abstract. The early warning technology of crop diseases and insect pests is a strong guarantee to respond to the increasingly dire situation of major pests and diseases and ensure national food security. At present, the method of the monitoring data collection about the information of diseases and insect pests mainly relies on a field visit carried out by the plant protection staff, sampling and analysis, which directly impacts on the accuracy of early warning analysis. Firstly, the research status of the early warning analysis technology of crop diseases and insect pests in the field of agricultural are investigated and studied thoroughly and deeply, and then the comparison and analysis of the mainstream technology for pest and disease warning and algorithms are done in detail. The final combination of the technology of the Internet of things, a better pest early warning solution of the facility agriculture is put forward in order to provide a useful reference for the study of the early warning of crop pests.

Keywords: diseases and insect pests, early warning, facility agriculture, Internet of things, plant protection.

1 Introduction

There are more than 1400 species crop diseases and insect pests in China. China is one of the countries that the plant diseases and insect pests occur frequently. The crop diseases and insect pests occur frequently each year and bring huge economic losses. At the same time, the problems of crop diseases and insect pests also exacerbated environmental pollution, food safety and other issues. Therefore, it is a particularly important work that ensuring the safety of China's agriculture to timely and effective monitoring, early warning and prevention and management. At present, China still lacks qualified experts and the traditional methods are still used by a majority of plant protection departments to detect the diseases and insect pests, the survey data is filled manually and the information can not be reported to the higher authorities until the diseases and insect pest occurs and then the management work begins. Most traditional approaches are inefficient, timeliness poor and can not make maximum use

---

* Corresponding author.

D. Li and Y. Chen (Eds.): CCTA 2013, Part II, IFIP AICT 420, pp. 177–187, 2014.
© IFIP International Federation for Information Processing 2014

of existing data[1]. It has become an urgent problem to apply existing computer technology and communication technology in the field of detecting the crop diseases and insect pests to build an early warning system of the crop diseases and insect pests. Combined the real-time data and historical data, the system will be able to make advance warning when the crop diseases and insect pests occur. And then the necessary protective measures are carried out in time, controlling the crop diseases and insect pests in the embryonic stage.

The paper summarized the existing theories and methods of early warning analysis for the crop diseases and insect pests, generalized existing early warning analysis algorithms and models and then made comparison and analysis in detail. Against a background of Internet of things, big data, cloud computing and other information technology, crop pest early warning analysis scheme based on real-time sensing data is put forward. The paper summarized various research methods to early warning analysis of the crop pests, providing a reference for the research and development of the early warning technology of the crop pests.

2   Warning Analysis Theory

2.1   The Basic Concept and Theory

Early warning technology of the crop diseases and insect pests relies on the principles of the biology, ecology and mathematics, and it analyzes a variety of correlative factors collected from the historical figures and the present figures about the diseases and insect pests and estimates theirs future changes and development trend to achieve the purpose of reduce the catastrophic losses. It is the cornerstone of pest diagnosis and treatment to make accurate and timely pest early warning and early warning can improve the ecological environment[2]. Pest warning is to test the environment, pathogeny, individuals themselves, analyze the monitoring data obtained from the test and then predict the incidence according to the relationship between the impact factors. The system will send out warning information if there is a possibility of disease or the forecast data exceeds a certain threshold value.

In accordance with the warning category pest early warning can be divided into several aspects crop warning, livestock and poultry warning, aquatic product warning and forestry warning. The paper focuses on the research methods of the crop warning.

The process of the pest early warning can be divided into several stages: Firstly it is to determine the alarm, which can be inspected from two aspects. One is alarm pheromones that are the component of the early warning indicators. The emergence period, the prevalence rate, the order of severity and the state of an illness are generally as the alarm pheromones of the crop pests. The other is the extent of the warning situation, which usually be divided into five levels that are without warning situation, light warning situation, moderate warning situation, Severe warning situation, the serious warning situation. The second is to look the police sources. Each police source can be subdivided and that a police source will be picked out as the analysis focus should be in accordance with specific conditions. The third is to analyze the warning signs. Different alarms are corresponding to different warning signs. The warning signs can be understood as the proliferation of warning situation and the phenomena generated during the period of the proliferation. Fourthly, forecast the warning degrees, which is the purpose of the early warning. For improving the warning degrees we should combine empirical methods and expert methods. Finally, it is to overcome the warning situation. Based on the principles of prevention first, integrated control and protection of the environment there are some recommendations of prevention for users.

2.2   Early Warning Methods

The early warning methods are a logical thinking process that is to collect data, sort out the data, generate the early warning information and then realize the early warning function. According to the statistical data there are more than 200 kinds of warning method so far.

On the basis of the methods and means of the early warning the methods can be divided into the expert experience method, the model method and the index method. The expert experience methods are mainly based on the past experience, adopt some mathematical models such as simulation experiments, the optimistic and pessimistic methods, competition theory and make use of the analogy methods to deduce the trend of development[3]. The model method is according to the dependent variables that are the state of an illness and the independent variables warning signs to establish early warning model. That is to say it processes the original data and gets the early warning information and expresses it.

According to the content of the early warning, the early warning method can be divided into the occurrence warning, the occurrence amount warning and the disaster degree warning .etc. The occurrence warning mainly alert the patent period or the dangerous period of the state and the level of plant diseases and insect pests. For the long distance migratory pests and the pests which have the ability of diffusion behavior, it can alert the period when the pests move out and move in this locality and takes the date as the basis of determining the prevent periods.

The occurrence amount warning method estimates that whether the tendency of pests appeared in the future reaches the threshold which indicates that it is the time to prevention and cure through the means of forecasting the quantity of the vermin and the pest density in the field. However, it will not be credible if there does not exist a large number of data collected for many years.

The disaster degree warning method relies on the two methods in front. Combined arable farming and the outbreak of the pests to early warning the most sensitive period one crop to the pests, it judges whether the period absolutely coincides with the destructive power of the pests or not and estimates whether the invasiveness of the pests is identical to the period that there emerge more and more pests or not and finally deduce the degree of the plague and the size of the loss caused by the pests[4,5].

According to the timeliness of early warning, the warning methods can be divided into four methods. They are short early warning, mid-term early warning, long-term early warning and extra long range early warning. Short Warning happens and warning the state after a few days when the pests emerge. The Mid-term early warning alerts the dynamics about a month later. Long-term warning makes warning about the dynamics of pests a few months later. Relying on the research of the law of development of the diseases and insect pests, Ultra-long warning explores the occurrence trend in the next year and ultra-long warning.

On the basis of the fundamental theory and early warning method of the agricultural pests and diseases, the paper builds the general framework[6] for early warning of the agricultural pest.

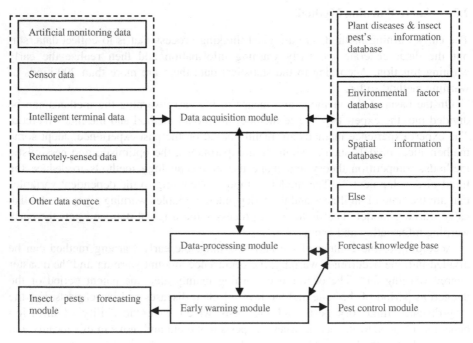

Fig. 1. The Overall Frame of the plant diseases and insect pests Warning

3   The Algorithm and Model of the Warning Analysis

3.1   The Main Early Warning Analysis Algorithm

There are many algorithms for the pest warning analysis, such as the genetic algorithm, BP neural network, radial basis function neural network, rough set theory algorithm, information fusion algorithms and statistical algorithms. The warning models are built on the basis of these algorithms[7]. The following content gives a very brief overview of the several typical algorithms.

3.1.1   Rough Set Theory Algorithm

Rough set theory algorithm is a new mathematical tool processing the vague and imprecise issues. For the border area thought the rough set is proposed and the individual that can not be confirmed are all belonged to the boundary line area. The area is defined as the difference set of the upper approximation set and the lower approximation set. Rough set theory processes and analyzes large amounts of data in the sets contained finite elements, gets rid of the compatibility information in accordance with the dependency relationship between two equivalence relations on the domain and then extracts potential and valuable rules and knowledge. Among them, the simplification and nuclear are two important concepts. The definitions of these basic concepts are given in the following.

Let U be the discourse domain, R is a indistinct relationship on the domain U. $X \subset U$, $R^*(X)$ shown in the formula(1)represents the upper approximation set of x and the lower approximation set $R*(X)$ is shown in the formula (2).

$$R^*(X) = \{x \in U, R(x) \cap X \neq \phi\} \tag{1}$$

$$R*(X) = \{x \in U, R(x) \subseteq X\} \tag{2}$$

ϕ is null set and $BN_R(X)$ represents the boundary of X and it is defined as formula (3).

$$BN_R(X) = R^*(X) - R*(X) \tag{3}$$

For knowledge R, $R*(X)$ is a set whose elements are not only contained in the domain U but must be included by X, $R^*(X)$ is a set whose elements are not only contained in the domain U but may be included by X and $B_{NR}(X)$ is a set composed of the remainder elements that not exit in the set X and $\overline{X}$ .The positive region of X is defined in the formula (4).

$$_{posR}(X) = R*(X) \tag{4}$$

R is a equivalence relation and r represents any one element, $r \in R$. R can be simplified through formula (6) if the formula (5) is true. The function of ind() expresses indistinct relation. If R can not be simplified further, the simplification of R can be expressed as red(R).

$$ind(R) = ind(R - \{r\}) \tag{5}$$

$$ind(R) = ind(R - \{r\}) \tag{6}$$

The R's nuclear core(R) is defined ass the intersection of all the red(R), such as (7).

$$core(R) = \cap red(R) \tag{7}$$

Nuclear is included in all simplified clusters and it is the characteristic set that is can not be eliminated. R and U is an equivalence relation on the domain U and the positive region R of the Q is defined as the formula (8). The equivalence relation is shown in formula (9).

$$P_{OSR}(Q) = \cup R*(X) \quad X \in Q \tag{8}$$

$$_{rR}(Q) = card(_{POSR}(Q)) / card(U) \quad 0 \leq _{rR}(Q) \leq 1 \tag{9}$$

Among them, the radix of B is defined as card (B).Taking advantage of the dependency relationship $_{rR}(Q)$, we can determine the compatibility of equivalence classes R and Q. R and Q are compatible if the value of $_{rR}(Q)$ equals 1. Otherwise, they are incompatible.

Information Fusion steps[8] through rough set theory are as follows

1) Draw up an information sheet contained condition attribute and conclusions attribute according to the collected sample information.
2) Taking advantage of the concepts such as simplification and nuclear, remove redundant condition attributes and duplicate information and finally draw s simplified information table.
3) Calculate the nuclear value table.
4) Obtain the simplified form of the information table according to the nuclear value table
5) Gather the corresponding minimum rules and educe the fastest fusion algorithm.

3.1.2   Neural Network Algorithm

BP neural network prediction model[11] makes use of the self-adaption and self-learning ability owned by BP neural network to study and analyze the data samples, and then it finds out the inherent laws and determines the connection weights and thresholds in the network. In the plant pests and diseases prediction method and an agricultural pest's prediction method, the optimization capability of a single BP neural network is weak. There are also studies that the genetic algorithm was introduced to BP neural network optimization and achieved good results. However, genetic algorithm makes arithmetic operators such as crossover, mutation, selection. Therefore, its convergence rate is slow and optimize efficiency is not high and it often expenses a lot of learning time to establish connection weights and thresholds.

The PSO optimization algorithm is good at handling optimization problems[10], and the study sets the connection weights and thresholds of the BP neural network as the position vector elements in the PSO optimization algorithm and uses it to achieve the initial optimization of the PB neural network, which improves optimization speed and accuracy of the connection weights and thresholds in PB neural network.

The idea of the PSO optimization algorithm is as follows:

1) It multiple calculates and optimizes position vector of itself until the fitness approaches the optimal. Solutions group tends to be stable if there are no significant changes. At this time, the position vector element is closer to the needs of the application;
2) Based on this, optimize the elements of the position vector further using of BP algorithm until the optimal connection weights and thresholds are found out.

3.2   Early Warning Model

Firstly, we usually establish pest warning indicator system and then combine some appropriate algorithms to build pest early warning model. The specific steps are as follows.

3.2.1   Establish a Warning Indicator System

(1) Determine the early warning indicators

Under the guidance of the experts in related fields, the early warning indicator whose affect is small and not easy to be detected is removed and several or a dozen indicators that are more important for pest control and prevention are selected as warning factor that is used to level evaluate police level.

(2) Sort the important degree of the early warning indicators

It is an important part to determine alarm level. Whether the indicators selected are appropriate or not directly affects the early warning results. The expert investigation method, matrix analysis method and the Delphi method are generally selected to sort.

(3) Identify the Warning limit of the warning indicators

According to the indicators and the order of the importance, the appropriate number of key indicators is selected as the basis for pest warning. For example, the formula of the police limits are determined using of the expert investigation and it is shown in formula (10).

$$D = \frac{\sum (Q(w_i) \times d_i)}{N \times Q} \tag{10}$$

D represents the alert level, Q is the expert weight and its range interval is from 1 to 10, d represents the alarm level set by experts, N represents the number of experts.

3.2.2   Build the Early Warning Model

There are some common forecasting models at home and abroad such as univariate decision judgment model, multivariate linear judgment model, multiple logical model, multivariate probability ratio regression model, artificial neural network model and combined forecasting model[11,12,13]. BP neural network model firstly predict the warning factor under the existing situation, and then identify the alarm size, according to the change speed of the monitoring value for a period predict whether the current state is abnormal or not.

Studies show that two hidden layer BP neural network can approximate any continuous function and it's any derivative with any precision and it can better approximate nonlinear function than the BP neural network that contains many hidden layer neural networks[11]. For BP neural network, fewer layers can also achieve good effect if the number of nodes in the hidden layer is designed reasonably. In order to reduce the computational complexity we design three-layer BP neural network and it is shown in the following picture 2.

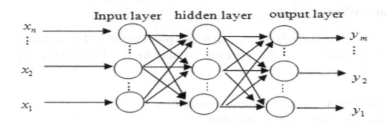

Fig. 2. Three-layer BP neural network

The early warning of the plant diseases and insect pest's three-layer BP neural network contains input layer, hidden layer and output layer. The number of neurons in input layer is determined by the input indicators. The BP neural network model needs optimizing through certain methods, the model needs training through the sample data and finally the prediction model is determined.

3.3   Early Warning Analysis Based on Real Sensory Data Analysis

Internet of things[9] is a constituent part of new generation of information technology. It makes real-time acquisition through the technology such as the sensor, radio frequency identification technology and global positioning system. The collection content includes any objects or process that needs monitoring, connecting and interacting and various information such as temperature, humidity and light. It can achieve the ubiquitous link among things and things, things and people through accessing to various possible networks and then realizes intelligent perception, identification and management to the objects and the process.

It will achieve the desired effect to apply the modern communications technology and the Internet of things technology to the field of agriculture and use of artificial neural network model to analyze the pest early warning. It takes advantage of varieties of sensors to accurately monitor facility environmental factors, which makes the facility environment be in real-time monitoring. Meanwhile, it collects and processes various data[14,15]. Therefore, the paper put forward real-time sensing data-based early warning analysis and its treatment flow is shown in Figure 3.

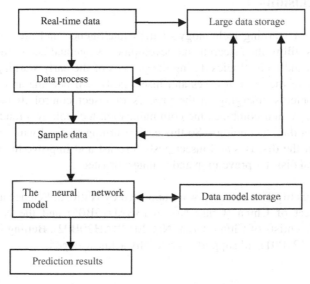

Fig. 3. The process flow of the early warning analysis based on real-time sensing data

4   Comparisons and Analysis

Based on the research of the pest warning algorithm model we make a contrastive analysis of various model and summarize their excellences and weaknesses. They are shown in the table 1.

There are many algorithm for pest early warning analysis, such as genetic algorithm, BP neural network, radial basis function neural network, rough set theory, information fusion algorithm and statistical algorithms. The pest early warning model is built based on these algorithms. At present, common forecasting model at home and abroad are univariate decision judgment model, multivariate linear judgment model, multiple logical model, multivariate probability ratio regression model, artificial neural network model and combined forecasting model.

Table 1. The Result of the Comparison and Analysis

the early warning algorithm	advantages	shortcoming
genetic algorithm	The global search capability is Strong.	Local search ability is weak.
BP neural network algorithm	Local search ability is strong.	Convergence speed is slow; local minimum; point instability.
Joint Algorithm	These algorithms complement each other.	There is not defect.

5   Conclusions

At present, pest warning technology of high intelligence and precision in the field agricultural is still in the research and development stage and needs constantly adding new practices and technologies during the process of the early warning research. The early warning of the crop diseases and insect pests will become an effective way to solve the problems emerging in the process of insect control .It takes Internet of things as its core and combines the communication technology, cloud computing, as well as the big data technology. So this early warning method will be more accurate and timely for the diseases and insect pests forecast and improve the overall level of the agricultural disaster prevention and counter-disaster.

Acknowledgment. This research was supported by National Science and Technology Support Project of China (Grant No. 2013BAD15B05) and the National Science Foundation Scientists of China (Grant No. 2012BAH20B02), Beijing Natural Science Foundation (4121001), all support is gratefully acknowledged.

References

1. Wang, M., Ma, Z., Jin, X.: The Construction of the Crop pests Early Warning Information System in Beijing. Journal of Plant Pathology 35(6), 67–70 (2005)
2. Alka, B., Munkvold, G.P.: Relationships of Environmental and Cultural Factors with Severity of Gray Leaf Spot in Maize. Plant Disease 86(10), 1127–1133 (2002)
3. Wu, X., Bao, S.: Buildup and application of multi-factor spatial interpolation model in the monitoring and warning system for crop diseases and insect pests. Transactions of the Chinese Society of Agricultural Engineering 23(10), 162–166 (2007)
4. Bao, R., Shen, Z.: The data acquisition system applied in the Agricultural pest forecasting. Plant Protection 29(5), 54–57 (2003)
5. Xia, B., Wang, J., Zhang, Y.: The Establishment and Application of crop pest monitoring information system in China. China Plant Protection Herald 26(12), 5–7 (2006)
6. Wang, M., Jin, X., Liu, Q.: The construction and application of the remote warning information system of the major crop diseases and insect pests. China Plant Protection Herald 26(7), 5–8 (2006)
7. Zhang, G., Zhu, Y., Zhai, B.: The warning system of Crop pests based on WebGIS. Transactions of the Chinese Society of Agricultural Engineering 23(12), 176–181 (2007)
8. Zhao, M., Liu, G., Li, M.: The management information system of apple pests based on GIS. Transactions of the Chinese Society of Agricultural Engineering 22(12), 150–154 (2006)
9. Liu, R.D., Zhang, N.: The monitoring and forecasting warning system of the meteorological disaster based on real-time data-driven in Yiyang. Anhui Agricultural Sciences 38(2), 788–789 (2010)
10. Wang, B., Wang, C., Liang, G.: D-S algorithm based on the particle swarm optimization. Sensors and Micro-Systems 26(1), 84–86 (2007)
11. Ji, H., Chaolun, B., Chen, S.: Ice prediction model of the BP network of the Fuzzy optimization based on genetic algorithm. China Rural Water and Hydropower (1), 5–7 (2009)

12. Liu, W., Wu, X., Ren, B.: The construction of crop pest digitized monitoring and early warning U.S. China Plant Protection Herald (8), 51–54 (2010)
13. Zhong, T., Liu, W., Huang, C.: Accelerate the construction of digital monitoring and early warning to provide support for the construction of a modern plant protection. China Plant Protections Herald (12), 05–03 (2012)
14. Yang, F., Yang, Z.: Disaster and Prevention Countermeasures leisure agriculture in Hunan. Agricultural Services 28(1), 85–86 (2011)
15. Hu, X., Li, Y.: Research of the agricultural disaster monitoring, early warning and prevention. Modern Agricultural Science and Technology (4), 25–37 (2012)

Study on the Way of Production, Life and Thinking of Farmers in Mobile Internet Era

Yang Yong, Wang Wensheng, Guo Leifeng, Sun Zhiguo, and Li Xiufeng

Agricultural Information Institute, Chinese Academy of Agricultural Sciences,
Beijing 100081, China
wheatblue@163.com, Wangwsh@caas.net.cn, guoleifeng@126.com,
sunbox.cn@qq.com, lifeng@caas.cn

Abstract. The mobile Internet and 3G mobile terminal has brought profound changes to people's life in the city, and also affected the way of production, life and thinking of farmers gradually. Due to the surveys conducted in various counties of China, Farmers are possible to obtain and share mass of external information and communicate with others far away from them whenever and wherever. However, the mobile internet just solved the hardwares of information acquirement. How to get the useful information, how to identify the information and how to use the information? All these questions have always baffled the farmers. Based on that, the author put forward suggestion as using cloud computing and cloud services to integrate and share agricultural and rural production and living information, equipping grass-root agricultural technician with 3G terminal to help farmers dealing with practical problems in agricultural production and using SNS to help farmers setting up mobile Internet social relationships for information sharing to solve all these problems in mobile internet era.

Keywords: Mobile Internet, Farmers, Cloud computing, agricultural technician, SNS.

1 Introduction

Mobile Internet is a whole body that combines mobile communication with Internet. With the rapid development of broadband wireless access technology and mobile terminal technology, people are eager to access to Internet for getting information and services whenever and wherever, especially leaving the computer. In recent years, mobile Internet developed rapidly, at the end of the September of 2012, the global mobile Internet users reached 1.5 billions.

In recent years, the China has been committed to product agricultural production and to develop sustainable agriculture in many rural areas. For the purpose, developing and training farmers is the most essential issue in China. Currently, experienced farmers are seriously exhausted and few of them prefer staying in remote areas. Because of the large population of farmers and complex agriculture industry structure from the western world, developing agricultural ICTs in Chinese farmers needs to adopt special way.

D. Li and Y. Chen (Eds.): CCTA 2013, Part II, IFIP AICT 420, pp. 188–197, 2014.
© IFIP International Federation for Information Processing 2014

Farmers as a special and large group in China, what they are effected and their response is an important issue related to the future and destiny of the country, and will directly affect the implementation of long-term sustainable development strategy of China's agriculture and rural areas. In view of this, the aim of this paper is to analysis the way of production, life and thinking of farmers in mobile Internet era, find the Find the crux of the problem, advantages and disadvantages of the mobile terminal that affected the farmers. Following that, combined with the characteristics of the application of mobile internet in recent years, this paper put forward some countermeasures, that is using cloud computing and cloud services to integrate and share agricultural and rural production and living information, equipping grass-root agricultural technician with 3G terminal to help farmers dealing with practical problems in agricultural production and using SNS to help farmers setting up mobile internet social relationships for information sharing to solve all these problems.

2   Survey and Methods

2.1   Experimental Samples

Since 2010, Agricultural Information Institute, Chinese Academy of Agricultural Science (AII-CAAS) have conducted an ICT demonstration program in Miyun county of Beijing(north of China), Xinhua county of Jiangsu Province(east of China), Linyin county of Henan Province(Middle of China) and Turpan region of Xinjiang Xinjiang Uygur Autonomous Region(North-east of China). That is to guide the demonstration farmers and agricultural technicians to use smart phones(3G) to improve their ability to get agricultural information and participate in online communication. Based on that, this survey was conducted in these four counties and 120 farmers was interviewed in September 2012.

2.2   Experimental Methods

The purpose of the research is to deeply probe if mobile internet and terminal better support farmers' long term agricultural products, agricultural development and rural life.

In the research, the author address farmers' background and problems through exploratory study that utilizes grounded theory approach and use triangulation method to developing and refining research questions and data collection. The epistemological perspective is oriented toward cataloging and developing explanatory frameworks for the variety of behavior patterns that are becoming increasingly obvious to all systematic observers.

While in interviewing, the author will collect data to describe and test the data through surveying/observing in the research process.The research will focus on Key informant interviews and a focus group to analyze and identify strengths, weaknesses, opportunities and threats (SWOT) of the research program. The interview is a conversation to get specific information and deep drawing research theme. During the interview, the author will encourage participant to describe his/ her background or successful experience to make a relax atmosphere, and it will design a questionnaire to probe the program and guide the topic around the purpose.

According to the surveys and data, the research put forward suggestion for the decision maker, technicians and farmers in mobile internet era.

3   Way of Production, Life and Thinking of Farmers in Mobile Internet Era

The mobile Internet and 3G mobile terminal has brought profound changes to people's life in the city, but also affected the way of production, life and thinking of farmers.

3.1   Effect on Way of Production of Farmers

Farmers usually adopt the very best picked seeds, fertilizer and pesticide to improve productivity. However, they are more eager to obtain more instruction on anti-pest management, updating planting techniques and changing conventional farming methods in the light of the market needs. With the rapid development of modern network and information technology, relationships between production information and farmers have been brought into closer connected.

3.1.1   Acquire Agricultural Production Information from Various of Channel

In recent years, many farmers are possible to use mobile internet to obtain mass of external information whenever and wherever, and no longer confined to traditional literature, newspapers, radio, telephone, television. However, in the past, they need to go to information center in their village or town to search information online. Many farmers have learned not only to browse the webpage, but also to use blog, micro-blog, wechat to communicate with others far from them, such as in another town, county or province. They actively joined or establish QQ group, circle of interest to share information and exchange agricultural production information, so they can introduce new varieties of crops or vegetables,or new technology, machine to improved the yield and their income. They have no longer played a passive role for acceptance of knowledge and information, but actively choose to get the information they need.

3.1.2   Change the Structure of Agricultural Production

Various channel for information acquirement have facilitated the farmers to be educated, and also significantly enhanced the farmers' creativity. Many farmers are no longer limited in planting traditional crops,livestock and Aquatic product, they started to plant and raise special, superior product to get more money. Some farmers are no longer limited to land management, they started to set up their own workshops and factories because they are easy to learn new technology and get help from professional people through mobile terminal and internet.

3.1.3   Increase the Risk of Agricultural Production

In mobile internet era, mass media provides us some inaccurate and false information all the time, that will make the farmers easy to lost money. Some farmers got the false information from micro-blog, wechat or webpages and trust it without field investigation, then suffered tremendous of losses. But on the other hand, in some underdeveloped rural areas, the asymmetry of information make the farmers in inferior position in market,the farmers are difficult to grasp market information timely for planting structure adjustment, so they can not produce marketable products and can not get good price.

3.1.4   Improve the Systematism Degree of Agricultural Production

Convenient information acquirement and communication will gradually lead the farmers to produce in a professional way. Variety of production cooperatives, farmers associations, professional and technical associations, leading agricultural enterprises and family farm have come into being. Farmers can obtain production and market information systematically whenever and wherever.When the farmer's individual rights and interests are violated, the organization agent on behalf of individuals will conduct the negotiations with counterpart to safeguard the interests of farmers. Before the production, farmers will plan and adjust the production structure according the analysis market supply and demand information, during the production, farmers can get the help from experts or massive information online, after the production farmers can release sales information of farm product by mobile terminal at any time. Farmers have swinged from individual management to intensive and scale management, and become the real competitive market players.

3.2   Effect on Way of Life of Farmers

The popularization of terminal and mobile Internet not only provided the farmers a variety of educational opportunities and sharing of information resources, but also changed their life style.

3.2.1   Enrich the Entertainment of Farmers

With the popularization of a variety of mobile phones, application software and network everywhere in rural areas, farmers have opportunity to enjoy more cultural exchanges and entertainment services, that have enriched their spiritual life, and improved their quality of life. In the past, while in slack season, farmers usually spend leisure time on playing cards and mahjong. But now the farmers can not only watch TV, but also can use intelligent mobile phone to contact with the outside for booker, buy stock, read the news online, watch movies and play games etc..

3.2.2   Facilitate Farmers' Life

Mobile internet facilitate farmers' communication with outside. They no longer have to worry about going out and lost outside. As for management of village affairs, meeting and discussion and marriage event in rural areas, they no longer need to be informed door by door and ask somebody to take a oral message, they can notice them at any time and everywhere. They can break the occlusion to understand domestic and foreign news and other useful information in the world. Although they work in their field, they still can have a direct dialogue face to face with agricultural experts in the institute through the mobile internet terminal. They can ask questions to the expert and get the advice through the screen. With the development of e-commerce in the rural areas, farmers can timely know agricultural products supply and demand in the fields.

3.2.3   Broaden the Employment Channel for Farmers

In the past, farmers often stick to their fields of agricultural production. Because of low efficiency of traditional agricultural production, the rural health care, children education have become a heavy burden for farmers. Now farmers can find out the demand according to their ability through the mobile internet, they can find way out of original narrow space to have new business or go to the city to find a new job for more money. These changes have sped up the integration process of rural urbanization in china.

3.2.4   Improve the Spirit and Outlook of Farmers

In the past, farmers have a weak legal concept, and lack of consciousness to use the law as weapons to protect their own legal rights.This asymmetry of information make the farmer's legitimate rights easy to be violated. Now, farmers can conveniently obtain the relevant legal knowledge to protect their own rights and interests, they can express their dissatisfaction or expose the problems through the website, micro-blog, weichat etc..In addition, since farmers have acquired new information actively, they have increased their knowledge, strengthened self-confidence and improved the social status gradually.

3.3   Effect on Way of Thinking of Farmers

Way of thinking is the program and method of people's thinking formed of factors in certain cultural background, knowledge structure and habits. In mobile Internet era, rapid and timely dissemination of information, knowledge popularization, information acquired by passive steering interaction change, have effected the concept and mode of thinking of farmers.

3.3.1   Excite the Creativity of Farmers to Become Rich

Farmer's way of thinking are influenced by region, the scope of activities, cultural atmosphere and other factors. Network and terminal of mobile Internet have been spread the external information to every corner of the rural, thus broken the constraints of region,the scope of activities and cultural atmosphere, and extremely triggered the farmers' ability of thinking and imagination. This kind of imagination can help the farmers to break away from the constraints of direct space and real space, and enter the possible and ideal space, then establish the target quickly. It can favor the innovation of knowledge and technology of farmers, so they will create new inspiration and new ways to get rich.

3.3.2   Improve the Capacity of Farmers to Accept Education

Mobile internet bring the farmers instant information communication and abundant information resources, make the farmers' understanding of the outside world far beyond what they can directly experience of their living world. So the interest of understanding external world of farmers was greatly stimulated. Acceptance of education of farmers will swing rapidly from passive to active,they have been infected by the surrounding farmers who got the information and became rich. They actively offered to accept knowledge and education, use modern information technology and equipment to link with the outside world as soon as possible. Based on that, under the improvement of basic education facilities, rich information resource of practical production technology, life knowledge and training in rural areas, farmers' ability to accept the education will be improved greatly.

3.3.3   Test the Farmers' Ability to Identify the Right Things

Large quantities of Agricultural and rural information resources were sent to the farmers through the terminal of mobile Internet. They are information of practical production technology, market and healthy entertainment information etc., in the same time, some of them are information of fraud, superstition, pornography and violence and negative decadent content. The latter will harm the farmers with low education. So it is a test for the farmers to correct the discrimination of things, and will force the farmers gradually to learn how to make a judgment and choice.

3.3.4   Get Rid of the Superstitious Feudal Idea of Farmers

The countryside is a place with high-rate occurrences of superstitious feudal ideas and rumors. The mobile internet and other advanced means of information dissemination have linked the rural and outside. Farmers gradually get and master more scientific knowledge, understand that they should consider all kinds of relationships with the things around when they deal with complex things. Thus they can break the shackles of traditional ideas and thinking, open to consider new problems, solve new contradictions and accept new things, gradually they will move away from the idea and the custom of feudal superstition.

4   Suggestions for Farmers in Mobile Internet Era

Mobile Internet brings farmers many conveniences, but also raises a series of new problems. In view of this, the author puts forward the following suggestions.

4.1   Using Cloud Computing and Cloud Services to Integrate and Share Agricultural and Rural Production and Living Information

In China, scattered agricultural production mode determines the agricultural information resources scattered, the cost of collection is high, and it is difficult for integration. At the same time, most of the integration are often for professional data resources from some scientific research institutions. In fact, large number of information kept by rural grassroots technicians, experts and managers and communicated in agricultural production activities can not be integrated and shared. This information plays an important role in the guidance of Agricultural production. As a sharing infrastructure, cloud computing can easily integrate various of information resources. Cloud storage make all types of users access to their own data and others by using a variety of terminals, establish their own information center without worrying about memory and servers. Low cost cloud storage can reduce high data collection costs, broaden the information service breadth, reduces information threshold and improve the information service benefit.

The requirements of information technology ability for terminal user are very low under cloud computing environment. It is what the groups of farmers needed. Farmers can use a variety of terminals to access the cloud, it can be a computer or mobile phone with Internet and browser. Do not need to install a wide variety of applications, do not need to care about data storage location and security of the data, whether it upgraded or not, even without antivirus software and firewall. All this work is finished by cloud computing center, professional cloud team will provide professional services to the user.

Under the cloud computing environment, agricultural information consulting becomes simple (as shown in Fig.1). The farmer put forward request, cloud services platform get the request, decide its types and characteristics, according to different agricultural information service advantage, it will schedule problem automobility and specify one agricultural service organizations to answer user's questions.Cloud service platform can be regarded as black box between the farmers and the agricultural information service institutions, it plays the role of scheduling, allocation of resources and assisting the advisory communication. Due to the adoption of joint service mode, farmers can get solutions soon.

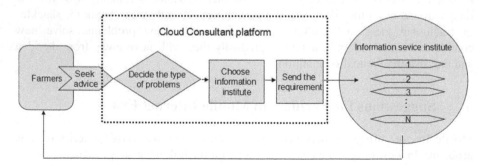

Fig. 1. Agricultural information consulting platform under the cloud computing

4.2   Equipping Grass-root Agricultural Technician with 3G Terminal to Help the Farmers to Deal with Practical Problems in Agricultural Production

Grass-root agri-technique extension related to food security, farmers' income and rural development. In mobile internet era, most of farmers are easy to get information from the internet and communicate with outside. However the mobile internet only solved the hardwares of information acquirement, how to get the useful information, how to identify the information and how to use the information still puzzled the farmers. Abundant information have broaden the knowledge of farmers, but a wide range of knowledge, if can not be combed, is often scattered and inefficient. As for farmers, the important is not to get the information for production activities, that is how to use it. In many countryside area of eastern coastal area, because of the fast economic development, information construction developed faster, the wireless broadband Internet access is not a problem, but the reality is, application level is very low. Since most of them are low educated, so they really need the grass-root agricultural technicians to help them.

Ministry of Agriculture, P.R.China have focused on enhancing the problem-solving ability of grass-root technicians by trainings and helping farmers solving the problem of agricultural production on the spot many years ago. However, limitation of timeliness and specificity of the information always puzzled the technicians and farmers. So it is necessary to equip the grass-root agricultural technician with 3G terminal for timely problem-solving of agricultural production for the farmers.

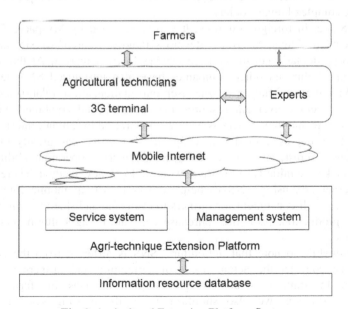

Fig. 2. Agricultural Extension Platform System

Through 3G terminal, technicians can educate farmers information awareness, enhance their ability to use information. Then they can educate the farmers about the new agricultural technology and ask for help from related institute and university by video conference system or submit the questions while they went to help the farmers. Moreover, technicians can be managed in a high efficiency by problem-solving record and GPS location. So the officers of agricultural management department can motivate them in a measurable way.

Now Chinese Academy of Agricultural Sciences have started to establish an Agri-technique Extension Platform System to provide the service for agricultural technicians with 3G terminal installed Android system, that will allow technician to get timely and detail problem-solving information or guidance to help the farmers(as shown in Fig 2).

4.3   Using SNS to Help Farmers Setting Up Mobile Internet Social Relationships for Information Sharing

In recent years, social networking (Social Networking Services, SNS) is developing rapidly in China and abroad, brings an Internet revolution, some comments regarded that as the signs of Web3.0 times, and it bring about an all-round experience for Internet users. The development of SNS based on the "six degrees of separation" theory and the "150 law". Six degrees of separation (Six Degrees) refers to a person with a stranger by interval not more than 6 person. SNS has its unique development space in the complex human society.

The SNS site, in foreign countries, it was represented by Myspace, Twitter and Facebook etc.. In China, SNS also develops rapidly. Renren, Kaixin, and 51.com makes SNS particularly conspicuous in China's Internet market. At the same time, Tencent, Sina, Sohu and other companies have also introduced SNS development strategy. With the advantages of true user information, clear relationship, content diversity, easy and convenient resource management, and combined with mobile communication technology, SNS has been widely praised by China's Internet users.

SNS with its unique attraction and high user viscosity, can greatly improve the farmer online interest, and further enhance its Internet operation ability, so that farmers can take the initiative to learn new technology, new method. Therefore in the mobile Internet era, using SNS to help farmers to set up mobile Internet social relationships, it will convenient for farmers to exchange and share the information of agricultural production, rural life, purchase and sale of agricultural products with others outside.

It is worthwhile to note that agricultural SNS is different from the commercial SNS, it must consider the actual needs of agriculture and farmers in the initial construction. In addition to rely on agricultural technicians, big farmers, experts exchange information, we also should rely on the existing scientific research resources to establish useful agricultural knowledge database. Using open API, the existing agricultural service concepts and technologies are integrated in the SNS platform, that can promote the agricultural SNS service capability and increase the users and resources, and mining the knowledge generated by users. With the rapidly growing of agriculture SNS user, the docking of supply and demand of agricultural products can bring huge profits for the website, such as the Taobao, 58 city and so on, that also can solve the practical problem of agricultural products sales for farmers.

Acknowledgment. Funds for this research was provided by the National Science and Technology Support Program(2011BAD21B01), that is Key Technologies for Agricultural Field Information Comprehensive Sensing and Rural extension.

References

1. Chen, X., Liu, Y.: Development and utilization of agricultural information resources under the network environment-The network development and construction should be suitable for farmers' use. Agricultural Library and Information Science 17(12), 56–58 (2005)
2. Wang, H.: Study on informational consciousness influencing on behaviors of farmers in Yizheng, Jiangsu Province, China as a case analysis. Dissertation of China Agricultural University, Beijing (2005)
3. Li, G., Zheng, H., Tan, C., et al.: Research on the Agricultural Information Services under the Cloud Computing Environment. Journal of Anhui Agri. Sci. 39(27), 16959–16961 (2011)
4. Wang, W.: Research on the key technology and integration of agricultural technology extension informationization based on 3G. China Science and Technology Achievements 12(8), 66 (2011)
5. Yi, L.: Profit model of SNS social networking site in China. E-commerce (7), 57–59 (2011)
6. Yin, G.W., Wang, W.S., Sun, Z.G.: Promote The Grass-root Agro-technique Extension Service and Management with Information Technology. Applied Mechanics and Materials, 263–266, 3298-3300 (2013)
7. Ford, S.A., Babb, E.M.: Farmer sources and uses of information. Agribusiness 5(5), 465–476 (1989)
8. Hua, C.: Source of sci-tech information acquirement for farmers. Agricultural Network Information (7), 50–52 (2005)

Studies on Domestic and Overseas in Research Progress of Agricultural Information Technologies

Ying Yu[1], Cunjun Li[1], Leigang Shi[1], Heju Huai[1], and Xiangyang Qin[2,*]

[1] Beijing Research Center for Information Technology in Agriculture, Beijng 100097, China
[2] Beijing Academy of Agriculture and Forestry Sciences, Beijing 100097, China
{yuy,licj,shilg,huaihj}@nercita.org.cn,
qinxiangyang@baafs.net.cn

Abstract. This paper is a comparative study of domestic and overseas in the aspect of agricultural information technologies research progress. In domestic, it is short of independent development of core technology in precision operation integrated system. The key problem that needs to be resolved in agricultural information acquisition and analysis technology is how to improve the accuracy, the agricultural products information collection technology is still a blank. Lacking of shared and integrated simulation platform in agricultural digital model and virtual technology, and agricultural productivity prediction and early warning technology still needs to be improved. In the technology of agricultural intelligent decision, China is blank in the field of knowledge grid, knowledge sharing cannot be achieved. The high technical cost and the low operation reliability are the main problems in agricultural intelligent equipment. In modern agricultural mechanical equipment technology, China is lack of uniformed standards and regulations. There is a big gap between domestic and overseas in rapid detection technology and bio-mechanical composite technology.

Keywords: Agricultural information, technology, domestic, overseas.

1   Introduction

The modern advanced agricultural information technologies have been much interest in not just precision agriculture (PA), but also in areas where precision operation technology integrated system, agricultural information acquisition and analysis technology, agricultural digital model and virtual technology, agricultural intelligent decision technology, agricultural intelligent equipment technology, modern agricultural mechanical equipment. And these technologies were fast developed in the world-wide, especially in some developed countries.

China has occupied the first positions in the list of countries with the fastest growing 100 cities. The implications of such dramatic shifts for economic development, urbanization and energy consumption are immense [1]. However, China

---

* Corresponding author.

D. Li and Y. Chen (Eds.): CCTA 2013, Part II, IFIP AICT 420, pp. 198–211, 2014.
© IFIP International Federation for Information Processing 2014

as a large agricultural country, the application of information technology in agriculture area is at the starting stage, but has achieved great accomplishments. The aim of this paper is to comparative analysis on agricultural information key technologies of domestic and overseas that will promote the communication and development of the technologies in the world-wide.

Rapid socio-economic changes in some developing countries, including China, are creating new scopes for application of agricultural information technology. PA is conceptualized by a system approach to re-organize the total system of agriculture towards a low-input, high-efficiency, and sustainable agriculture. Many researchers applied the precision technology to manage farmland product system and agricultural product traceability system etc. FH Wang [2] pointed that there were many soft and hardware, such as 3S, farmland information acquisition system, and intelligent field variable-rate operation system etc. were studied and applied in China. Besides, four key technological components, i.e., field information collection, field information management, variable-rate decision-making, and variable-rate technologies were analyzed and evaluated; the impacts of precision agriculture on farm input, crop yield and quality were summarized; the barriers for development and adoption of precision agriculture were also pointed out by CJ Zhao etc. [3, 4]. NQ Zhang etc. [5] have provided the impact of precision agriculture technologies on farm profitability and environment, worldwide applications and adoption trend of precision agriculture technologies, and potentials of the technologies in modernizing the agriculture in China.

Related research has overwhelmingly showed that rural residents have an extensive range of information needs, with agricultural technological information, market information, income generation information and policy information being the most needed; and that they rely very much on interpersonal relationships for acquiring both general information and information for agriculture production [6]. Therefore, the agricultural information acquisition and analysis technology is urgent demand in rural area. Lake [7] reported that studies of Internet adoption in the general U.S. population had produced estimates of 60 to 100 million households using the Internet. Within the U.S. farm population, Internet adoption in 1999 was estimated at 43% for U.S. farms with sales over $100,000 [8]. Computer adoption on U.S. farms has been studied by many researchers. Business characteristics such as experience with other technology, the use of farm records services, the use of consultants, the size of the farm business, the complexity of the farm business, the level of farm income, the type of commodities produced by the farm, and the number of employees have been hypothesized to affect PC adoption [9, 10, 11]. These studies have taken the perspective that the computer is an integral tool for turning farm data into information upon which management can act. In other words, the computer is a key element of the managerial information system (MIS) on most farms [12].

Worldwide, farmers are increasingly purchasing and using on-farm computers to provide decision support information and assist in meeting their tax and other reporting commitments. While, having purchased, the farmers clearly believe the investment is justified, there is some before and after data to support this conclusion. [13, 14] Nuthall[15] pointed out that, on average, the profit has tended to increase after purchasing a computer. However, the wide variations and involvement of many factors make categorical conclusions difficult. It leaves open the question whether computer technology makes it easier for good managers to make decisions that in the past were of a similar quality but took longer to obtain, perhaps with subconscious observation and intuition, in contrast to the computer based decision systems actually creating valuable new information.

In agriculture production, we have a well-established range of instruments for measuring variables such as mass, volume, temperature, relative humidity, gas and fluid flow. All are capable of working reliably in the agricultural environment, with sufficient accuracy for most purposes. Usually they are based on a sensor in direct contact with the solid, liquid or gas concerned [16, 17].

Drabenstott [18] has identified five challenges that will be critical in shaping the rural economic outlook in the USA, but they applied to many other settings as well: tapping digital technology, encouraging entrepreneurs, leveraging the new agriculture, improving human capital, and sustaining the rural environment. Three of the five are connected rather directly to the digital economy and the use of information and communication technologies (ICTs): digital technology, entrepreneurship, and human capital. The new agriculture also exploits new technologies and assembles them (GPS, databases and geographic information systems) into precision agriculture to optimize agricultural inputs to specific locations, perhaps with little involvement of farmers' themselves [19].

Fountas etc. [20] described the development of a system based model to characterize farmers' decision-making process in information-intensive practices, and its evaluation in the context of precision agriculture. A participative methodology was developed in which farm managers decomposed their process of decision-making into brief decision statements along with associated information requirements. A knowledge-based intelligent e-commerce system for selling agricultural products was found out by W Wen [21]. The KIES system not only provided agricultural products sales, financial analysis and sales forecasting, but also provided feasible solutions or actions based on the results of rule-based reasoning.

Use of computers and sensors for real-time decisions in cropping systems is increasing rapidly. Yet, the value of technology can be best realized when integrated with agronomic knowledge, resulting in a seamless process of assessment, interpretation, and targeted operation [22]. Henten [23] etc. indicated a procedure and the results of an optimal design of the kinematic structure of a manipulator to be used for autonomous cucumber harvesting in greenhouses.

Research activities concerning automatic guidance of agricultural vehicles have led to various solutions. Sensors, including mechanical ones, global navigation satellite systems (GNSS), machine vision, laser triangulation, ultrasonic and geomagnetic, generate position, attitude and direction-of-movement information to supply control algorithms. In America, a conceptual framework of an agricultural vehicle guidance automation system includes navigation sensors, navigation planner, vehicle motion models, and steering controllers [24]. In Europe, Keicher [25] etc. indicated that it was depending on who is funding the project, the systems range from a PC, with a frame grabber or a GNSS receiver used to guide an implement along a predefined path with speeds up to 3 m/s, to a multiprocessor bifocal road recognition system for autonomous cars driving on motorways with a speed of 130 km/h.

All the development of agricultural information technologies in the world-wide is as to improve agriculture production efficiency and farmer income, reduce the damage to environment, and promote the sustainable development of agriculture.

2   The Key Technologies

2.1   Precision Operation Technology Integrated System

In China, it is immature of the precision control technology, and the precision operation guidance system is not very reliable. The theoretical research on zoning management is not enough, and its system application cost is very high. Generally, the precision operation integrated system is lack of core and key technologies.

Table 1. Precision operation technology integrated system

Key technology	Comparison
Precision soil preparation and seeding	Overseas: the application of precision seeding operation has been widespread
	Domestic: laser has been applied in soil preparation
Precision operation technology	Overseas: variable-rate fertilization, precision irrigation, intelligent spray, mechanical weeding have been achieved
	Domestic: variable-rate fertilization of single fertilizer has been accomplished
Precision zoning management and prescription map technology	Overseas: the application of zoning management and prescription map technology has been popular
	Domestic: variable-rate fertilization on single plot has been used in the demonstration stage
Precision guidance system	Overseas: auto and assistant guidance system has been used on farmland operation
	Domestic: assistant operation guidance has been used on regional demonstration
Precision integrated system	Overseas: the technology has been applied in field, orchard, greenhouse, animal husbandry, fishery and forestry
	Domestic: only the main field crops have been demonstrated

Gao Liangzhi [26] accounted that agriculture system is a complex system which composed of agricultural biology, agricultural environment, agricultural technology and agricultural economy—the four subsystems, with a certain internal relations. Luo Xiwen [27] introduced the precision agriculture technology system from the thought of technology, supporting technology, operation process and application examples. Zhao Chunjiang [3] analyzed and discussed the technology of precision agriculture system from the four key technological components of precision agriculture (information acquisition, information analysis, decision making and decision implementation). Cao Hongxin et al. [28] discussed the digital cultivation framework and technical system.

Precision agriculture cannot work without the construction of agricultural information infrastructure, and the basic of infrastructure construction is the network connection device. According to the statistics of USDA in 2009, digital subscriber line (DSL) was the most common method of accessing the Internet. The second is Satellite and wireless, and cable is the third method. In China, the infrastructure building in Beijing had certain scale, and the network coverage in rural counties and the rate of network into villages almost reached 100% [29].

2.2   Agricultural Information Acquisition and Analysis Technology

In the international, sensor is developing rapidly but in China, it is still in the primary stage. The cost of sensor research in agriculture is high and its accuracy is much low, that is the same difficulty as agricultural information acquisition technology. China had no research inthe key technology of agricultural products information collection. Lack of sharing platform,model design and low monitoring accuracy which are the significant problems in remote sensing.

Table 2. Agricultural information acquisition and analysis technology

Key technology	Comparison
Growth information acquisition technology	Overseas: crop growth, physical and chemical properties, disease pest and weed have been detected intelligently
	Domestic: sensor detection of plant growth has been accomplished
Soil information acquisition technology	Overseas: sensor system online measurement of the main soil physical and chemical properties has been achieved
	Domestic: sensor system achieved the measurement of soil moisture and hardness
Agricultural products information collection technology	Overseas: nanotechnology and biochip technology etc advanced technologies have been widely used
	Domestic: the traditional methods have mainly been adopted
Remote sensing information acquisition technology	Overseas: remote sensing have been applied in determination of soil physical and chemical properties and crop growth, and fertilization recommendation on internet
	Domestic: remote sensing has been used in crop classification, farmland area estimation, growing of large-range detection

The key information technologies in agricultural inforamtion acquisition, it mainly included of some aspects as below: the rapid detection technology, auto identification technology, the system and platform of agricultural products, network technology and database technology.[30]

In university of California, Davis, a new test scanner was developed which used in testing the wine whether deterioration without opening the packaging. The development of the scanner referred to the magnetic resonance imaging (MRI) technology which is widely used in the field of medicine; it can detect acetic acid in wine. In Germany, there was a light, rapid and effective PEN2 electronic nose system, it was used in inspecting and testing gaseous matter and steamy, now all the scent of meat, fruit, yogurt, milk, alcohol and coffee etc could be tested by it. [31].

Beijing CapitalBio Corporation and Beijing Entry-Exit Inspection and Quarantine Bureau cooperated to develop 'Protein Chip Veterinary drug residues in protein microarray platform', multiple samples in a variety of veterinary drug residues could be detect on the same chip [32]. Changchun Jilin University Instrument Co., Ltd developed a serial of measuring instruments, one of them is 'Pesticide Residues in Food Supplies fast detector', it can be applied to flour, rice, soybeans, green beans surface-site rapid detection of pesticide residues [33]. Chinese Military Academy of Medical Sciences used indicator paper and the light reflection sensors to develop equipment: 'portable multi-function devices for food safety rapid testing'. It was the first time to blend the following technologies together, such as electro-optical technology, sensor technology, microelectronics, micro-mechanical technology, computer technology and food safety testing technology, the testing covered all kinds of food daily, each sample test only 15 seconds to 30 minutes [34].

Near infrared spectroscopy in the soil inspection has a good application, to determine soil moisture, organic matter, total nitrogen and available nitrogen, organic carbon and total carbon [35]. Laser-induced breakdown spectroscopy (LIBS) has emerged in the past ten years as a promising technique for analysis and measure [36]. A Laser induced breakdown spectroscopy (LIBS) system used for detecting heavy metals in polluted soil was established. Samples containing various heavy metals such as Cd, Cu, Pb, Cr, Zn, Ni and soil sample were analyzed by this system, and main spectral lines of heavy metals and main elements were recognized [37].

Shen Guanglei etc designed and developed a beef quality and safety traceability system via internet technology. The system adopted JSP to design the Object-Oriented dynamic pages and adopted the MySQL to design the database. The B/S (Browser/Server) structure was used to put the management system on the internet, which implemented the management of beef traceability system via internet and made the beef traceability system be more networking and popularizing [38].

Based on analyzing agricultural products traceability and investigating the producing enterprises of agricultural products, agricultural products archive management system (FPAMS) with B/S (Browser/Server) structure was founded by using database technology [39].

2.3   Agricultural Digital Model and Virtual Technology

In the digital agriculture area, there is seldom application in simulation object and dynamic model in China. The R&D of platform and model is lack of sharing and integration. Time, space productivity prediction and early warning need to be strengthened.

Table 3. Agricultural digital model and virtual technology

Key technology	Comparison
Digital modeling technology of bio-environmental system	Overseas: the prediction of plant and animal growth, output and quality has been achieved
	Domestic: simulation of the growth and output of main crops has been accomplished
Agricultural biologic form virtual representation technology	Overseas: the morphological change process of main agriculture and forestry plant has been dynamic displayed
	Domestic: static virtual of the main agriculture and forestry plant on growth period has been accomplished
Digital visual bio-mimetic platform	Overseas: function and structure coupled crop growth simulation system and components have been formed
	Domestic: crop growth and environment simulation system is still in build
Agricultural biologic system virtual design technology	Overseas: agronomy system design and environmental effect assessment have been achieved based on scenario simulation
	Domestic: part of the virtual design and regulation has been used based on simulated evaluation
Agricultural productivity prediction and early warning technology	Overseas: model has been applied on region productivity prediction and management under climate change
	Domestic: productivity prediction and analysis under site level has been achieved

The digital resource of 'National Cultural Information Resources Sharing Project' has reached 105.28TB, an accumulative total of 890 million people who got services. 'E-home' is the primary public information service place in rural areas, its number had reached 824; China has more than 250 rural digital cinemas, nearly 40,000 digital cinema projection equipment in rural areas [40].

The potential for rural areas to benefit from telecommunications technology is a persistent question [41]. Rural America is digital: rural communities and people are connected to the Internet. The definitive 2000 report, Falling through the Net [42], and its 2002 counterpart, A Nation Online [43] found that rural households, which historically trailed those in central cities and urban areas, have shown significant gains in Internet access.

2.4   Agricultural Intelligent Decision Technology

The low utilized information and non-shared knowledge is the problem of agricultural information fusion and knowledge discovery technology in domestic. It is still a blank of the application of knowledge grid technology in the field of agriculture. In the decision support technology, the self-learning ability needs to be improved.

Table 4. Agricultural intelligent decision technology

Key technology	Comparison
Information fusion technology	Overseas: multi-information fusion technology has been achieved
	Domestic: information is scattered and not integrated
Knowledge discovery technology	Overseas: massive knowledge base building has been achieved
	Domestic: the structure of knowledge base is single and not uniform
Knowledge grid technology	Overseas: knowledge grid calculation technology has been achieved
	Domestic: the knowledge grid technology is just starting
Decision support technology	Overseas: intelligent agriculture decision service has been achieved
	Domestic: the application of expert system is very well

At the moment, the AGROVOC thesaurus is mapped to the following recourses: EuroVoc [44], NALT [45], GEMET [46], LCSH [47], STW-Thesaurus for Economics [48], and RAMEAU [49]. With the launch of the AGROVOC linked open data [50], the stage is set for organizations around the world to start publishing their agriculture knowledge models by linking them to AGROVOC, as well as utilizing AGROVOC for resource management. Here in MIMOS, we are publishing four knowledge models in agriculture. The first being a generic crop ontology, followed by ontology each for tomato, corn and chili. Dickson Lukose briefly described the evolution of the LOD, the emerging world-wide semantic web (WWSW), and explore the scalability and performance features of the service oriented architecture that forms the foundation of the semantic technology platform developed at MIMOS Bhd., for addressing the challenges posed by the intelligent future internet[51].

2.5   Agricultural Intelligent Equipment Technology

In China, the cost of network building and maintenance is very high of intelligent monitoring technology, especially the cost of sensor which applied in detection. It is weakness of the study foundation of the model intelligent control technology. Single species and low reliability are the problem of intelligent robot.

At present, the farm computer equipment is mainly used to conduct online transactions. In 2009, 81 percent of U.S. farms with sales and government payments had access to a computer, 79 percent owned or leased a computer, 69 percent were using a computer for their farm business, and 76 percent had Internet access [52]. In Japan, about 34% households had PC, of which 12.2% had access to the internet [53].

Table 5. Agricultural intelligent equipment technology

Key technology	Comparison
Intelligent monitoring technology	Overseas: the multi-parameter detection simultaneously and wireless network have been achieved
	Domestic: single point monitoring and wireless transmission has been accomplished
Intelligent control technology	Overseas: linkage control based on growth and environment has been achieved
	Domestic: control technology has not been combined with growing
Intelligent detection technology	Overseas: the detection is intelligent and the operation is easy
	Domestic: intelligent environment monitoring hasn't been accomplished
Intelligent robot technology	Overseas: intelligent robot has been developed in multi-domain
	Domestic: intelligent robot is still in developing and no mature products

Gloya [54] found that in the USA, the producers were not sure as to how the information technology can best be used in their farm to create value. Furthermore, in India Raju [55] concluded that organizational linkages and networking capacities are to be strengthened for 'digital unity' to provide multiple opportunities to the agriculture communities to exploit local resources for their self-development. In New Zealand rural areas, farmers are increasingly purchasing and using on-farm computers to provide decision support information and assist in meeting their tax and other contracts management. While the farmers purchase on Internet, they clearly believe the investment is justified, although there is not enough data to support this conclusion [56].

For China, the issue of agricultural communication development has been conventionally examined under labels such as universal service, digital, divide, broadband deployment, and e-government, which generally fall into two seemingly distinct categories—access and applications. In China, these concepts are currently incorporated into a single program, if not a single term—'Village Informatization Program' ('VIP') [57, 58].

2.6   Modern Agricultural Mechanical Equipment

It is short of uniformed standards and regulations of large-scale agricultural operations control system in domestic. The intelligent equipment control level is lower than developed countries and lack of common platform. The rapid detection technology of agricultural products processing equipment still needs to be improved. There is no research in the area of bio-mechanical composite technology.

Table 6. Modern Agricultural mechanical equipment

Key technology	Comparison
Large tractor operation control system	Overseas: onboard communication, bus control, condition monitoring and self-walking function have been achieved
	Domestic: traditional integrated circuit and electric relay control has been used
Large harvest equipment	Overseas: intelligent cotton picker, grain combine harvester has been developed
	Domestic: there is medium-sized combine harvester, no intelligent cotton picker
Hills operation equipment	Overseas: hills operation has been mechanization
	Domestic: traditional walking tractor operation is still used
Agricultural products processing equipment	Overseas: fruit and vegetable classification, fisheries automation have been accomplished
	Domestic: fruit mechanical classification has been achieved
Bio-mechanical composite technology	Overseas: human-machine coordination and object tracking has been achieved
	Domestic: no research in this area

Holland, Italy, France, UK, Spain, USA, Canada, Israel, Turkey, Japan, Korea, Australia and other countries started earlier in agricultural technology research facility, which developed fast and in the highest level of the world currently[59,60]. Xu Fang et al. summarized the survey and development of agricultural mechanical equipment in protected agriculture. It included machinery for seeding, transplanting, automatic engrafting, cultivating, harvesting and transporter for vegetable production. Flower-product ion machines and agricultural robots were also introduced. Some problems about the technical development of agricultural machinery in protected agriculture in the future are discussed [61].

3   Conclusions

The key agricultural information technologies comparison of domestic and overseas shows that, although the beginning of China is late the development is very fast. But in some key technologies, there is still a large gap with the developed countries. The main problem of Chinese agricultural information technology is that the basic research and deep development is not enough, lacking of independent core technology, reliability and accuracy is low. The development of high accuracy sensor is the main technical difficulty, and the high cost is also a problem. In digital agriculture, it is short of deeply theoretical research and technology development, and the agricultural prediction and early warning ability still needs to be improved. There is no uniformed standards and regulations in mechanical equipment development, the low information utilization is a serious factor to restrict knowledge sharing and hinder intelligent decision support system development. It is still a blank in some technology research fields in China, such as agricultural information collection technology, knowledge grid technology and bio-mechanical composite technology etc.

China as a large developing country, there is no advantage to compare with developed countries in resource and environment etc. Currently, China is confronted with huge competitive pressure on the quality of agricultural products and benefit of agricultural production. There is much practical value and strategic significance in researching and developing agricultural information technology, to reduce agricultural production costs, improve the output and quality of agricultural products.

Acknowledgment. Funds for this research was provided by the construction of integrated information "three dimensional rural" service platform in National Modern Agricultural City for Science and Technology Projects (D121100003212003).

References

1. Mondal, P., Basu, M.: Adoption of precision agriculture technologies in India and in some developing countries: Scope, present status and strategies. Progress in Natural Science 19, 659–666 (2009)
2. Maohua, W.: Thinking through the experiment, demonstration and development research on precision agriculture. Journal of Agricultural Science and Technology 5(1), 7–12 (2003)
3. Zhao, C., Xue, X., Wang, X., Chen, L., Pan, Y., Meng, Z.: Advance and prospects of precision agriculture technology system. Transactions of the Chinese Society of Agricultural Engineering 19(4), 7–12 (2003)
4. Zhao, C.: Progress of agricultural information technology. International Academic Publishers. International Academic Publisher (2000)
5. Zhang, N., Wang, M., Wang, N.: Precision agriculture—a worldwide overview. Computers and Electronics in Agriculture 36, 113–132 (2002)
6. Yao, Z., Yu, L.: Information for social and economic participation: A review of related research on the information needs and acquisition of rural Chinese. The International Information & Library Review 41(2), 63–70 (2009)
7. Lake, D.: Spotlight: how big is the U.S. net population? The Industry Standard (November 29), http://www.thestandard.com/metrics/display/0,2149,1071,00.html (accessed January 6, 2000)
8. National Agricultural Statistics Service (NASS), Agricultural Statistics Board, U. S. Department of Agriculture, Farm Computer Usage and Ownership (1999)
9. Lewis, T.: Evolution of farm management information systems. Computers and Electronics in Agriculture 19, 233–248 (1998)
10. Amponsah, W.A.: Computer adoption and use of information services by North Carolina commercial farmers. Journal of Agricultural and Applied Economics 27, 565–576 (1995)
11. Baker, G.A.: Computer adoption and use by New Mexico nonfarm agribusinesses. American Journal of Agricultural Economics 74, 737–744 (1992)
12. Gloya, B.A., Akridge, J.T.: Computer and internet adoption on large U.S. farms. International Food and Agribusiness Management Review 3, 323–338 (2000)
13. Lazarus, W.R., Strecter, D., Jofré-Girando, E.: Management information system: impact on dairy farm profitability. North Central Journal of Agricultural Economics 12, 267–277 (1990)

14. Verstegen, J.A.A.M., Huirne, R.U.M., Dijkhuizen, A.A., King, R.P.: Quantifying economic benefits of sow—herd management information systems using panel data. American Journal of Agricultural Economics 77, 387–396 (1995)
15. Nuthall, P.L.: Case studies of the interactions between farm profitability and the use of a farm computer. Computers and Electronics in Agriculture 42, 19–30 (2004)
16. Cox, S.: Measurement and Control in Agriculture. Blackwell Science, Oxford (1997)
17. Cox, S.: Information technology: the global key to precision agriculture and sustainability. Computers and Electronics in Agriculture 36, 93–111 (2002)
18. Mark, D.: New policies for a new rural America. International Regional Science Review 24, 3–15 (2001)
19. Tsouvalis, J., Seymour, S., Watkins, C.: Exploring knowledge cultures: precision farming, yield mapping, and the expert—farmer interface. Environment and Planning A 32, 909–924 (2000)
20. Fountas, S., Wulfsohn, D., Blackmore, B.S., Jacobsen, H.L., Pedersen, S.M.: A model of decision-making and information flows for information-intensive agriculture. Agricultural Systems 87, 192–210 (2006)
21. Wen, W.: A knowledge-based intelligent electronic commerce system for selling agricultural products. Computers and Electronics in Agriculture 57, 33–46 (2007)
22. Newell, R.K.: Emerging technologies for real-time and integrated agriculture decisions. Computers and Electronics in Agriculture 61(1), 1–3 (2008)
23. Van Henten, E.J., Van't Slot, D.A., Hol, C.W.J., Van Willigenburg, L.G.: Optimal manipulator design for a cucumber harvesting robot. Computers and Electronics in Agriculture 65(2), 247–257 (2009)
24. Reid, J.F., Zhang, Q., Noguchi, N., Dickson, M.: Agricultural automatic guidance research in North America. Computers and Electronics in Agriculture 25, 155–167 (2000)
25. Keicher, R., Seufert, H.: Automatic guidance for agricultural vehicles in Europe. Computers and Electronics in Agriculture 25, 169–194 (2000)
26. Gao, L.: Fundamentals of Agricultural Systems. Jiangsu science and Technology Press, Nanjing (1993)
27. Luo, X., Zhang, T., Hong, T.: Technical System and Application of Precision Agriculture. Transactions of the Chinese Society of Agricultural Machinery 32(2), 103–106 (2001)
28. Cao, H., Zhang, C., Jin, Z., et al.: The discussion of the framework and technical system of digital cultivation. Tillage and Cultivation (3), 4–7, 40 (2005)
29. Yu, Y., Qin, X., Zhang, L.: Development status of rural informatization in Beijing, China. International Journal of Agriculture & Biological Engineering 4(4), 59–65 (2011)
30. Yu, Y., Li, J., Qin, X.: The Information Key Technologies for Quality & Safety Monitor and Management of Agricultural Products. Advanced Materials Research 634-638, 4004–4010 (2013)
31. Wang, X., Cao, Y., Ma, J., Huang, F., Yu, A.: New advances in fast detection techniques for quality and security control of farm produce. Modern Scientific Instruments 1, 121–123 (2006)
32. Wu, L., Dong, J.: Simultaneous detection of a variety of veterinary drug residues, 'protein chip for detection of veterinary drug residues platform'. Jiangxi Feed 3, 41 (2005)
33. Feng, G., Wang, X., Xie, F., Yu, A.: The in situ detection of residual pesticide in rice, flour, fruit and vegetable. Modern Scientific Instruments 1, 86–88 (2005)
34. Wang, X., Wu, Z.: Food safety on-site rapid detection equipment comes out. Family & Traditional Chinese Medicine 12, 13 (2005)

35. Chen, P., Liu, L., Wang, J., Shen, T., Lu, A., Zhao, C.: Real-time analysis of soil N and P with near infrared diffuse reflectance spectroscopy. Spectroscopy and Spectral Analysis 28, 295–298 (2008)
36. Wang, J., Zhang, N., Hou, K., Li, H.: Application of LIBS technology to the rapid measure of heavy metal contamination in soils. Progress in Chemistry 20, 1165–1171 (2008)
37. Yu, L., Zhao, H., Ma, X., Liu, Y., Zhang, M., Liao, Y.: Research of LIBS method for detection of heavy metals in polluted soil. Laser Journal 5, 66–67 (2008)
38. Shen, G., Zan, L., Duan, J., Wang, L., Zheng, T.: Implementation of beef quality and safety traceability system via internet. Transactions of the Chinese Society of Agricultural Engineering 23, 170–173 (2007)
39. Yang, X., Qian, J., Sun, C., Liu, X., Han, X.: Implement of farm product archives management system based on traceability system. Chinese Agricultural Science Bulletin 22, 441 (2006)
40. Yu, Y., Qin, X.: Study on Rural Cultural and Entertainment Informatization Status in China. In: 2011 International Conference on Future Information Technology, vol. 13, pp. 402–406 (2011)
41. Malecki, E.J.: Digital Development in Rural Areas: Potentials and Pitfalls. Journal of Rural Studies 19, 201–214 (2003)
42. NTIA. National Telecommunications and Information Administration. Falling Through the Net: Toward Digital Inclusion (2000), Information on,
    http://www.ntia.doc.gov/ntiahome/fttn00/contents00.html
43. NTIA and ESA. National Telecommunications and Information Administration and Economics and Statistics Administration. A Nation Online: How Americans Are Expanding Their Use of the Internet. NTIA, Washington, DC (2002)
44. Euro Voc. Euro Voc: multilingual thesaurus of the European Union (2011),
    http://eurovoc.europa.eu/drupal/
45. NALT. Agriculture thesaurus and glossary, national agriculture library, United States department of agriculture, United States of America (2011),
    http://agclass.nal.usda.gov/agt.shtml
46. GEMET. General multilingual environmental thesaurus, European topic center on catalogue of data sources, European environmental agency (2011),
    http://www.eionet.europa.eu/gemet
47. LCSH 2011, Library of Congress Classification Outline, The library of Congress, United States of America. (2011), http://www.loc.gov/catdir/cpso/lcco/STW
48. STW Thesaurus for economics, Leibniz information center for economics,
    http://zbw.eu/stw/versions/latest/about
49. RAMEAU. Répertoire d'autorité-matière encyclopédique et alphabétique unifié, National Library of France (2011), http://rameau.bnf.fr/
50. ALOD. AGROVOC linked open data. Food and Agriculture Organization of the United Nations (FAO) (2011),
    http://aims.fao.org/website/Linked-Open-Data/sub
51. Lukose, D.: World-Wide Semantic Web of Agriculture Knowledge. Journal of Integrative Agriculture 11(5), 769–774 (2012)
52. Farm computer usage and ownership. Washington: National Agricultural Statistics Service, Agricultural Statistics Board, U.S. Department of Agriculture (2009)
53. Zhao, Y.: The characteristics of agriculture informatization in developed countries. Chinese Rural Economy (7), 74–78 (2002)
54. Brent, A., Gloya, J.T.: Computer and internet adoption on large U. International Food and Agribusiness Management Review (3), 323–338 (2000)

55. Raju, K.A.: A case for harnessing information technology for rural development. The International Information & Library Review 36(3), 233–240 (2004)
56. Nuthall, P.L.: Case studies of the interactions between farm profitability and the use of a farm computer. Computers and Electronics in Agriculture (42), 19–30 (2004)
57. Xia, J.: Linking ICTs to rural development: China's rural information policy. Government Information Quarterly 27(2), 187–195 (2010)
58. Xia, J., Luingjie: Bridging the digital divide for rural communities: The Case of China. Telecommunications Policy 32, 686–696 (2008)
59. Von Elsner, B., Briassoulis, D., Waaijenberd, D., et al.: Review of structural and functional characterist ics of greenhouses in European Union countries, Part I: design requirements. Part II: typical design. Journal of Agricultural Engineer Research 75(1-16), 111–126 (2000)
60. Wang, M.: The development of industrialized agriculture and Engineering Science and technology innovation. Publishing house, Beijing (2000)
61. Xu, F., Zhang, L., Ji, S., Wan, Y.: Mechanical equipment in protected agriculture and its technical development. Journal of Zhejiang University of Technology 29(2), 136–141 (2001)

Research and Design of Peanut Diseases Diagnosis and Prevention Expert System

Kun Zhang, Benjing Zhu, Fengzhen Liu, and Yongshan Wan[*]

College of Agronomy, State Key Laboratory of Crop Biology,
Shandong Agricultural University, Taian 271018, China
{kunzh,yswan}@sdau.edu.cn

Abstract. Based on the analysis of characteristics and rules of the peanut diseases, this article sums up the table for peanut diseases diagnose, presents an expert system for peanut diagnose, information service. The architecture, functions, knowledge base and inference engine are designed. The system will greatly improve the prevention of peanut diseases.

Key words: Peanut diseases, Diagnose, Expert system.

1   Introduction

There are many types of peanut diseases, happen frequently, which seriously affected the yield and quality of peanut. The occurrence of disease is affected by many factors. It is complex, boundary fuzzy, uncertainty, poor controllability (Wan, 2003; Wang, 1999; Sun, 1998). Due to the shortage of basic technical personnel and lack of knowledge, when peanut diseases occurs, the correct diagnosis and proper treatment can not be given at the first time, which bring about a large area of peanut diseases and seriously impact on the yield and quality of peanut. Therefore, the scientific and timely for the diagnosis and treatment of the diseases of peanut would be improved through studying the method of diagnosis and treatment of diseases of peanut and setting up peanut diseases diagnosis and prevention expert system. It's especially good for the diagnosis and prevention of major peanut diseases, which would greatly enhance the level and efficiency of peanut diseases prevention and control.

There are lots of research and application for plant protection expert system abroad, and there are many successful cases. Such as PLANT/ds soybean disease diagnosis expert system developed by the University of Illinois is the earliest development for agricultural expert system in the world. The tea (Camellia Sinensis) pest expert system (TEAPEST) developed by Ghosh (2003) is object-oriented expert system based on rules. It can identify the most diseases of tea plant, and achieved very good results in practical application. Apple disease management decision-making expert system developed by S. Haley can help farmers by providing advice, how to improve the apple

---

[*] Corresponding author.

D. Li and Y. Chen (Eds.): CCTA 2013, Part II, IFIP AICT 420, pp. 212–221, 2014.

production, and can give proper means to diagnosis after identify the species, and this system is suitable for the less experienced personnel.

In recent years, research and application of agricultural expert system in China developed quickly. Many agricultural expert system has been applied in practice and played an important role, and also has made great economic benefits and social benefits. Such as "corn diseases and insect pests prevention expert system" implemented by Jilin University and Jilin Academy of Agricultural Sciences, is one for the farmers based on uncertainty reasoning, multimedia and friendly interface technology (1999). Important vegetable pest diagnosis and consultation system developed by Zheng Yongli (2004) in Zhejiang Province is programmed by the object-oriented relational database Visual FoxPro 8.0, and it adopted the event-driven model such as properties, methods and events programming. While as for peanut, there is no expert system of peanut pest identification, which is systematic and helpful for users. Therefore, we would develop a peanut diseases diagnosis and prevention expert system.

2   The System Development Tools

The software is designed based on the operating system Windows XP, using VB6.0 as the expert system programming tool, Access 2003 as the background database, Dream-weaver as HTML browsing software, Photoshop6.0 to process the images.

3   Structure and Function Design of the System

3.1   Structure of the System

The system is developed using the modular design thought. It's composed of inference engine, database, knowledge base, system maintenance and user interface, etc. The core of the system is the knowledge base and inference engine. System has a modular structure, each module is independent, and the knowledge database and inference engine is separated, so it is easy to maintain and update.

3.2   Function of the System

In the principle of usability, the system realized the function of peanut disease diagnosis, disease information querying, scientific application of pesticides consultation and knowledge base maintenance, etc. Each functional module is described as follows:

3.2.1   Disease Diagnosis

Diagnose the diseases which have occurred in the current field, according to the characteristics of symptoms of disease the user supplied; then all the information of the diseases diagnosis will be given, including symptoms, etiology, pathogenesis, prevention measures and the picture for the characteristics.

3.2.2   Disease Information Querying

The module provides advisory function for the user. It mainly provides disease symptoms, occurrence, control methods and other aspects of the information queried. There are five types of disease included, namely, fungal, bacterial diseases, viral diseases, nematodes disease, and noninfectious diseases. Users can query the information, learnt identify, improve the correct rate of the recognition only a simply click of the mouse.

3.2.3   Advisory Scientific Medication

Mainly to provide service of retrieval, query, consulting and learning about relevant information about commonly agricultural chemicals used in peanut. For example, characteristics of pesticide, the commodity name, English name, Chinese name, and control object toxicity, control method and characteristics of the function and so on.

3.2.4   System Maintenance

This part mainly used for maintenance of system database, knowledge base, picture library editing, updating, deleting, expansion and system program, etc.. Also it can detect the consistency and integrity of knowledge, to avoid redundant or the contradictions of old and new knowledge in the knowledge base, and make certain limits of user authority to ensure the normal operation of the system.

4   Realization of System Functions

4.1   Establishment of Database

The system contains disease database, pesticide database and disease diagnosis database. The structure and examples of the three databases are as follows (table 1., table2., table 3.).

Table 1. Structure and examples of diseases database

Fields	Field type	Field length	Example
Disease name	text	14	leaf spot of peanut
alias	text	30	early leaf spot of peanut
Symptoms	remarks	automatically	omit
Pathogens	remarks	automatically	Groundnut Cercospora, belonging to ascomycotina fungi
Disseminated cycle	remarks	automatically	omit
Pathogenic factors	remarks	automatically	The optimum temperature 25-28 •, the onset heavy in rainy and humid climate,
Prevention method	remarks	automatically	Selection of resistant varieties, spraying pesticides
Disease picture	OLE object		omit

Table 2. Structure and examples of pesticide database

Fields	Field type	Field length	Example
Chinese name	text	10	多菌灵
English name	text	20	carbendazim
Trade name	text	30	MBC
Categorical attribute	text	20	bactericide
Dosage	text	100	25%, 50%carbendazol wettable powder
Control object	text	200	peanut seedling blight, stem rot, root rot, leaf spot, net blotch
Toxicity	remarks	automatically	Mouse acute oral LD50 8000-10000 mg / kg
Matter need attention	remarks	automatically	omit

Table 3. Structure and examples of diagnose of diseases database

Fields	Field type	Field length	Example
Diseased location	text	10	leaf
Symptom 1	text	50	spot round or irregular in shape
Symptom 2	text	50	spot with halo around
Symptom 3	text	50	Spots of dark brown, large, yellow halo, Grey Mildew account raw when wet
Disease name	text	20	cercospora brown spot of peanut

4.2   Realization of Disease Diagnosis

4.2.1   Organization Structure of Knowledge

Built diagnostic tree about disease (occurred in root, stem, leaf and plant) and characteristics of the damage, by collecting the characteristics of peanut diseases, summarizing these data and sorting out the peanut disease retrieval table, finally the making the retrieval table into diagnostic tree for computer program reasoning, as shown in figure 1. Each node in Fig. 1 corresponds to a group of knowledge. The method of diagnosis tree analysis will promote an unknown problem layer by layer, finally get the disease name.

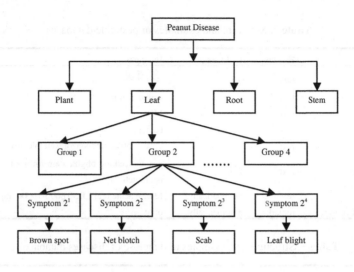

Fig. 1. The knowledge structure of peanut disease diagnose

Note :

Group 2 : Lesion round or irregular in shape.

Symptom 21 : Lesion tea brown or dark brown, with yellow halo, gray powder mildew in the wet spot surface

Symptom 22 : Stellate black spot at the beginning, chlorotic with halo, then reticulate brown spots, with small black protrusions.

Symptom 23 : The lesion sag middle, uplift at the edge, adaxially yellow-brown, caused perforation.

Symptom 24 : A dark brown irregular moire spot appears first in the tip and edge of the leaf, leaf rot off, when humidity is big, with white mycelium and sclerotium.

The first layer: Onset location, there are four parts, the whole plant, leaf, root and stem.

Second layer: Symptom group, which is the common features reorganized and extracted by a part of a disease experts and knowledge engineers.

Third layer: Characteristics of disease symptoms, is the further detailed description. This system is distinct uniformly by the symptoms of 1, 2, 3. If give the conclusion directly by the symptoms of nodes, can get the final point, otherwise continue to establish the node, and so on, until all nodes come to the final node.

4.2.2   Knowledge Representation
The system adopts two levels including the rule frame and the body of the rule to describe the knowledge of disease diagnosis through the way of thinking and the hierarchical structure model that analysis of disease diagnosis. Rule frame reflects the

logic relationship between the premise and conclusion factor, which is a reasoning framework; the rule body reflects the value knowledge between premise factors and conclusion factors, it can contain formulas, may also have a set of rules. Its general form is:

IF Ei(I) then A

The Ei (1, 2, 3,...... , n), as the logic relationship of "and", Ei and A are collectively referred to "fact", "concept" or "factors" in "rule frame + rule body". Take the peanut leaf disease as an example:

RSn:IF lesion mildew layer

    THEN Name of peanut diseases

RB: IF gray mold on lesion

    THEN peanut brown spot

    IF lesions with white mycelium and sclerotium

    THEN peanut leaf blight

    IF lesion dark brown, in concentric rings of spots, with dark brown mildew

    THEN peanut brown spot

    IF lesion brown by yellow, dark brown edge with a yellow halo, rupture, like anxiety, with small black spots

    THEN peanut focal spot

Among them, RS represents the rule frame identifier, n is rule frame number, and is also the rule group number. RB is the rule body identifier, and the downward to the next rule frame identifier are all the rule body content in the rule group. In the system, the rule groups are mutual independence, the group number can be identified arbitrary, as long as not to repeat, that is place of the rule group is arbitrary. System will put similar knowledge in the same group rules, such as the diagnosis of diseases based on leaf symptoms, rule body expresses the solution process to the conclusion set organically and naturally, avoiding redundant and dispersion in writing rules, knowledge base is compressed greatly, and the speed of reasoning is accelerate greatly.

4.2.3 Realization of Inference Engine

The system adopts depth-first method to perform comprehensive reasoning, by analyzing the process of peanut disease diagnosis, summarizing the experts' advice and experience. The search tree for disease diagnosis established, and the search is from the top to down along the hierarchical structure of knowledge. The reasoning process of the system can be seen as a search process for the results along the search tree.

    The basic idea of depth-first search is: start from the initial node S, select a node to inspect among the child nodes, if not the target node, then select another node to inspect, search down so on and so on. When reach some child node, if the node is not the destination node also cannot continue to expand, and then select its sibling nodes to inspect. Take peanut leaf disease as an example: firstly, using reverse reasoning, to determine the position of peanut diseases is in peanut leaf, and then using the depth-first method forward reasoning, confirm the diagnosis group (i.e. initial node) for lesion plaque surface mildew layer, select a node in its child node to inspect, namely specific symptoms contained, if not the target node, then inspect in the next level of symptoms, search down until the diagnosis results are obtained. If the diagnostic results are inconsistent with the actual, this disease is not the target node and also cannot continue to expand, then reselect the symptoms to inspect, repeat the process, until the results are obtained.

4.2.4   Main Program

Programming method to realize the function of disease diagnosis is as follows.

```
Option Explicit
Public db1 As Database
Public str1 As String, str2 As String
Public rs1 As DAO.Recordset
Dim cn As Connection
Dim rs As Recordset
Private Sub Form_Load()
Text1.Text = Date
Text2.Text = "Taian in Shandong Province"
Text12.Text = " Adult onset "
Text3.Text = czqzd.Combo1.Text
Text4.Text = czqzd.Combo2.Text
Text5.Text = czqzd.Combo3.Text
Text6.Text = czqzd.Combo4.Text
Set db1 = OpenDatabase(App.Path & "\database.mdb")
'Match knowledge in rule base according to information the user selected
str1 = "select * from bhzd where zz1 like '" + czqzd.Combo2.Text + "' and zz2
like'" + czqzd.Combo3.Text + "' and zz3 like '" + czqzd.Combo4.Text + "'"
Set rs1 = db1.OpenRecordset(str1, dbOpenDynaset)
If rs1.BOF And rs1.EOF Then
      MsgBox " There's no symptoms according to your diagnosis, please choose
the symptoms correctly!", vbExclamation, "diagnosis Suggestion"
      Else
      Text7.Text = rs1.Fields("bhmc")
      rs1.close
      db1.close
End If
   Set adocon = New Connection
   adocon.CursorLocation = adUseClient
   adocon.open "Provider=Microsoft.Jet.OLEDB.4.0;Data Source=" & App.Path
& "\database.mdb"
   Set rs = New Recordset
   rs.open "binghai", adocon, adOpenDynamic, adLockOptimistic
   rs.MoveFirst
   On Error Resume Next
   Do Until rs.EOF
     If rs.Fields("Disease name ") = Text7.Text Then
       Text8.Text = rs.Fields("alias ")
```

```
            Text9.Text = rs.Fields("Pathogens ")
            Text10.Text = rs.Fields("Disseminated cycle and Pathogenic factors ")
            Text11.Text = rs.Fields("Prevention method ")
         Image1.Picture = LoadPicture(App.Path & rs.Fields("The image path "))
        Exit Do
          Else
            rs.MoveNext
          End If
        Loop
          rs.MoveFirst
        End Sub
```

The operation interface of peanut diseases diagnosis is as follows.

Firstly, select disease symptoms according to the local actual situation (Fig.2). Then we get the diagnosis results after the "start diagnosis" button been clicked (Fig.3).

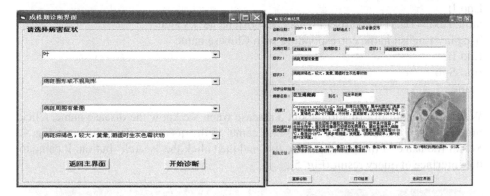

Fig. 2. Interface of the symptom diagnoses   **Fig. 3.** Result of the disease diagnose

4.3   Realization of the Query Function

The system will add the corresponding Chinese name exists in the peanut disease inquiries to the list of user interface box, gain the corresponding diseases of picture and text information, by selecting one of the Chinese name or calling the query function internal system by inputting known Chinese name. Retrieval function by English name or Chinese name is basically the same in design method. Take the disease Chinese name search as example; the programming method of realizing the retrieval functions is as follows:

```
Private Sub cmdfind_Click ( )
' 'Button after the user complete entering in the text box, to trigger the retrieval event,
program starts'
Data1.RecordSource = "select * from" & binghai & "where disease name =" & """ &
txtname.Text & """
```

' 'Find the corresponding Chinese name in the library'
If Data1.Recordset.recordcount <= 0 Then
MsgBox "sorry, there is no information about what you find, make sure you typed it correctly"
Data1.RecordSource = "select*from" & binghai
End If
''There is not the corresponding Chinese name to find in the library, display a message'
If Data1.Recordset.recordcount > 0 Then
''If find the corresponding Chinese name in the library'
Data1.Recordset.movefirst
richtextbox1.loadfile Str & txtname.Text
''Call the corresponding text information'
image1.Picture = LoadPicture(Str & "data\picture\" & Data1.Recordset.Fields ("picture" & m)". Value)
''Call the corresponding picture information'
End If
latin.Text = Data1.Recordset.Fields ("English name").Value
"display English name corresponding with Chinese name '
End If
End Sub
''End'

We can get more information about a disease when we know the disease name. Click button of "query according to the disease name" in the query interface shown in Fig. 4. Select the name of the disease in the list, and then click the "search" button. It came to the interface of query result (Fig. 5).

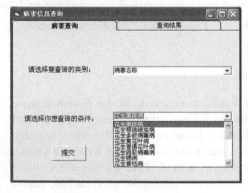

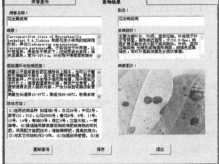

Fig. 4. Disease query interface   **Fig. 5.** Disease query result interface

5   Conclusion

The system adopts the modular design, which only needs to improve and expand the specific modules. It can effectively provide the details of peanut diseases and the comprehensive control technical point of diseases, through collecting the massive peanut disease relevant knowledge, and analysising comprehensively the knowledge base. The system can provide perfect consult service for agricultural technicians and farmers against peanut diseases, change the present situation of blind use of pesticides, and protect the ecological environment. So it can play an important role, and also can alleviate the situation of seriously lack of the advanced agricultural technology personnel. So it has practical significance to promote peanut diseases diagnosis and prevention expert system. The accuracy of the model and fitting in with the actual production are the critical factors to decide the success of the system, therefore it also need continuous improvement to verify the system in future.

Acknowledgment. The project received financial support from the China Agriculture Research System (CARS-14).

References

1. Wan, S.: Peanut cultivation in China. Shanghai Science and Technology Press, Shanghai (2003)
2. Wang, Z., Gai, S.: Shandong Peanut. Shanghai Science and Technology Press, Shanghai (1999)
3. Sun, H., Ding, J., Li, Y.: Expert system (ES) and its application in agriculture. Journal of Shandong Agricultural University 29(2), 270–276 (1998)
4. Ghoshi, S.: Teapest: An expert system for insect pest management in tea. Applied Engineering in Agriculture 19(5), 619–625 (2003)
5. Liu, D., Zhuang, T.: Maize diseases and pests control expert system. Research and Development of Computer 36(1), 36–41 (1999)
6. Zheng, Y., Cheng, J., Zhang, Q.: A preliminary study on the main diseases and insect pests of vegetables consulting system in Zhejiang Province. Zhejiang Journal of Agricultural Sciences 16(4), 186–191 (2004)
7. Shao, G., Li, Z., Wang, W., et al.: Study on VPRDES of vegetable diseases remote diagnosis expert system in Beijing area. Plant Protection 32(1), 51–54 (2006)
8. Ma, Q., Kong, J.: Utility color atlas on economic crops diseases and insect pests. Henan Science and Technology Press, Zhengzhou (1998)
9. Tu, Y.: Pesticide application technology standardization. China Standard Press, Beijing (2001)
10. Niu, Z., Yang, X., Shou, S., et al.: Research and design of cucumber disease diagnosis expert system knowledge organization. System Sciences and Comprehensive Studies 20(1), 33–36 (2004)
11. Lian, S.: Introduction to artificial intelligence technology, pp. 34–42. Xi'an Electronic and Science University Press, Xi'an (2001)
12. Giarratano, J., Riley, G.: Principles and programming on expert systems, pp. 67–75. Machinery Industry Press, Beijing (2000)
13. Black, R., Sweetmore, A., Holt, J.: Plant clinic: a training system for decision-making and resource management in plant disease diagnosis. 4, 52–55 (1995)

Filling Holes in Triangular Meshes of Plant Organs

Zhihui Sun, Xinyu Guo*, Shenglian Lu, Weiliang Wen, and Youjia Chen

Beijing Research Center for Information Technology in Agriculture, Beijing 100097, China
guoxy@nercita.org.cn

Abstract. Data missing was always when scanning plant organs by using 3D laser scanner, and this could lead to holes appearing in the mesh surface generated from the scanned point cloud. Basing on the analyzing to the geometrical features of different types of plant organs, a hole-filling algorithm was proposed which based on the normal vector of the mesh to make the repair area as close as possible to the original model of the organs mesh. The algorithm firstly extracts the hole boundary in the surface mesh, and compute the positions of new points based on the normal vector and normal curvature of the mesh boundary, then connects the new points and boundary points to generate a new triangular mesh. Experiments on different plant organs of geometric shapes were taken and new triangular mesh qualities were evaluated. Results show that the algorithm can fill holes in several kinds of scanned plant organs mesh quickly and efficiently.

Keywords: mesh surface, hole-filling, normal vector estimation, normal curvature, plant organs.

1   Introduction

With the maturity of three-dimensional (3D) laser scanning technology and the popularity of related equipment, the researches of 3D reconstruction from scanned point cloud has became spotlight in recent years. For the measurement of plant morphology, traditional manual method for data collection has shortcomings, such as low speed, low accuracy of the measured data and so on. On the other hand, the 3D laser scanner can measure any object in non-touch approach in the advantages of high scanning speed, real-time, high accuracy *etc.* [1]. As such laser scanner has been used more and more frequently for plant modeling in recent years. The reconstructed plant model based on 3D laser scanned data not only can present real plant morphology, but also provide accurate structural model for the calculation of plant physiological function, and also has important impact on the display and outreach of the new variety. However, due to the complex in the structural system of plant, such as shelter between organs and the complex optical property of organ surface, the quality of measured data by using 3D scanner could be easily interrupted, result in the data missing and formation of holes in the constructed 3D model of plant organs. This will

---

* Corresponding author.

D. Li and Y. Chen (Eds.): CCTA 2013, Part II, IFIP AICT 420, pp. 222–231, 2014.

bring difficulty for further analysis such as calculating area and volume of the plant organ, finite element analysis, textures *etc*. Therefore, the hole-filling work is very important in 3D plant modeling based on the laser scanning technology.

Many works have been done in hole-filling. Davis *et al.* [2] applied the voxel diffusion to fill the holes, and displayed the mesh model using MC (Marching Cube) algorithm. The application of this algorithm could get good results in hole-filling, but its shortcoming is that the original mesh model was changed. Liepa [3] performed triangulation to the hole and adjusted the new vertices to make the new mesh matching the original one. Wei *et al.* [4] presented a method which firstly filled the holes, then refined and optimized the hole area to get new points inside the hole, and finally fitted the weighted points to create the surface. The new mesh could be adjusted by adjusting inside points of the hole. Zhang *et al.* [5] performed triangulation to the hole and refined it and adjusted its geometry by few times of iteration. In short, all above methods were based on the mesh refinement, which firstly mesh the hole, then adjust the new mesh in order to match the whole mesh model. The difference among them is the application of the algorithms to adjust the new mesh.

Zhao *et al.* [6] combined wave front method and Poisson equation to adjust the vertices of the new mesh. Similarly Wang *et al.* [7] combined wave front method and normal of boundary with Laplacian coordinate to fill the holes. Du *et al.* [8-9] applied RBF to fill the holes, without considering the consistency of the morphology between the hole and surrounding regions. Qian *et al.* [10] suggested a mesh-repairing algorithm through recovering the missing sharp features. The edges and corners of the sharp features could be obtained through establishing parabola functions around the sharp features, then filled the holes using the extend Marching Cube algorithm. Levy [11] flattened the whole mesh model to fill the holes. But the algorithm has low efficiency if the hole is much smaller than the whole mesh. Brunton *et al.* [12] flatten the hole boundary into the reference plane to fill the holes, then embed back to the spatial mesh.

The above algorithms can obtain good results while repairing flat areas and small holes. However, due to the complex and diverse morphology of plant organ, the mesh-repairing for plant organ is still a difficult task to address. This paper proposed a filling hole algorithm by modifying the traditional repairing method with considering the curvature of the surface as the constraint, which is more suitable to the surface mesh-repairing of complex plant organ. In the proposed algorithm, the mesh hole boundaries were extracted and the angles between the adjacent boundary edges were calculated. Then the positions of the new points were calculated according to the bisector, the normal vector and the vertex curvature direction. Finally the adjacent vertices were connected to finish the mesh-repairing.

2   Filling Holes Method

Our proposed filling holes method includes four steps. Based on an input triangular mesh model, the method firstly finds holes and extracts the boundary of each hole. Then the normal vector of hole boundary was estimated, and the boundary angle was calculated. These information were used at the last step for generating new points to fill the hole.

2.1   Extracting the Boundary of Mesh Hole

For the triangular mesh model $M = (V, \ F, \ E)$, in which V, F, E indicates the set of vertex, surface and edge, respectively. If M is a closed mesh, the boundary edge of a hole in the mesh should belong to only one triangle. So the boundary edge could be found by traversing the whole mesh in this principle. If M is a non-closed mesh, due to the non-closed mesh has its own boundary, when detecting the hole boundary, one edge of the hole should be interactively selected so as to traverse the whole mesh to find its adjacent edge. The detailed explanation of the algorithm is in reference [13]. This work applied interactive approach to select the boundary triangles. The procedures of the algorithm were as follows:

1) interactively selecting one hole triangle F; finding the edge E which belong to only one triangle (assumed the two vertices of E are v_1 and v_2) as the first hole edge; adding v_1 and v_2 into array V, which includes vertices of boundaries.

2) searching the boundary of the hole along edge E from v_1 or v_2; traversing the whole triangle mesh from v_2 to find the set of edge connected with the hole boundary. Traversing the set of edge to find the edge belong to only one triangle as the adjacent hole edge E'; adding the vertices of E' into V; deleting duplicate vertices.

3) if one vertex in E' is v_1, then stop searching, else, follow step 2 to continue searching the vertex of the hole boundary.

2.2   Estimating the Normal Vector of Hole Boundary

The normal estimation of vertex on the mesh usually use a normal weighted average on 1-ring of the vertex, the simplest weighted way is area weighted. This method is relative accurate for the estimation of the normal vector for the vertex inside the mesh. But for the boundary points without the complete ring structure, the estimated normal vector is not accurate. To obtain the accurate normal vector of the hole boundary, first patching the missing ring mesh, then calculating the weighted area of ring triangular surface normal, as the normal vector of the vertex. We assumed that the vertex ring region has K numbers of triangular facets, the normal vector of the triangular facets inside the ring region could be calculated from the vector product of both sides of the triangle. The normal vector of triangular facets is assumed as ($i=1, 2, ..., k$), the area is assumed as ($i=1, 2, ..., k$), equation 1 shows how to calculate the normal vector of vertex. Figure 1 presents the calculation of the normal vector of the vertex of the missing triangle. The dashed line indicated the patched edge.

$$n_v = \frac{\sum_{i=1}^{k} A_i n_i}{\| \sum_{i=1}^{k} A_i n_i \|} \tag{1}$$

2.3   Calculation of Boundary Angle

To more convenience calculate the boundary angle, firstly, we should orient the repairing mesh and sort the boundary edge clockwise, the detail explanation of the algorithm is in reference [13]. We assumed that each two adjacent edges of the hole polygon corresponding one angle (Fig. 1 mid and right). The a, b, and n are the unit normal vectors of $\overrightarrow{v_i v_{i-1}}$, $\overrightarrow{v_i v_{i+1}}$ and vertex v_i, respectively. The angle θ is determined by the normal vector of the vertex and the vector angle. If the vector product of a and b have the same direction with the normal vector of the vertex, the angle θ is smaller than 180 degree. Else, the angle θ is larger than 180 degree. Figure 1 (mid and right) shows the two situations of the hole boundary angle. Equation 2 shows the calculation of the hole boundary angle.

$$\begin{cases} if((a \times b) \cdot n > 0) \\ \theta = \arccos(a \cdot b) \\ else \\ \theta = 360 - \arccos(a \cdot b) \end{cases} \tag{2}$$

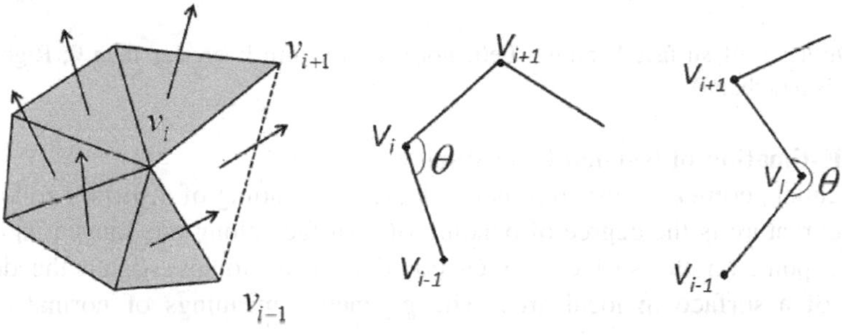

Fig. 1. Boundary normal vector estimation of and boundary angle calculation. Left: Estimation of boundary normal vector. Mid: angle less than 180 degrees. Right: Angle more than 180 degrees.

2.4   Calculation of New Points

2.4.1   Calculation of Initial New Points

We used the principle of the smallest angle to fill the hole. Firstly, the smallest boundary angle was found, then doing the following steps: if $0 < \theta_i \leq 75$, then directly connect $v_{i-1} v_{i+1}$, as shown in figure 2 left, if $75 < \theta_i \leq 135$, then the new points were created from the bisector, as shown in figure 2 mid,   if $135 < \theta_i < 180$, v_{new} and v_{new2} were created from the three times bisector, as shown in figure 2 right.

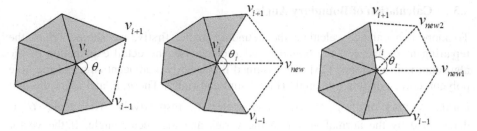

Fig. 2. Calculation rules of initial new point. Left: $0 < \theta_i \le 75$, Mid: $75 < \theta_i \le 135$, Right: $135 < q_i < 180$

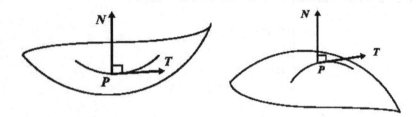

Fig. 3. Direction of surface bending. Left: normal curvature is greater than 0; Right: normal curvature is less than 0.

2.4.2   Estimation of Normal Curvature

In differential geometry, curvature is the degree of bending of a surface in local area. Normal curvature is the degree of bending of a surface along one tangential direction from one point on the surface, which could be used to investigate the degree of bending of a surface in local area. The geometric meanings of normal curvature include:

(1) the absolute of the normal curvature reflects the bending degree of the surface from one point along one direction;

(2) the positive/negative sign of the normal curvature reflects the bending direction of the surface along one tangential direction.

If the normal curvature is larger than zero, the surface bends in the direction of the normal vector of the point, as shown in figure 3 left; if the normal curvature is less than zero, the surface bends in the reverse direction of the outer normal vector of the point, as shown in figure 3 right. Taubin[14] estimated the curvature as:

$$k_p(T) \approx \frac{2\langle N, q - p \rangle}{\| q - p \|^2} \tag{3}$$

Where p is one point on the surface, N is the normal vector, q is another point on the surface around p, $k_p(T)$ is the curvature along the direction T from the point p. T is the projection unit vector of q-p on the tangent plane of point p. In this work, the curvature along the T direction of v_i point is:

$$k_{v_i}(T) \approx \frac{2 \langle N_{v_i}, v_{new} - v_i \rangle}{\| v_{new} - v_i \|^2} \tag{4}$$

2.4.3   Adjusting New Points

The initial new points were created from the bisector of the hole boundary angle. Filling hole by connecting the new points cannot reflect the local feature of the surface. So we need to adjust the new points. We assume that the new point v_{new} was created from the angle bisector vector of $\angle v_{i-1} v_i v_{i+1}$. Figure 4 shows the schematic diagram for adjusting new points.

The point v_{new} was adjusted on the flat π. The direction of adjusting was determined by the sign of $k_{v_i}(T)$. If $k_{v_i}(T) > 0$, the new point will bend in the direction of the normal vector, that is, the surface is convex in the direction of T. The adjusting angle is θ. If $k_{v_i}(T) < 0$, the new point will bend in the reverse direction of the normal vector, that is, the surface is concave in the direction of T. The adjusting angle is also θ. The angle θ is determined by the absolute of the normal curvature and the threshold η, which should set manually.

$$\begin{cases} v' = v + \eta \cdot | k_{v_i}(T) | \cdot n_i \\ \theta = arc\, cos(v, v') \end{cases} \tag{5}$$

The adjusted new points are calculated as:

$$v_{new} = v_i + \frac{1}{2} (\| v_{i-1}, v_i \| + \| v_i, v_{i+1} \|) \cdot \frac{v'}{\| v' \|} \tag{6}$$

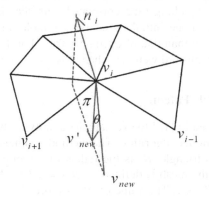

Fig. 4. Adjustment of new point

3   Results and Discussion

3.1   Experimental Results and Analysis

To test the effect of this algorithm for filling holes in triangular mesh, we used this algorithm on several scanned mesh models of plant organs. A handheld 3D laser scanner (Polhemus Ins. FastScan) was used to scan real plant organs. Mesh models were constructed from the obtained 3D point cloud using RBF software attached by the scanner. The fruits of green pepper and cucumber were scanned directly from the plants in greenhouse. Mustard and ginger were removed to the lab for scanning, but without additional treatment like spray white. Figure 5 presents the mesh models with holes, the holes patching model, the enlarge effect of the holes patching (from left to right) of green pepper fruit, cucumber fruit, mustard and ginger (from the top down). Table 1 shows the number of new vertices, the number of new triangles, and the repairing time in holes patching processing for green pepper fruit, maize leaf, cucumber fruit, mustard and ginger.

The experimental results show that this algorithm is fast and effective for patching holes in mesh model. The patching results matched well with the initial density and the morphology of the mesh. For the flat area with little variation of curvature, like the mesh model of cucumber fruit (Fig. 5), a good result obtained by setting a smaller threshold. However, for the area with large variation of curvature, like the mesh model of ginger (Fig. 5), a good result could be obtained by setting different thresholds. The new point is determined by the length of the boundary of the hole, so the effect of patching is related with the initial mesh. If the initial mesh is even distributed, and the length of each edge of the hole boundary equal to the mean length of the edges of the whole mesh, the patching results could match well with the density of the initial mesh. If the initial mesh is uneven distributed, the patching result is not good. From table 1, the patching time correlated with the size of the hole, but not correlated with the size of the initial mesh. For example, the number of the initial mesh of mustard was more than that of cucumber fruit, but the number of new points and the patching time of mustard were less than those of cucumber fruit.

3.2   Evaluation of Hole Patching

We use a method proposed in [15] to evaluate the quality of the filled triangular mesh by using the above mentioned algorithm. The method assumed that R_a is the radius of the inscribed circle of a triangle, R_b is the radius of the circumscribed circle of the triangle. The quality of the mesh is defined as $\mu = 2R_a / R_b$. If the triangle is regular triangle, the value of μ is 1. The value of μ is more closer to 1, the quality of the created triangular mesh is better. Usually, the quality of the triangular mesh is deemed as good when the value of μ larger than 0.5. The figures of the quality of the mesh patching were automatic created by the program of algorithm (Fig. 6), in which the horizontal axis indicates the value of μ, while the ordinate indicates the number of triangles. We retained two decimal digits in the data analysis for better counting the number of triangles. Figure 6 presents the quality of the mesh patching for green pepper fruit, cucumber fruit, ginger and mustard, respectively.

The quality of the patching triangular mesh shown in Fig. 6 is high. The value of μ were mostly larger than 0.6, concentrated in the interval (0.9, 1.0), which indicates the created triangle meshes closer to regular triangle. The mean values of μ was larger than 0.85 for all organ models, which indicates that the proposed algorithm in this work could well patch the mesh holes of different plant organs.

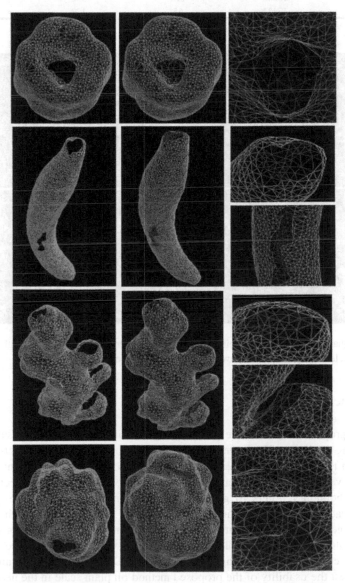

Fig. 5. The holes patching results of plant organs of five variety: the mesh model with holes, the holes patching model, the enlarge effect of the holes patching (from left to right)

Table 1. The number of new meshes and the patching time in holes patching processing

Model name	Original mesh number (Np)	New added vertex number (Np)	New added triangle number (Nm)	Repairing time (s)
Green pepper	8931	15	50	4.234
Leaf of maize	6725	18	75	5.485
Cucumber	11498	68	194	8.922
Ginger	13608	116	356	15.781
Mustard	12434	65	190	8.813

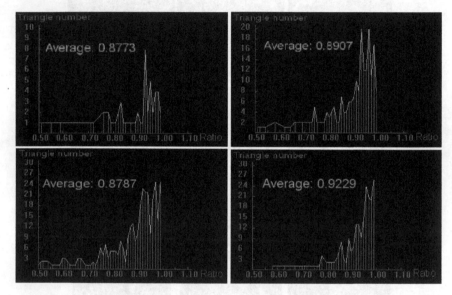

Fig. 6. The quality statistics of patching mesh. Up left: green pepper model; Up right: cucumber model; Down left: ginger model; Down right: mustard model.

4   Conclusions

This work proposed a mesh hole patching algorithm due to the data missing during the scanning in using 3D laser scanner. The algorithm firstly extracts the hole boundary in the triangular mesh, and compute the positions of new points based on the normal vector and normal curvature of mesh boundary, then connects the new points and boundary points to generate a new triangular mesh. The experimental results indicates that the algorithm could effectively patch the mesh with large hole area and large variation of curvature. However, all the evaluations in this work were in organ scale. For the whole plant, the shelter between organs is more serious. It means more complex holes will appear when using 3D laser scanner to scan the plant. We would test the usability of the proposed method on plant scale in the near future.

Acknowledgment. This work is supported by National Natural Science Foundation of China Grant No. 31171454, China National Science and Technology Support Program Grant No. 2012BAD35B01 and Special Fund for Agro-scientific Research in the Public Interest Grant No. 201203026.

References

1. Zhao, C.J., Lu, S.L., Guo, X.Y., et al.: Exploration of digital plant and its technology system. Scientia Agricultura Sinica 43(10), 2023–2030 (2010)
2. Davis, J., Marschne, S.R., Garr, M., et al.: Filling holes in complex surfaces using volumetric diffusion. In: Proceedings of the First International Symposium on 3D Data Processing Visualization, and Transmission, pp. 428–438. IEEE Computer Society Press, Los Alamitos (2002)
3. Liepa, P.: Filling holes in meshes. In: Proceedings of the Eurographics, ACM SIGGRAPH Symposium on Processing, Aachen, Germany, pp. 200–205 (2003)
4. Zhang, J., Yue, W.N., Wang, N., et al.: Anisotropic Hole Filling Algorithm for Triangle Mesh Models. Journal of Computer-Aided Design & Computer Graphics 19(7), 892–897 (2007)
5. Wei, Z.L., Zhong, Y.X., Yuan, C.L., et al.: Research on Smooth Filling Algorithm of Large Holes in Triangular Mesh Model. China Mechanical Engineering 19(8), 949–954 (2008)
6. Zhao, W., Gao, S.M., Lin, H.W.: A robust hole-filling algorithm for triangular mesh. The Visual Computer 23(12), 987–997 (2007)
7. Wang, X.C., Cao, J.J., Liu, X.P., et al.: Advancing Front Method in Triangular Meshes Hole-Filling Application. Journal of Computer-Aided Design & Computer Graphics 23(6), 1048–1054 (2011)
8. Du, J., Zhang, L.Y., Wang, H.T., et al.: Hole Repairing in Triangular Meshes Based on Radial Basis Function. Journal of Computer-Aided Design & Computer Graphics 17(9), 1976–1982 (2005)
9. Wang, H.T., Zhang, L.Y., Li, Z.W., et al.: Repairing Holes in Triangular Meshes Based on Radial Basis Function Neural Network. China Mechanical Engineering 16(23), 2072–2079 (2005)
10. Qian, G.P., Pan, R.F., Tong, R.F.: Feature-preserving Mesh Completion. Journal of Image and Graphics 15(2), 334–339 (2010)
11. Lévy, B.: Dual domain extrapolation. In: Computer Graphics Proceedings, Annual Conference Series, ACM SIGGRAPH, pp. 364–369. ACM Press, New York (2003)
12. Brunton, A., Wuhrer, S., Shu, C., et al.: Filling hole in triangular meshes by curve unfolding. In: Proceedings of IEEE International Conference on Shape Modeling and Applications, pp. 66–72. Institute of Electrical and Electronics Engineers Press, Beijing (2009)
13. Zhang, L.Y., Zhou, R.R., Zhou, L.S.: Research on the Algorithm of Hole Repairing in Mesh Surfaces. Journal of Applied Sciences 20(3), 221–224 (2002)
14. Taubin, G.: Estimating the tensor of curvature of a surface from a polyhedral approximation. In: Proceedings of Fifth International Conference on Computer Vision (ICCV 1995), pp. 902–907 (June 1995)
15. Field, D.A.: Qualitative measures for initial meshes. Numerical Methods in Engineering 47(4), 887–906 (2000)

Semantic-Based Reasoning for Vegetable Supply Chain Knowledge Retrieval

Xinyu Liu, Lifen Hou, and Yonghao Wang

Yantai Automobile Engineering Professional College
YanTai, 265500
xinyu8898858@sina.com

Abstract. Aiming at problems such as backward management, low informationization level etc. in China vegetable supply chain management, ontology theory is introduced. Ontology model of vegetable supply chain knowledge retrieval is constructed firstly, and then the ontology model is formalized by RDF(S) in order to make it can be identified by computer. After confirming inference axiom rules, the inference model of vegetable supply chain knowledge retrieval is created. Finally, the validity of the inference model is tested by the experiment.

Keywords: supply chain, semantic, inference rules.

1 Introduction

China is a large agricultural country, also the largest producer of vegetables. Vegetable industry plays an important role in China agriculture. However, compared with developed countries, China vegetable industry exists some problems such as backward management, low informationization level, inefficient production and circulation process, high circulation cost, product quality safety in question etc., which affect the market competitiveness of china vegetable industry in international market.

Using information technology and Internet technology to manage vegetable supply chain is considered as one of the most important means to improved supply chain management function[1]. Compared with developed countries, agricultural information level in china is still relatively low. In Germany more than 85% farmers have a personal computer and agricultural products are processed using specific software. In the UK the information technology has also plays a very important role. The knowledge will be transferred to each other better in different countries and regions by getting help from information technology.

Although China's agricultural information level is relatively low, and the basic information infrastructure is not perfect, but the information technology and Internet technology in China is growing rapidly. According to China Internet Development Statistics Report in July 2012, the total Internet users in China has reached about 538 million people, the penetration rate of Internet has arrived at 39.9%, and the scale of the netizen is 146 million which increases 14.64 million compared with the end of 2011.

D. Li and Y. Chen (Eds.): CCTA 2013, Part II, IFIP AICT 420, pp. 232–238, 2014.

2   Related Works

Semantic reasoning is achieved on ontology theory. The study on ontology focusing on theory research mostly instead of application research presently. Currently, the application projects on ontology abroad mainly includes Gene Ontology (GO) [2], Business Process Management Ontology (BPMO) [3], Drug Ontology Project for Elsevier (DOPE) [4], Ontology-based Environmental Decision Support System (OntoWEDSS)[5], Agricultural ontology services (AOS) [6].

Presently, the main projects on knowledge acquisition include Mindnet and Advanced Knowledge Technology (AKT). Mindnet is responsible of natural language processing research group of Microsoft, and AKT is charged by U.K. Engineering and Physical Sciences Research Council, which is a knowledge technology research project funded since October 2000[7].

The human knowledge can be understood better on semantic layer by using semantic retrieval and semantic tagged content retrieval. The most direct means to achieved semantic retrieval based on keywords retrieval is to introduce dictionary ontology in specific steps of retrieval process, and the most common used dictionary ontology is WordNet currently. In the work of Moldovan[8] and Buscaldi[9], Boolean operation of current search engines is adopted to expand terms to polysemy or synonym situation. Kruse[10] select a specific meaning of a word in WordNet firstly, and then the specific meaning will be combined with the original retrieval keyword by Boolean operation. Guha[11] uses keywords-based retrieval methods mentioned above not only in text database but also in keywords matching with terms in Resource Description Framework. Rocha etc. [12] proposed an algorithm, which begins the retrieval from a starting point obtained from the original text retrieval. In the retrieval system proposed by Airio etc. that the retrieval is carried on in ontology browser mode[13].

The semantic retrieval model based on system domain ontology includes seven function modules: resource annotation module, labels recommending module, questions processing module, semantic retrieval module, results processing module, user feedback processing module and ontology construction and management module[14]. The ontology-based reasoning can be divided into term-oriented reasoning and instance-oriented reasoning[15].

Ontology formalization and developing tools consist of XML and OWL etc. besides RDF(S). RDF is a data model about objects (or resources) and their relationship, which also provides its basic semantic expression. RDF Schema is a glossary to describing properties and classes of RDF objects (or resources), which also provides its semantic hierarchical structure expression. XML lacks a reasoning system comparing with XML although RDF data model can be described by XML. Comparing with OWL, OWL formalization is built on RDF(S) top layer, which is used for information dealt with application instead of human beings reasoning[16]. This paper focuses on ontology based data model and semantic reasoning, RDF(S) is selected as formalization tool.

3   Semantic-Based Retrieval of Vegetable Supply China Knowledge

3.1   Ontology Model of Vegetable Supply Chain Knowledge Retrieval

Ontology model consists of class and relationship. And the name of class generally consists of a noun concepts or a verb concept. The relationship represents interactions among classes (concepts). There are four relationships between two classes which are: "part_of", "kind_of", "attribute_of" and "instance_of". By which, the ontology model of vegetable supply chain knowledge retrieval system is constructed. The ontology model describes all classes and relationships of the system, which is shown in fig.1.

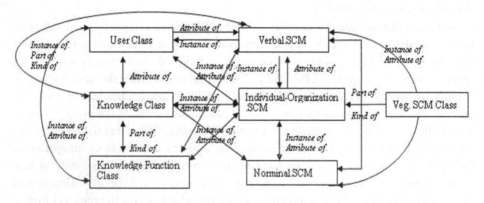

Fig. 1. Ontology model of Vegetable Supply Chain Knowledge Retrieval

3.2   RDF(S) Formalization of System Ontology Model

Ontology of vegetable supply chain knowledge retrieval system is formalized by RDF(S) to carry on concept-acquisition reasoning. The vegetable producer class of ontology model is taken as an instance, which is the first layer subset of vegetable supply chain class. The vegetable producer class is divided into three the second layer subsets, which are no-contact-producer, cooperative-producer and contract-producer. The second layer subset of the first layer subset (noun concepts subset) is defined as people class in the end of ontology formalization, of which the instances take instances of vegetable supply chain producer class.

```
<?xml version="1.0"?>
<rdf:RDF
Xmlns rdf=http://www.w3c.org/1999/0222-rdf-syntax-ns#>
```

```
<rdf:Description rdf:ID="VEG.SC">
<rdf:type rdfresource="http"//www.w3c.org/2000/01/rdf-schema#Class"/>
</rdf:Description>

<rdf:Description rdf:ID="Vegetable Producer_VEG.SC">
<rdf:type rdfresource="http"//www.w3c.org/2000/01/rdf-schema#Class"/>
<rdfs:subclassOf rdf:resource="#VEG.SC"/>
</rdf:Description>

<rdf:Description rdf:ID="No-Contract-Individual_Vegetable Producer_VEG.SC">
<rdf:type rdfresource="http"//www.w3c.org/2000/01/rdf-schema#Class"/>
<rdfs:subclassOf rdf:resource="# Vegetable Producer"/>
</rdf:Description>

<rdf:Description rdf:ID="Cooperative Producer_Vegetable Producer_VEG.SC">
<rdf:type rdfresource="http"//www.w3c.org/2000/01/rdf-schema#Class"/>
<rdfs:subclassOf rdf:resource="# Vegetable Producer"/>
</rdf:Description>

<rdf:Description rdf:ID="Contract Producer_Vegetable Producer_VEG.SC">
<rdf:type rdfresource="http"//www.w3c.org/2000/01/rdf-schema#Class"/>
<rdfs:subclassOf rdf:resource="# Vegetable Producer"/>
</rdf:Description>

<rdf:Discription rdf:ID="Person_Nnominal Concept">
<rdf:type
rdfresource="http"//www.w3c.org/2000/01/rdf-schema#Class"/Property>
<rdfs:domain rdf:resource="# Vegetable Producer_VEG.SC "/>
<rdfs:range rdf:resource="# Person_Nnominal Concept"/>
</rdf:Description>

</rdf:RDF>
```

4   Construction of Reasonging Rules

4.1   Inference Axiom

Creating Inference Axiom

Ontology inference rules should obey the inference axiom rules, two inference axiom rules are adopted as below.

[Equivalence Relationship: (?a Equate ?c), (?b Equate ?c),notEqual(?a, ?b)->(?a Equate ?b)]

[Synonymic Relationship: (?a Near-synonyms ?b),(?a Near-synonyms ?c), notEqual(?b, ?c)->(?b Near-synonyms ?c)]

Creating Inference Model

Based on inference axiom rules, an inference model for retrieval system will be created.

An external file is used to define reasoning rules firstly, which will be introduced to attributes which is thought as retrieval resource.

myresource.addProperty(ReasonerVocabulary.PROPruleMode, "hybrid");

myresource.addProperty(ReasonerVocabulary.PROPruleSet,   "reasoning rules file");

And then an instance of retrieval inference machine is created as below.

Reasoner reasoner = GenericRuleReasonerFactory.theInstance().create(myresource);

Finally, combine the instance of retrieval inference machine and ontology model to create reasoning model.

infModel=ModelFactory.createInfModel(reasoner, data);

4.2   Experiment Results and Analysis

There are 637 records in the knowledge database. Five concepts (agri_product、 fruit 、 inventory、 logistics、 transportation) relevant with vegetable supply chain process are selected. The reasoning rules are carried on the ontology model. The retrieval result is shown in table 1.

Table 1. Experiment Results

	agri_product	fruit	inventory	logistics	transportation	average recall rate
Kyewords-based retrieval	56 records	16 records	18 records	149 records	48 records	0.268
Ontology-based retrieval	91 recordss	91 records	286 records	350 records	316 records	0.987

In general, the Precision Ratio and Recall Ratio are two most basic evaluated targets for retrieval. The goal of information retrieval pursues not only a higher Precision Ratio but also a higher Ratio. Date of knowledge base just for vegetable supply chain domain have been analyzed and processed in their collection, therefore results for keywords-based retrieval and ontology-based retrieval all have high Precision Ratio. Recall Ratio for two different retrieval ways will be emphasized. It is obviously that average recall rate of ontology-based retrieval is 0.987 much higher than 0.268, which is average recall rate of keywords-based retrieval.

Because the ontology model and semantic expansion are achieved offline, the main retrieval time is to read ontology model into memory and retrieval reasoning. The retrieval can arrive at real-time response for the current amount of date in knowledge base.

5   Conclusions

The ontology model of vegetable supply chain knowledge retrieval is constructed and formalized by RDF(S) in this paper. After confirming inference axiom rules, knowledge retrieval inference model is put forward. Finally, its validity is proved by the experiment. This work will enhance knowledge transferring in different countries and regions, and improve the international competitiveness of China vegetable supply chain.

Aknowledgement. This work was supported by the Natural Science Foundation project of Shandong Province (ZR2012FM008), National Natural Science Foundation project of China (61100115,61170161), National Science and Technology Support Program project (2011BAD21B01, 2011BAD21B06) and Higher Educational Science and Technology Program Project of Shandong Province (J13LN68).

References

[1] Lancioni, R.A., Smith, M.F., Oliva, T.A.: The role of the Internet in supply chain management. Industrial Marketing Management 29, 45–56 (2000)
[2] http://www.geneontology.org
[3] http://www.bpiresearch.com/Resources/RE_OSSOnt/re_ossont.htm
[4] Broekstra, J., Fluit, C., et al.: The Drug ontology Project for Elsevier. In: Proceedings of the WWW 2004 workshop on Application Design, Development and Implementation Issues in the Semantic Web, New York (May 2004)
[5] Ceccaroni, L., Cortés, U., Sànchez-Marrè, M.: OntoWEDSS: augmenting environmental decision-support systems with ontologies. Environmental Modelling and Software 19(9), 785–797 (2004)
[6] http://www.fao.org/agris/aos
[7] Advanced Knowledge Technologies, http://www.aktors.org
[8] Moldovan, D.I., Mihalcea, R.: Using wordnet and lexical operators to improve internetsearches. IEEE Internet Computing 4, 34–43 (2000)
[9] Buscaldi, D., Rosso, P., Arnal, E.S.: A wordnet-based query expansion method for geographical information retrieval. In: Working Notes for the CLEF Workshop (2005)
[10] Kruse, P.M., Naujoks, A., Roesner, D., Kunze, M.: Clever search: A wordnet based wrapper for internet search engines. In: Proceedings of the 2nd GermaNet Workshop (2005)
[11] Guha, R., McCool, R., Miller, E.: Semantic search. In: WWW 2003: Proceedings of the 12th International Conference on World Wide Web, pp. 700–709. ACM Press (2003)

[12] Rocha, C., Schwabe, D., de Aragão, M.P.: A hybrid approach for searching in the semantic web. In: Proceedings of the 13th International Conference on World Wide Web, pp. 374–383 (2004)

[13] Airio, E., Järvelin, K., Saatsi, P., Kekäläinen, J., Suomela, S.: Ciri - an ontology-based query interface for text retrieval. In: Hyvönen, E., Kauppinen, T., Salminen, M., Viljanen, K., Ala-Siuru, P. (eds.) Web Intelligence: Proceedings of the 11th Finnish Artificial Intelligence Conference (2004)

[14] Wang, H., Huo, J., et al.: Ontology model structure design for semantic-based retrieval. System Engineering and Electronic Technology 32(1), 166–174 (2010)

[15] Dou, Y., He, J., Liu, D.: Study on ontology-based semantic retrieval model for people tagged system. Journal of the China Society for Scientific and Technical Information 31(4), 381–389 (2012)

[16] http://www.docin.com/p-224295860.html

Spectral Characteristics of Tobacco Cultivars with Different Nitrogen Efficiency and Its Relationship with Nitrogen Use

Taibo Liang, Jianwei Wang, Yanling Zhang, Jiaqin Xi, Hanping Zhou,
Baolin Wang, and Qisheng Yin*

Zhengzhou Tobacco Research Institute of CNTC, Zhengzhou 450001, China
taibol@163.com

Abstract. To investigate the relationship between spectral characteristics of tobacco cultivars and their nitrogen use characters, four tobacco cultivars with different nitrogen use efficiency were used in a ^{15}N pot experiment. The result showed that, in the visible light range, the spectral reflectance was lower in higher nitrogen level (N2) than N1, while opposite in near infrared range. The spectral reflectance of K326 and HD were lower than ZY100 and NC89 in visible light range, which closely related to their higher chlorophyll content. Both the nitrogen utilization rate of basal fertilizer and topdressing fertilizer in HD and K326 were higher than that in ZY100 and NC89 under N1 and N2 levels. The basal fertilizer use efficiency was negatively correlated with ρ550, and significantly positively correlated with RVI (800, 550), DVI (800, 550) and NDVI (800, 550) both in N1 and N2 level. The top dressed fertilizer use efficiency was significantly positively correlated with RVI (800, 550), DVI (800, 550) and NDVI (800, 550) in N1 level. Therefore, spectral characteristics can be an important method for diagnosing tobacco nitrogen metabolism characteristics.

Keywords: tobacco (*Nicotiana tabacum L.*), nitrogen use efficiency, spectral characteristics, nitrogen utilization.

1   Introduction

Tobacco is an important economic crop in China, but also an important model plant in science research. In the tobacco production, in order to get higher yield and income, excessive application of nitrogen fertilizer occurred from time to time, which caused potential threat to field ecosystem. Therefore, how to improve the tobacco nitrogen use efficiency has become a hot topic in tobacco nutrition study. Some studies showed that, crop nitrogen use efficiency closely related to cultivars, nitrogen levels and soil conditions and so on [1-3]. In certain ecological conditions, tobacco nitrogen use efficiency have significant differences among different genotypes, which closely related to plant nitrogen metabolism and nitrogen uptake and use efficiency [4-6].

In recent years, the use of remote sensing technique for real-time monitoring and plant nutrition diagnosis has become a hot topic in the application of remote sensing in

D. Li and Y. Chen (Eds.): CCTA 2013, Part II, IFIP AICT 420, pp. 239–246, 2014.
© IFIP International Federation for Information Processing 2014

agriculture. Hyperspectral technology, has some advantages such as wide spectrum range, high resolution, and large amount of data [7]. Through spectral changes monitoring, the relationship between spectra reflectance of crops and the leaf area index, aboveground biomass, chlorophyll content can be studied, which can provide basis for crop growth monitoring and yield estimation. [8-10]. Study showed that, crop nutrition condition closely related to spectral characteristics, and the crop spectra reflectance had significant differences under different nitrogen levels [11-12]. The relationship between leaf nitrogen accumulation and ratio of near infrared and green band R_{810}/ R_{560} was established, to monitor nitrogen nutrition of rice [13]. Tobacco nitrogen use efficiency have significant difference among different cultivars, however, the relationship between spectral characteristics and nitrogen utilization of different varieties was rarely reported. In this study, pot experiment was carried out to study the spectral characteristics of different nitrogen efficiency tobacco cultivars under different nitrogen levels and their relationship with nitrogen utilization, so as to provide theoretical basis for breeding and nutrient management.

2   Materials and Methods

2.1   Experimental Design

A ^{15}N tracer pot experiment was performed in the greenhouse of Zhengzhou Tobacco Research Institute of China National Tobacco Corporation, Zhenzhou, China, from April to August (growing season) in 2012. Four cultivars, K326, Zhongyan 100 (ZY100), Hongda (HD) and NC89 were employed in the experiment under two nitrogen levels, N1 (1.0 g pot^{-1}) and N2 (3.0 g pot^{-1}), respectively.

Isotope fertilizer were ammonium nitrate ^{15}N double labeled (abundance of 10%) and ^{15}N potassium nitrate (abundance of 10%), which provided by Shanghai Research Institute of Chemical Industry. In ten pots, ^{15}N ammonium nitrate fertilizer were applied before transplanting, and common potassium nitrate fertilizer were top dressed. In other ten pots, the order of fertilizer was opposite. Seventy percent N, all P$_2$O$_5$ and part of the K$_2$O fertilizers were applied before transplanting. Other parts of the N and K$_2$O was top dressed as potassium nitrate at the resettling growth stage. Each treatment had 20 pots, wherein each pot contained 15 kg soil.

2.2   Measure Items and Methods

At the vigorous growth stage (60 days after transplanting) and maturity stage (85 days after transplanting), the spectral characteristics was determined with ASD FieldSpec Hand-Held (AnalyticalSpectral Device, USA). Meanwhile, the tobacco plants were sampled for determination of dry weight and nitrogen content.

The chlorophyll content was measured by the method as described by Porra et al [14]. The ^{15}N samples were analyzed by ZHT$_2$O$_2$ mass spectrometer in Academy of Agriculture and Forestry of Hebei Province.

2.3   Data Analysis

Using Microsoft Excel 2010 and DPS (Data Processing System) for data processing and statistical analysis. RVI(λ1, λ2)=$\rho\lambda$1/$\rho\lambda$2; DVI(λ1, λ2)= | $\rho\lambda$1-$\rho\lambda$2 | ; NDVI(λ1, λ2)= | $\rho\lambda$1-$\rho\lambda$2 | /($\rho\lambda$1+$\rho\lambda$2); among them, ρ was reflectance, λ was wave length.

3   Results and Analysis

3.1   Spectral Characteristics of Tobacco Cultivars with Different Nitrogen Efficiency

Figure 1 showed the spectral reflectance of different tobacco cultivars in two nitrogen levels. In the visible light range, the chlorophyll absorption of visible light formed obvious "green peak" at 500-600nm. From 670-760nm, the spectral reflectance increased rapidly with the increasing wavelength, and formed high reflection platform in the near infrared range 780-1050 nm. Tobacco leaf spectral reflectance showed significant differences under different nitrogen levels. In the visible light range, the spectral reflectance was lower in higher nitrogen level (N2) than N1, while opposite in near infrared range. Compared among different cultivars, the spectral reflectance of K326 and HD were lower than ZY100 and NC89 in visible light range, which closely related to their higher chlorophyll content. In the near infrared range, the spectral reflectance of four cultivars showed the different order in N1 and N2, which showed the difference of response to nitrogen level among different cultivars.

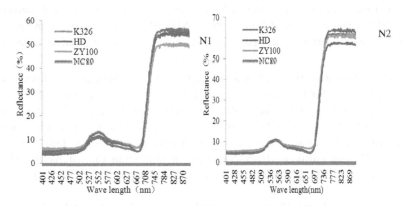

Fig. 1. Changes of spectral characteristics of four tobacco cultivars with different nitrogen efficiency

3.2   Nitrogen Use Efficiency of Different Tobacco Cultivars

[15]N tracer test results (Table1) showed that, 15.54%~21.72% nitrogen accumulation amount derived from fertilizer, and 78.28%~84.46% from soil at N1 level; while at N2 level, they were 45.04%~48.29% and 51.71%~54.96%, respectively. Therefore, tobacco plant absorbed more nitrogen from soil under lower nitrogen level, while more

nitrogen from fertilizer under higher nitrogen level. Compared among different cultivars, the proportion of nitrogen accumulation from fertilizer of ZY100 was higher than other cultivars under both N1 and N2 levels.

According to nitrogen accumulation amount, all the nitrogen from fertilizer, soil and total amount were higher in HD and K326 than that in ZY100 and NC89 under both N1 and N2 levels.

Table 1. Nitrogen use efficiency of different tobacco cultivars

Nitrogen level	Cultivar	TNUA (mg·plant^{-1})	NDFF						NDFS	
			NDFBF		NDFTF		Total			
			NUA (mg·plant^{-1})	NUR %	NUA (mg·plant^{-1})	NUR %	NUA (mg·plant^{-1})	NUR %	NUA (mg·plant^{-1})	NUR %
N1	HD	2808.96	307.31	10.94	199.02	7.09	506.33	18.03	2302.63	81.97
	K326	2409.47	302.21	12.54	162.21	6.73	464.42	19.27	1945.06	80.73
	ZY100	2046.39	278.28	13.60	166.15	8.12	444.43	21.72	1601.96	78.28
	NC89	2314.79	212.10	9.16	147.60	6.38	359.71	15.54	1955.09	84.46
N2	HD	2917.23	839.79	28.79	513.06	17.59	1352.84	46.37	1564.38	53.63
	K326	2784.23	822.12	29.53	431.77	15.51	1253.89	45.04	1530.34	54.96
	ZY100	2419.85	740.06	30.58	428.61	17.71	1168.67	48.29	1251.19	51.71
	NC89	2279.63	655.36	28.75	418.94	18.38	1074.30	47.13	1205.32	52.87

TNUA: total nitrogen uptake amount; NUA: nitrogen uptake amount; NUR: nitrogen uptake ratio; NDFF: nitrogen derived from fertilizer; NDFBF: nitrogen derived from basal fertilizer; NDFTF: nitrogen derived from topdressing fertilizer; NDFS: nitrogen derived from soil.

The nitrogen use efficiency was calculated by ^{15}N technique in different treatments (Table 2). Under N1 level, the utilization rate of basal fertilizer were 30.30%~43.90%, and topdressing fertilizer were 49.20%~66.30%; while under N2 level, that were 31.21%~39.99% and 46.55%~57.01%, respectively. The nitrogen utilization rate of topdressing fertilizer was significantly higher than that of basal fertilizer. Compared among different cultivars, both the nitrogen utilization rate of basal fertilizer and topdressing fertilizer aligned as HD and K326> ZY100 and NC89 under N1 and N2 levels.

Also it showed that, the nitrogen residual rate of topdressing fertilizer were lower than that in basal fertilizer. Compared among different cultivars, both the soil residual rate and loss rate were lower in HD, which may be related to the well root growth and higher nitrogen use efficiency in this cultivar.

Table 2. Nitrogen use efficiency of different tobacco cultivars

Nitrogen level	Cultivar	Fertilization time	Uptake (%)				Residual ratio (%)	Loss ratio (%)
			Root	Stem	Leaf	Total		
N1	HD	BF	3.96	5.36	34.58	43.90	45.13	10.97
		TF	6.00	9.87	50.47	66.30	20.76	12.94
	K326	BF	5.85	4.87	32.45	43.17	46.22	10.61
		TF	10.37	10.26	33.44	54.07	22.35	23.58
	ZY100	BF	3.50	6.61	29.65	39.75	35.37	24.88
		TF	5.05	7.65	42.68	55.38	20.74	23.87
	NC89	BF	4.06	3.56	22.68	30.30	52.99	16.71
		TF	7.57	6.18	35.46	49.20	13.83	36.97
N2	HD	BF	3.03	6.14	30.82	39.99	33.05	26.96
		TF	5.79	10.03	41.18	57.01	14.20	28.79
	K326	BF	3.81	5.05	30.29	39.15	43.11	17.74
		TF	6.82	8.38	32.78	47.96	18.55	33.49
	ZY100	BF	3.10	4.93	27.20	35.24	40.03	24.73
		TF	6.09	8.02	33.51	47.62	17.56	34.82
	NC89	BF	4.97	4.20	22.04	31.21	39.53	29.26
		TF	8.71	7.88	29.96	46.55	14.28	39.17

3.3   The Correlation Analysis between Vegetation Index and Nitrogen Metabolism Index

It showed that (Table 3), the chlorophyll content was significantly negatively correlated with ρ550, and significantly positively correlated with RVI(800,550), DVI(800,550) and NDVI(800,550) in N1 level. The total nitrogen uptake amount was significantly negatively correlated with ρ550, and significantly positively correlated with RVI (800,550), DVI (800,550) and NDVI (800,550) in N2 level. The BFUE was negatively correlated with ρ550, and significantly positively correlated with RVI (800,550), DVI (800,550) and NDVI (800,550) both in N1 and N2 level. The TFUE was significantly positively correlated with RVI (800,550), DVI (800,550) and NDVI (800,550) in N1 level.

Table 3. The correlation index between some vegetation index and Chlorophyll, TNUA, BFUE, and TFUE

Nitrogen level	Index	ρ550	ρ800	RVI(800,550)	DVI(800,550)	NDVI(800,550)
N1	Chlorophyll content	-0.98**	0.81	0.97*	0.89*	0.97**
	TNUA	-0.85	0.4	0.77	0.53	0.73
	BFUE	-0.84	0.96**	0.90*	0.97**	0.93*
	TFUE	-0.84	0.84	0.90*	0.87*	0.88*
N2	Chlorophyll content	-0.44	0.65	0.58	0.63	0.55
	TNUA	-0.87*	0.92*	0.92*	0.92*	0.91*
	BFUE	-0.96**	0.97**	0.98**	0.97**	0.97**
	TFUE	0.59	0.78	0.72	0.77	0.7

TNUA: total nitrogen uptake amount; BFUE: basal fertilizer use efficiency; TFUE: topdressing fertilizer use efficiency.

4 Conclusion and Discussion

Tobacco leaf spectral reflectance showed significant differences under different nitrogen levels, which showed the difference of response to nitrogen level among different cultivars. Studies showed that the spectral reflectance closely related to chlorophyll content [15-16]. In the visible light range, the tobacco cultivar with higher nitrogen use efficiency showed lower spectral reflectance, which closely related to their higher chlorophyll content.

The canopy spectral characteristics under different nitrogen conditions were analyzed in previous studies [7,11]. Tang *et al* [17] had illustrated that the spectral difference were clear for the canopy and leaves of rice under different nitrogen levels. In this study, the spectral reflectance in the visible light range was lower in higher nitrogen level, which consistent with previous studies. Some studies showed that crop nitrogen nutrition condition can determined using canopy spectral, however, the relationship between canopy spectral characteristics and plant nitrogen use was rarely reported. In this study, the basal fertilizer use efficiency was negatively correlated with ρ550, and significantly positively correlated with RVI (800, 550), DVI (800, 550) and NDVI (800, 550) both in N1 and N2 level. The topdressing fertilizer use efficiency was significantly positively correlated with RVI (800, 550), DVI (800, 550) and NDVI (800, 550) in N1 level. Therefore, spectral characteristics can be an important method for diagnosing tobacco nitrogen metabolism characteristics.

Acknowledgement. This study was supported by the National Natural Science Foundation of China (31101120) and the key Science and Technology Program of Henan Province, China (102101110600).

References

1. Serret, M.D., Ortiz-Monasterio, I., Pardo, A., et al.: The Effects of Urea Fertilisation and Genotype on Yield, Nitrogen Use Efficiency, 15^N and 13^C in Wheat. Annals of Applied Biology 153(2), 243–257 (2008)
2. Gallais, A., Hirel, B.: An Approach to the Genetics of Nitrogen Use Efficiency in Maize. Journal of Experimental Botany 55(396), 295–306 (2004)
3. Gill, S., Ahid, M., Azam, F.: Root Induced Changes in Potential Nitrification and Nitrate Reductase Activity of the Rhizospheric Soil of Wheat (triticum aestivum L) and Chichpea (cicer arietinum L. Pakistan Journal of Botany 38(4), 991–997 (2006)
4. Juan, M.R., Rosa, M.R., Luis, M.C., et al.: Grafting to Improve Nitrogen-use Efficiency Traits in Tobacco Plants. Journal of the Science of Food and Agriculture 86(6), 1014–1021 (2006)
5. Kruse, J., Kopriva, S., Hansch, R., et al.: Interaction of Sulfur and Nitrogen Nutrition in Tobacco (Nicotiana tabacum) Plants: Significance of Nitrogen Source and Root Nitrate Reductase. Plant Biology 9(5), 638–646 (2007)
6. Tang, W.J., He, F.F., Zhou, J.H., He, W., Yang, Z.Y., Yang, H.Q., Xiao, Z.X.: Transformation Law of Different Formal Nitrogen Fertilizer in Paddy Field and its Effect on Growth and Nicotine Content of Flue-cured Tobacco. Crop Research 23(1), 30–34 (2009) (in Chinese)
7. Wang, L., Bai, Y.L.: Correlation Between Corn Leaf Spectral Reflectance and Leaf Total Nitrogen and Chlorophyll Content under Different Nitrogen Level. Scientia Agricultura Sinica 38(11), 2268–2276 (2005)
8. Feng, W., Zhu, Y., Tian, Y.C., et al.: Monitoring Canopy Leaf Pigment Density in Wheat with Hyperspectral Remote Sensing. Acta Ecologica Sinica 28(10), 4902–4911 (2008) (in Chinese)
9. Huang, C.Y., Wang, D.W., Yan, J., Zhang, Y.X., Cao, L.P., Cheng, C.: Monitoring of Cotton Canopy Chlorophyll Density and Leaf Nitrogen Accumulation Status by Using Hyperspectral Data. Acta Agronomica Sinica 33(6), 931–936 (2007) (in Chinese)
10. Royo, C., Aparicion, N., Villegas, D., Casadesus, J., Monneveux, P., Araus, J.L.: Usefulness of Spectral Reflectance Indices as Durum Wheat Yield Predictors Under Contrasting Mediterranean Conditions. International Journal of Remote Sensing 24(22), 4403–4419 (2003)
11. Dai, H., Hu, C.S., Cheng, Y.S., Song, W.C.: Correlation Between Agronomic Parameters and Spectral Vegetation Index in Winter Wheat Under Different Nitrogen Levels. Agricultural Research in the Arid Areas 23(4), 16–22 (2005) (in Chinese)
12. Sun, H., Li, M.Z., Zhang, Y.E., Zhao, Y., Wang, H.H.: Spectral Characteristics of Corn Under Different Nitrogen Treatments. Spectroscopy and Spectral Analysis 30(3), 715–719 (2010) (in Chinese)
13. Xue, L.H., Cao, W.X., Luo, W.H., Jiang, D., Meng, Y.L., Zhu, Y.: Diagnosis of Nitrogen Statues in Rice Leaves with the Canopy Spectral Reflectance. Sci Agric Sinica 36(7), 807–812 (2003) (in Chinese)

14. Porra, R.J., Thompson, W.A., Kriedemann, P.E.: Determination of Accurate Extinction Coefficients and Simultaneous Equations for Assaying Chlorophylls a and b Extracted with Four Different Solvents: Verification of the Concentration of Chlorophyll Standards by Atomic Absorption Spectroscopy. Biochimica. Et. Biophysica. Acta. 975, 384–394 (1989)
15. Li, X.Y., Liu, G.X., Yang, Y.F., Zhao, C.H., Yu, Q.W., Song, S.X.: Relationship Between Hyperspectra Parameters and Physiological and Biochemical Indexes of Flue-cured Tobacco Leaves. Scientia Agricultura Sinica 40(5), 987–994 (2007)
16. Meng, Z.Q., Hu, C.S., Cheng, Y.S.: Study on Correlation Between Chlorophyll Density of Winter Wheat and Hyperspectral Data. Agricultural Research in the Arid Areas 25(6), 74–79 (2007) (in Chinese)
17. Tang, Y.L., Wang, R.C., Huang, J.F., Kong, W.S., Cheng, Q.: Hyperspectral Data and Their Relationships Correlative to the Pigment Contents for Rice Under Different Nitrogen Support Level. Journal of Remote Sensing 8(2), 185–192 (2004) (in Chinese)

Research on Agricultural Products Cold-Chain Logistics of Mobile Services Application

Congcong Chen[2], Tian'en Chen[1], Chi Zhang[1], and Guozhen Xie[1]

[1] National Engineering Research Center for Information Technology in Agriculture,
Beijing 100097, China
[2] Southwest University, College of Computer and Information Science,
Chongqing 400715, China
Chencc198702@163.com, {chente,zhangc}@nercita.org.cn,
xieguozhen1990@hotmail.com

Abstract. Real-time monitoring of agricultural products cold-chain logistics and transport can effectively ensure the quality and safety of agricultural products, reducing logistics cost. This paper analyzes three functional architectures, including the agricultural application of mobile terminal data acquisition, logistics warning and mobile payment. The cold-chain logistics of mobile service application system process and module structure are designed on the basis of the hardware environment of mobile device and wireless network environment. The system can monitor and manage the process of storage and transportation of agricultural products cold-chain through the mobile terminal, improving the efficiency of logistics.

Keywords: cold-chain logistics, agricultural products, mobile terminal.

1 Introduction

With the rapid development of agricultural economy in our country, the demand for agricultural products and the volume is also growing rapidly. Agricultural products themselves have some characteristics, such as saving cycle short, perishable, low temperature storage and so on, therefore, they need to be effective oversight and management in the production, storage, and transportation[1]. Each link in any tiny mistake could cause huge losses, and even lead to food safety hidden trouble. Improvement of information level makes the cold-chain logistics have a comprehensive development, the use of relevant technology such as RFID and WSN make traditional cold-chain logistics monitoring mode cannot adapt to the agricultural information level, therefore, cold-chain logistics monitoring need to meet the current and the development of information technology and user needs[2].

In this paper, by introducing the agricultural cold chain logistics present situation, the three major functional modules and practical case analysis, can effectively improve the current application limitation of cold-chain logistics and logistics efficiency, it can improve the current deficiencies and defects existing in the cold chain logistics of agricultural products, has a certain application value.

D. Li and Y. Chen (Eds.): CCTA 2013, Part II, IFIP AICT 420, pp. 247–254, 2014.

2   Agricultural Products Cold-Chain Logistics Present Situation

Agricultural products cold-chain logistics is refers to the perishable agricultural products production, processing, transportation, distribution of behavior such as a series of logistics activities, the whole process of agricultural products needs to be in a state of low temperature[3]. As the attention of the customers to the agricultural product quality and food safety, and security of preservation of agricultural products gradually becomes a necessary. The main flow of Cold-chain logistics chart as shown in the Fig.1.

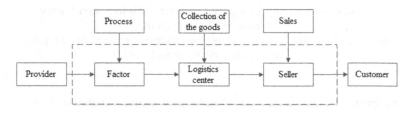

Fig. 1. The Main Flow of Cold-Chain Logistics Chart

In foreign countries, cold chain logistics develop more mature. Mostly adopted in the process of cold collection transportation automatic temperature control device, can real-time monitor the temperature of the cold box changes and ensure that transport goods qualitative change will not occur. Cold chain facilities and cold-chain equipment relatively backward in our country, the original old equipment, development and distribution is not balanced, is unable to provide a low temperature for perishable food circulation system, as for the lagged far behind in the cold chain technology application abroad.

In terms of logistics information system, service network and logistics information system of cold chain logistics in our country plays an important role in the development, service network and imperfect, incomplete information system, the influence on the quality of agricultural products logistics, accuracy and timeliness, at the same time, the cost of agricultural products cold chain and the degree of loss of goods is also high.

Through the research on China's cold-chain logistics development present situation, mainly exist the following problems:

➢   Cold-chain logistics information utility ratio is low.
➢   Data is not timely and effective feedback.
➢   Transportation of agricultural products quality safety exist in the process of great hidden trouble.
➢   Between each node of supply chain information flow is not smooth

Can be seen from the above problem, the current agricultural products cold-chain logistics information utilization rate is very low, as a direct result of the entire 00agricultural product logistics process and the information chain of security monitoring is blocked[4]. Meanwhile, Agricultural product logistics in our country has a long time, low efficiency, high cost, poor security monitoring, it is difficult to meet the needs of the market.

Therefore, this paper in view of the present agricultural products logistics existing problem, combining with the consumer in the process of transportation of agricultural products such as real-time monitoring and mobile service requirements, to build a mobile agricultural products cold-chain logistics service system.

3   Cold-Chain Logistics of Mobile Service System Overall Design

Mobile service on patterns of supply of agricultural products, logistics mainly includes three phases: the phase is perishable agricultural products from the production processing vendor to the distribution center of agricultural products, The second stage is the transportation of agricultural products, The third stage is agricultural products delivered to the user, by user sign for it.

In this paper, based on Internet of things and web service technology, the production and processing of agricultural products as a node, with the help of third party logistics complete cold-chain logistics mobile service system. The system structure as shown in the Fig.2.

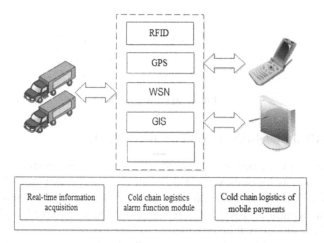

Fig. 2. The System Structure Chart

Cold-chain logistics of mobile services for consumers and the driver provides a convenient and quick service[5]. Complete database system based on agricultural products. Build the real-time data acquisition, cold-chain logistics alarm function module and mobile payment three parts content.

3.1   Real-Time Information Acquisition of Agricultural Products

Agricultural products real-time information acquisition through the electronic label record raw agricultural products production and processing all information, provide information of origin traceability data base. Using RFID technology, store the key to influence the quality and safety of agricultural products processing information. Information of housing environment was collected through wireless sensor. Through the above information can provide mobile services data base.

Consumers through the electronic label can be directly traced back to the information in the process of production and processing of agricultural products, ensure food safety. System data flow diagram is shown in Fig.3.

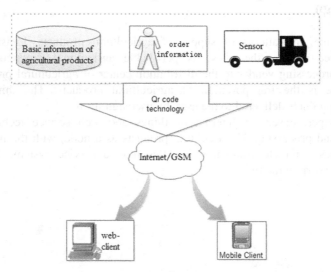

Fig. 3. System data flow

3.2   Cold-Chain Logistics Alarm Function Module

Wireless sensors in the box body can accurately obtain the current situation of the agricultural products logistics vehicle driving on the way, such as body temperature, humidity etc.. Combining different agricultural products quality safety evaluation model, When monitoring data in the box body reaches or exceeds safety threshold, intelligent alarm service will be provided to the truck driver.

The system is based on Web Services Technology, and establishes a shared information system architecture based on Internet. This technology allows users to call web service of the complex agricultural information platform without restraint under the environment of TCP/IP, in order to realize the information share in Internet. Warning module flow chart is as follows Fig.4.

According to the shipping order number, the function obtains the real-time temperature monitoring of the current WSN, and gets the threshold temperature of the agricultural products from Web Service. Every 5 seconds the temperature is compared with the temperature threshold. if it is more than the threshold temperature, rings or SMS alerts are sent to transport driver through the mobile terminal, to timely adjust the environment, and ensure the quality and freshness of the transport agricultural products[6]. Mobile warning mainly has the following advantages: the user does not need to carry too much hardware device box alarm to save cost; through the use of mobile devices, real-time data of agricultural products can be monitored effectively, greatly improving the utilization rate of agricultural product data[7][8].

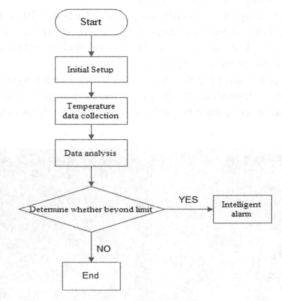

Fig. 4. Warning module flow chart

4 Cold-Chain Logistics of Mobile Payments

Cold-chain logistics mobile service applications provide complete information interaction mode, from sales to transport for the user to provide a convenient and efficient service. The rise of mobile electronic commerce brings development prospects for application in agriculture[9].

Cold-chain logistics mobile payment uses two-dimensional code technology[10]. When the goods are delivered to the users, users can realize online payment by scanning two-dimensional code, and transfer the payment information to the server, implementing the interaction with the server. Ultimately, the users and vendors are identified to ensure the normal sales of goods. At the same time, the entire transport chain is presented transparent to users, ensuring that food safety, sales, transportation can be traced back.

Through the above three modules, the user can use the mobile terminal real-time monitor product transport of live, at the same time, the vehicle positioning, to check the vehicle information in time, convenient for users to manage the goods. Mobile agricultural cold chain logistics services to improve the efficiency of logistics, has solved the traditional user must be in the specified environment view logistics vehicles, have certain application value.

5   Case Analysis

Cold-chain logistics mobile system can provide complete backtracking on meat from farm to consumers throughout the supply chain.

The RFID radio frequency identification and electronic label technology are combined as the solution, which integrates the beef growth with complete processing data in electronic tag. The mobile terminal can use mobile network to access the database, which is convenient for user's inquires.

Fresh meat are processed by cold-chain logistics monitoring system to ensure its quality and freshness. Meanwhile, the quality of the environment is strictly controlled in the transportation process, once beyond the safety threshold, issue a warning to the driver(Fig.5).

Fig. 5. Casing monitoring view

On the other hand, remote monitoring can monitoring logistics vehicle current location in real time in the electronic map , transportation routes, local condition and so on to provide navigation guidance for transport drivers. Production of beef are transported to vendors and users(Fig.6).

Fig. 6. Vehicle service view

Providing the mobile terminal distribution services can realize mobile sign, mobile place an order, mobile source query and other functions.

6   Conclusions

This paper analyzed the current cold-chain logistics problems and users' need for cold-chain logistics, proposing the key technology and the design of cold-chain logistics service system. Through the cold-chain logistics service system, agricultural products from field to table can be fully in management while the navigation path is provided for Logistics. This system not only ensures the quality and safety of agricultural products, but also saves the cost of logistics, improves the logistics efficiency, and has a broad application prospect.

Acknowledgment. This research was supported by National Science and Technology Support Project of China. (Grant No.2013BAD15B05) and the National Science Foundation Scientists of China(Grant No. 2012BAH20B02) ,all support is gratefully acknowledged.

References

1. Zhu, C.C.: Study on Farm Products Cold-chain Logistics System of China. Journal of Anhui Agri. Sci. 39(4), 2317–2318 (2011)
2. Wang, Y., Gu, Y.N.: The Study of Agricultural Products Cold Chain Logistics. Loglstlcs Englneering and Management 32(9), 4–5 (2010)

3. Liu, G.M., Sun, X.D.: Monitoring and Tracking System Agricultural Products Cold Chain Logistics Based on WSN and RFID. Journal of Agricultural Mechanization Research 4(4), 179–182 (2011)
4. Qi, L., Han, Y.B., Zhang, X.S., et al.: Real Time Monitoring System for Aquatic Cold-chain Logistics Based on WSN. Transactions of the Chinese Society for Agricultural Machinery 43(8), 134–140 (2012)
5. Zhou, X.M., Qian, J.P., Yang, X.T., et al.: Review the Application of Information Technology in Agricultural Products Logistics and Distribution. Chinese Agricultural Science Bulletin 26(8), 323–327 (2010)
6. Yang, X.T., Qian, J.P., Fan, B.L., et al.: Establishment of Intelligent Distribution System Applying in Logistics Process Traceability for Agricultural Product. Transactions of the Chinese Society for Agricultural Machinery 42(5), 125–130 (2011)
7. Wang, H.J.: Research on Cold Chain Temperature Monitoring System Based on Wireless Sensor Network, pp. 1–45. Harbin University of Science and Technology, Harbin (2011)
8. Alessio, C., Simone, C., Marco, P., et al.: A Wireless Sensor Network for Cold-Chain Monitoring. IEEE Transactions on Instrumentation and Measurement 58(5), 1405–1411 (2009)
9. Zhang, L., Pang, Y.: Comparative Study on Mode of Agricultural Products Cold Chain Logistics. Logistics Engineering and Management 32(10), 1–6 (2010)
10. Fu, X.X., Zhou, S.Q., Xie, X.P.: Smart Monitoring and Tracking Technology for Transportation Equipment of Agricultural Products Logistics. Journal of Agricultural Mechanization Research 8(8), 166–169 (2010)

The WSN Real-Time Monitoring System
for Agricultural Products Cold-Chain Logistics

Chen Liu[2], Ruirui Zhang[1,3], Tian'en Chen[1,*], and Tongchuan Yi[4]

[1] National Engineering Research Center for Information Technology in Agriculture,
Beijing 100097, China
[2] College of Mechanical and Electronic Engineering, Northwest A & F University,
Shaan xi, Yangling 712100, China
[3] College of Information and Electrical Engineering, China Agricultural University,
Beijing 100083, China
[4] College of Information Engineering, Capital Normal University, Beijing 100048, China
xmyliuchen@126.com, {zhangrr,chente}@nercita.org.cn,
y_t_ch@aliyun.com

Abstract. In order to reduce the loss of fresh agricultural products and to ensure the food safety in the process of cold chain logistics, a real-time monitoring wireless sensor network system, based on Zigbee technology for agricultural products cold-chain logistics is proposed. The sensor node is designed with CC2530 SOC and SHT15 sensor, in which the communication stack is optimized and a sleep-wake up mechanism is designed. By analyzing the experimental test and actual application, it shows that the system has a good performance in low power consumption and is satisfied with cold chain logistics application.

Keywords: cold chain logistics monitoring, transportation of agricultural products, wireless sensor network.

1 Introduction

In China, agricultural products consumption is 400 million each year. It accounts for over 50% of the total consumption [1]. As a special commodity, agricultural products including fruit, meat, vegetables, eggs and aquatic products is fresh and perishable. In order to ensure the safety of agricultural products and to reduce logistics losses, agricultural products have to be handled under controlled environmental quantities such as temperature and humidity. So cold chain logistics is used in agricultural products transportation. The chain that brings the temperature-sensitive products from the factory to the consumer through an uninterrupted series of steps under a controlled temperature is usually called the cold chain [2]. Some thermal requirements have to be installed in the refrigerated vehicles to avoid the situation that may occur could cause a significant change in the product temperature during the transportation. Wireless sensor network is one of the best ways to solve this problem [3-4].

A wireless sensor network consists of spatially distributed autonomous sensors to monitor physical or environmental conditions, and to cooperatively pass their data

---

* Corresponding author.

D. Li and Y. Chen (Eds.): CCTA 2013, Part II, IFIP AICT 420, pp. 255–261, 2014.

through the network to a main location. The main characteristics of a WSN include low-power consumption and mobility of nodes [5-7]. On the one hand, it can improve the reliability and flexibility of the system. On the other hand, nodes using batteries can be worked for a long time.

In that case, this paper proposes a wireless sensor network system of cold chain logistics based on Zigbee technology. The sensor node is designed with CC2530 SOC and SHT15 sensor, in which the communication stack is optimized and a sleep-wake up mechanism is designed. The testing results shows that the system is suitable for cold chain logistics.

2   Hardware Design of Wireless Sensor Node

The WSN is built of nodes. According to the function, wireless sensor node can be classified as sensor node and coordinator. The hardware structure of the coordinator is similar and more simple than the sensor node, so this article put emphasis on the hardware design of the monitoring node.

Each sensor node is typically composed of five parts: a radio transceiver with an internal antenna or connection to an external antenna, a microcontroller, an electronic circuit for interfacing with the sensors and an energy source, usually a battery or an embedded form of energy harvesting [8]. Considering the practical requirements of the agricultural products cold chain logistics, this paper chooses CC2530 chip as a processor and wireless communication solution. Sensor module is included SHT15 temperature and humidity sensor. Two batteries can be the enough power source. Fig.1 shows the structure of sensor node.

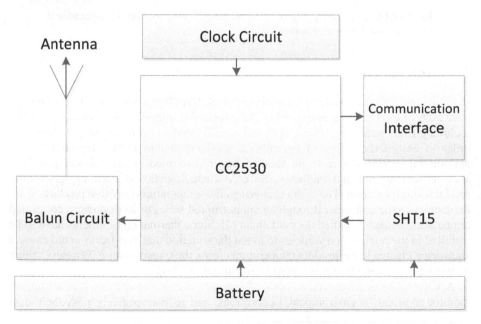

Fig. 1. The structure of sensor network node

CC2530 is a true system-on-chip (SoC) solution for IEEE 802.15.4 and Zigbee applications. It combines the excellent performance of a leading RF transceiver with an industry-standard enhanced 8051 MCU, in-system programmable flash memory, 8-KB RAM, and many other powerful features. The operating ambient temperature range is from 40 ℃ to 125 ℃ and the operating supply voltage is from 2.0 ~ 3.6 V [9]. The CC2530 has various operating modes, making it highly suited for systems where ultralow power consumption is required. Short transition times between operating modes further ensure low energy consumption. The core current consumption is about 1 μA in power mode 2 which need only 1 ms to be active [10].

SHT15 is Sensirion's family of surface mountable relative humidity and temperature sensors. The sensors integrate sensor elements plus signal processing on a tiny foot print and provide a fully calibrated digital output. A unique capacitive sensor element is used for measuring relative humidity while temperature is measured by a band gap sensor. Both sensors are seamlessly coupled to a 14bit analog to digital converter and a serial interface circuit. The operating Range is from 40 ℃ to 123.8 ℃ and the resolution is ± 2.0 % and ± 0.3 ℃. The average power consumption is only 90 μW [11].

3   Software Architecture

The software architecture is based on Z - Stack operating system in IAR Embedded Workbench development environment with C language. Z-Stack operating system is semi-open systems based on priority developed by TI for CC2530 SoC [12]. Monitoring functions can be realized by adding definition pins of CC2530 and SHT15 in hardware abstraction layer and writing appropriate programs in application layer.

System nodes complete initialization including system clock, stack, hardware, operating system and so on. After the initialization system start to query task events with priority followed by entering into MAC layer, network layer, hardware layer, application support sub-layer and application layer.

If any event of binding time happens in application layer, sensor node reset SHT15 sensor and start to collect data. Data collection includes two parts which are the temperature and the humidity. The monitoring node sends data to coordinator after collecting. Then, the monitoring node enters into sleep mode to save power and refresh the timer.

CC2530 has three kinds of sleep modes: PM1, PM2 and PM3. Sensor node works in PM1 mode normally. PM2 mode is used to save power consumption and can be awakened periodically. PM3 mode has the least power consumption but it can be awakened by external interrupt only. In this paper, the function of "DRFD_RCVC_ALWAYS_ON" is defined as false. Sleep-wake up mechanism program can be realized by adding function of "POWER_SAVING" at the end of each application program. The flow chart of execution is shown in fig 2.

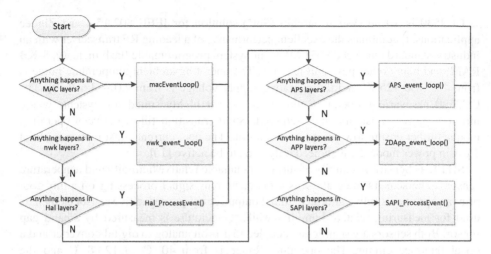

Fig. 2. Flow chart of monitoring node data acquisition

4   Testing and Results Analysis

For verification the system has been tested with two sensor nodes measuring the temperature and humidity at two different positions. SmartRF Studio 7 software developed by TI company is used for testing RSSI of sensor node and packet loss rate. The communication radio frequency is 2405 MHz. Sensor node sends 300 data in eight different distances with ten levels of emission power. The averaged results are showed in Table 1 and Table 2.

Table 1. Testing results of RSSI (dBm)

Power /dBm	Distance/m							
	1	5	10	15	20	30	40	50
4.5	-54.9	-64.0	-69.9	-72.7	-80.0	-85.5	-87.1	-94.9
2.5	-57.0	-66.2	-71.7	-74.6	-85.6	-86.5	-89.5	-98.7
1.0	-61.6	-66.5	-72.3	-77.7	-89.3	-92.4	-92.0	-99.4
-0.5	-63.8	-68.1	-73.7	-81.5	-92.5	-93.3	-92.7	-98.0
-1.5	-66.5	-69.7	-75.4	-83.1	-94.0	-93.6	-95.1	-100.3
-4.0	-68.0	-74.1	-79.8	-86.8	-94.4	-97.1	-98.5	
-8.0	-68.9	-76.1	-83.7	-89.3	-95.2	-98.4	-99.3	
-12.0	-72.2	-84.5	-86.8	-92.2	-98.0	-99.4		
-16.0	-77.0	-84.1	-92.0	-95.4	-100.2			
-20.0	-80.3	-87.8	-95.0	-97.2				

Note: the blank space means node could not receive the signal.

Table 2. Testing results of packet losing rate (%)

Power	Distance /m							
/dBm	1	5	10	15	20	30	40	50
4.5	0	0	0	0	0.5	0.7	1.3	2.9
2.5	0	0	0	0	1.0	1.1	2.3	53.2
1.0	0	0	0	0.2	1.7	2.7	3.4	51.6
-0.5	0	0	0.2	0.6	3.8	7.2	8.7	61.8
-1.5	0	0	0.1	0.8	7.5	7.3	9.2	92.5
-4.0	0	0	0.3	0.8	14.0	27.2	42.0	100
-8.0	0	0	0.7	1.6	20.0	50.7	78.8	100
-12.0	0	0.8	0.7	1.8	33.1	71.5	100	100
-16.0	0	1.1	4.2	4.6	96.7	100	100	100
-20.0	0.3	4.2	7.1	20.0	100	100	100	100

Test results show that RSSI attenuation trend is obvious related with the decrease of transmitted power. When the distance becomes longer, RSSI becomes weak and the packet loss rate begins to rise. Packet loss phenomenon is obvious when the RSSI reduced to - 100 dBm. Within the scope of 15 m, even - 16 dBm transmitted power can guarantee less than 5 % packet loss rate. But only transmitted power is 4.5 dBm sensor node can guarantee the communication reliable. The wireless link is more reliable only if the communication distance is limited in a reasonable scope with a strong transmitted power. The comparison chart of RSSI and packet loss rate is showed in Fig.3 and Fig.4.

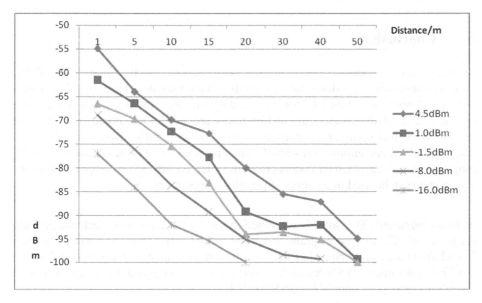

Fig. 3. Comparison chart of RSSI

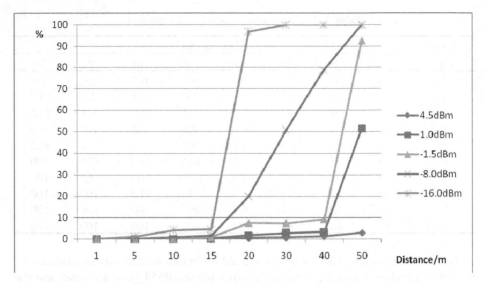

Fig. 4. Comparison chart of packet losing rate

The comparison results show that RSSI is above - 95 dBm and packet loss rate is controlled within 8% when the distance limited to 30 m and the transmitted power above - 1.5 dBm. According to the actual situation, the distance between sensor nodes is no more than 20 m. It proves that the system can be applied in cold chain logistics and the communication is reliable.

5   Conclusions

In this paper, a measuring system that is conceived for the monitoring of the temperature-sensitive products during their distribution has been proposed. The system consists of nodes based on CC2530 SoC and SHT15 sensor that are able to carry out measurements of the temperature and humidity around the monitored products. The testing shows that system performed well when the transmitting power is above - 1.5 dBm. The RSSI can ensure over - 95 dBm and packet loss rate is less than 10 % in the range of 30 m. The obtained results prove the effectiveness of the proposed solution. The system can be used in a variety of practical applications.

Acknowledgment. This work is supported in part by Research Key Technology R&D Program of China under grant No. 2012AA101901, Beijing Natural Science Foundation under grant No. 4121001, and Research Foundation for Young Scholar of BAAFS under grant "WSN research and equipment development for its application in greenhouse group monitor"(No. QNJJ201217).

References

1. Zhu, C.: Study on Farm Products Cold-chain Logistics System of China. Journal of Anhui Agriculture Science 39(4), 2317–2318 (2011)
2. Ruiz-Garcia, L., Barreiro, P., Robla, J.I., et al.: Testing ZigBee motes for monitoring refrigerated vegetable transportation under real conditions. Sensors 10(5), 4968–4982 (2010)
3. Strazdins, G., Elsts, A., Nesenbergs, K., et al.: Wireless Sensor Network Operating System Design Rules Based on Real-World Deployment Survey. Journal of Sensor and Actuator Networks 2(3), 509–556 (2013)
4. Lakshmil, V.R., Vijayakumar, S.: Wireless Sensor Network based Alert System for Cold Chain Management. Procedia Engineering 38, 537–543 (2012)
5. Heidmann, N., Janßen, S., Lang, W., et al.: Implementation and Verification of a Low-Power UHF/LF Wireless Sensor Network as Part of the Intelligent Container. Procedia Engineering 47, 68–71 (2012)
6. Sun, Y., Yang, H., Liu, Z., et al.: An Intelligent Logistics Tracking System Based on WSN. Journal of Computer Research and Development 48, 343–349 (2011)
7. Guo, B., Zhang, T., Qian, J., et al.: Design of Zigbee-based Wireless Information Collection Node For Cold Chain. Chinese Agricultural Science Bulletin 26(5), 318–321 (2010)
8. Ruiz-Garcia, L., Lunadei, L., Barreiro, P., et al.: A review of wireless sensor technologies and applications in agriculture and food industry: state of the art and current trends. Sensors 9(6), 4728–4750 (2009)
9. Hwang, J., Yoe, H.: Study of the ubiquitous hog farm system using wireless sensor networks for environmental monitoring and facilities control. Sensors 10(12), 10752–10777 (2010)
10. Sharma, P.: Socio-Economic Implications of Wireless Sensor Networks with Special Reference to its Application in Agriculture. African Journal of Computing & ICT 6(2) (2013)
11. Park, D.H., Park, J.W.: Wireless sensor network-based greenhouse environment monitoring and automatic control system for dew condensation prevention. Sensors 11(4), 3640–3651 (2011)
12. Carullo, A., Corbellini, S., Parvis, M., et al.: A wireless sensor network for cold-chain monitoring. IEEE Transactions on Instrumentation and Measurement 58(5), 1405–1411 (2009)

The Construction of Agricultural Products Traceability System Based on the Internet of Things—The Cases of Pollution-Free Vegetables in Leping of Jiangxi Province

Fang Yang

School of Business Administration, Jiangxi University of Finance & Economics,
Nanchang, 330013, China
Yfang118@163.com

Abstract. With the problems of food safety happening increasingly, the demand of the public for agricultural products becomes stronger and stronger. This paper declared the status of the production process of pollution-free vegetables in Leping city of Jiangxi Province in detail. At the same time, it put forward the existing problems in the current process and the recommend for new system. In the end, based on the internet of things, it designed the Agricultural Farm System, including the process of purchasing, storage and outbound, planting, processing, sale and distribution.

Keywords: the internet of things, agricultural products traceability, information system.

1 Introduction

Agricultural product is the necessity for people to survive, however, in recent years, the quality problem of agricultural product has exposed frequently in China, which has seriously influenced the healthy bodies and daily lives of consumers. The agricultural products with higher safety and higher quality are popular in consumer markets. Especially with the development of economy, the production and sale of agricultural product are increasingly separating, which made it more difficult to get safe information of agricultural product. So, in order to realize a traceable market requirement, which asks for the development and establishment of an Agricultural Product Traceable Information System directly, more effective measures must be taken to strengthen the management and supervision of agricultural production.

Since the 1990s, many developed countries already have traceability systems. The countries in European Union (EU) first apply agricultural products traceability system in the product of live cattle and beef. EU issued a White Paper on Food Safety in January 2000, and declared all relevant production operators' responsibility in the process of product circulation from farm to table. In the regulation No 179/2000 of EU, companies are stipulated to provide assurance measures on materials and date to ensure their safety and traceability in the process of production, processing and sales. In USA, government could farther to promote management functions based on

D. Li and Y. Chen (Eds.): CCTA 2013, Part II, IFIP AICT 420, pp. 262–268, 2014.

agricultural products traceability system. In 2003, U.S. food and drug administration (FDA) published the food safety tracking regulations, in which all enterprises, involved in food transportation, distribution and import, are required to establish and preserve the whole process of food distribution records. In order to promote the traceability of agricultural products, a series relevant laws and regulations are established in the field of agricultural products traceability. Domestic scholars emphasis the importance of establishing agricultural products traceability system to the quality and safety of agricultural products, and put forwards many countermeasures against the current problems.

2   The Problems and Present Situations of Vegetable Production in Leping

There are 20,000 hectares for vegetable planting in Leping in 2007. Its total output has broken through 600,000 tons; the vegetable planting area of the city is 17,000 hectares, and the total output is 620,100 tons. It covers the ten vegetable varieties of 100 major categories, of which more than 80% of the export of vegetables, by the end of 2010 the city's vegetable planting area has reached 22,000 hectares with an annual average increase of 4%; the total output of vegetable is 960,000 tons with an annual average increase of 12%; and the total output value of about 1180,000,000 RMB with an annual average increase of 17%. Vegetable wholesale market transaction volume reaches 701,000 tons with an annual average increase of 1.5%; turnover is expected to reach 1380,000,000 RMB, an increase of 17.4%. By 2011, the city's vegetable planting area is more than 25,000 hectares, and total output is 990,000 tons. Leping became the largest vegetable base and vegetable distribution center, price and formation center, information and Communication Center in Jiangxi province. It became the important distribution center of vegetables in the Yangtze River Basin, and has long enjoyed a good reputation of national dish country of the Yangtze River and the national pollution-free vegetables Demonstration County.

2.1   Existing Problems of the Vegetable Production Business Pattern

Problem 1: Pick up seeds. Due to the lack of the information of seed selection and source, the production of seeds planting cannot be adjusted to response to market demand in time, and cannot fit for the customer's demand.

Problem 2: Breeding. The planning of breeding is not reasonable that leads the lack of production or too much of production.

Problem 3: Decide planting. In planting decisions, the processes cannot be precisely controlled due to non-complete historical information.

Problem 4: Field management. Don't have complete record. So the field management cannot provide statistical information, and predict crop's needs of fertilizer, water and sunshine to provide reference standard for plant operation.

Problem 5: Harvest. Harvest does not have the unity of records due to individual records. People cannot statistical the output of fields.

2.2   Existing Problems of the Current Mode of Vegetable Processing

Problem 1: Process management. In vegetable processing, such information cannot be recorded, including time when vegetables reach the machining center, weight, operation personnel. So the formation cannot be reversely traced.

Problem 2: Warehousing operations. No record the related information of each storage products, including the name, batch, quantity, storage time and warehousing operations et al. The information of the products in storage cannot be traced back.

Problem 3: The inventory query. Due to lack of inventory information, the inventory situation are present cannot been known in any time, so cannot provide and reference information to production, procurement and sales.

Problem 4: Scrap processing. Without a complete waste treatment process, at present the warehouse keeper will statistics the information about the scrapped productions according to scrap processing requirements, information will be summarized and reported to the superior leadership.

Problem 5: Outbound processing. No record about the reason of the outgoing products, including name, product batch number, storage products, delivery time, delivery etc. The detailed information of the outgoing products cannot be traced reversely.

2.3   The Problems Existing in the Sales Distribution Nowadays

Problem 1: Distribution management. The plan of distribution lacks of the distribution automatic plan system, nowadays the formulate of the plan about distribution is related to the forms and records of distribution, but very little of the members information can complete the tasks at the required time, so as you can see, it will be very difficult to complete the tasks when the member scale expanded.

Problem 2: Distribution statistics. We just have each piece of distribution plan, if you want to statistics one of the project targets, you need to do it by yourself.

Problem 3: Satisfactory of the statistics. Don't set completed satisfactory evaluation system, cannot give a completed statistics and analysis to the customer's evaluation, so it is very difficult to improve the operation of the product, plant, and distribute.

3   The Design of Function Modules

Through the analysis of the production process, at the same time, in order to make system exchange and share data with the related system better, especially in order to complete the system after the long and arduous system maintenance tasks in the operation, reduce the maintenance work in the development, on this basis, we can propose the component development according to the function of the system, as shown in Fig. 1. It is the core modules and the basic functions of the system.

According to the business process, we can design the parts of procurement process, storage process, planting process, processing flow, flow distribution in the agricultural products traceability system. In the follow sections, the design of each part and links of those parts will be respectively described in detail.

3.1   Procurement Design

Purchase link is the initial part of the production and processing. Through setting up a rigorous procurement process, we can avoid that inventory source become a problem. Purchasing module can be divided into four sub-modules, including the procurement plan, procurement process, sourcing and purchasing statistics. The node sets realize that the whole purchasing process can more be standardized, and the authentic data can be obtained.

In the procurement procedure of setting up purchase plan, each procurement plan number can corresponds to multiple purchase batches, and the information of each batch includes corresponding product and the relations information. Because a product's bar code corresponding to sole one purchase batches, we can quickly find the source of kinds of problems by inversely information tracing.

3.2   In-Out Stock

In-out stock represents the link of the raw materials' procurement and the recipient of warehouse. In order to guarantee the accuracy of entering warehouse and alleviate the press on the audit, a link of audit can be set. Thus, not only can guarantee the accuracy of the data in stock, but also will make the process of structure more clear and easy. The efficiency of all works can be improved based on the accuracy of data. No matter what kind of storage types, including procurement warehousing and unplanned put in stock, all need to be handled in storage applications. This is the goods cannot be put in stock until they pass the audit.

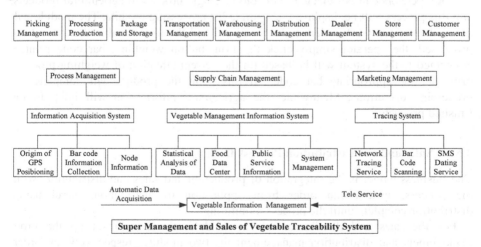

Fig. 1. Construction of vegetable traceability system

3.3   Planting Process

According to the demand analysis report, the main processes for planting include seeding, sowing, the field of management and harvesting link. Although the types of field management are far too many, these are just some of the operating records in the system, namely the choice of the state. Therefore, the planting processes designed in the system sequentially include fielding, seeding, managing, and harvesting.

The module settings of the planting management in the system includes the planting of structure module, the field of management module, the record of harvest module, and the temperature field management module, etc. When users begin to run the system, the planting module chooses the large field or basement to plant and gets the seeds to be sowed. At the same time, the system will automatically calculate the planting area according to the basic data of the large field, while associating the information about sowing seeds of purchasing batch and supplier, etc. in the module of field management, the user can record the daily maintenance and the operation management. The later product traceability can be supported by the data provided from system module.

3.4   Processing Flow

With regard to fruit and vegetable products, the processing of agricultural products includes picking, weighting, packaging and printing barcodes. The detailed processes can be shown in Fig.2.

In view of the processing flow, there are three sub-modules in the system: the product to be processed, the processing order management and processing lists. The product information in process module comes from the module of picking operation. After completing pick operation in user management module, the system will automatically jump to the agricultural information management module and continued to process in cultivation of agricultural products.

The operators can be selected to develop a single process to agricultural products, and this process will be displayed in a single list processing module. The module will detail the various work orders to be processed and the total number that have been processed, the operator simply click the Print button weighing, bar code printers connected to the system will be based on the current situation of weighing products print out the appropriate bar code, At this point the product will complete the processing operations, Meanwhile, the agricultural information will jump to the finished products.

3.5   Sales and the Distribution Process Design

The business model is the integration of production and sales and distribution, so in the process, system can order from application to distribution, warehousing, distribution complete until the process monitoring.

For the sales and distribution of the module, the system set up the order management and distribution management the two modules respectively, an order,

order review, order tracking, order record four modules including order management module, and distribution management module in distribution, distribution and delivery records specified on the three modules. Orders for information input module has two types, one is the company clerk keyboard added, another is the customer place an order on the official site, and then automatically displayed in the modules of the system. When there is an order, staff and click "submit" button, the order will be submitted to the order review, for review staff review, through the audit, a database management module "plan library" module will have the invoice, the invoice is generated automatically according to the orders, the staff can do the picking operation the order here, open distribution interface, using scanning gun scanning distribution product bar code, distribution is completed, the invoice status displays distribution is completed, the message appeared "outbound" button, click the button to open the library, library editing interface, enter the appropriate information from the storage, outbound order, the product he completed the warehouse operation. Then, information of the order is shown in the distribution of the specified module, here, the staff want to edit the order distribution list, specify the distribution of personnel and vehicles, while the distribution process of the vehicle real-time monitoring, until the completion of the income distribution in the signature and date of receipt, the monitoring end.

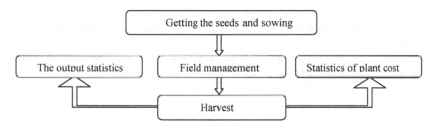

Fig. 2. The flow chart of planting process

3.6   Ascend Ways

There are two kinds of query methods of agriculture products traceability, the direct use of mobile phone or PDA to scan the bar code. Two-dimensional code affixed on the product packaging can query to the relevant information of the product. A product bar code on the official website can also be traced back to the relevant information of the product.

4   Conclusions

In the paper, on the background of frequent food safety issues, some related policy and technology application in the current domestic system is summarized firstly. The facing problem, the command of information system and agricultural production, processing and sales process flow are detailed analyzed. Finally, according to the recommend analysis, it built the Agricultural Information Tracking System which shows the optimization and improvement of production process, especially in the aspects of procurement and field management. The design of multi-node audit guarantees the authenticity of the data and the executive supervision.

Acknowledgment. This work was supported by a grant from the 2012 Science Foundation for Youths of the Department of Education of Jiangxi Province (No. GJJ12277), a grant from the 2012 Humanities and Social Science foundation of College of Jiangxi Province (No.GL1216), a grant from the 2012 Humanities and Social Science foundation for Youths of Education Department of China (No. 12YJC630263), a grant from the 2010Humanities and Social Science foundation for Youths of Education Department of China (No. 10YJC630255) a grant from the 2011 The national natural science fund projects of China (No. 71162012).

References

1. Ruiz-Garcia, L., Steinberger, G., Rothmund, M.: A model and prototype implementation for tracking and tracing agricultural batch products along the food chain. Food Control 21(2), 112–121 (2010)
2. Zhang, X., Lu, S., Xu, M., et al.: Applying evolutionary prototyping model for eliciting system requirement of meat traceability at agribusiness level. Food Control 21(11), 1556–1562 (2010)
3. Breeding, M.: An Analytical Approach to Assessing the Effectiveness of Web-Based Resources. Computers in Libraries 10(1), 20–22 (2008)

Key Technologies and Alogrithms' Application in Agricultural Food Supply Chain Tracking System in E-commerce

Lijuan Huang[1] and Pan Liu[2]

[1] Guangzhou University, Guangzhou 510006, Guangdong, China
[2] Shanghai Business School, Shanghai 201400, China
huanglijuan66s@126.com, pan1008@163.com

Abstract. In nowadays' world, frequently occurred agricultural food safety events have severely done harm to people's health and directly initiated great trust crisis to governments and agricultural food enterprises, and the E-commerce as virtual economy can enlarge the trust crisis terrifically. As a result, to establish unified traceability information sharing platform of agricultural supply chain has become an urgent task for the related organizations. This paper begins with an exploratory research on design and realization of Agricultural Food Supply Chain Tracking System (AFSCTS) in E-commerce environment, which can provide a powerful technical support for timely tracing and finding the link where a agricultural food safety problem happens in the supply chain. This paper has 5 sections: the first to briefly introduce the concepts relevant to AFSCTS and the significance of this research; the second to analyze three key technologies (RFID technology, database integration technology and data security technology) utilized in AFSCTS; the third to elaborate on two key algorithms (traceability algorithm and data encryption algorithm) designed in AFSCTS; the fourth to design and implement AFSCTS based on three key technologies and two key algorithms, and the overall system frame and function structure are narrated detailedly in the paper; the last to apply the AFSCTS in Nanfeng county (the hometown of famous Nanfeng Orange in China) for tracing agricultural food, and the system implementation has achieved experimental success.

Keywords: agricultural food, tracking system, supply chain, E-commerce.

1   Introduction

1.1   Related Concepts

Currently, there is no a unified authority definition of Food Supply Chain Tracking System (FSCTS). According to ISO standards (ISO22005, 2013), tracking system can identify and track object position change by recording items code. Food traceability can be defined as tracking all stages (from production, processing to sales) about food products (J. A. Monahan, 2012; T. Simpson, 2013) [1]. The European

D. Li and Y. Chen (Eds.): CCTA 2013, Part II, IFIP AICT 420, pp. 269–281, 2014.
© IFIP International Federation for Information Processing 2014

commission defines that FSCTS is the Management Information System (MIS) to trace material used in animals or feeds in any specified stages of food supply chain(M. Reynolds, 2012; R. Angeles 2012) [2-3]. The definition of Agricultural Food Supply Chain Tracking System (AFSCTS), provided by the United Nations Codex Alimentations Commission (UNCAC), is a computer system to trace agricultural food in production, processing, distribution and sale process of agricultural supply chain (F. Fisher, 2013) [4].

In western nations, more and more attention is focused on the research of Tracking System, such as Tracking System based on products supply chain (May Tajima, 2013), Tracking System based on hospital applications (T. Monahan, 2012; P. Stanfield, 2013), and children's tracking system for large amusement parks (Xiaodong Lin, etc., 2012) [5-7]. These systems have achieved traceability function by database technology and RFID technology. With continuous development of RFID technology and continuous decline of its cost, research on Tracking System is getting increasing attention, especially in monitoring the quality and safety of food (MinBo Li, ZhuXu Jing, Chen Chen, 2012) [8]. These applications based on RFID have accelerated RFID technology development in E-commerce environment (Feng Huang, Peng Hao, HuaRui Wu, 2012) [9].

European Bovine Spongiform Encephalopathy (BSE) crisis in EU countries represented as the global scope vicious food safety incidents' outbreak. CAC Biotechnology Intergovernmental Task Force Meeting divided FSCTS into five parts (records management, inquiring management, marking management, liability management and credit management). According to the features differences of agricultural food Tracking System, Elise Golan, an American scholar, sets three standards to measure AFSCTS: breadth, depth and precision (X. L. Kwan, 2010) [10]. Among them, breadth means the information scope that the system contained, depth means the distance that traceability information goes forward or backward, and precision means the ability that can determine unsafe food source or certain features of goods (Elise Golan, 2008) [11].

Generally, AFSCTS is divided into two kinds: one is oriented to the consumers, this means tracing from the upstream node of food supply chain to downstream node (e.g. consumers); the other is oriented to the suppliers, this means tracing from the downstream node of food supply chain to the upstream node (e.g. food production source). In view of the importance of food safety, there are more researches focused on the latter tracking system in E-commerce environment.

1.2   Background and Significance

In recent years, frequently occured food safety events (such as SanLu milk powder, Foot-and-mouth disease, Avian flu, PRRS, etc) have been severely doing harm to consumers' health, causing huge economic losses in many countries, and directly initiating great trust crisis to governments and enterprises in food supply chain (N. K. Porter, 2011). In order to help the governments and the related organizations timely find food safety problems in the process of production or circulation and make the responsibility of enterprises or related departments clear, it has become an urgent and important task for domestic and foreign governments and the related organizations to establish unified traceability information sharing platform in food supply chain.

Developed countries have issued a series of relevant laws and regulations, which require that all enterprises engaged in production, processing, packaging or managing food of human or animal must provide traceability information platform (C.W. Lau, 2011). In the United States, food enterprises must trace and find food problem within 48 hours,or face large sum of astonishing fines; In Japan, only through many hurdles, then food could be on dining-table; In France, if a store is found to sell expired food, the store will be shut; In Canada, 80 percent of food can be traced to source, and the country is realizing "the brand Canada" strategy (Stanford, 2010).

At present, China's frequent malignant events of food safely (such as toxic powder event, Sudan red event, turbot fish event, and bonny big event) directly shocks the Chinese Central government. Food safety problem has become the focus of attention of the whole China society (C. J. Tao, 2011). Researching and developing AFSCTS is imminent in China, and it can provide powerful technical support to crack down on food crime and safeguard people's health. In this paper, we planned and designed AFSCTS based on   three key technologies and two key alogrithms.

2   Key Technologies

There are three key technologies (RFID technology, database integration technology and data security technology) utilized in AFSCTS under the environment of E-commerce.

2.1   RFID Technology

RFID (Radio Frequency Identification), arises in recent years, is a kind of automatic identification technology. It has become one of indispensable support technologies in Internet of Things today. RFID has some characteristics such as unique identifier, fast read and write, untouched identification, mobile identification, and multi-target recognition (1000 tags can be simultaneously identified per second) make it possible to realize efficient traceability for supply chain system. In RFID System, digital memory chips with unique electronic commodity codes can be pasted on single food product, and receiving equipment can activate RFID tags, read and change data, and transmit data to host computer for further processing the data. Operating principle of RFID technology is shown in Fig.1.

At present, the method based on traditional bar codes can not realize traceability management for the whole food supply chain. Compared with bar code technology, RFID tags are more suitable for managing the whole process of food supply chain from farmland to dining-table. Because there are unique identification codes, data erase duplication, tag with large storage for data, and fast response for identifying tag with a long service life, RFID can be used in bad condition such as high temperature and humidity. Applying RFID technology can not only identify every unit of goods but also may effectively identify each node of supply chain to track and trace each link of supply chain. These links include processing, packaging, storage, transportation and sales. This may timely find out existing problems and properly handle them.

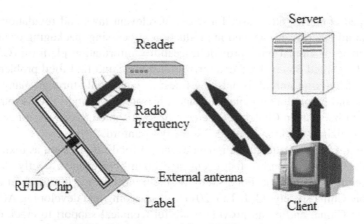

Fig. 1. Operating principle of RFID technology

2.2   Database Integration Technology

Heterogeneous database integration technology can merge and share data resources or hardware resources from different databases. Through database integration technology, information of monitoring product from databases of each enterprise in supply chain will be integrated. This can achieve to trace food in whole supply chain, and each enterprise can use its database without redesigning or rebuilding new databases. This can save costs for tracing food supply chain.

In recent years, technologies applied to integration of databases mainly include middle ware, mobile agent and XML. The features of mobile agent is autonomy, mobility and collaboration. There are also distributed self-learning and self-reasoning about mobile Agent. So, the rapid development of mobile agent can give a new way for research on the heterogeneous database integration technology. XML has gradually become industry standard of data representation and information exchange, and it may provide a platform for transferring data of relational databases of heterogeneous and different platforms. The combination with mobile agent and XML could improve the efficiency of user queries in process of heterogeneous database integration. XML is a model used to describe data structure. It utilizes a mechanism to describe contents of related table in database and standardized combination form in files. XML also defines description rules such as file structure, data type and so on. XML has been widely used to access to heterogeneous data sources and transmission of middle ware. As network technology and software technology continue to evolve, distributed and heterogeneous topology has become remarkable characteristics of all kinds of calculation and development environment of system. Heterogeneous database integration technology has been considered to be hotpot and difficult issues in Database fields by domestic and foreign academia and industry.

3   Key Algorithms

There are two key algorithms (Traceability algorithm and data encryption algorithm) designed in AFSCTS under the environment of E-commerce.

3.1   Traceability Algorithm

In China, the traditional traceability algorithm can only trace a enterprise but can not trace all node in the supply chain, and the traceability procedure only starts after unsafe food have done harm to the society. As a result, this traceability is often too late and not so successful.

In order to establish tracking and tracing model in Tracking System, we need to make sure the EPC coding about the products in the supply chain and the mapping of the data information in Tracking System. In the actual application system, mainly information of product coding contains two parts: identification information and record information. The identification information can be used to set the only logo to the product, and it gets the EPC logo through multi-level to make sure the continuity of tracing process and tracking process. Generally, the record information contains main information of the product. It mainly is used by supply chain enterprises to communicate between internal and external. The tracking information of product comes from the object events, and it changes tracing event information into tracing unit information for realizing tracing each unit back on the supply chain.

Product traceability unit can be defined as formula (5):

$$S_i = \{s_1, s_2, \cdots, s_n\} \tag{5}$$

S_i is used to describe object products set in the supply chain, and it also consists other objects in same supply chain. The change of each object is corresponding to the corresponding events. So each object E_{ij} contains event set eik, and $k \in \{1,2,\ldots,j\}$. E_{ij} can be expressed as formula (6):

$$E_{ij} = \{e_{i1}, e_{i2}, \cdots, e_{ij}\} \tag{6}$$

Where, object i contains events j. Each event is corresponding to the change of information, such as product status, position and properties, and tracing events can effectively ensure continuity of traceability. Therefore, Object Si in time t runs through a supply chain node, and then its status Ct(Si) can be described by several parameters as   formula (7):

$$C_t(S_i) = \{M, E, t, ID_t, Type_t(S_i), Loc(S_i), Step(S_i), Real_t(S_i)\} \tag{7}$$

Where, Time t is used to express the status of time utility.

IDt is the corresponding labels about the object Si at moment t.

Type$t$ (Si) is the business operation types in the process of the supply chain about the object Si at moment t.

Loc(Si) is the position in the process of supply chain about the Si at moment t.

Step(Si) is the business step at supply chain node about Si at moment t.

Real$t$(Sj) is used to describe the association contact between object Si and other object Sj, such as the relationship between the product object and the packing bag object, and it realizes the binding and split between different objects.

Object value M is serial number about the object Si   associated with the other object Sj. M can be expressed as formula (8), where i is not equal to j:

$$M = \sum_{j=1}^{n} Mat_t(S_i) \tag{8}$$

Event value E is sum of all Function Ej with parameter Si in given time t, and it can be computed as formula (9):

$$E = \sum_{j}^{n} E_j(S_i) \tag{9}$$

AFSCTS designed in this paper can not only trace a node, but also trace the whole supply chain. It can not only support traceability after the event, but also support traceability before the event. The AFSCTS with real-time monitoring can eliminate the potential unsafely of food and to protect people's life safety. Traceability algorithm can obtain and express corresponding tracing results through EPCIS interfaces. Pseudo-code of main tracing algorithm in the AFSCTS is described as follows :

```
Retrospect (Epc_code, eventlist, event *curr, track *pro ,&n)
 { eventlist = getQueryNames();
    epc_code = getSubsIDs(*curr, &n);
   int i; string e_track, e_stander_track;
   for(i ; i<=n;  i++)
     { if (eventlist. epc_code == epc_code)
          e_track = getVendorVer () ;
          e_stander_tracl == getStanVer ();}
       if (eventlist ==null)
               output(getSubsIDs(*curr, &n)) ;
       else {
           for(i=1,i<=length(eventlist),i++)
           {output(getSubsIDs(queryName: string))}}
```
The main interfaces of EPCIS in the ASCTS have been specified in Table I.

Table 1. The main interfaces parameter of EPCIS

No.	Name	Parameter	Type
1	Subscribe	Query Name	string
2	Subscription Controls	Subscription_ ID	string
3	unsubscribe	Subscription_ID	string
4	getQueryNames		list
5	getVendorVersion		string
6	getVendorVersion		string

3.2   Encryption Algorithm

There are some important information (e.g. money or assets information and consumer information) in AFSCTS. It is hard for enterprise to define potential intruders, because information resources are sharing by all firms in supply chain. If food supply chain information system is illegally invaded, that may cause huge financial losses for the firm which is invaded, or even all node enterprises in the whole supply chain may suffer huge losses. Data encryption is very important for the system because integrated database is used in the system. It is essential to use electronic tag with identity authentication and encryption system in order to protect commercial secrets of each node enterprise in AFSCTS. At present, the research about data security of RFID is paid more attention. Advanced Encryption Standard (AES) is post by NIST in 1997, and now it is considered to be a replacement for DES that represents traditional symmetric encryption algorithms. After several years, NIST sifts many algorithms and eventually determines to take Rijndael algorithms as AES, which is invented by Vincent Rijmen and Joan Daemen as the replacement of DES. NIST have established a new standard of AES in May, 2002. Rijndael uses a very limited storage of RAM and ROM, this makes Rijndael to be a excellent candidate when it is applied in resource-constrained environment. Therefore, it is possible to realize encryption of RIFD technology. So this research proposes that AES algorithm is transported to the food supply chain Tracking System for guaranteeing data security in shared database. But taking into account such factors as the cost of technology, successfully applying Rijndael in RFID needs to be studied; this article raised the idea of the data encryption.

Main part of the Rijndael is Mix Column transform, and it has mainly accessed knowledge of polynomial arithmetic. Here are the domain F2n of arithmetic, which can be calculated as formula (10) :

$$f(x) - a_{n-1}x^{n-1} + a_{n-2}x^{n-2} + \cdots \mid a_1 x \mid a_0 \qquad (10)$$

Where, coefficient a_i is from Polynomial coefficients as the n-tuple, and the tuple can correspond to the number of binary computer. So, it is convenient to realize program to apply this algorithm in computer. Rijndael algorithm includes: S-box, shiftRow transform, MixColumn transform, and AddRoundKey. AddRoundKey is regarded as the last step, and each round key is combined with the message by XOR operation, which may have formula (11) to represent, and $j=0,\ldots,L_b$-1:

$$\left(b_{0,j},b_{1,j},b_{2,j},b_{3,j}\right) \leftarrow \left(b_{0,j},b_{1,j},b_{2,j},b_{3,j}\right) \oplus \left(k_{0,j},k_{1,j},k_{2,j},k_{3,j}\right) \qquad (11)$$

AES uses a loop structure to realize iterative encryption and input data by permutations and substitutions. Permutations are rearranging data. Substitutions are replacing another with a unit of data. ASE uses all kind of different technologies to achieve permutations and substitutions. In our AFSCTS, the key can be produced by data encryption algorithm. The pseudo-code is as follows:

```
Keyproduction (byte Key1, byte Key2)
{   for (i=0; i<Nk; i++)
        Key2[i]=(Key1[4*i],Key1[4*i+1], Key1[4*i+2],Key1[4*i+3]);
        for (i =NK; i< Nb*(Nr+1); i++)
          { temp = Key2[i-1];
            if ( i%Nk ==0 )
            tempx=RotBytes(Rcon[i/Nk])
        temp=SubBytes(tempx);
            else if ((Nk ==8) && (i%Nk ==4))
                temp =SubBytes(temp);
                Key2[i] = Key2[i-Nk] ;}}
```

3.3   Data Security Algorithm

With rapid development of network technology, the degree of openness, sharing and interconnection of Internet are continuously improved. It is more and more important for information safety to construct AFSCTS, which should take reliable measures to guarantee data safety. At present, there are several methods of security design about AFSCTS based on RFID technology.

(1) Data of RFID tags is encrypted. EPC code is encrypted by RSA algorithm and then the code is written down in tags. Literacy device with private-key read labels cipher text can encode and decrypt to get EPC gold-digging. RSA algorithm is the evolution of RICE which was published by Ron Rivest, Adi Shamir and Len Adleman from MIT in 1987. RSA is a kind of block cipher. The text passwords and cipher text of the RSA are integer between 0 and n-1. Based on block M of text passwords and block C of cipher text, the process of encryption and decryption could be described as formula (1) and formula (2):

$$C = M^e \tag{1}$$

$$M = C^d \bmod n = M^{e.d} \bmod n \tag{2}$$

Based on the above formulas, both sender and receiver have known the variable n and variable e, so the only requirement is that receiver has to know the variable d. The equation of public key is PU={e,n}, and the equation of public key is PR={d,n}.

(2) RFID reader approves the authenticity of labels by HB Protocol, and HB Protocol is introduced by Hoper and Blum (2012). HB Protocol can be used to solve personal identity authentication, and it can also effectively solve message of asking for reply in process of eavesdropping and authentication.

(3) Though database decipherments to protect the security of system data. During the process of encrypting AFSCTS data, encrypted data have been stored in the server. The data in the server is safe, because there are not keys on the server. Even if someone can make an incursion to the server, and he has no access to the data encrypted. The client system has a copy of encryption key, so users can obtain the data through corresponding method. During the process of encrypting AFSCTS data, each record of the database is encrypted by block. Each line named Ri can be taken as

a continuous block which can be expressed as formula (3) and the entire line can be expressed as formula (4):

$$B_i = \left(x_{i1} \| x_{i2} \| \ldots \| x_{iM} \right) \tag{3}$$

$$E\left(k, B_j\right) = E\left(k, \left(x_{i1} \| x_{i2} \| \ldots \| x_{iM}\right)\right) \tag{4}$$

(4) Using CA technology can realize RFID data safety in the process of transmission on the net. User's identity is identified between RFID client and data center in tracking server rooms by digital signature technology. In the process of transmission, message authentication code is added into confidential RFID data so as to realize packet encryption and data integrity checks.

According to safety requirements of AFSCTS, we can independently or collectively apply those safety technologies. Eclectically considering between cost investment and security degree, we realized storage and encryption algorithms of security system by software. Certainly, they can rely on hardware to achieve, but with lower processing power and higher cost of tags encrypted than the former.

4   Application and Realization

At present, AFSCTS is designed and applied in Nanfeng county. Nanfeng county is located in the southeast of Jiangxi Province, China. It is famous orange hometown, where Nanfeng Orange was regarded as "Orange King" by the president of former Soviet Union Stalin and also praised as "Gold Orange" by Chinese Premier Wen Jiabao.

Beautiful Nanfeng is abundant not only in orange but also in other characteristic agricultural foods (e.g. rice, pickles, turtle, bean milk skin and beating tea). Although international market demand is very large for these agricultural foods, actual exports cannot meet them, even Nanfeng Oranges was unsalable in domestic market at one time, partly because there is no tracking system of agricultural foods, so it is very difficult to track agricultural foods and to find the reasons when food unsafely events happen. Some unsafely events occurred in recent years have seriously done harm to the world brand of Nanfeng Orange and shares of international market. So, to ensure quality safety of agricultural foods is very meanful work for striving to expand the international market. Based on the above three key technologies and two key algorithms, overall framework and function structure of Nanfeng county's AFSCTS are described as follows.

4.1   Overall Framework

General idea of the system design is described as follows: Integrate the data processed by Event handler, then store the above data in a public database of AFSCTS. Because the integrated data usually involves business secrets, supply chain enterprises can encrypt the data for protecting business secrets. The user who has permission can access public database to query the corresponding data. Integrated database involves the whole supply chain of each enterprise's monitoring data, so there is huge amount of data in this shared database, and it needs to store the previous data regularly in historical database.

When a tracing request involves historical data, it can use tracing query by calling the historical database or archiving the data in the historical database to the current data. Integration database of AFSCTS generally stores each enterprise's monitoring data, and it adds a historical database for relieving the pressure of integration database and increasing query sensitivity of current product status information. Based on supply chain integration database, we can design a special food product traceability platform conveniently for the users who have permissions to trace products, and feedback the result of retroactive to users directly. Here the user usually refers to the customer who has key, because it involves the data security and the data on all aspects of the product may be encrypted. The AFSCTS for Nanfeng county can also track the flow of products, and it's more convenient for enterprises   to analyzing the market.

Overall framework of Nanfeng county's AFSCTS is shown in fig.2, where this framework includes RFID data flow based on complex event handling mechanism of EPCIS. According to the EPCIS standards, AFSCTS can be divided into two parts (Master data and Event data), and the two events include simple events and complex events. According to the business function of complex events, it can be divided into Object Event, Quantity Event, Transaction Event and Aggregation Event. Filtering repeated data and classifying into various incidents based on its function, EPCIS has special events capture operation module, and it will capture events to handle the

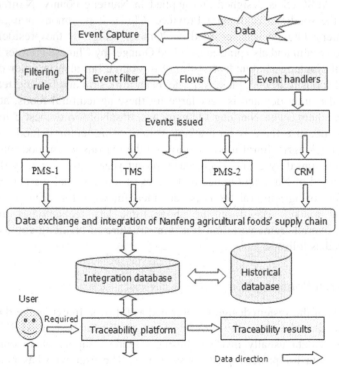

Fig. 2. Overall framework of Nanfeng county's AFSCTS

processing center in the stream of event, then issue these events. The capture interface can be defined as Capture (event:List< EPCISEvent >): void.In the whole of this operation, data is employed by RFID technology, then deposited in each link database of the subsystems in AFSCTS respectively.

4.2   Function Structure

Based on the above design thought, monitoring information of products in the whole Nanfeng agricultural food supply chain is integrated, and then stored in integration database of AFSCTS. Hence, any authorized nodes in the supply chain (e.g. enterprises, Governments, or consumer) can trace the sources and whereabouts of Nanfeng agricultural foods conveniently from this database. In this system, it's worth saying that Nanfeng agricultural foods can be reversely tracked and positively tracked in the supply chain.

There are main four functions of AFSCTS to be designed for Nanfeng agricultural food supply chain in the paper, namely planting monitoring, processing monitoring, transportation monitoring and sales monitoring. Its function structure is shown in Fig. 3.

(1) Planting monitoring: through RFID technology, this system mainly real-timely records the detail of Nanfeng agricultural products' planting information, such as plant varieties, seeding records, irrigation records, fertilization records, pest control records, and manage the information about pesticide purchase, storage, use and safety period. If pesticide expired or banned pesticide appears, it will give off warning information. After Nanfeng agricultural foods are picked, picking date, plot numbers and picking sequence will be batched by the times number. Put all basic information including management personnel's name together in a label card written by the times number, thus make each participating object on tracing chain of agricultural foods.

(2) Processing monitoring: because information about Nanfeng agricultural foods can be easily added to labels, once these foods enter the processing link, the related processing enterprise can read the label information of the products first, and then according to the needs in itself, change or add the corresponding information (such as processing enterprise name, processing time, processing address and packed weight) to the labels.

(3) Transportation monitoring: when preparing Nanfeng agricultural foods for loading, it uses fixed literacy to detect the products in loading area. AFSCTS first reads the label card of transport vehicles automatically, this label card records basic information of the vehicles such as nameplate and owner'name. If the label card is read with illegal data, it will alarm automatically. After loading finished, loading personnel needs to add loading information to label card, such as loading time, start time, destination and so on.

(4) Sales monitoring: Manager can read label by reader automatically to establish special sales monitoring system (such as monitoring system of supermarkets) based on RFID technology. If the customers pay online, AFSCTS can not only trace information record of this customer and know final whereabouts of Nanfeng agricultural foods, but also it can trace these products sourcing in the upper supply chain.

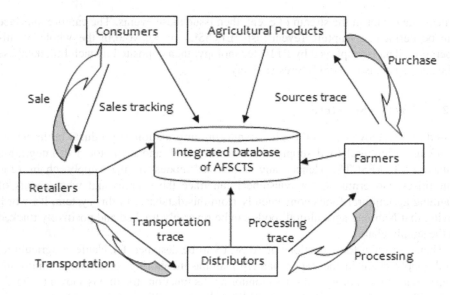

Fig. 3. Flow-process diagram of AFSCTS

5   Conclusion

There is an old saying "Food is the paramount necessity of the people." In recent years, frequently occurred agricultural food safety events all over the world have been severely doing harm to people's health, and also directly initiating great trust crisis to governments and the society. In order to timely trace and find unsafe agricultural food source and provide a powerful technical support to crack down on food crime and safeguard people's health, designing and developing AFSCTS is of great practical value and social significance and is supported by the government and all walks of life. Based on three key technologies and two key alogrithms, our research groups begins with exploratory research for designing and planning an AFSCTS in E-commerce environment, which is applied in Nanfeng county (the hometown of famous Nanfeng Orange in China) for tracing food safety of Nanfeng agricultural Supply Chain, and the experimental result of this system implementation is successful. As supply chain Tracking System has seldom been applied to agricultural food supply chain in China, the research and application is an innovation, and its fruit can be used in pharmaceutical and dangerous goods industry as well as agricultural food supply chain.

Acknowledgment. The authors thank the anonymous reviewers for their valuable remarks and comments. This work is supported by Shanghai Natural Science Fund (No.13ZR1429600), 2012 Teaching reform Project of Colleges and Universities of Jiangxi Province (Grant No. JXJG-12-3-16), the 11th Five-year Social Science Planning Project of Jiangxi Province in 2010 (Grant No. 10GL35), Social Science

Research Project of Guangzhou University, Innovation Program of Shanghai Municipal Education Commission (No. 13YZ141), and Young teacher training scheme of Shanghai Universities (No. SXY12014).

References

1. Tajima, M.: Value of RFID in supply chain management. Journal of Purchasing & Supply Management 69(12), 261–273 (2013)
2. Kshetri, N.: Barriers to e-commerce and competitive business models in developing countries: A case study. Electronic Commerce Research and Applications 6(4), 443–452 (2013)
3. Angeles, R.: Supply-Chain Applications and Implementation Issue. Information Systems Management 23(10), 51–65 (2012)
4. Nath, B., Reynolds, F.: Want R, RFID technology and applications. IEEE Pervasive Computing 5(1), 22–24 (2012)
5. Knospe, H., Pohl, H.: Information Security Technical Report. RFID security 9(4), 39–50 (2012)
6. Ko, J.M., Kwak, C., Cho, Y., Kim, C.O.: Adaptive product tracking in RFID-enabled large-scale supply chain. Expert Systems with Applications 38(3), 1583–1590 (2012)
7. Fisher, J.A., Monahan, T.: Tracking the social dimensions of RFID systems in hospitals. International Journal of Medical Informatics 77(3), 176–183 (2013)
8. Huang Li-juan, A.: personalized recommendation system for e-supply chain based on improved fat-growth algorithm. Journal of Applied Mathematics and Statistics 45(13), 135–141 (2013)
9. Simpson, L.K.T., Stanfield, P.: A model for quantifying the value of RFID-enabled equipment tracking in hospitals. Advanced Engineering Informatics 25(8), 23–31 (2013)
10. Jing, M.X., Chen, C.: The application of RFID in tracing system. Computer Integrated Manufacturing Systems 126(10), 241–253 (2012)
11. Huang, F., Hao, P., Wu, H.R.: Application of RFID middleware in food product safety Tracking System. Transactions of the CSAE 24(5), 177–181 (2012)
12. Wang, F.S., Liu, S.R., Liu, P.Y.: A temporal RFID data model for querying physical objects. Pervasive and Mobile Computing 6(3), 382–397 (2012)
13. Huang, L., Yu, P., Luo, Q., Zou, C.: E-Tourism Supply Chain Evaluation Based On Ahp And Fce Method. Journal of Theoretical and Applied Information Technology 45(2), 702–709 (2012)
14. Hopper, N.J., Blum, M.: Secure Human Identification Protocols. In: Boyd, C. (ed.) ASIACRYPT 2001. LNCS, vol. 2248, pp. 52–66. Springer, Heidelberg (2001)
15. Lee, C.K.M., William, H., Ho, G.T.S., Lau, H.C.W.: Design and development of logistics workflow systems for demand management with RFID. Expert Systems with Applications 38(5), 428–437 (2011)
16. Saidi, K.S., Teizer, J., Franaszek, M., Lytle, A.M.: Static and dynamic performance evaluation of a commercially-available ultra wideband tracking system. Automation in Construction 29(3), 453–458 (2013)

Evaluation of EPIC Model of Soil NO₃-N in Irrigated and Wheat-Maize Rotation Field on the Loess Plateau of China

Xuechun Wang [1,*], Shishun Tao [1], Jun Li[2], and Yongjun Chen[1]

[1] School of Life Science and Technology, Southwest University of Science and Technology,
Mianyang, Sichuan, 621010, China
[2] College of Agronomy, Northwest A & F University, Yangling, Shaanxi 712100, China
xuechunwang@swust.edu.cn

Abstract. EPIC model has been evaluated and used world wide, however there is still some disagreements on the simulation results of nitrogen cycle. Based on field experimental data, simulation results of soil NO_3-N was evaluated and the parameter sensitivity for simulated NO_3-N was analyzed in irrigated winter wheat / summer maize field on the Loess Plateau of China. Results showed 1) EPIC model estimated soil NO_3-N content and its movement among different soil layers well, with the mean RRMSE value of 0.46, for irrigated winter wheat / summer maize cropping system in the semi-humid region of the Loess Plateau. 2) Simulation results of soil NO_3-N was more sensitive to soil parameters, compared with crop parameters and meteorological parameters. 3）To improve the parameter value of BN2, HI, TB, WA, CNDS, BD and FC was better to the EPIC model to simulate soil NO_3-N on the Loess Plateau of China.

Key word: NO_3-N, Parameter sensitivity, EPIC model, The Loess Plateau.

1   Introduction

Nitrogen is one of the macronutrients necessary for plant growth[1] and plays an important role in increasing crop yield during the past century[2], however it also causes NO_3-N contamination of groundwater in areas of intensive agriculture, due to the over use of nitrogen by farmers on the Loess Plateau of China[3]. In general, any downward movement of water through the soil profile will cause the leaching of NO3-N, with the magnitude of the loss being proportional to the concentration of NO3-N in the soil solution and the volume of leaching water[4-5]. There is potential for NO3-N to be leached wherever rainfall or water supply exceeds evapo-transpiration. A common conclusion reported by many researchers is that proper fertilizer, crop, water, and soil management can minimize leaching of NO3-N and increase crop yield[6-7]. However, it is difficult to make a decision considering so many factors at the same time by field experiment. With the developing of computer, crop model has been used to solve this kind of problem by simulation method.

---

* Corresponding author.

D. Li and Y. Chen (Eds.): CCTA 2013, Part II, IFIP AICT 420, pp. 282–289, 2014.

As an multi-crop model, EPIC model (Environmental Policy Integrated Climate Model) has been evaluated and used world wide to simulate crop yield[8], soil water[9] and nutrition cycle[10] for different cropping system with different fertilizer and irrigation managements. Gaiser et al. (2010) reported that EPIC model can explain about 80% of the variance in crop yield in tropical sub-humid West Africa and semi-arid Brazil [11]. Wang and Li (2010) showed that EPIC model can estimate soil water and crop yield well with the new database built up for the Loess Plateau [8]. Several validation studies (Cavero et al. 1998 and 1999) found that EPIC satisfactorily simulated measured soil nitrogen (N) and/or crop N uptake levels [12-13]. However, less accurate soil N and crop N uptake results were reported in EPIC validation studies by Chung et al. (2002) [14]. Therefore its necessary to evaluate EPIC model before using EPIC model to simulate soil NO₃-N in irrigated field on the Loess Plateau of China.

Objectives of this paper were to evaluate EPIC model of simulated soil NO3 in irrigated field with different fertilizer levels and to point out some advices for the better application of EPIC model in the world.

2   Material and Method

2.1   Site Description

Yulin (E108.07°, N34.26°), as a typical intensive agriculture place of semi-humid region on the Loess Plateau, is located at the southern part of Shanxi providence. It is an arid continental monsoon climate zone and is a main wheat and maize area of China. Its annual precipitation is 550~600mm and 50% of them dropped in July, August and September. Its mean annual temperature is 10°C with the mean frost-free period of 152 d, and the mean sunshine hours of 2158h. Predominant soil used for winter wheat/summer maize in this area is Lou soil, with the field capacity and wilting point of 210~222 g/kg and 110-120 g/kg respectively.

2.2   Field Experiment

The field experiment, for irrigated winter wheat / summer maize cropping system, was carried out at Yangling eco-agriculture station from 1994 to 1997. Its fertilizer treatments were as follows: (1) no nitrogen fertilizer (N0), (2) 130 kg pure N/hm² (N130), (3) 260 kg pure N/hm² (N260), (4)390kg pure N/hm² (N390), (5) 520kg pure N/hm² (N520). Phosphorous fertilizer applied for each treatment was the same, 52 kg/hm² of P₂O₅. All experiments were established in 15 plots of 10.26m×6.5m (with a buffer zone of 1m between plots). Plots were arranged in a Randomized Complete Block Design (RCBD) with three replications. According to the requirement of crops, 150mm underground water were used to irrigate winter wheat and summer maize each year.

2.3   Method

2.3.1   Soil NO_3^-

In this research, soil samples for different soil layers (0~0.1m, 0.1~0.2m, 0.2~0.4m, 0.4~0.6m, 0.6~0.8m, 0.8~1.0m, 1.0~1.4m, 1.4~1.8m, 1.8~2.2m, 2.2~2.6m, 2.6~3.0m, 3.0~3.5m, 3.5~4.0m) were collected by core break method, before planting and after harvest in each experiment plot. NO_3 –N was abstracted using 1mol/L KCl solution, the content of NO_3-N was measured by Continuous Flow Analyzer.

2.3.2   Evaluation Method

Relation index (R), root mean square error (RMSE), relative root mean square error (RRMSE)and relative error (RE) [8-9] was used to evaluate how well the EPIC model simulated soil NO_3-N in irrigated winter wheat / summer maize cropping system on the Loess Plateau of China.

In order to evaluate the simulation results of NO_3-N movement in different soil layers, equation 1 and 2 wer used to calculate the changing of NO_3-N in each soil layer.

$$\Delta O = OH - OS \ (1) \ ; \qquad \Delta S = SH - SS \qquad (2)$$

Where ΔO and ΔS was observed and simulated, respectively, changing of NO_3-N in one soil layer from planting to harvesting. OH and SH was observed and simulated, respectively, NO_3-N of one soil layer before planting. OS and SS was observed and simulated, respectively, NO_3-N of one soil layer after harvest.

Sensitivity of model parameters on simulated soil NO_3-N was analyzed using Extended Fourier Amplitude Sensitivity Test (EFAST) [15-16]. Model parameters and their interconnection together impacted the Variation of Simulation Results (VSR), therefore we use equation 3 to divide the variation of simulation results, based on EFAST method.

$$V = \sum_i V_i + \sum_{i \neq j} V_{ij} + \sum_{i \neq j \neq m} V_{ijm} + \cdots + V_{12 \cdots k} \qquad (3)$$

Where V_i indicated the contribution of parameter X_i to VSR by itself, V_{ij} indicated the contribution of parameter X_i to VSR through parameter X_j; V_{ijm} indicated the contribution of parameter X_i to VSR through parameter X_j and X_m. $V_{12 \ldots k}$ indicated the contribution of parameter X_i to VSR through parameter X_1, X_2 .....and X_k.

After dividing the variation of simulation results, a normalization processing was made to got the first order sensitivity index of parameter X_i, by equation 4, and the second order sensitivity index by equation 5, the third order sensitivity index by equation 6.

$$S_i = \frac{V_i}{V} \qquad (4) \qquad S_{ij} = \frac{V_{ij}}{V} \qquad (5) \quad S_{ijm} = \frac{V_{ijm}}{V} \qquad (6)$$

Where S$_i$ was the first order sensitivity index of parameter X$_i$. S$_{ij}$ was the second order sensitivity index of parameter X$_i$, S$_{ijm}$ was the third order sensitivity index of parameter X$_i$. Then the total sensitivity of parameter X$_i$ can be calculated using equation 7. Where S was the total sensitivity of parameter X$_i$.

$$S = S_i + S_{ij} + S_{ijm} + \cdots + S_{12\cdots i\cdots k} \qquad (7)$$

3   Results

3.1   Content of NO$_3$-N in different Soil Layers for different Treatments

EPIC model simulated soil NO$_3$-N content well   with the mean R^2 value of 0.82 (figure 1). RMSE value of N0, N130, N260, N390 and N520 treatment was 1.016, 1.217, 2.781, 2.749, 3.873 mg/kg respectively, and the RRMSE value was 0.46, 0.43, 0.58, 0.33 and 0.39 respectively. Observed mean value of soil NO$_3$-N content was 3.96, 4.39, 8.63, 10.41 and 11.67mg/kg respectively, simulated mean value was 3.77, 4.22, 7.64, 10.15, 11.21 mg/kg respectively. simulated value was lower than observed value with the RE value of -0.05, -0.04, -0.11, -0.02 and -0.04 for N0, N130, N260, N390 and N520 respectively. Except N260 treatment, relationship between simulated and observe value of soil NO$_3$-N was significant, with the R value of 0.89, 0.86, 0.84 and 0.83 respectively(figure 1),

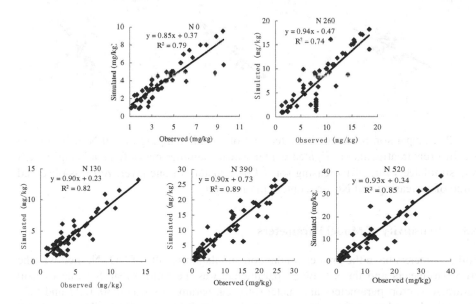

Fig. 1. Comparison between simulated and observed soil NO$_3$-N content for different fertilizer treatments in irrigated winter wheat / summer maize field at Yangling

3.2   Changing of Soil NO₃-N Content for different Treatments

Changing direction (increase or decrease) of $\triangle$S (Simulated value of increasing NO₃-N conteny) agreed well with $\triangle$O (Observed value of increasing NO₃-N content) except N260 treatment (figure 2). Mean value of $\triangle$O for N0, N130, N260, N390 and N520 was 1.95, 2.01, 4.46, 4.44, 5.59 mg/kg respectively, and mean value of $\triangle$S was 1.96, 1.92, 3.09, 4.79, 6.12 mg/kg. $\triangle$S was higher than $\triangle$O, with the RE value 1%, 8% and 9% for N0, N390 and N520 respectively,   and was lower than $\triangle$O with the RE value of -4% and -31% for N130 and N260 respectively. The EPIC model estimated well the variation of soil NO₃-N in irrigated winter wheat / summer maize field, with the RMSE value of 0.99, 1.447, 3.611, 3.014, 4.604 mg/kg, for N0, N130, N260, N390 and N520 respectively, the RRMSE value of 0.51, 0.75, 1.17, 0.63 and 0.70 respectively.

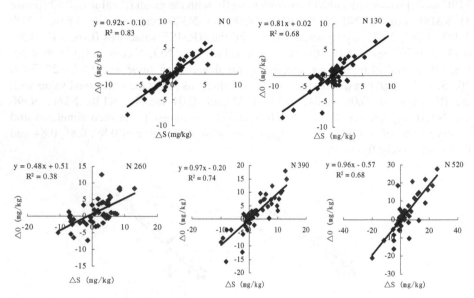

Fig. 2. Comparison between simulated and observed changing of soil NO₃-N content for different treatments in irrigated winter wheat / summer maize field at Yangling $\triangle$S was simulated value of increasing soil NO₃-N content in one layer, $\triangle$O was observed value of increasing soil NO₃-N content in one layer

3.3   Sensitivity of Model Parameters

Soil parameters have the greatest impact on simulation results of NO₃-N, then was the crop parameters and the metrology parameters was the third. Mean S value for soil parameters, crop parameters and meteorological parameters was 0.30, 0.27 and 0.23 respectively, and mean S1 value was 0.16, 0.13 and 0.11 respectively. The fact that S value of BN2, HI, TB and WA was higher than other crop parameters means that crop parameters of BN2, HI, TB and WA can be used to improve the simulation results of

NO₃-N of EPIC model. The same results was founded for soil parameters of CNDS, BD, and FC; for meteorological parameters of PRCP and RAD. Though the crop parameters HI and TB got an higher value of S, their value of S1 was lower comparing with WA, this means that though using HI and TB can improve the simulation results of NO₃-N in EPIC model, using other related parameters may be the better way. Similar results was founded for soil parameters of SLMX and CBN.

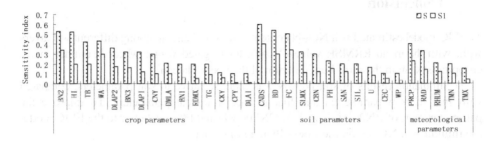

Fig. 3. Sensitivity of model parameters for simulated soil NO₃-N in irrigated winter wheat / summer maize field at Yangling. S was the total sensitivity of parameter X. S_1 was the first order sensitivity index of parameter X.

4   Discussion

For N260 treatment, the RRMSE value between simulated and observed soil NO₃-N content was higher than 0.5, and it was higher than 1.0 between $\triangle$O and $\triangle$S. This fact indicated that a far distance existed between simulation results and observed results. However, checking observed results carefully, we found that soil NO₃-N content was the same (7.97mg/kg) from top soil to deep soil after the harvest of winter wheat in 1995. If not considering these error values, the RRMSE value between $\triangle$O and $\triangle$S was 0.46, and it was 0.34 between simulated and observed soil NO₃-N content. Therefore, it is the error of observed value that cause the higher value of RRMSE value for N260 treatment.

Engelke and Fabrewitz (1991) found that EPIC estimates of denitrification and mineralization were plausible[17]; however, Richter and Benbi (1996) described EPIC's mineralization predictions as very poor[18]. Edwardset al. (1994) found that annual EPIC estimates of nutrient losses were significantly correlated with measured values, except for nitrate-N [19]. Chung et al. (2002) found that EPIC model estimated N loss in tile flow in Iowa region for corn and cotton[14]. Our results found that Except N260, simulated soil NO₃-N content was consistent with observed (figure 1), and simulated changing of soil NO3-N was similar with observed (figure 2). So EPIC model is an effective tool to simulate NO₃-N in irrigated cropping system on the Loess Plateau of China.

A sensitivity analysis by Benson et al. (1992) showed that EPIC N leaching estimates can be very sensitive to choice of evapo-transpiration routine and soil moisture estimates[20]. Niu et al (2010) found that most of model uncertainties introduced by input data that are not site-specific but commonly used[10]. Our results found that 1) to improve soil parameters was better for EPIC model to got a better simulation results of soil NO_3-N. 2) Simulation results of NO_3-N was sensitive to crop parameters of BN2, HI, TB and WA, soil parameters of CNDS, BD and FC, meteorological parameters of PRCP and RAD.

5   Conclusion

1) EPIC model estimated soil NO_3-N content and its movement among different soil layers well, with the mean RRMSE value of 0.46, for irrigated winter wheat / summer maize cropping system in the semi-humid region of the Loess Plateau.

2) Simulation results of soil NO_3-N was more sensitive to soil parameters, compared with that to crop parameters and that to meteorological parameters. To improve the parameter value of BN2, HI, TB, WA, CNDS, BD and FC was better to the EPIC model to simulate soil NO_3-N on the Loess Plateau of China.

Acknowledgement. This study was sponsored by the Foundation of SWUST university for new teachers with doctor degree (No. 11zx7140) and the foundation of Sichuan provincial Education Department (13zd1132). We are grateful to Shu-Lan Zhang who had provided experimental data of irrigated winter wheat / summer maize cropping system at Yangling.

References

[1] Bouman, O.T., Mazzocca, M.A., Conrad, C.: Soil NO3-N leaching during growth of three grass–white-clover mixtures with mineral N applications. Agriculture Ecosystems and Environment 136, 111–115 (2010)

[2] Wang, Q.J., Bai, Y.H., Gao, H.W., He, J., Chen, H., Chesney, R.C., Kuhn, N.J., Li, H.W.: Soil chemical properties and microbial biomass after 16 years of no-tillage farming on the Loess Plateau, China. Geoderma 144, 502–508 (2008)

[3] Zhao, B.Z., Zhang, J.B., Flury, M., Zhu, A.N., Jiang, Q.A., Bi, J.W.: Groundwater contamination with NO3-N in a wheat-corn cropping system in the north china plain. Pedosphere 17(6), 721–731 (2007)

[4] Guo, S.L., Zhu, H.H., Dang, T.H., Wu, J.S., Liu, W.Z., Hao, M.D., Li, Y., Keith, S.J.: Winter wheat grain yield associated with precipitation distribution under long-term nitrogen fertilization in the semiarid Loess Plateau in China. Geoderma 189, 442–450 (2012)

[5] Dimitroulopoulou, C., Marshm, A.R.W.: Modeling studies of NO3-N nighttime chemistry and its effects on subsequent ozone formation. Atmospheric Emironment 31(18), 3041–3057 (1997)

[6] Liu, X.J., Ju, X.T., Zhang, F.S., Pan, J.R., Peter, C.: Nitrogen dynamics and budgets in a winter wheat–maize cropping system in the North China Plain. Field Crops Research 83, 111–124 (2003)

[7] Yvonne, O., Yvonne, K., Roland, B., Wolfgang, W.: Nitrate leaching in soil: Tracing the NO3 -N sources with the help of stable N and O isotopes. Soil Biology & Biochemistry 39, 3024–3033 (2007)

[8] Wang, X.C., Li, J., Muhammad, N.T., Fang, X.Y.: Validation of the EPIC model and its utilization to research the sustainable recovery of soil desiccation after alfalfa (Medicago sativa L.) by grain crop rotation system in the semi-humid region of the Loess Plateau. Agriculture, Ecosystems and Environment 161, 152–160 (2012)

[9] Wang, X.C., Li, J.: Evaluation of crop yield and soil water estimates using the EPIC model for the Loess Plateau of China. Mathematical and Computer Modeling 51, 1390–1397 (2010)

[10] Niu, X.Z., William, E., Cynthia, J.H., Allyson, J., Linda, M.: Reliability and input-data induced uncertainty of the EPIC model to estimate climate change impact on sorghum yields in the U.S. Great Plains. Agriculture, Ecosystems & Environment 129(1-3), 268–276 (2009)

[11] Gaiser, T., de Barros, I., Sereke, F., Lange, F.M.: Validation and reliability of the EPIC model to simulate maize production in small-holder farming systems in tropical sub-humid West Africa and semi-arid Brazil. Agriculture, Ecosystems & Environment 135, 318–327 (2010)

[12] Cavero, J., Plant, R.E., Shennan, C., Williams, J.R., Kiniry, J.R., Benson, V.W.: Application of EPIC Model to Nitrogen Cycling in Irrigated Processing Tomatoes under Different Management Systems. Agriculture System 60, 123–135 (1997)

[13] Cavero, J., Plant, R.E., Shennan, C., Williams, J.R., Kiniry, J.R., Benson, V.W.: Application of EPIC Model to Nitrogen Cycling in Irrigated Processing Tomatoes under Different Management Systems. Agriculture System 56(4), 391–414 (1998)

[14] Chung, S.W., Gassman, P.W., Gu, R., Kanwar, R.S.: Evaluation of EPIC for Assessing Tile Flow and Nitrogen Losses for Alternative Agricultural Management Systems. Transactions of the ASABE 45(4), 1135–1146 (2002)

[15] Wu, J., Yu, F.S., Chen, Z.X., Chen, J.: Global sensitivity analysis of growth simulation parameters of winter wheat based on EPIC model. Transactions of the CSAE 25(7), 136–142 (2009)

[16] Saltelli, A., Tarantola, S., Chan, K.P.S.: A quantitative model-independent method for global sensitivity analysis of model output. Technometrics 41(1), 39–56 (1999)

[17] Engelke, R., Fabrewitz, S.: Simulation Runs with the EPIC Model for Different Data Sets. In: Soil and Groundwater Report II: Nitrate in Soils, pp. 288–299 (1991)

[18] Richter, J., Benbi, D.K.: Modeling of Nitrogen Transformations and Translocations. Plant Soil 181, 109–121 (1996)

[19] Edwards, D.R., Benson, V.W., Williams, J.R., Daniel, T.C., Lemunyon, J., Gilbert, R.G.: Use of the EPIC Model to Predict Runoff Transport of Surface-Applied Inorganic Fertilizer and Poultry Manure Constituents. Transactions of the ASABE 37(2), 403–409 (1994)

[20] Beckie, H.J., Moulin, A.P., Campbell, C.A., Brandt, S.A.: Testing Effectiveness of Four Simulation Models for Estimating Nitrates and Water in Two Soils. Canadian Journal of Soil Science 75, 135–143 (1992)

Three-Dimensional Reconstruction and Characteristics Computation of Corn Ears Based on Machine Vision

Jianjun Du, Xinyu Guo[*], Chuanyu Wang, Sheng Wu, and Boxiang Xiao

Beijing Research Center for Information Technology in Agriculture,
Beijing Academy of Agriculture and Forestry Sciences, Beijing, 100097, China
{dujj,guoxy,wangcy,wus,xiaobx}@nercita.org.cn

Abstract. Three-dimensional shape descriptors of corn ears are important traits in corn breeding, genetic and genomics research, however it is difficult to accurately and consistently measure 3D features of corn ears by hand or traditional tools. This study presents a 3D modeling method based on machine vision to reconstruct the 3D model of corn ears for quantitative feature computation and analysis. Firstly, a simple machine vision system is designed to capture images of corn ears from different angles of view. The corn ears in these images are then registered in the uniform coordinate system using a rapid process pipeline which consists of image processing, object detection, distortion correction and registration in pixel level etc. After the registration, the point sets in edge contours and center skeletons of corn ears are used to reconstruct the surface model based on resample and interpolation techniques. The experimental results demonstrate that the presented method can not only build realistic 3D models of corn ears for visualization, also be used to accurately compute geometric characteristics.

Keywords: corn ear, reconstruction, modeling, machine vision.

1 Introduction

Three-dimensional shape descriptors of corn ears are important traits for corn breeding, genetic and genomics research. However, it is difficult to measure 3D shape descriptors by manual operations or traditional tools. Machine vision (MI), which includes methods for acquiring, processing, analyzing, and understanding images, has been applied to automatic inspection, process control, and robot guidance in industry [1]. In agriculture applications, considerable research was reported using machine vision and image processing to identify features of agricultural products [2]. Hugue et al. described an approach for automatic assessment of crop and weed area in images of cereal crops using a tractor mounted camera [3]. Zhang et al. implemented fruits classification using computer vision and a multiclass support vector machine, and achieved classification accuracy of 88.2% [4]. Ni et al. designed a prototype machine vision system for inspecting corn kernels with random orientation, and respectively

---

[*] Corresponding author.

D. Li and Y. Chen (Eds.): CCTA 2013, Part II, IFIP AICT 420, pp. 290–300, 2014.

obtained classification rates of 91% and 94% for whole and broken kernels [5]. In general, the detection system using machine vision only utilizes two-dimensional information of detected objects, and it is difficult to directly obtain the three-dimensional shape measurements from images.

Recently, a measurement method for 3D geometric features of corn ears based on image processing was developed [6]. Moreover, analysis models of corn ears are also established to analyze corn threshing process using the Discrete Element Method [7, 8]. The above methods assumed the corn ear as segmented truncated cones, thus the computation accuracy was limited owing to the deficiencies of the individual differences in its transverse profiles. The objective of this study is to develop a three-dimensional reconstruction method of corn ears based on machine vision for computing and visualizing the geometric characteristics of corn ears. This method mainly consists of five steps: image acquirement, image analysis (object detection, distortion correction), skeleton registration in pixel level, surface reconstruction, 3D characteristics computation and visualization.

2   Experiments and Methods

2.1   Image Acquirement

A traditional vision system is designed as a controllable device to capture consecutive images of corn ears from different profiles, as shown in Fig. 1(a). This system consists of digital camera, turn-plane, needle, stepping motor and LED light source etc. A simple turn-plane with a steel needle is used to fix the corn ear, and driven to rotate by stepping motor. At each specified rotate angle, a picture is captured by the digital camera and store as JEPG format once the turn-plane stays stable. The size of corn ear in each image is related with the position of cameras, the distance between camera and corn ear, the offset orientation of ears and the inner parameters of the camera. It is difficult to guarantee the consistence of the rotate axis and the center axis of corn ear in the real application, therefore the corn ear in the images captured from different rotate angles will have different heights. Fig. 1(b) shows the schematic diagram of this machine vision system. We take the height of corn ear in two images with 180 degrees as $H1$ and $H3$, so the standard height of corn ear can be calculated as $H = (H1 + H3)/2$. In the next step, the corn ear will be respectively extracted from serial images, and scaled to the same height according to the standard height.

The number of images is determined by the included angle between each two adjacent imaging positions. In order to accurately reconstruct the three-dimensional skeleton of corn ear, at least two orthogonal images are prerequisite in the image sets, and two images with 180 included angles are also integrant. Therein, the orthogonal images with 90 degrees will be used to determine the center axis of corn ears in three-dimensional space, and the two images with 180 degrees are used to compute the standard height of corn ear. Thus, at least four images need to be captured for each corn ear. Fig. 1(c) shows four images among which each image has included angle of 90 degrees with its adjacent images.

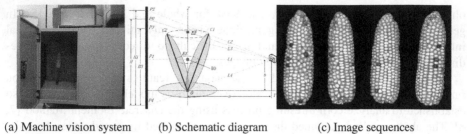

(a) Machine vision system   (b) Schematic diagram   (c) Image sequences

Fig. 1. Image acquirement based on machine vision system

In the image sequence, each image of corn ears only provides the local information of the entire corn ear. Due to the position, dip angle and shape differences of corn ears, the distortion of corn ears in different images manifests as inconsistencies of length, width and texture etc. To build the three-dimensional model of corn ears, these images with local information must be registered to the uniform coordinate system. In this study, the center axis of corn ear and the captured angle of images are used to determine the coordinate system of corn ears. Therefore, three continuous processes are implemented to register these images. Firstly, the valid region of corn ears in these images is extracted by threshold, contour extraction and classification techniques. The OBB of the corn ear in each image is then calculated and rotated to parallel to the (Cartesian) coordinate axes, and further zoomed to the standard height with the same proportion in length and width orientation of images. Finally, the contours of corn ears are extracted and split into three types of point sets. Combined with the captured angle of images, these point sets can be registered in the uniform coordinate system.

2.2   Object Extraction

The captured image generally contains corn ear and some background units, such as steel needle, turn-plane etc. In order to robustly extract the objects of interest, a simple method is used to clean the background information. A calibrated image is firstly obtained before placing corn ears, and then the subtraction result between this image and the image captured after placing corn ears will generate a cleaned image which only contains the corn ear. Furthermore, the cleaned image is converted into gray-level image for image segmentation and object extraction.

Since the image only contains the corn ear, Otsu's method [9] is used to calculate the threshold which converts the gray level image into the binary image. This method can automatically generate the optimum threshold by minimizing the intra-class variance or maximizing inter-class variance. Due to the color and shape varieties of corn ears, the islands or holes in binary images are difficult to be fully cleaned. Therefore, the contour of corn ear is more effective to represent the valid region of corn ear. Moreover, in order to obtain the smoothed edges of corn ears in binary images, iterative morphology operations with increasing radius of structure element will be used to clean the image until the image only contain the object of interest. Therein, Open operations are used to remove the islands, and Close operation to fill the holes. The maximum radius of structure element is taken as 10 using trial and error procedure to avoid the excessive smoothness. In most cases, the radius with 5 can effectively smooth the edges and fill holes in binary images. After each iterative morphology operation, the object number is counted using the connected-component labeling algorithm [10].

The edge contour of corn ears is determined by contour extraction and classification. A simple method is also used to rapidly extract the contours in the binary image. The binary image is firstly eroded one pixel, and then subtracted by this image. Valid pixels in the resulted image are then linked into closed contours according to their connectivity. Among all the contours, the contour with maximum pixels is taken as the final contour of corn ear. This contour is further filled with uniform label value using region growing method which utilizes the contour center as the growing point. Valid regions of corn ears in binary images were represented as B_{11}, B_{12}, ..., B_{1N}. Accordingly, regions of corn ears in RGB images can also be obtained according to the corresponding relation of pixel coordinates between the binary image and captured image, and expressed as S_{11}, S_{12}, ..., S_{1N}.

2.3   Distortion Correction

As mentioned above, the distortion of corn ears in images must be firstly corrected. The rotation and scaling of corn ears are implemented to uniform the corn ear to the standard height and the same proportion. To rotated corn ears in images to parallel to the (Cartesian) coordinate axes, the oriented bounding box (OBB) is used to determine the center axis of corn ears. OBB as a two-dimensional rectangle containing the object is the most simple and appropriate shape descriptor for the valid regions of corn ears. Because of the inconsistency between the rotate axis and center axis of corn ear, the major axis of OBB always offset to the rotate axis to some extent. The OBB of corn ear will be rotated to be aligned with the axes of the Cartesian coordinate system in the image. The offset angle between the major axis of corn ear and vertical orientation can be computed according to the vertices position of OBB, and is simultaneously used to rotate the binary and color images with the same degree.

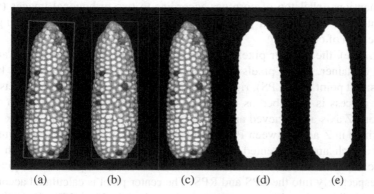

(a)   (b)   (c)   (d)   (e)

Fig. 2. Object extraction and distortion adjustment of corn ears (a) Captured RGB image; (b) Rotated RGB image and the extracted contour; (c) Rotated gray-level image; (d) Rotated binary image (B1); (e) Scaled binary image

The heights of OBBs in serial images are inconsistent, so an image resample method based on a central differencing scheme in combination with a nearest neighbor interpolation [11] is used to scale OBBs to the standard height. During the distortion correction, the valid regions of corn ears are rotated and scaled for generating new serial binary images B21, B22 … B2N and RGB images S21, S22 … S2N. Therein, binary images are used to build the skeleton of corn ear, and RGB images to provide the rendering texture.

Since the reconstruction results are susceptible to the edge of input images, we test and verify the availability of edge maintain. For images of different ears, we use the above image analysis methods to obtain the contours of corn ears. Fig. 2 shows the process to extract and correct objects from a given input image. Fig. 2(a) computes the OBB of corn ear, and obtain the angle between the rotate axis and the longitude axis of OBB is 0.038(2.162 degree). Therefore, the corn ear is rotated to vertical orientation as Fig. 2(b). In Fig. 1(b), the interval angles between the center axis and rotate axis of four images with 90 intervals are respectively 2.16, -1.28, 2.15 and 3.96 degrees. These corn ears are rotated to parallel with vertical orientation, and the heights of corn ear are respectively 594, 576, 564 and 591 pixels. Corn ears are further zoomed to the standard height with scales (1.02, 0.99, 0.97 and 1.02). The pixel resolution in this vision system has been measured as 0.02887cm/pixel, therefore the maximum height difference of corn ears come up to 30 pixels (0.87cm, about 1.5 seed height). Fig. 2(d) and (e) shows the rotated and scaled corn ears. Compared with the manual measurement, the standard height of corn ear is consistent within 0.5% relative deviation. Accordingly, the contours of corn ears can accurately represent the edges of ears, and RGB region surrounded by the contours can also be used for the texture image.

2.4   Skeleton Registration

The surface model of corn ear originates from ear contours of serial captured images. After the distortion correction of corn ears, a registration method in pixel-level is used to reset the contour pixels in three-dimensional space. This method firstly built the three-dimensional skeleton of corn ear according to two orthogonal images (with 90 degree interval). And then all pixels in contours will be registered according to the center skeleton of corn ears.

In Fig. 2 (e), the contour pixels of corn ears are extracted and stored to an initial point set container. These pixels in the contours are firstly split into three kinds of point sets: left point sets (LPS), right point sets (RPS) and the center point sets (CPS). The split process is described as follows: For each initial point set, the two extreme points along Z axis are retrieved and respectively set as the Pzmin and Pzmax. In each pixel position in Z axis between Pzmin and Pzmax, the pixels are collected into a sub point set which are represented as Pixels (z). Therefore, two pixels with longest distance can be extracted from each Pixel (z) as the left and right contour pixels, and pushed respectively into the LPS and RPS. The center pixel is calculated according to the corresponding left and right contour pixels, and pushed into CPS. Once the above processes are executed at each position between P(zmin) and P(zmax), LPS, RPS and CPS will be filled with these feature pixels in the same number.

In each image, CPS can be taken as the two-dimensional skeleton of corn ear. Therefore, we use two orthogonal images to determine the three-dimensional skeleton of the corn ear throughout coordinate exchange. The global coordination of corn ears is defined according to two orthogonal images, and the first image is assumes in X-Z plane (its orthogonal image locates in Y-Z plane). The origin point is set in the intersection point between the short edge and the longer center axis of OBB. The skeleton of corn ears is determined as follows: the coordinates of all pixels in the first image is defined as (X, 0, Z), and the coordinates of pixels in its corresponding orthogonal image is defined as (0, Y, Z). Therefore, the points in the three-dimensional skeleton of the ear can be represented as (X, Y, Z). The global coordination of corn ears is determined according to the OBBs and their camera angles. Each input image provides a longitudinal profile. In each valid position along Z axis, the transverse profile of corn ears consists of contour pixels from serial images, as shown in Fig. 3. The point O is assumed as the skeleton point of the corn ear. We assume images C1 and C2 as two orthogonal images. Therein, C1 has two edge points (C11, C12), and C2 is (C21, C22). The three-dimensional center of corn ears in Z coordinate is determined as follows: the X value in image C2 was modified as X value of the center point O, and the Y value in image C1 was also modified as the Y value of the center point O. Likewise, the coordination of points in the other images will be determined by the rotation angle of the images. After center point sets in two orthogonal images are brought into coincidence, point sets of the skeleton can be generated.

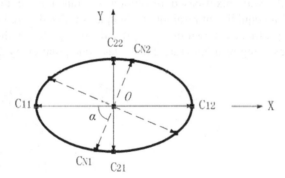

Fig. 3. A transverse profile of corn ears generated from serial images

2.5   Surface Reconstruction

The surface contour points can be classified into longitudinal or transverse profile points. The contour points in the longitudinal profile are corresponding to the contour obtained from one image, and the points in the same transverse profile possess the same Z coordinate. Each image only provides two contour points for each transverse profile, as shown in Fig. 3. Once the number of serial images is enough to describe the edge features of corn ears in pixel-level, the elaborate three-dimensional surface model can be generated. However, it is low effective and unnecessary to capture lots of images to reconstruction the surface model of corn ears. Therefore, to fit the three-dimensional model based on several images becomes an essential method for improving the efficiency of modeling. Here, B-spline curve is used to interpolate and resample the surface points of corn ears.

The B-spline curve is typically specified in terms of n+1 control points $(P_0, P_1, ..., P_n)$ a knot vector $U = \{u_0, u_1, ..., u_m\}$, and a degree p, as shown in the following:

$$C(u) = \sum_{i=0}^{n} N_{i,p}(u) P_i \qquad (1)$$

Where, $N_{i,p}(u)$ is B-spline basis function of degree p, and $m = n + p + 1$. To wrap control points or knots can construct closed B-spline curve. The construction procedure: (1) Add a new control point $P_{n+1} = P_n$; (2) Find an appropriate knot sequence of $n+1$ knots $u_0, u_1, ..., u_n$; (3) Add p+2 knots and wrap around the first $p+2$ knots: $u_{n+1} = u_0, u_{n+2} = u_1, ..., u_{n+p} = u_{p-1}, u_{n+p+1} = u_p, u_{n+p+2} = u_{p+1}$. (4) The closed curve with C^{p-1} continuity has n+p+2 knots, and $C(u_0) = C(u_{n+1})$.

In each position of center skeleton of corn ear, a closed B-spline curve based on several edge points from images is generated to fit the transverse contour, and is sampled according to the assigned number of points. Since the points in the longitudinal profile with pixel-based resolution are much more than ones in the transverse profile, an equidistant resample technique is utilized to reduce the points in each longitudinal profile, as shown in Fig. 4(a) and (b). Fig. 4(c) is the surface model with 2500 triangles using resample rate 24 (i.e. take one point every 24 points) in the

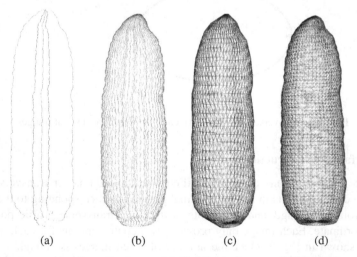

(a)   (b)   (c)   (d)

Fig. 4. Surface reconstruction of corn ears (a) Skeletons of corn ears; (b) Contour points after interpolation and resample; (c) Triangle meshes of 2500; (d) Triangle meshes of 5900

longitudinal profile; and Fig. 4(d) is the surface model with 5900 triangles using resample rate 10. Obviously, different resample rates of contour points will produce surface models with different detail characteristics. The point interpolation and resample techniques can regulate the number of points in each transverse and longitudinal profile. The polygonal surface model of corn ears can further be generated by linking the corresponding points between the adjacent contours.

3   Results and Discussion

The length and width indices are the most fundamental descriptors of size. For each image, the oriented bounding box can be considered to truly represent the size of corn ear in the corresponding angle of view. However, the serial images of the same corn ear will result in different sizes which vary within a larger-scope. As described in section 1, the length and width of corn ears can be computed according to serial images. In this study, the length and width of 168 corn ears were respectively measured and computed by hands and by the presented method to verify the computation effectiveness based on machine vision. Since the measured results by hand may be little different due to subjective judgment for a given corn ear, we only check whether the computation result falls into the measuring range by hand. The comparison of results shows the presented method can accurately obtain the size (height and width) of corn ears, even though the center axis of corn ear has an obvious angle deviation with the rotation axis.

The reconstructed model of corn ears consists of sets of triangles which are collected together to form the surface of corn ear. This surface model can be used to calculate the 3D shape parameters, such as volume, surface area, shape index etc. Generally, these descriptors of corn ears are difficult to be measured by hand and tools. Although the volume can be measured using draining method, the experimental results always lead to large deviations due to surface concave-convex, water permeability, the thickness and tightness of wrappings etc. From the geometric modeling perspective, if the width and height of corn ears is enough accurate, the volume and surface area can also be accurately computed. The experiments have shown that the computed length and width can represent the true sizes of corn ears, thus other indices computed using three-dimensional model can be deemed to locate in high confidence level.

Based on three-dimensional models of corn ears, both volume and surface area are calculated using numerical integration. Other shape features of interest can also be identified and each feature is quantified by one or more indices. For example, the non-sphericity index (NSI) can express how closely the shape of the corn ear resembles a sphere [12]. The NSI is defined as:

$$NSI = 1 - (18\pi)^{\frac{1}{3}} \frac{V^{\frac{2}{3}}}{S} \tag{2}$$

Where V is the volume and S is the surface area.

The number of resample points in the surface model is an important parameter to balance the efficiency and accuracy of three-dimensional modeling. We use the volume variance to determine the optimum resample rate. The source model in pixel-level resolutions can be taken as the standard model which provides the most accurate description for surface shape of corn ears. In different resample level, the three-dimensional model of corn ears can be generated and used to compute their surface area, volume and *NSI*. The computed error between the current model and the standard model is computed as:

$$Val_{error} = \frac{Val_{s\tan dard} - Val_{current}}{Val_{s\tan dard}} \tag{3}$$

Where, $Val_{s\tan dard}$ is the computed value (the surface area, volume or *NSI*) of the standard model, and $Val_{current}$ is the value of the current model in the designated resample rate. Fig. 5 shows the relationship between the number of triangles in the current model and computation errors. The volume error is the most relatively insensitive to the number change of triangles. Meanwhile, the surface area varies significantly with the decrease of triangles. For the models with 2500, 5900 and 29100 triangles, the errors of the volume, *NSI* and surface area are respectively (0.05%, 1.75% and 3.50%), (0.49%, 3.44% and 7.08%) and (1.93%, 3.75% and 8.56%). Generally, NSI can be taken as an effective shape descriptor of corn ears to determine the triangle number in the three-dimensional modeling. Therefore, the resample rate 10 (i.e. reserve one per ten points) can dynamically improve the modeling efficiency, and the shape error is controlled within 5%.

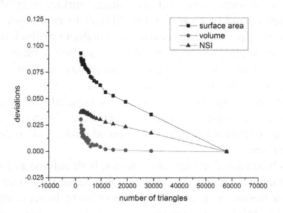

Fig. 5. The error analysis of surface area, volume and NSI

There-dimensional models of corn ears can not only be used for the shape characteristic calculation, but for visualization analysis. Three-dimensional models of six corn ears are respectively reconstructed using the presented method, and then serial two-dimensional images are projected onto the surfaces of three-dimensional models to obtain visualization results, as shown in Fig. 6.

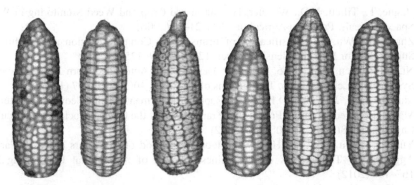

Fig. 6. Three-dimensional visualization of corn ears

4   Conclusions

To measure the surface, volume and shape descriptors of corn ears is very difficult by hand or traditional tools, therefore we presented an automatic reconstruction method of corn ears based on machine vision for features computation and visualization. This method can automatically build the three-dimensional models of corn ears based on serial image of corn ears, thus it is suitable for breeders to study the geometrical, morphological and visual symptoms characteristics in quantitative and qualitative analysis. A successive processing pipeline has been developed to extract the effective corn regions, adjust the imaging distortion to unify the region sizes, classify the pixels in contours and skeletons of corn ears into different point sets, and register all points in pixel-level to the uniform coordinate system. The advantages of the proposed method contain: a simple system based on machine vision is designed to automatically capture image sequences of corn ears from different angles of view; the distortion correction of valid regions provides more precise computation results of sizes and 3D shape descriptors; and the registration and resample in pixel level can accurately reconstruct three-dimensional corn ears and improve the modeling efficiency.

Acknowledgment. The research work was supported by Special Fund for Agro-scientific Research in the Public Interest under Grant No. 201203026, National Science & Technology Pillar Program under Grant No. 2012BAD35B01, and Special Fund for S&T Innovation of Beijing Academy of Agriculture and Forestry Sciences under Grant No. KJCX201204007.

References

1. Shapiro, L.G., Stockman, G.C.: Computer Vision. Prentice Hall (2001)
2. Costa, C., Antonucci, F., Pallottino, F., et al.: Shape Analysis of Agricultural Products: a Review of Recent Research Advances and Potential Application to Computer Vision. Food and Bioprocess Technology 4(5), 673–692 (2011)

3. Hague, T., Tillett, N.D., Wheeler, H.: Automated Crop and Weed Monitoring in Widely Spaced Cereals. Precision Agriculture 7(1), 21–32 (2006)
4. Zhang, Y., Wu, L.: Classification of Fruits Using Computer Vision and a Multiclass Support Vector Machine. Sensors (Basel) 12(9), 12489–12505 (2012)
5. Ni, B., Paulsen, M.R., Liao, K., et al.: Design of an Automated Corn Kernel Inspection System for Machine Vision. Transactions of the ASAE 40(2), 401–497 (1997)
6. Ma, Q., Jiang, J.T., Zhu, D.H., et al.: Rapid Measurement for 3D Geometric Features of Maize Ear based on Image Processing. Transactions of the Chinese Society of Agricultural Engineering 28(suppl. 2), 208–212 (2012)
7. Yu, Y.J., Zhou, H.L., Fu, H., et al.: Modeling Method of Corn Ears based on Particles Agglomerate. Transactions of the Chinese Society of Agricultural Engineering 28(8), 167–174 (2012)
8. Fu, H., Lv, Y., Li, Y.S., et al.: Analysis of Corn Threshing Process based DEM. Journal of Jilin University (Engineering and Technology Edition) 42(4), 997–1002 (2012)
9. Otsu, N.: A Threshold Selection Method from Gray-Level Histograms. IEEE Transactions on Systems, Man and Cybernetics 9(1), 62–66 (1979)
10. Wu, K.S., Otoo, E., Suzuki, K.: Optimizing Two-Pass Connected-Component Labeling Algorithms. Pattern Analysis and Applications 12(2), 117–135 (2009)
11. Parker, J., Kenyon, R.V., Troxel, D.E.: Comparison of Interpolating Methods for Image Resampling. IEEE Trans Med Imaging 2(1), 31–39 (1983)
12. Lauric, A., Miller, E.L., Baharogluet, M.I., et al.: 3D Shape Analysis of Intracranial Aneurysms Using the Writhe Number as a Discriminant for Rupture. Annals of Biomedical Engineering 39(5), 1457–1469 (2011)

Application of RMAN Backup Technology
in the Agricultural Products Wholesale Market System

Ping Yu[1,2] and Nan Zhou[1]

[1]Network Center, China Agricultural University, Beijing 100083, China
[2] School of Economics & Management, China Agricultural University, Beijing 100083, China
{yuping,zhnan}@cau.edu.cn

Abstract. Based on Oracle Database RMAN backup technology, this paper designs and implements a database backup subsystem for the agricultural products wholesale market system. Through full backup, incremental backups and cumulative incremental backup, it ensures the timeliness, integrity and efficiency of data recovery, when data mistake happens in the agricultural products wholesale market system database or the database collapses.

Keywords: agricultural products wholesale market system, Oracle database backup, RMAN.

1   Introduction

The development of computer technology and network technology makes database technology to be widely applied in enterprise management information systems, e-commerce systems, e-government systems and other application systems. With 随the continuous propelling of informatization , the amount of data in application systems grows sustainably. Data security issue in database systems have become increasingly important.

RMAN is Oracle tools which is used to backup, restore and restore Oracle database. RMAN can only be used in ORACLE8 or later versions. Researchers have applied Oracle backup technology to different areas. ZHANG Yun-fan discusses the application of backup and recovery technology of Oracle database in oil field exploration project [1]. Feng ke et al. gives a brief summary for database administrators who pay more attentions to the usability of backup data usability and auto-monitoring technique, the technique is applied in database backup of university and proved to be effective   [2]. ZHU Youcun et al. realized rapid backup and recovery based on oracle 10 and applied it in the database of hospital [3]. Following the database backup technology of Daqing logging production database wag still in the stage of the logical export manually and has not yet had an effective data backup technology to ensure data security, LIU Yubin uses RMAN backup technology on the features of the production database to deploy a perfect backup program in practice [4].

The agricultural products wholesale market system database has stored the price information and quality testing data of agricultural products. The safe storage of these data plays a crucial role in price management, quality inspection and macro decision-making of the agricultural products wholesale market. Therefore, when it appears that

D. Li and Y. Chen (Eds.): CCTA 2013, Part II, IFIP AICT 420, pp. 301–308, 2014.

data is accidentally deleted or database crashes, the requirement for its data recovery is zero-loss and to recovery to a designated time point. In addition, as the amount of data in this database is very large, and the allowed downtime of this system is 10 minutes, it requires the backup and recovery efficiency is as high as possible. Common logical backup will cause some data cannot be recovered, and the recovery time is far from the actual needs, therefore it needs to design a real-time, fully synchronized and strategic reasonable data backup subsystem.

RMAN can back up the entire database or database components, such as table spaces, data files, control files, archive files, and Spfile parameter file [5]. RMAN also allows you to conduct block-level incremental backup, incremental RMAN backups are time and space efficient because they only backup those data blocks which have changed since the last backup [6]. Therefore, it can well meet the data recovery needs of the agricultural products wholesale market system, whether from the respect of real-time or recovery efficiency.

In this paper, we used Oracle Database RMAN backup technology to design and implement a backup subsystem for the agricultural products wholesale market system.

2   Agricultural Products Wholesale Market System

The agricultural products wholesale market system is the first phase of Golden Agriculture Project, it uses the agricultural products wholesale market as the main datasource to collect the prices and the quality inspection data of agricultural products, it goes through data collection segment, the data processing segment, data

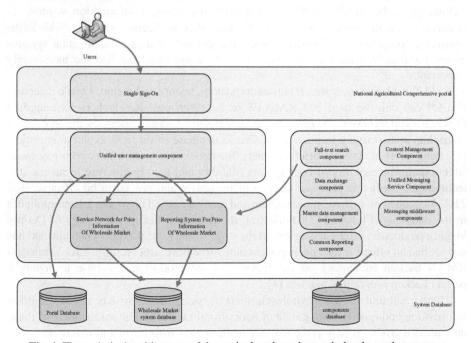

Fig. 1. The technical architecture of the agricultural products wholesale market system

analysis and display segment. The technical architecture of the agricultural products wholesale market system is shown as Fig.1.

As shown in Figure 1, System database includes portal database, the wholesale market system database and components database, it is an important support for the running of the agricultural product wholesale market system, as all of the system's data changes have been recorded in the database, if these data is lost, it will directly affect the normal running of the agricultural products wholesale business, so how to ensure data not to be lost, is essential. So it is important to choose a reasonable backup mode and develop a reasonable backup strategy to design and implement the agricultural products wholesale market system backup subsystem, so as to ensure the data not to be lost.

We choose Oracle10g for windows to meet the system's requirement for the capacity, response speed, database security, stability and scalability of database.

3   Database Backup Subsystem Based on RMAN Technology

3.1   The Architecture of RMAN Technology

The architecture of RMAN can be expressed in Fig.2 [7].

Through the above architecture diagram, we can know oracle RMAN is actually through the queries on the basis table of target database and the system views to conduct online hot backup for the database data files ,control files and spfile files and

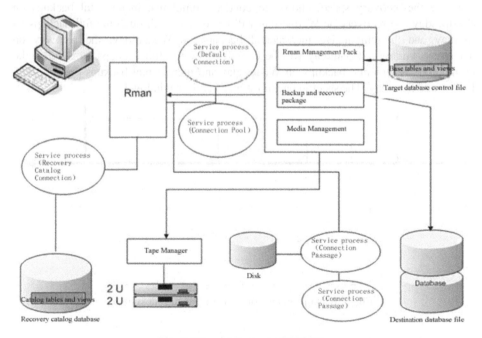

Fig. 2. The architecture of RMAN

so on, so as to achieve the recovery of target database after a crash [8]. In addition, in order to avoid bringing the performance loss to the operating system by backing up all files in each backup, RMAN also provides a backup of the archive logs, so that we can develop an appropriate strategy, namely conducting incremental archive logs backup after full backup. When the database crashes, we first restore full backup, and then apply the incremental backup archive to achieve full recovery or the specified point in time recovery. At backup time, RMAN will store the backup information of each time in the recovery catalog database, and establish a connection channel, and then through media management, transfer the backup sets to the backup media. When restoring database, RMAN will search in the recovery catalog database according to our specified conditions, when it finds the corresponding backup piece, it will use the oracle recovery package to conduct the corresponding recovery.

3.2   The Design of Database Backup Subsystem

By analyzing the characteristics of the agricultural products wholesale market system, we know that the amount of data in the database will be very large, and every Monday to Friday is the peak of the business, the data volume has increased significantly more than on Saturday and Sunday; the importance of its data is very high, it does not allow missing data and its recovery time should be as short as possible. Therefore, its backup strategy is as follows: Choose full backups on Sunday to reduce impact of full backups on the database performance; the data recovery on Monday, Tuesday will be completed through the full backups on Sunday and the incremental backup on the two days; in order to improve the recovery speed, choose to conduct cumulative incremental backups on Wednesday, the recovery on Wednesday will be completed through the full backups on Sunday and the cumulative incremental backups on Wednesday; the recovery on Thursday, Friday and Saturday will be completed by the full backups on Sunday, the cumulative incremental backups on Wednesday and incremental backups on Thursday, Friday and Saturday. The specific backup strategy is shown in Fig.3.

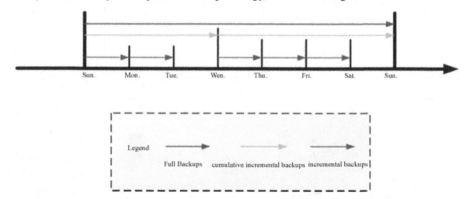

Fig. 3. Backup strategy of database backup subsystem

3.3   The Implementation of Database Backup Subsystem

1.   Application Preparation
Before using RMAN backup for the first time, we need to do some preparatory work, including set the target database to archive database, because RMAN needs the

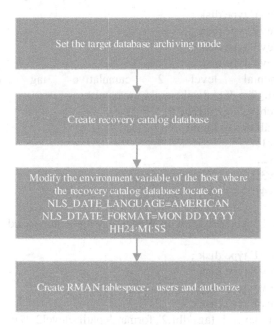

Fig. 4. Preparation before application

archive log file support when doing the time-based recovery or the full recovery or other recoveries. In addition, we need to create a database and create a table space and user in the database to store the recovery catalog information. These works can be broadly described in Fig.4.

2.   The design and implementation of backup and recovery program
    The full backup program on every Sunday are summarized as below:
run {
allocate channel 'dev1' type disk ;
allocate channel 'dev2' type disk ;
allocate channel 'dev3' type disk ;
allocate channel 'dev4' type disk ;
backup incremental level 0 tag 'dbL0' format '<path>\level0_%u_%p_%c' database skip readonly;
sql 'alter system archive log current' ;
backup archivelog all delete input;
release channel dev1;
release channel dev2;
release channel dev3;

release channel dev4;
}

The incremental backup program on Monday, Tuesday, Thursday, Friday and Saturday are summarized as below:

```
run {
allocate channel 'dev1' type disk ;
allocate channel 'dev2' type disk ;
allocate channel 'dev3' type disk ;
allocate channel 'dev4' type disk ;
backup    incremental    level    2    cumulative    tag    'dbL2'    format
'<path>/level2c_%u_%p_%c' database skip readonly;
sql 'alter system archive log current' ;
backup archivelog all delete input;
release channel dev1;
release channel dev2;
release channel dev3;
release channel dev4;
}
```

The incremental backup program on Wednesday are summarized as below:

```
run {
allocate channel 'dev1' type disk ;
allocate channel 'dev2' type disk ;
allocate channel 'dev3' type disk ;
allocate channel 'dev4' type disk ;
backup incremental level 1 tag 'dbL2' format '<path>/level2_%u_%p_%c' database
skip readonly;
sql 'alter system archive log current' ;
backup archivelog all delete input;
release channel dev1;
release channel dev2;
release channel dev3;
release channel dev4;
}
```

These backup program are set to be executed through the timed job system of windows.

3.   Backup and recovery joint adjusting

After making these preparations, we can officially start using RMAN for data backup. This includes running RMAN Manager to connect to the target database and recovery catalog database, register the target database, synchronize the information between the target database, the directory database, and the backup pieces, distribute and establish channels to a connection, run the script for backup and recovery, and finally release the connecting channels and other steps. These steps can be simplified represented in Fig.5.

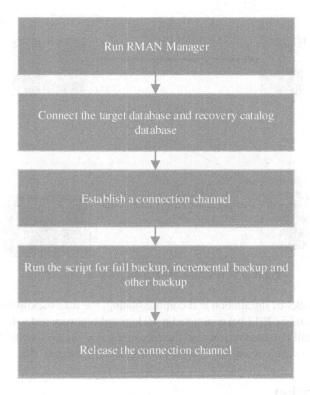

Fig. 5. The backup and recovery processes of RMAN

4   Test and Run

The agricultural products wholesale market system was officially launched in the end of December 2011, its database system was Oracle 10G installed under WINDOWS, and Oracle practical cluster systems was installed to ensure the high availability of the database. The physical structure of its operating environment is shown as Fig.6.

As can be seen from the figure, the database realized Oracle utilities (RAC) via two IBM p690 cluster systems.When there is a data error operation, the data can be restored through the database backup system.

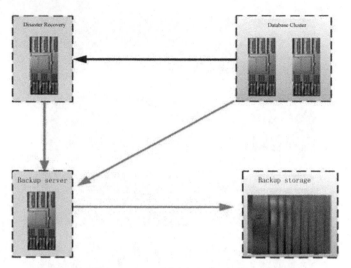

Fig. 6. The structure of database backup subsystem

After two weeks of installation and commissioning, the data backup functionality has been achieved, and in 2012 we conducted a backup and recovery test, which successfully implement the full database restore and the recovery based on the specified point in time, and ensured zero data loss and the recovery of accidentally deleted data.

5   Conclusion

This paper designed and implemented a database backup subsystem based on the Oracle Database RMAN backup technology, and successfully deployed it into the agricultural products wholesale market system, which ensures the security, reliability and recoverability of the back-end database of the system. It is also able to adapt to the security requirement of the slimily system.

References

1. Zhang, Y.-F.: Backup and Recovery Strategy of Oracle Database. Computer Engineering 35(15), 85–87 (2009)
2. Feng, K., Liu, N.-J., Wang, Q., Li, T.: Research and practice of auto-monitoring for Oracle database backup. Experimental Technology and Management 28(4) (2011)
3. Zhu, Y.-C., Luo, D., Yang, X.-R.: RMAN Backup, Recovery and optimization Based on oracle 10g. Medical Equipment 33(10), 45–46 (2012)
4. Liu, Y.-B.: The Research and Implementation of Database Backup Program Based on Oracle. Programming Skills & Maintenance 14, 025 (2009)
5. Freeman, R.G., Hart, M.: Oracle9i RMAN backup & recovery. McGraw-Hill/Osborne (2002)
6. Hart, M., Freeman, R.G.: Oracle Database 10g RMAN Backup and Recovery (2007)
7. Oracle RMAN Pocket Reference. O'Reilly Media, Inc. (2009)
8. Alapati, S.R., Kuhn, D., Nanda, A.: RMAN recipes for Oracle database 11g: A problem-solution approach. Apress (2007)

Effects of Water and Nutrition on Photoassimilates Partitioning Coefficient Variation[*]

Jianhua Jin and Yang-ren Wang

Department of Water Conservancy, Tianjin Agriculture University, Tianjin300384, China
{Jinjh2010,wyrf}@163.com

Abstract. Photoassimilates partitioning was studied using the experimental data of winter wheat in 2008-2009 at irrigation experimental station of Tianjin Agriculture University.The Research showed that when relative development stage (RDS) was greater than 0.35 fr(root) ,fr (stem and sheath) and fr(leaf) decreased ,fr(Ear) increased with the increase of RDS,and turned into negative When RDS were 0.59,0.67,0.7 respectively.It was due to the existence of repartitioning of photoassimilates, and photoassimilates storaged in root , stem and sheath and leaf began to transport to ear.Water and nutrition had a certain influence on the partitioning of photoassimilates among organs.Excessive water and nutrition were unfavourable for the transportation of photoassimilates to the growth center.

Keywords: winter wheat, photoassimilates, partitioning coefficient.

1   Introduction

Photoassimilates partitioning, known as dry matter partitioning, mainly refers to the partitioning of organic matter producted through photosynthesis among organs[1]. The status of water and nutrient in soil have great influence on photoassimilates accumulation and partitioning[2].   Variation of photoassimilate partitioning coefficient and the effects of diffrent water and nutrition treatment on winter wheat photoassimilate partitioning were discussed ,which will   provide a theoretical basis for determining the amount of water and nutrition of the high yield for winter wheat.

2   Experiments and Methods

2.1   Experimental Design

The experiment were conducted at irrigation experimental station of Tianjin Agriculture University in winter wheat growing season, 2008-2009, located in Xiqing district, Tianjin,longitude $116^{0}57$ ', latitude $39^{o}08$ ', height of 5.49 m above sea level.

---

[*] Tianjin Nature Science Foundation （10JCBJC09400）.

Xiqing district belongs to the warm temperate semi-humid continental monsoon climate zone. The soil of 0-2.7m was medium loam;The level of NO_3-N, NH_4-N of 0-0.6m were 3.87g/kg ,3.05g/kg respectively. Winter wheat varieties was 6001. Winter wheat were sowed in October 7, 2008, harvested in June 15, 2009.

The test consisted of 5 treatments,2 of which were choosen for analysis in this paper, high water and nutrition ,medium water and nutrition(on behalf of the local normal water and nutrition level).Base manure of high nutrition treatment was compound fertilizer 750kg/hm^2.After manuring was urea 225 kg/hm^2. Base manure of medium water and nutrition was compound fertilizer 450kg/hm^2,After manuring was urea 150 kg/hm^2. Irrigation times of high water treatment (the whole growth period) were 4, winter water(December 2, 2008), jointing water(April 9, 2009), esring water(May 7, 2009) , filling water (May 18, 2009) respectively; Irrigation times of medium water treatment was 3, winter water(December 2, 2008), jointing water(April 9, 2009), filling water (May 18, 2009) respectively;Irrigation quota was 60mm for all treatments.Each treatment contained 3 replicates, the area of a plot was 40m^2.the total area of two treatment was 240m^2.There was no isolation zone between replicates, .Three plots of one treatment were adjacent. There was protection zone in test area,with width of more than 3m.

2.2   Test Items and Methods

The test index included dry matter, soil moisture,nutrition, plant height, leaf area, temperature, wind speed, radiation,etc. The measuring of dry matter: 10 representative plant samples were chosed and cut against ground.The root were got through a stratified sampling method with root drill, 10cm a layers,washed with tap water, dived into several groups according to plant organ subsequently.Each groups dryed to constant weight under the conditions of 80°C after fixxing 30 minute under the conditions of 105°C, then weighted.

Soil moisture was determined by neutron instrument, with measuring depth was 160cm and testing one time each 20cm .

Temperature, wind speed and radiation, etc meteorological data were from meteorological station of Xiqing , Tianjin city.

2.3   Data Analysis

Dry matter of Every organs were measured for 8 times from March 27,2009 to June 15,2009.The data were analyzed by Excel2003 .The variation of dry matter of every organs with time were simulated to determine every organs daily dry matter . The difference between daily dry matter weight and the previous day's was the dry matter increment.the ratio of dry matter increment to the total of each organ was definded as partitioning coefficient of an organ.

3   Results and Analysis

3.1   The Photoassimilates Partitioning Coefficient of Organ of Medium Water and Nutrition

Photoassimilates from the plant photosynthetic organ is distributed to each organ.Plant organs(root,stem,leaves and storage organs) receive photoassimilates for the maintenance of their physiological activity and growth.The rate of photoassimilates partitioning to   each organ is in connection with RDS, the amount of which depends on photoassimilates partitioning coefficient. The photoassimilates   partitioning coefficient is as follows:

$$fr(org) = \frac{GAA(org)}{Fgass}$$

Where org is the plant organ, $fr(org)$ is photoassimilates   partitioning cocfficient, $GAA(org)$ is the rate of photoassimilates partitioning to various organs $(kg/ha^{-1} \cdot d^{-1})$, $Fgass$ is the total rate of assimilation of the field-crop plants $(kg/ha^{-1} \cdot d^{-1})$。

photoassimilates   partitioning coefficient of medium water and nutrition   was shown in figure 1.It can be seen from Figure 1 that   fr(root) ,fr（stem and sheath）and fr(leaf) decreased ,fr(Ear) increased with the increase of RDS when RDS was greater than 0.35. It was the reason that the absorption and using process of sinks had significant effects on the partitioning of assimilates. The   photoassimilates partitioning of each sink dependde on the competition ability of each sink mainly.The stronger the competition ability was,the more photoassimilates.Competition ability have two meanings,   strength and priority.Strength is defined as the potential demand or potential capacity   to photoassimilates of sink[2]. Priority refers to the   order of meet the need to photoassimilates of sinks when photoassimilates supply is not sufficient.According to the previous studies, the   priority order of plant organs from strong to weak were seed, fruit (stem and leaves), cambium layer, root and storage structure[3-4].Fr(ear) was the bigest because the ear competitiveness to photoassimilate was greater than   root, stem and sheath, leaf when RDS was greater than 0.59, and increased with increasing RDS.

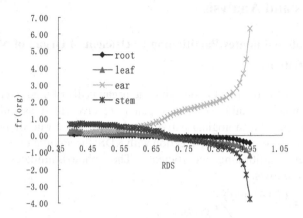

Fig. 1. Photoassimilates partitioning coefficient of winter wheat of medium water and nutrition

It can be seen from Figure 1 that fr(root) ,fr(stem and sheath) and fr(leaf) turned into negative When RDS were 0.59,0.67,0.7 respectively . It was due to the existence of repartitioning of photoassimilates.Component of plant body with the exception of cell walls ,even the inorganic substance, can be redistributed to other organs.For example, when the leaf senescence, most of the sugar and N, P, K etc evacuated and redistribute to new organs nearby.When RDS was greater than 0.59 ear become the growth center of winter wheat,so photoassimilates storaged in root , stem and sheath and leaf began to transport to ear. The amount of photoassimilates of root , stem and sheath and leaf decreased with time,so the corresponding photoassimilates   partitioning coefficient were negative.Photoassimilates supply to ear mainly,so fr(ear) was the   maximum.when RDS=0.95, partitioning coefficient of ear reached 6.35.The ratio of photoassimilates transported   from stemandsheath was the largest,followed by leaves and roots.

3.2   Effects of Water and Nutrient on Photoassimilates Partitioning Coefficient

The data of high water and nutrition ,medium water and   nutrition treatments were analysized to detect whether the nutrition and water had influence on the partitioning coefficient.It can be seen from Figure 2 that photoassimilates partitioning coefficient of root,stem and sheath, leaf of high water and nutrition   treatment were greater than that of medium water and nutrition. That was, vegetative growth of high water and nutrition was luxuriantly than that of   medium water and nutrition.When RDS was 0.90 fr(root) ,fr（stem and sheath）, fr(leaf), fr(ear) of medium water and nutrition were -0.18,-0.47,-0.89,2.54 respectively . Fr（stem and   sheath）, fr(leaf), fr(ear) of medium water and nutrition were -0.13,-0.25,-0.45,1.83 respectively.It showed that excessive water and nutrition made against the transportation of photoassimilates to the growth center.The ratio of photoassimilates transported to ear from root , stem and sheath and leaf of high water and nutrition treatment was lower than   that of medium water and nutrition treatment.So water and nutrition should be controled appropriately in later growth period in order to realize high yield of winter wheat.

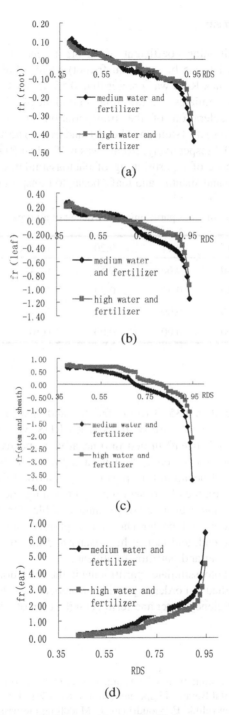

Fig. 2. Photoassimilates partitioning coefficient among organs

3.3   Data Comparison

Photoassimilates partitioning coefficient of all organs measured by P.M.Driessen[5] were shown in Table 1.It can be seen that fr(root) fr(stem) and fr(leaf) were 0.00, and fr(ear) was 1.00 when RDS was greater than 0.56. In his study, the minimum and maximum of Photoassimilates partitioning coefficient of all organs were 0.00 and 1.0 without the consideration of the transportation of photoassimilates among organs.However fr(root) ,fr（stem and sheath）and fr(leaf) turned into negative When RDS were 0.59,0.67,0.7 respectively, fr(ear) was 6.35 when RDS=0.95 in the paper.It was due to the existence of repartitioning of photoassimilates, and photoassimilates storaged in root , stem and sheath and leaf began to transport to ear.

Table 1. Photoassimilates partitioning coefficient of all organs of winter wheat

RDS	0	0.11	0.20	0.35	0.47	>0.56
fr(leaf)	0.50	0.66	0.56	0.34	0.10	0.00
fr(root)	0.50	0.34	0.23	0.09	0.04	0.00
fr(stem)	0.00	0.00	0.21	0.57	0.86	0.00
fr(s.o.)	0.00	0.00	0.00	0.10	0.00	1.00

4   Conclution

When RDS was greater than 0.35   fr(root) ,fr（stem and sheath）and fr(leaf) decreased, fr(Ear) increased with the increase of RDS.When RDS were 0.59,0.67,0.7 fr(root), fr（stem and sheath） and fr(leaf) turned into negative respectively .It was due to the existence of repartitioning of photoassimilates,and photoassimilates storaged in root , stem and sheath and leaf began to transport to ear. It showed that water and nutrition had a certain influence on the photoassimilates partitioning coefficient among organs, and excessive water and nutrition were unfavourable for the transportation of photoassimilates to the growth center.The ratio of photoassimilates transported to ear from root , stem and sheath and leaf of high water and nutrition treatment was lower than that of medium water and nutrition treatment .

In future research photoassimilates partitioning in the conditions of limited water and nutrition should be studied , in order to provide theoretical basis for the determination of high yield and high efficiency water and ferlilizer utilization of winter wheat.

Reference

1. Marcelis, L.F.M.: Sink strength as a determinant of dry matter partitioning in the whole plant. Journal of Experimental Botany 47(special issue), 1281–1291 (1996)
2. Marcelis, L.F.M., Hewvelink, E., Goudriaan, J.: Modelling biomass production and yield of horticultural crops: a review. ScientiaHorticulturae 74, 83–111 (1998)

3. Minchin, P.E.H., Thorpe, M.R.: What determines carbon partitioning between competing sinks? Journal of Experimental Botany 47(special issue), 1293–1296 (1996)
4. Wardlaw, I.F.: The control of carbon partitioning in plants. New Phytologist 116, 341–381 (1990)
5. Yu, Z.-R., Wang, J.-W., Qiu, J.-J.: Land-use systems analysis, vol. 195. China Agricultural Science and Technology Publishing House, Beijing (1997)

Effect of Water and Nitrogen Stresses
on Correlation among Winter Wheat Organs

Zhou Xin-yang and Wang Yang-ren[*]

Hydraulic Engineering Department, Tianjin Agricultural University, Tianjin, China, 300384
xinyang87@126.com, wyrf@163.com

Abstract. Correlation among root, stem, leaf and grain is the basis for the establishment of photosynthate partitioning model, and is of great significance for rational regulation of plant morphology and population structure and improvement of crop economic production. The paper is based on experiment results of dry mater weight of root, stem, leaf, and grain of winter wheat of 5 treatments with different water and fertilizer level, and experiment results of soil moisture and soil nitrogen (NH^{4+}-N和NO^{3-}+N). The experiment was carried out in the Farmland Water Cycling Experiment Station of Tianjin Agricultural University in 2008-2009. The relationships between leaf to root ratio, stem to leaf ratio, and grain to stem ratio with root nitrogen uptake was established, respectively. The results indicted that the leaf to root ratio, and stem to leaf ratio was more sensitive to water and nitrogen stresses than that of grain to stem ratio. Both the leaf to root ratio and stem to leaf ratio increased with the root nitrogen uptake, and had positive correlations with root nitrogen uptake with the correlation coefficients above 0.91. The correlation between grain to stem ratio and root nitrogen uptake was not significant within the given range of water and nitrogen stresses. The ratio of grain to stem under water and nitrogen stresses was approximately equal to that without water and nitrogen stresses.

Key words: growth correlation, water and nitrogen stresses, winter wheat, leaf to root ratio, root nitrogen uptake.

1   Introduction

Various parts of winter wheat (including root, stem, leaf and grain) integrate to a whole, where each organ has its unique function with close links between different organs. This phenomenon is known as correlation among plant organs (Zhang, 2007). Correlation among root, stem, leaf and grain is the basis for the establishment of photosynthate partitioning model, and is of great significance for rational regulation of plant morphology and population structure and improvement of crop economic

---

[*] Corresponding author.

D. Li and Y. Chen (Eds.): CCTA 2013, Part II, IFIP AICT 420, pp. 316–325, 2014.

production. Davison (1969) republished correlation model between root and shoot in dry matter, the model included nitrogen uptake in previous results. The other previous results presented only correlation models among plant organs and these models are all without both water stress and nutrient stress. For example Michaelis-Menten's formula (Liu, et al. 2010), Thomas, et al. (1997), in additional biomass partitioning indexes (Liu, et al. 2010), or partitioning coefficients (P.M. Driessen, et al. 1997), which imply correlation relationships among plant organs, between crop organs. Based on field experiment results, this paper republished the correlation models between the ratio (the leaf to root ratio, stem to leaf ratio and grain to stem ratio) with nitrogen uptake of winter wheat under water and nitrogen stresses conditions.

2   Materials and Methods

The experiment of water and fertilizer in winter wheat double factors was carried out at the water cycling experiment bases of Tianjin Agricultural University in 2009. The station located at 116°57'E and 39°08'N, elevation 5.49m, groundwater table 3.70-2.06m. The type of winter wheat is 6001, Growth period was from October 7, 2008 to June 15, 2009. The soil bulk density in 0-100cm is 1.43 g / cm^3. The number of experiment treatments was 11 with 3 replications, and the area of each experiment plot is 66.7 m^2. The dry matter data of roots, stems, leaves, grain, soil moisture and nutrient (include ammonium nitrogen NH_4^+ and nitrate NO_3^+) were measured for all treatments.

These treatments include the combinations of high water (4 times water during all growth period, 60mm each time) and high fertilizer (base fertilizer: 750kg/ha compound fertilizer with potassium sulfate type with 15% nitrogen, 15% phosphorus, 15% potassium, and over 45% total nutrients, topdressing: 225kg/hm^2 urea with 46.2% total nitrogen), medium water (3 times) and medium fertilizer (450kg/ha compound fertilizer, 150kg/ha urea), medium water and low fertilizer (150kg/hu compound fertilizer, 75kg/ hm^2 urea), and without water and fertilizer.

The test depth of soil moisture and nitrogen is 0-160cm which is divided into eight layers, 20cm each layer. From sowing to harvest, eight times was sampled to test nitrogen in indoor method, and soil moisture is tested 22 times using the neutron probe. The root was sampled using root drilling with the inner diameter of 5cm and the drill bit length of 10 cm, cleaned, oven-dried and weighted to obtain the dry weight. Root samples were taken in the ridge and between two ridges, and their mean value is used to calculate the dry weight of the root. Sampling depth is 0-60cm before jointing and 0-100cm after jointing stage, 10cm a layer. Dry matter weight was tested nine times.

3   The Calculation of Nitrogen Uptake by Winter Wheat Root

Nitrogen uptake amount is calculated from the root water uptake multiplied by the soil water nitrogen concentration in different soil layers, which is

$$x(t) = 0.1 \cdot \sum_{i=1}^{t} \sum_{j=1}^{4} S_{ij} \cdot \Delta z_j \cdot C_{ij} / \theta_{ij} \tag{1}$$

Where $x(t)$ = cumulative nitrogen uptake by crop root from sowing to the day t, kg/hm²; S_{ij} = root water uptake on day i and soil layer j, 1/d; C_{ij} = soil nitrogen content (the sum of nitrate and ammonium nitrogen) on day i and soil layer j, mg/kg; Δz_j = soil depth of layer j, cm, and Δz_j = 20cm in this study; i = days number from sowing date, d; j = soil layer numbers, which is j = 1, 2, 3 and 4 for 0-20cm, 20-40, 40-60cm, 60-80cm soil layers, respectively. The root water uptake can be calculated by Equation (2)(Kang, et al. 1994).

$$S_r(z,t) = 2.1565 \cdot T_p(t) \cdot \frac{e^{-1.8z/z_r}}{z_r} \left(\frac{\theta(z,t) - \theta_{wp}}{\theta_{cr} - \theta_{wp}} \right)^{0.6967} \tag{2}$$

Where $T_p(t)$ = potential crop transpiration rate at time t, mm/d; z = the soil depth from the surface, cm; $\theta(z,t)$ = soil volumetric water content on day t and at depth z, cm³/cm³; z_r = crop root depth at time t, cm; θ_{wp} = soil moisture content at the wilting point, cm³/cm³; θ_{cr} = critical soil moisture content affecting the root water uptake, cm³/cm³. In this $\theta_{cr} = 0.75\theta_f$, where θ_f = soil moisture content at field capacity and θ_f = 36 cm³/cm³ within the 0-100cm soil layer. Following Driessen and Konijn (1997), z_r can be approximated by

$$z_r = \begin{cases} 0 & t \leq 14 \\ 0.3623t + 1.3768 & 14 < t \leq 221 \\ 80 & t > 221 \end{cases} \tag{3}$$

Potential transpiration rate can be expressed as (Kang, et al. 1994),

$$T_p(t) = ET_p(t) \cdot (1 - e^{-K \cdot LAI(t)}) \tag{4}$$

Where $LAI(t)$ = Leaf area index; K = extinction coefficient which is taken as 0.6 in this paper; $ET_p(t) = K_c(t) \cdot ET_o(t)$, $ET_p(t)$ = Crop potential evapotranspiration, mm/d; $ET_o(t)$ = Reference crop evapotranspiration, mm/d, which can be calculated from the Penman-Monteith equation (Richard, et, al. 1998); $K_c(t)$ = crop coefficient.

In days without LAI measurement, LAI can be estimated from linear interpolation of two adjacent measurements. $K_c(t)$ can be calculated from (5) (Kang, et al. 1994),

$$K_c(t) = \begin{cases} aLAI(t) + b & LAI(t) < LAI_0 \\ K_{co} & LAI(t) \geq LAI_0 \end{cases} \tag{5}$$

Where a, b, LAI_0, and K_{co} are constants, which are 0.21, 0.42, 4.2, and 1.30, respectively.

In Eqs. (1) and (2), soil moisture content and soil nitrogen content can be obtained from measured values in the field test. In days without measurement, soil moisture content and soil nitrogen content of different soil layers can be obtained from linear interpolation of measurement. Then the amount of daily root water uptake and nitrogen uptake can be calculated. Using Eq. (1), the cumulative nitrogen uptake of winter wheat can be obtained for different irrigation and fertilization treatments. From Eq.(1), the cumulative nitrogen uptake considers both root water uptake and soil nitrogen content, and can reflects soil moisture and soil nutrient status to some extent. Therefore, cumulative nitrogen uptake can be used for the quantitative representation of nutrient and water stresses effects on plant growth. In this paper, cumulative nitrogen uptake is taken as the moisture-nutrient stresses index to analyze the effect of water and nitrogen stresses on correlation among winter wheat organs, and to establish the relationship between stem to leaf ratio ($K_{sl}(t)$), grain to stem ratio ($K_{s\theta}(t)$), leaf to root ratio ($K_{lr}(t)$) and root nitrogen uptake. The above ratios are defined as $K_{sl}(t) = W_s(t)/W_l(t)$, $K_{\theta s}(t) = W_\theta(t)/W_s(t)$, $K_{lr}(t) = W_l(t)/W_r(t)$, where $W_l(t)$, $W_s(t)$, $W_\theta(t)$ and $W_r(t)$ are dry weights of leaf, stem, ear and root, kg/hm^2. Because the grain test is more difficult, especially before maturity, and ear dry weight test is easier; ear dry weight rather than grain dry weight was used. This is appropriate because the correlation between the maturity grain dry weight and ear dry weight is significant. For example, sampling data on June 15, 2009 resulted in y=0.7872x-117.79, R^2=0.981, where y= grain dry weight, x= ear dry weight, kg/hm^2.

4   Results and Discussion

4.1   Effect of Water and Nitrogen Stresses on Winter Wheat Root Water Uptake and Nitrogen Uptake, Yield and Total Dry Matter Weight

Amount of winter wheat root water uptake and nitrogen uptake are calculated from Eqs (2) and (1) (Table 1) by using 5 treatments. For different treatment, amount of winter wheat root water uptake and nitrogen uptake have great differences, which includes not only the impact of water shortage on the amount of root water uptake, but also the effect on leaf area index difference on winter wheat root water uptake. As can be seen from Figure 2, due to water, nitrogen double stresses, leaf area indexes have greater difference under different treatments. Leaf area indexes of high water high fertilizer and medium water and fertilizer treatments are greater, and their maximum reach 6.53 and 4.96 which appeared in about 0.7 (relative growth period). LAI is smaller for medium water and low fertilizer treatment, and smallest for the medium water and zero fertilizer treatment. Thus, to some extent, root nitrogen uptake amount is a better representation for the degree of water and nitrogen stresses.

Table 1. Amount of winter wheat root water uptake and nitrogen uptake in the whole growing period under different treatments

Treatment	Root water uptake /mm	Root nitrogen uptake /kg/hm^2	Grain yield /t/hm^2	Total dry matter /t/hm^2
High water and high fertilizer	299.4	496.7	6.362	20.355
Medium water and fertilizer	231.8	187.4	6.297	16.245
Medium water and low fertilizer	195.4	95.1	5.522	17.895
Medium water zero fertilizer	166.9	85.5	4.461	14.085
Without water and fertilizer	151.9	67.8	3.491	12.015

4.2   The Relationship between Leaf to Root Ratio, Stem to Leaf Ratio, Stem to Grain Ratio and Root Nitrogen Uptake of Winter Wheat

4.2.1   The Relationship between Leaf to Root Ratio and Root Nitrogen Uptake

The relationship between leaf to root ratio and root nitrogen uptake can be expressed as the power function $K_{lr}(t) = ax(t)^b$, where a and b are two regression coefficients, and the regression results is given in Table 2 using 5 treatments. The regression equations for some testing time do not reach a significant level, but the leaf to root ratio increases with root nitrogen uptake and its power index is around 0.5 with the average value of 0.5460. The coefficient a decreases with the relative growth period. By fixing b=0.5, the regression analysis of $K_{lr}(t) = ax(t)^{0.5}$ results in $a = 0.0584t^{-1.9884}$. Using 2 to approximate the power index 1.9884, the relationship between leaf to root ratio and root nitrogen uptake can be expressed by equation (5).

$$K_{lr}(t) = 0.0584x(t)^{0.5} \cdot t^{-2} \tag{5}$$

Table 2. Relationships of leaf to root ratio and root nitrogen uptake at different growing period

Time/year-month-day	Growth days	Relative growth period	a	b	Squared correlation coefficient R^2	F(1,2)
2008-12-17	71	0.268	1.3258	0.4164	0.3191	0.937
2009-03-27	171	0.367	0.5652	0.3754	0.3741	1.793
2009-04-16	191	0.488	0.1642	0.4875	0.5933	4.376
2009-05-07	212	0.655	0.0864	0.5658	0.7373	8.420
2009-05-14	219	0.722	0.0990	0.5196	0.7929	11.486
2009-05-21	226	0.789	0.0467	0.6612	0.8591	18.292
2009-05-27	232	0.852	0.0271	0.7575	0.7894	11.245
2009-06-04	240	0.936	0.0645	0.5022	0.5864	4.253
2009-06-15	251	1.055	0.0281	0.6285	0.6702	6.096

Note: significant $F>F_{0.05}=18.5$, not significant $F<F_{0.1}=8.53$.

4.2.2   The Relationship between Stem to Leaf Ratio and Root Nitrogen Uptake

The relationship between $K_{slm}(t)/K_{sl}(t)$ and $1/x(t)$ is found to be linear (Fig. 1) , where $K_{slm}(t)$ is the stem to leaf ratio under high water high fertilizer treatment. From the linear regression results, the relationship between stem to leaf ratio and root nitrogen uptake can be expressed as

$$K_{sl}(t) = \frac{K_{slm}(t)}{1.0123 - 27.496/x(t)} \qquad (6)$$

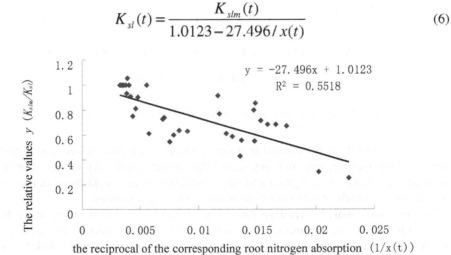

Fig. 1. Relationship between stem to leaf ratio and root nitrogen uptake

4.2.3   The Relationship between Grain to Stem Ratio and Root Nitrogen Uptake

Relative grain to stem ratio is defined as the grain to stem ratio under varying conditions divided by the ratio under high water and high fertilizer condition. The relationship between relative grain to stem ratio and root nitrogen uptake is shown Fig. 2. As seen from Figure 5, the relative grain to stem ratio ($y_r = K_{\theta s}(t)$) remains constant for different root nitrogen uptake. Linear regression of $y_r = a_2 x(t) + b_2$ results in $a_2 = -0.00005$, $b_2 = 0.98$ and the corresponding correlation coefficient $R = 0.0245 < R_{0.05}(2,33) = 0.349$. It shows that the effect of water and nitrogen stresses on grain to stem ratio is not significant. For winter wheat, in a certain water and nitrogen stresses range (root nitrogen uptake range in 57-306kg/hm^2), the relative grain to stem ratio is not related to root nitrogen uptake, and the value is 0.98 and very close to 1. It indicates that the grain to stem ratio is not affected by water and nitrogen stresses and its value is equal to grain to stem ratio without water and nitrogen stresses.

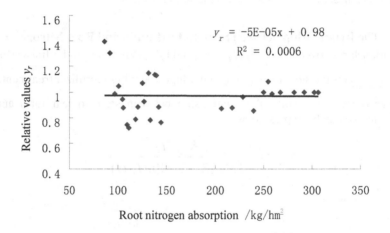

Fig. 2. Relationship of grain to stem ratio and root nitrogen uptake

The above results do not mean that grain growth and stem growth were not related to water and nitrogen stresses, and only shown that the ratio of the dry weight of grain to stem did not change or the grain and stem weights increase or decrease in the same degree for winter wheat suffering from water and nitrogen stresses.

In the above analysis, relative ratios with respect to with the high water high fertilization treatment (regarding as no water and nitrogen stresses) were used. As a result, data from different treatment and time can be combined in models, which can be used for correlations analysis of water and nitrogen stresses on the whole growth period of winter wheat.

4.3   Validation

4.3.1   Validation of the Relationship between the Leaf to Root Ratio with Root Nitrogen Uptake

Leaf to root ratio can be calculated from Eq.(5) for other different treatments. The scatter plot of measured and calculated values (Fig 3) shown that the scatter points lies around the 1:1 line, with the correlation coefficient $R = 0.9164 >$ $R_{0.01}(2,33) = 0.449$ and reached a significant level. This shown that Eq. (5) can be described the relationship between the leaf to root ratio and root nitrogen uptake.

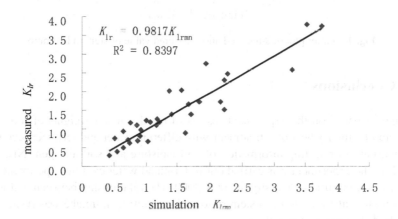

Fig. 3. Scatter plot of simulated and measured values of leaf to root ratio

4.3.2   Validation of the Relationship between Stem to Leaf Ratio with Root Nitrogen Uptake

Stem to leaf ratio under various root nitrogen uptake conditions can be obtained from Eq. (6). The scatter plot of measured and calculated stem to leaf ratios (Fig 4) shown that the simulated and measured values are close to the 1:1 line with the correlation coefficient $R = 0.9167 > R_{0.01}(2,33) = 0.449$ and reached a very significant level. This shown that the formula (6) expressed the effect of water and nitrogen stresses on stem to leaf ratio of winter wheat. The ratio increases with root nitrogen uptake.

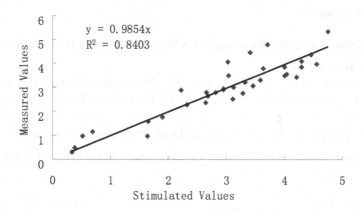

Fig. 4. Scatter plot of simulated and measured value of stem to leaf ratio

5   Conclusions

The paper is based on the experiment results of the dry mater weights of root, stem, leaf, grain of winter wheat of 5 treatment with different water and fertilizer levels, and the synchronization testing information of soil moisture and soil nitrogen (NH_4^+-N和 NO_3^-+N). The experiment was carried out in farmland water cycling experiment bases of Tianjin Agricultural University in 2008-2009. The relationship between leaf to root ratio, stem to leaf ratio, grain to stem ratio and root nitrogen uptake was respectively established. The results show that:

(1) The water and nitrogen uptake by roots, to a certain extent, can be reflected the degree of water and nitrogen stresses and can be taken as the index to describe the effect of water and nitrogen stresses on crops growth.

(2) Leaf to root ratio and stem to leaf ratio are more sensitive to water and nitrogen stresses than grain to stem ratio, and have significant correlations with root nitrogen uptake, and increase with root nitrogen uptake. For the range of water and nitrogen stresses in this study, grain to stem ratio is not correlated with root nitrogen uptake.

(3) In this paper, the range of water and nitrogen stresses is greater, with the root water uptake varying from 151.9 to 299.4mm, root nitrogen uptake from 67.8 to 496.7 kg/hm^2, and corresponding crop yield from 3.491 to 6.362 t/hm^2, the total dry matter from 12.015 to 20.355 t/hm^2.

Acknowledgment. Funds for this research was provided by the Natural Science foundation Projects (50679055), the Tianjin Science and Technology Development project (10JCYBJC09400).

References

1. Xu, K.-Z.: Plant physiology. China Agriculture Press, Beijing (2007)
2. Davidson, R.I.: Effect of root/leaf temperature differentials on root/shoot ratios in some pasture grasses and clover. Ann. Bot. 33, 561–569 (1969)
3. Liu, T.-M., Xie, G.-S.: Agricultural System Simulation, pp. 145–148. Science Publishers, Beijing (2010)
4. Kätterer, T., Eckersten, H., Andrén, O., Pettersson, R.: Winter wheat biomass and nitrogen dynamics under different fertilization and water regimes: application of a crop growth model. Ecological Modelling 102, 301–314 (1997)
5. Shi, J.-Z., Wang, T.-D.: A mechanistic model describing the photosynthate partitioning during vegetative phase. Acta Botanica Sinica 36(3), 181–189 (1994)
6. Wang, Y.-R.: Water, Heat Transfer and Crop Growth Simulation in SPAC with Water and Nutrient Stress. Northwest A&F University (2004)
7. Allen, R.G., Perein, L., Raes, D., Smith, M.: Guidelines for computing crop water requirements. FAO Irrigation and Drainage 56 (1998)
8. Zhang, Y.-F., Zhang, W.-Z.: The Present Status of Study on Uptake of Nitrogen by Crop Roots. Irrigation and Drainage 15(3), 35–39 (1996)
9. Kang, S.-Z., Liu, X.-M., Xiong, Y.-Z.: Soil-Plant-Atmosphere Continuum Water Transmission Theory and Its Application. Hydraulic and Electric Power Press, Beijing (1994)
10. Driessen, P.M., Konijn, N.T., Yu, Z.-R., Wang, J.-W., Qiu, J.-J., et al.: Land use system analysis. China Agricultural Science and Technology Press, Beijing (1997)

Application of a Logical Reasoning Approach Based Petri Net in Agriculture Expert System

Xia Geng, Yong Liang, and Qiulan Wu

Institute of Information Science and Engineering,
Shandong Agricultural University, Taian Shandong 271000, China
{gx,yongl,zxylsg}@sdau.edu.cn

Abstract. First of all, a goal-guiding graphic reasoning approach that based on the predicate/transition system has been proposed for the first-order predicate logic. In process of reasoning, the premise is separated from the conclusion, which has been taken as the beginning of the backward reasoning that is purposeful and effective as well. Next, this reasoning approach has been applied in the agriculture expert system to present a method of solving problem, providing a new way for studying the reasoning mechanism of the agriculture expert system.

Keywords: first-order predicate logic, Predicate/transition system, goal guiding, graphic reasoning, backward reasoning, agriculture expert system.

1 Introduction

Agriculture expert system in agriculture is to widely apply the accumulated knowledge and experiences of agricultural experts by using computer techniques which can overcome the limit of time and space, so as to turn these knowledge and experiences into productivity[1].

To construct a good reasoning mechanism is the basis of agriculture expert system. Reasoning means the process of searching the answers from knowledge base for the given problems when domain-specific knowledge has been stored into the base in a certain form. A reasoning process is to determine whether the given proposition is contained in the selected sets of the first-level of facts and clause rules[2-3]. In the reasoning processes of agriculture expert system there already have applied many approaches. Recent years, the expert system based Petri net and its application have become one of the research hotspots in the field of intelligent control and intelligent system[2]. Yet, it is not common that the Petri net model has been employed by agriculture expert system, and, theoretical system has not been established.

The first-order predicate logic has already enjoyed a wide application in computer science. For its safety and reliability, the practical results of some other logic theories turn out that, the applications in which these logic theories had been employed can get essential conclusions no more than the application in which only the first-order predicate logic had been employed, and can not get a better intelligent system[7].

D. Li and Y. Chen (Eds.): CCTA 2013, Part II, IFIP AICT 420, pp. 326–341, 2014.

Petri net offers a new way to study the first-order predicate logic reasoning. Not only the Petri net itself can provide an intuitive semantic frame for traditional logic symbols, but also its properties can be used for the randomness of logic reasoning, which finds a way for realizing the machine reasoning, and which increases the chances for dealing with reasoning problems by using different and effective ways.

For the first-order predicate logic reasoning, predicate/transition system (Pr/T system) of the high-level Petri net can be utilized to build the model. Based on the model, researches of logic reasoning are divided into two types. ① For the set of Horn clauses, the algorithm of computing T-invariants had been proposed[9,15-16]; four reasoning algorithms that can support the conclusions obtained and can be popularized to the first-order predicate logic had been proposed, through the improved strategies of resolution refutation [17]. ② For the set of non-Horn clauses, an efficient algorithm and the backward and forward approaches of analyzing T-invariants had been put forward in accordance with the necessary and sufficient condition of contradiction which is contained in the set of non-Horn clauses[18-19]; for propositional logics, the reasoning process is turned to solve the non-negative integer solutions of linear equations of the incidence matrix, and this principle can be applied for predicate logic[8].

However, the existing reasoning methods of the first-order predicate logic that based on Pr/T system are equal to the traditional resolution refutation method, in which premise and conclusion are put together to make up the reasoning. In such a way, some heuristic information are not easily to be used for reasoning process, where a large number of useless steps may exist and the reasoning processes are inefficient. Therefore, this paper proposed, by borrowing ideas from the and/or resolution refutation reasoning and based on the Pr/T system, a goal guiding graphic reasoning approach of realizing the backward reasoning, which is applied in the agriculture expert system.

2   Basic Concepts

We assume that our readers know well the knowledge of agriculture expert system, Petri net, the first-order predicate logic and reasoning. For simplifying the description, we just list some related concepts and terms here.

Definition 1[8]: Given that D is the nonempty finite set, and V is the nonempty finite symbol set.

If all of the symbols of V set are representatives of the elements of D set, then V can be deemed as a variable set of D, and these symbols are the variables of D.

(1)Both the elements and variables of D are called the D-terms of D. If $f^{(n)}$ is an equation of D with n unknowns, and $v_1, v_2, ..., v_n$ are terms of D, then $f^{(n)}(v_1, v_2, ..., v_n)$ is also a term of D, and, no other terms exist.

(2)The n-ary vector $<v_1, v_2, \cdots, v_n>$ of which the components of terms of D is called the n-ary tuple of D, where n $>$ 1.

(3)The sum that multiple n-ary tuples of D are connected by "+" is called as the n-ary symbolic sum of D, symbolic sum for short. When $n = 0$, it is called as the empty symbolic sum, which is represented with "$NULL$" or "$< >$".

(4)The symbol "+" is commutative.

Definition 2[8]: Given that $\Sigma = (S,T;F,D,V,A_S,A_T,A_F,M_0)$ is the Pr/T system, which meets the following:

(1) $(S,T;F)$ means the directed net, which is the basic net of Σ.

(2) D is the nonempty finite set which is called as the individual set of Σ, and there are operative symbols set Ω of D.

(3) V is a variable set of D.

(4) $A_S : S \rightarrow \pi$, where, π is the dynamic predicate set of D, for $s \in S$, if $A_S(s)$ is a n-ary predicate, then s is called as the n-ary predicate.

(5) $A_T : T \rightarrow f_D$, where, f_D is the formulary of D, for $t \in T$, $A_T(t)$ can contain only the static predicates and operative symbols of Ω.

(6) $A_F : F \rightarrow f_S$, where, f_S is the symbolic sum set of D. For a n-ary predicate $s \in S$, if $(s,t) \in F$ or $(t,s) \in F$, then $A_F(t,s)$ or $A_F(s,t)$ is the n-ary symbolic sum. For $t \in T$, free variables in formula $A_T(t)$ must be the free variables of the directed arc with the end of t.

(7) $M_0 : S \rightarrow f_S$, for a n-ary predicate $s \in S$, $M_0(s)$ is the n-ary symbolic sum.

When describing a logic problem, the first-order predicates can be divied into two types: describing the premise and describing the conclusion.

In general, this paper adopts the method proposed in literature [14] to build the Pr/T system model for the first-order predicate.

Definition 3: Assume that P and Q are predicates that describe the premise and conclusion respectively. Given that $\Sigma_1 = (S,T;F,D,V,A_S,A_T,A_F,M_0)$ and $\Sigma_2 = (S,T;F,D,V,A_S,A_T,A_F,M_0)$ are Pr/T net systems corresponding to P and Q respectively, then Σ_1 is called as the premise Pr/T net system, or called as the premise net for short, and Σ_2 is called as the conclusion Pr/T net system, or called as the conclusion net.

Definition 4: Assume that $\Sigma = (S,T;F,D,V,A_S,A_T,A_F,M_0)$ is the conclusion net, then $\forall t \in T$ means the target transition.

Definition 5: The two-ary tuple of $N = (S,F)$ which meets the following conditions is called as Predicate-and/or graph, or called as Pre-and/or graph for short.

(1) $|S| \geq 2$

(2) $F \subseteq (S \times S)$

(3) $dom(F) \cup cod(F) = S$

where,

$dom(F) = \{x \in S \mid \exists y \in S : (x, y) \in F\}$

$cod(F) = \{x \in S \mid \exists y \in S : (y, x) \in F\}$

(4) $\forall s \in S$ represents an atomic predicate formula.

(5) $s_1, s_2, \dots s_n (n \geq 2)$ means that the relation between predicates is "or", if and only if:

$\exists s \in S$ and $(s, s_i) \in F (i = 1, 2, \dots, n)$

(6) $s_1, s_2, \dots s_n (n \geq 2)$ means that the relation between predicates is "and", if and only if:

$\exists s \in S$ and $(s_i, s) \in F (i = 1, 2, \dots, n)$

$(s_i, s) \in F (i = 1, 2, \dots, n)$ is connected with semicircle.

(7) If $\exists x \in S, {}^{\bullet}x = \varnothing$, then the node x is called as the end-node.

Definition 6: Assume that $\Sigma = (S, T; F, D, V, A_S, A_T, A_F, M_0)$ is a Pr/T net system, and $\Sigma' = (S_1 \cup S_2, T'; F', D, V, A_{S_1 \cup S_2}, A_{T'}, A_{F'}, M_0')$ is a two-level place Pr/T subnet of Σ, if and only if all of the following conditions are met:

(1) $S_2 = \{s_1, s_2, \cdots, s_m\} \subseteq S(m \geq 0)$;

(2) $T' = \{t \mid t \in T, t^{\bullet} = S_2\}$, and T' is not null set;

(3) $S_1 = \{s \mid s \in S, \exists t \in T', which\ makes\ s^{\bullet} = \{t\}\}$;

(4) $F' = \{(x_1, x_2) \mid x_1, x_2 \in S_1 \cup S_2 \cup T', (x_1, x_2) \in F\}$, for $\forall (x_1, x_2) \in F'$, $A_{F'}(x_1, x_2) = A_F(x_1, x_2)$, $A_{S_1 \cup S_2} : S_1 \cup S_2 \to \pi$, where π is a dynamic predicate set of D ; $A_{T'} : T' \to true$, for $t \in T'$, $A_{T'}(t)$ is a static predicate, $M_0' : S' \to <>$. S_2 is the output place set of Σ' , and S_1 is the input place set of Σ' .

If Σ itself is a two-level place Pr/T subnet, then this Σ is called as a two-level place Pr/T net.

For example, fig.1 shows the basic net of a two-level place Pr/T net. In particular, both S_2 and S_1 in definition 7 can be null set, but because that no independent node shall exist in the net, they can not be null set at the same time. Besides, provisions (2) and (3) of definition 7 do not require T' and S_1 to be the maximal set. Therefore, even S_2 has been determined, the obtained two-level place Pr/T subnet may not be the only one.

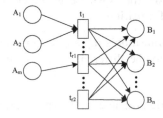

Fig. 1. The basic net of a two-level place Pr/T net

Definition 7: Assume that $\Sigma = (S,T;F,D,V,A_S,A_T,A_F,M_0)$ is a Pr/T net system, for $\forall t \in T$, if $t^{\bullet} \neq \varnothing$, then make $t^{\bullet} = \{s_1,s_2,\cdots,s_m\}(m \geq 1)$; define $\{s(X_1),s(X_2),\cdots,s_m(X_m)\}$ as the output predicate set of the transition t (where, $X_i(i=1,2,\cdots,m)$ is a term of $A_F(t,s_i)$). Similarly, if $^{\bullet}t \neq \varnothing$, then make $^{\bullet}t = \{s_1',s_2',\cdots,s_n'\}(n \geq 1)$; define $\{s_1(Y_1),s_2(Y_2),\cdots,s_n'(Y_n)\}$ as the input predicate set of the transition t (where, $Y_i(i=1,2,\cdots,n)$ is a term of $A_F(s_i',t)$). Elements of input/output predicate set are called the input/output predicate of transition t.

Definition 8: Assume that $\Sigma = (S_1 \cup S_2,T;F,D,V,A_S,A_T,A_F,M_0)$ is a two-level place Pr/T subnet, where, S_1 is the input predicate set of Σ and S_2 is the output predicate set. For any atomic predicate formula set, $P = \{P_1(X_1),P_2(X_2),\cdots,P_n(X_n)\}$, the formula set is successfully matched with the two-level place Pr/T subnet Σ, if and only if both of the following conditions are met:

(1) The place set corresponding to the set P meets $\{P_1,P_2,\cdots,P_n\} = S_2$;
(2) Exist a replacement θ, which makes
$\forall t \in T$, $X_i\theta = A_F(t,P_i)\theta(i=1,2,\cdots,n)$.

For the purposes of simplifying discussion, we assume that:

(1) In definition 8, predicate symbols of the predicate formula set P are different;
(2) Each directed edge of the Pr/T net system has a $n(n \geq 1)$-try tuple, and no symbolic sum forms that consist of $m(m \geq 2)$ n-try tuples.

3   Pre-and/or Graph Description of the Pr/T Net System

Assume that Σ is a two-level place Pr/T net which contains $r(r \geq 1)$ transitions. According to the definition of the two-level place Pr/T net, Σ can be shared composition of r two-level place Pr/T nets, each of which contains only one transition[22].

3.1   Pre-and/or Graph of the Two-Level Place Pr/T Net System Which Contains Only One Transition

The ratio of the numbers of transition's input place and output place is represented by $m:n(m,n \geq 0)$. According to the different value of $m:n$, the discussion can be divided into the following:

(1) $1:n(n \geq 1)$, as shown in fig.2(a), and its Pre-and/or graph is shown in fig.2(b).

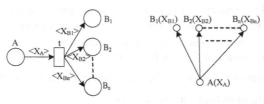

(a) $1:n(n\geq 1)$ (b) Pre-and/or graph corresponding to fig.(a)

Fig. 2. $1:n(n\geq 1)$ and its Pre-and/or graph

(2) For $0:n(n\geq 1)$, its Pre-and/or graph is similar to the fig.2(b), only that the atomic predicate formula $A(X_A)$ is replaced by "*NULL*".

(3) $m:1(m\geq 2)$, as shown in fig.3(a), and its Pre-and/or graph is shown in fig.3(b).

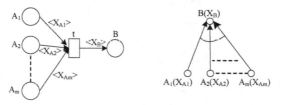

(a) $m:1(m\geq 2)$ (b) Pre-and/or graph corresponding to fig. (a)

Fig. 3. $m:1(m\geq 2)$ and its Pre-and/or graph

(4) For $m:0(m\geq 1)$, its Pre-and/or graph is similar to the fig.3(b), only that the atomic predicate formula $B(X_B)$ is replaced by "*NULL*".

(5) $m:n(m,n\geq 2)$, as shown in fig.4(a), and its Pre-and/or graph is shown in fig.4(b).

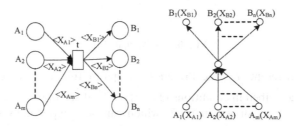

(a) $m:n(m,n\geq 2)$ (b) Pre-and/or graph corresponding to fig.(a)

Fig. 4. $m:n(m,n\geq 2)$ and its Pre-and/or graph

3.2   Pre-and/or Graph of the Two-Level Place Pr/T Net

Theorem 1: Given that $\sum$ is a two-level place Pr/T net which contains $r(r\geq 1)$ transitions, and its Pre-and/or graph can be shared composition of the five Pre-and/or graphs described in section 3.1.

Proof of Theorem 1: (1) when $r = 1$, then the conclusion is always true.

(2) when $r > 1$, we need to prove the following three situations according to the total number (n) of the output places of $\sum$ and definition 7.

1) When $n = 1$, as shown in fig.5, where, $P_i(i = 1, 2, \cdots, r)$ is the output place set for t_i, and $|P_i| \geq 0$ (explanation for P_i is the same hereinafter). The graph of fig.5 can be made up with r subgraphs that shown in fig.6.

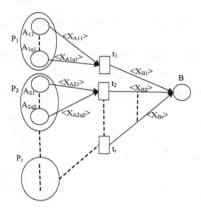

Fig. 5. $n = 1$

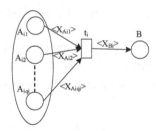

Fig. 6. A subgraph of fig.5

According to different $|P_i|$, the Pre-and/or graph of fig.5 can be composed by Pre-and/or graphs described in conditions (1), (2) or (3) of section 3.1.

2) When $n > 1$ as shown in fig.7, which can be composed by r subgraphs that shown in fig.8.

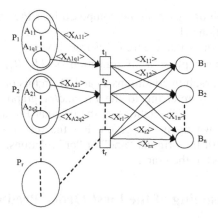

Fig. 7. $n > 1$

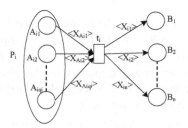

Fig. 8. A subgraph of fig.7

According to different $|P_i|$, the Pre-and/or graph of fig.7 can be composed by Pre-and/or graphs described in conditions (1), (2) or (5) of section 3.1.

3) When $n = 0$, as shown in fig.7 (where $|P_i| > 0$), which can be composed by r subgraphs that shown in fig.10.

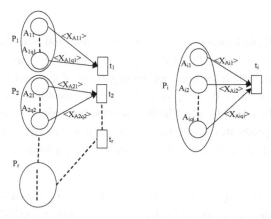

Fig. 9. $n = 0$ **Fig. 10.** A subgraph of fig.9

The Pre-and/or graph of fig. 9 can be composed by r Pre-and/or graphs described in conditions (4) of section 3.1.

From the above ,we know that theorem 1 is true. (over)

When $r > 1$, the output place B of transition or $\{B_1, B_2, \cdots, B_n\}$ will appear r times in the Pre-and/or graph in conditions 1) and 2). In this paper, we regulate that in the reasoning process by referring to the Pre-and/or graph, only one (set) output place inherits the "and" or "or" relations of the original Pre-and/or graph, and these relations should be drawn up in the graph. For the rest $r - 1$ (set) output places, although they also inherit the original "and" or "or" relations, they do not need to be drawn up for conciseness of the graph.

4   Graphic Reasoning of the First-Order Predicate Logic

4.1   The Goal Guiding Graphic Reasoning Approach of the First-Order Predicate Logic

We assume that the general form of the first-order predicate logic that needed to be proved is: $P: A_1, A_2, \cdots, A_m \rightarrow B$, in which, both the premise $A_i (i = 1, 2, \cdots, m)$ and conclusion B are the first-order predicate formulas, and B should be represented by prenex normal form without universal quantifiers. Specific steps of the goal guiding graphic reasoning approach of the first-order predicate logic are shown as follows:

Step 1. Build the Pr/T net systems for $A_1, A_2, \cdots, A_m$, respectively, and merge the same places to obtain the premise net, which is assumed to be $\Sigma_1 = (S_1, T_1; F_1, D, V, A_{S1}, A_{T1}, A_{F1}, M_{01})$. Build the Pr/T net systems for predicate formulas which are after the quantifiers of B to obtain the conclusion net, which is assumed to be $\Sigma_2 = (S_2, T_2; F_2, D, V, A_{S2}, A_{T2}, A_{F2}, M_{02})$ [14]. Rename some variables to so that the same variables will not appear on the input/output arcs of different transitions;

Step 2. For each target transition t_i ($i = 1$ to $|T_2|$), initialize the corresponding Pre-and/or graph $G_i (i = 1, 2, \cdots, |T_2|)$ to be null;

Step 3. for ($i = 1$ to $|T_2|$){

Step 3.1

(1) If the output predicates of t_i belong to a non-null set, then we assume the predicate set to be $Q_i = \{P_1(X_1), P_2(X_2), \cdots, P_n(X_n)\} (n \geq 1)$, and these n atomic predicate formulas, which are taken as the beginning of the reasoning, are represented by n end-nodes of G_i. If $t \in T_1$ and $t^* = \varnothing$, then another end-node marked with "$NULL$" should be added to G_i, which is also the beginning of reasoning;

(2) If the output predicates of the target transition t_i belong to a null set, then only one end-node that marked with "$NULL$" should be established as the beginning of reasoning;

Step 3.2

If (any subset of Q_i can not successfully match with any two-level place Pr/T subne in the premise net Σ_1), we consider that B is not an effective conclusion of $A_1, A_2, \cdots, A_m$, then goes to Step 5.

else{

While (a subset of Q_i is successfully matched with a certain two-level place Pr/T

subnet Σ_1' in the premise net Σ_1), do{

Step 3.2.1. Add the Pre-and/or graph of Σ_1' into G_i (if a node is involved in $n(n \geq 2)$ times of successful matching processes, then copy it for n times in G_i to make the Pre-and/or graphs adding into G_i independent with each other);

Step 3.2.2. Obtain a new set of end-nodes;

Step 3.2.3. If the atomic predicate formula of an end-node is the input predicate of the target transition t_i after replacement, then this end-node is marked as "terminational node";

Step 3.2.4. Given that Q_i is the atomic predicate formula set of non- terminational end-nodes.

}

};

Step 3.3

If there is in G_i a subgragh G_i', which meets:

(1)the atomic predicate formula set of the output predicate set of the target transition t_i is equal to that of the reasoning beginnings of G_i';

(2)the atomic predicate formula set of all the "terminational nodes" in subgraghs is equal to the input predicate set of the target transition t_i after a certain replacement.

then, it proves that the reasoning of the target transition t_i is successful. Otherwise, the reasoning is unsuccessful, which proves that B is not an effective conclusion of $A_1, A_2, \cdots, A_m$, then goes to step 5.

};

Step 4. If the reasoning of each target transition $t_i (i = 1, 2, \cdots, |T_2|)$ is successful, and replacements in reasoning process are coincident, then it proves that B is an effective conclusion of $A_1, A_2, \cdots, A_m$, else, it is not an effective conclusion.

Step 5. End the reasoning.

In the above-mentioned approach, if the atomic predicate formula which is added to the Pre-and/or graph $G_i (i = 1, 2, \cdots, |T_2|)$ is equal to an existing atomic predicate formula of G_i after replacement, then the merging should be made under the condition that no new end-node is added to G_i. When making merging, the following should be met:

(1) If both nodes are marked as the "terminational nodes", then these two nodes needed to be merged, otherwise not;

(2) In order to keep the original reasoning relations, the two nodes are not really merged into one node, but connected through a dotted line.

Step 1 regulates that the same variable shall not appear on the input/output arc of different transitions in Σ_1 and Σ_2. Therefore, if a replacement exists in Step 3.3, then it must be consistent with the replacement in Step 3.2.3. It can be known from section

3.1 that, when Pre-and/or graph $G_i(i=1,2,\cdots,|T_2|)$ has $n(n \geq 2)$ "terminational nodes", then there are n terminational nodes that represent the "and" relation between predicates, and there are directed path from these "terminational nodes" to the beginning of reasoning. Thus, for the target transition t_i, we assume its input predicate set to be $I_i = \{P_1(X_1), P_2(X_2), \cdots, P_n(X_n)\}(n \geq 0)$, and output predicate set to be $O_i = \{Q_1(X_1), Q_2(X_2), \cdots, Q_m(X_m)\}(m \geq 1)$. When the two conditions in Step 3.3 are met, then the following proposition is true: $P_1(X_1) \wedge P_2(X_2) \wedge \cdots \wedge P_n(X_n) \rightarrow Q_1(X_1) \vee Q_2(X_2) \vee \cdots \vee Q_m(X_m)$, i.e. the reasoning of the target transition t_i is successful, and Step 3.3 is correct. For other steps, it is obvious that they are effective and reasonable. Hence, for the first-order predicate logic, the proposed goal-guiding graphic reasoning approach is also effective and reasonable.

4.2   Application Example

Example
if $\forall x(C(x) \rightarrow W(x) \wedge R(x)) \wedge \exists x(C(x) \wedge Q(x))$,
prove that $\exists x(Q(x) \wedge R(x))$.

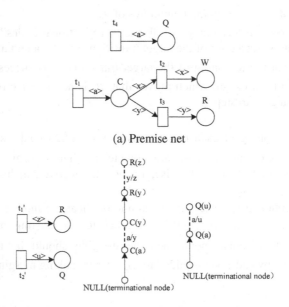

(a) Premise net

(b) Conclusion net   (c) Reasoning process of the target transitions t_1' and t_2'

Fig. 11. Premise and conclusion nets and the reasoning processes

Proof of the Example: According to literature [14] and Step1 of the goal-guiding graphic reasoning approach, we get the premise and conclusion nets, as shown in fig.11(a) and 11(b), respectively. The reasoning process of the target transitions t_1'

and t_2' is shown in fig.11(c) from the top to bottom. Because the two conditions in Step3.3 of the goal-guiding graphic reasoning approach are satisfied, the reasoning of t_1' and t_2' can be considered to be successful, and replacements in the reasoning process are consistent, thus the conclusion is true.

In fig.11(c), a dotted line represents a predicate formula of the premise net which is obtained by replacing the predicate formula of the corresponding node. In essence, the two nodes connected by the dotted line represent one node.

5   Application of the Goal-Guiding Graphic Reasoning Approach in Agriculture Expert System

At present, production rule has become a knowledge representation mode which enjoys the most artificial intelligent application, and which has been employed in many successful expert systems to represent knowledge[2]. In this paper, we assume that the agriculture expert system use the production rule representation. for instance:
IF   wz= suburbs, and nyhxptr=large
   THEN it means a large quantity of carbon emission per unit area
The goal-guiding graphic reasoning approach can not only prove the already known results, but also solve questions in agriculture expert system. Specific steps as follows:

Step 1. Build proper predicate formulas for production rules of the knowledge base and conclusions to be solved;
Step 2. According to literature [14] and Step1 of the goal-guiding graphic reasoning approach, build the Pr/T net model for production rules of the knowledge base and conclusions to be solved, to get the premise net Σ_1 and the conclusion net Σ_2 ;
Step 3. Do Step2-Step4 of the goal-guiding graphic reasoning approach;
Step 4. If the question that need to be solved is an effective conclusion, then the value of the variable in Σ_1 that obtained by replacement in the reasoning process is just the answer of the question; if the question is not an effective conclusion, then there is no answer for the question.

Example. Take the judgment of several common pest and disease damages during the cotton seedling period. In such an agriculture expert system knowledge base, the production rule representation of syndromes and diseases is described as follows[20]:
Diseases during the cotton seedling period:

IF systemic and cotyledon foliage and appear one of syndromes of yellow net, purple plague, ralstonia solanacearum and yellows.
THEN the cotton would be withered.
IF local and cotyledon foliage and appear water-soaked dots or small spots and turbid juice when wiped on the glass.
THEN it shows the angular.

IF local and cotyledon foliage and appear pyorrhea or incrustation on the extended scab.
THEN it shows the angular.
IF local and rhizome and appear tawny and annular constriction.
THEN it shows the seedling blight.
IF local and root and appear arachnoid tomentum with soil particles but not the cotton fiber.
THEN it shows the seedling blight.
IF local and rhizome and burst of long-thin spindle-shaped fibers.
THEN it shows the anthracnose.
IF local and cotyledon foliage and appear small dots with ashen in the center and dull-red in outer area.
THEN it shows the anthracnose.
IF local and rhizome and appear dark brown long round spot and constriction.
THEN it shows the redroot.
Here we do not list the diseases during cotton budding, blossing and boll opening periods.
Table 1 lists the relations between the predicates and syndromes, in which x represents a disease.

Table 1. Relations between the predicates and syndromes

Predicate	Syndrome	Predicate	Syndrome
Q1(x)	Systemic	P4(x)	pyorrhea or incrustation on the extended scab
Q2(x)	Local		
Q3(x)	Rhizome	P5(x)	tawny
Q4(x)	cotyledon foliage	P6(x)	annular constriction
Q5(x)	root	P7(x)	arachnoid tomentum with soil particles
Q6(x)	seeding period		
Q7(x)	budding period	P8(x)	cotton fiber
Q8(x)	blossing period	P9(x)	burst of long-thin spindle-shaped fibers
Q9(x)	Boll-opening period		
P1(x)	yellow net, purple plague, ralstonia solanacearum and yellows	P10(x)	small dots with ashen in the center and dull-red in outer area
P2(x)	water-soaked dots or small spots	P11(x)	dark brown long round spot
		P12(x)	constriction
P3(x)	turbid juice when wiped on the glass	R(x,A1)	x is blight
		R(x,A2)	x is anthracnose
		R(x,A3)	x is red rot

Given: during the seeding period, some rhizomes appear burst of long-thin spindle-shaped fibers, and the disease is not systemic but local, question: what is the conclusion?
Step1: Build corresponding predicate formulas for production rule representation of the knowledge base, here lists parts of them:

$$Q_6(x) \wedge Q_2(x) \wedge Q_3(x) \wedge P_9(x) \to R(x, A2)$$
$$Q_6(x) \wedge Q_2(x) \wedge Q_4(x) \wedge P_{10}(x) \to R(x, A2)$$
$$Q_6(x) \wedge Q_2(x) \wedge Q_3(x) \wedge P_5(x) \wedge P_6(x) \to R(x, A1)$$

Build predicate formulas for the conclusion that needs to be solved:
$Q_6(x) \wedge Q_2(x) \wedge Q_3(x) \wedge P_9(x) \to R(x, y)$, where, $R(x, y)$ represents that the disease x is y.

Step2: Part of the Pr/T net Σ_1 of corresponding production rules is shown in fig.12(a), and Pr/T net Σ_2 of the conclusion is shown in fig.12(b).

Step3: The reasoning process is shown in fig.12(c), from which we know that the question that needs to be solved is an effective conclusion.

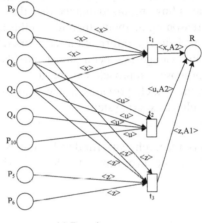

(a) Premise   net

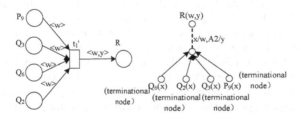

(b) Conclusion net   (c) Reasoning of the target transition t_1'

Fig. 12. Premise and conclusion nets and the reasoning processes

Step4: In the reasoning process, replacement A2/y had been used, which indicates that A2 is the answer of this question, namely, anthracnose is the conclusion.

6 Conclusions

In this paper, for the first-order predicate logic, a goal guiding graphic reasoning approach that based on the predicate/transition system has been proposed, and the approach has been applied in the reasoning process of agriculture expert system. Compared to other previous work,the paper has the following significance and innovation:

(1)The reasoning process of the approach proposed in this paper is started with the conclusion, so the approach is purposeful and effective with reducing many useless steps.

(2)For the approach in this paper,in process of reasoning, the premise is separated from the conclusion, avoiding the disadvantage that the causal relationship will be covered by traditional reasoning methods, therefore in this approach, the knowledge is highly readable and some heuristic information can be used in the reasoning process.

(3) When agriculture knowledge has been stored into the base in the form of the production rule, the proposed approach can be used to answer questions of the agriculture expert system, which provides a new way for studying the reasoning mechanism of the agriculture expert system.

Acknowledgment. This research was funded by the National High Technology Research and Development Program of China (863 Program) under Grant 2013AA102301.

References

1. Hui, Z.: The development of expert system and its application in the agriculture. Southwest China Journal of Agricultural Science 16(3), 117–121 (2003)
2. Ji, L.: Study and Realization on Chemical HAZOP Expert System Based on Petri Net Modeling. Beijing University of Chemical Technology, Beijing (2007)
3. Linfeng, Z.: Citrus Cultivation and Management Expert System for Construction and Implementation. Hunan Agricultural University, Changsha (2008)
4. Lihua, H., Liping, C., Yanjun, W.: Study on knowledge and model combination of agriculture expert system tool. Journal of Hebei Agricultural University 26(3), 93–96 (2003)
5. Li, H., Jiao, Y.: Technologies for developing the national 853 IAES. Sci/Tech Information Development & Economy 15(18), 236–238 (2005)
6. Wenke, J., Liping, C., Mei, Z.: Generating tools of agricultural expert system based on knowledge and the mathematical model. Computer and Agriculture (2), 4–8 (2000)
7. Yuping, Z.: Deduction property of some logics applied to computer science. Chinese Journal of Computers 22(6), 571–576 (1999)
8. Chongyi, Y.: PetriNets Theory and Application. Electronic Industry Press, Beijing (2005)
9. Peterka, G., Murata, T.: Proof procedure and answer extraction in Petri net model of logic programs. IEEE Transactions on Software Engineering 15(2), 209–217 (1989)
10. Murata, T., Zhang, D.: A Predicate-transition net model for parallel interpretation of logic programs. IEEE Transactions on Software Engineering 14(4), 481–497 (1988)

11. Lin, C., Wu, J.: Logical inference of clauses in Petri net models using fixpoint. Journal of Software 10(4), 359–365 (1999)
12. Fang, H., Wu, Z., Cui, H.: Method extraction based on reachability tree of Pr/T net for Horn clauses set. Journal of System Simulation 17(suppl. 1), 163–165 (2005)
13. Fang, H., Yin, Y., Xu, Y.: First-order predicate logic proposition proved by using Predicate/transition net. Computer Engineering 32(23), 191–198 (2006)
14. Geng, X., Wu, Z., Zhang, J.: Modeling of first-order predicate expression by using Predicate/transition system. Journal of System Simulation 19(suppl. 1), 9–15 (2007)
15. Lin, C., Chandhury, A., Whinston, A.B., et al.: Logical inference of Horn clauses in Petri net models. IEEE Transactions Knowledge and Data Engineering 5(3), 416–425 (1993)
16. Lin, C.: Application of Petri nets to logical inference of Horn clauses. Journal of Software 4(4), 32–37 (1993)
17. Yi, Z., Shilin, W.: New methods of logic inference of Petri net based on resolution refutation. Journal of Computer 20(3), 213–222 (1997)
18. Lin, C., Chanson, S.T., Murata, T.: Petri net models and efficient T-invariant analysis for logical inference of clauses. In: 1996 IEEE International Conference on Systems, Man and Cybernetics, Beijing, China, October 14-17, pp. 3174–3179 (1996)
19. Lin, C., Wang, D.: Logical inference of clauses using T-invariant of Petri nets. Journal of Computer 19(10), 762–767 (1996)
20. Qiuhong, L., Renpu, J., Yu, Z., et al.: Discussion on the agricultural expert system based on the production rule and principle attributed. Journal of Anhui Agricultural Sciences 36(10), 4307–4309 (2008)
21. Zuo, X., Li, W., Liu, Y.: Discrete Mathematics, pp. 2–79. Shanghai Scientific and Technical Publishers, Shanghai (1982)
22. Changjun, J.: Dynamic invariance of Petri net. Science in China (series E) 27(5), 567–573 (1997)
23. Murata, T., Subrabmanian, V.S., Wakayama, T.: A Petri net model for reasoning in the presence of inconsistency. IEEE Trans on Knowledge and Data Engineering 3(3), 281–292 (1991)

Fermentation Condition Optimization for Endophytic Fungus BS002 Isolated from Sophora Flavescens by Response Surface Methodology

Na Yu and Lu He

School of Chemical Engineering, University of Science and Technology LiaoNing,
Anshan 114051, China
yuna662007@126.com, 87478@163.com

Abstract. The endophytic fungus BS002 with good effect of disease-resistant has been isolated from the medicinal Sophora flavescens. The conditions of the fermentation medium were explored in this paper. On the basis of the single factor experiments, CH_3COONa, potato and glucose were defined to be the main factors by Plackett-Burman Design; Response surface methodology (RSM) with Box-Behnke Design was used to optimize fermentation conditions of endophytic fungi BS002. The optimal conditions are defined as follows: 100 mL in the container 250mL, potato 246.47 g/L, glucose 27.81 g/L, CH_3COONa 1.95 g/L, concentration of fungus 10^7 cfu/ml, temperature 25 °C, yeast extract 1.0 g/L, speed 150 rpm, fermentation for 4 d, dry weight of mycelium can reach 10.398 g/L.

Keywords: response surface methodology, Plackett-Burman, endophytic fungi, fermentation condition, Sophora flavescens.

1   Introduction

Endophytic fungi, which lives within healthy plant tissues or organs, do not bring some diseases to plants and the whole or part of the life cycle exists in the plant [1-3]. Studies have shown that, many endophytic fungi were isolated from plant roots, stems, leaves, etc. Part endophytic fungi can produce the same or similar pharmaceutically active ingredients with the host[4,5]. Medicinal plant Sophora with functions of heat-clearing, detoxifying, eyesight-improving, diuretic, rheumatism removing, desinsection, etc. is the dried root of plant, it is cold, bitter, affecting the heart, spleen, kidney. Matrine and flavonoids are contained in Sophora, and have better pharmacologically active and medicinal value[6-9]. Endophytic fungus BS002 strain was isolated from Sophora seeds. The test proved strong antifungal activities against the growth of *Botryosphaeria berengriana f.sp. piricola*, *Physalospora piricola*, *Cladosporium cucumerinum Ell. Arthur.*, *Fusarium oxysporum f.sp. cucumerinum*, *Fusarium moniliforme*, etc. The inhibition to *physalospora piricola* was the biggest with the antibacterial diameter of 45 mm.

D. Li and Y. Chen (Eds.): CCTA 2013, Part II, IFIP AICT 420, pp. 342–350, 2014.

In this paper, Plackett-Burman design[10-12] was used to define to the main factors and response surface methodology (RSM) with Box-Behnke design was used to optimize fermentation medium and fermentation conditions[13,14]. RSM, a mathematical and statistical method, is a way of finding the best conditions in the multi-factor systems. The accurate and effective results can be achieved by the experiment. The best combination of various factors and the optimal response value can be determined by this methed on the entire inspection area[15-19].

2   Experiments and Methods

2.1   Preparation

The endophytic fungus BS002 isolated from the seeds of S. flavescens, was provided by bio-pharmaceutical technology laboratory, College of Biological Engineering, University of Science and Technology LiaoNing, Liaoning Province, China.

Endophytic fungus BS002 (Penicillium sp. M-01, Penicillium variants), was inoculated in PD medium (potato 200 g, sugar 10 g, glucose 10 g, sodium acetate 1.66 g, peptone 1.02 g, water1000 mL) at 25 °C, 150 r / min training for 4 d, then filtrated.

2.2   Methods of Test

Liquid fermentation process parameters with inoculative dose, liquid volume, temperature, Metabolism factor, potatoes, carbon source, nitrogen source, etc. are optimized by single factor experiment. Each factor is set high and low level, denoted by 1 and -1, the high level was 1.5 times of low level in Plackett-Burman Design[19,20](Table 1). The significant factors can be analyzed.

Table 1. The measured values of Placket-Burman Design

No.	inoculative dose (cfu/mL)	medium volume (mL)	speed (rpm)	Temper -ature (℃)	CH₃COONa (g/L)	potato (g/L)	glucose (g/L)	yeast extract (g/L)	biomass (g/L)
1	1	-1	+1	-1	-1	-1	+1	+1	5.500
2	1	+1	-1	+1	-1	-1	-1	+1	3.335
3	-1	+1	+1	-1	+1	-1	-1	-1	5.410
4	+1	-1	+1	+1	-1	+1	-1	-1	5.350
5	+1	+1	-1	+1	+1	-1	+1	-1	6.625
6	+1	+1	+1	-1	+1	+1	-1	+1	7.260
7	-1	+1	+1	+1	-1	+1	+1	-1	6.410
8	-1	-1	+1	+1	+1	-1	+1	+1	8.995
9	-1	-1	-1	+1	+1	+1	-1	+1	6.600
10	+1	-1	-1	-1	+1	+1	+1	-1	8.080
11	-1	+1	-1	-1	-1	+1	+1	+1	6.365
12	-1	-1	-1	-1	-1	-1	-1	-1	3.200

RSM with Design expert 8.0 was used to optimize fermentation conditions of the endophytic fungus BS002. Optimal response can be analyzed and obtained by the significant factors were considered the independent variables and the mycelium dry weight was considered the response value. The design method was showed (Table 2).

Table 2. Level of response surface methodology experiment factors

Level	X_1 : $CH_3COONa(g/L)$	X_2 : glucose(g/L)	X_3: potato(g/L)
-1	1.5	20	200
0	1.88	25	250
+1	2.25	30	300

The mycelium was collected, washed with deionized water three times, and dried in vacuum oven to constant weight at 150℃ to obtain the biomass of each group(formula 1.).

$$\text{biomass} \ (g/L) = \text{dry cell weight} \ (g)/\text{fermentati on} \ (L) \qquad \text{(formula 1.)}$$

3 Results and Discussion

3.1 The Results of Plackett-Burman Design Test

The results of Plackett-Burman Design are shown in Table 3.

Table 3. Levels and effects of variables

NO.	Factors	level -1	level 1	Estimate	t-Value	P-Value	Rank
1	inoculative dose(cfu/mL)			-0.055	-0.417	0.704	8
2	medium volume(mL)	80	120	-0.010	-1.167	0.328	6
3	speed(rpm)	120	180	0.013	2.374	0.098	4
4	$CH_3COONa(g/L)$	1.5	2.25	2.847	6.443	0.008	1
5	temperature(℃)	22	33	0.025	0.754	0.505	7
6	potato (g/L)	200	300	0.012	3.521	0.039	3
7	glucose(g/L)	20	30	0.180	5.442	0.012	2
8	yeast extract(g/L)	0.8	1.2	1.242	1.449	0.231	5

According to the Plackett-Burman test, CH_3COONa, potato, glucose were proved to be significant factors. p-values were less than 0.05, and the order: $CH_3COONa >$ glucose > potato. Inoculative dose and medium volume had negative effect on experimental results, that Estimate < 0; the other six fators had positive effect, that Estimate > 0. Three significant factors would be the basis of the following experiments. According to the single factor experiment results, inoculative dose 10^7 cfu/mL, medium volume 100 mL/250 mL erlenmeyer flask, speed 150 rpm, temperature 25 ℃, yeast extract 1 g/L.

3.2   The Results of RSM

The test results are shown in Table 4. and Table 5.

Table 4. Experimental design and result of response surface methodology as N=17

No.	CH_3COONa (g/L)	glucose (g/L)	potato (g/L)	biomass (g/L)
1	1	-1	0	6.898
2	0	-1	1	5.385
3	0	0	0	9.953
4	0	0	0	8.638
5	0	1	1	9.160
6	1	1	0	9.270
7	-1	1	0	8.980
8	0	-1	-1	5.840
9	0	1	-1	8.858
10	-1	0	-1	7.390
11	0	0	0	10.178
12	0	0	0	9.725
13	1	0	1	7.630
14	-1	0	1	7.890
15	0	0	0	9.770
16	-1	-1	0	5.520
17	1	0	-1	9.120

The maths model (this test response value Y from the encoded variables A, B, C) was set up based on the above data by quadratic regression analysis with Design expert 8.0.6 : $Y=9.65+0.39A +1.58B - 0.14C-0.27AB-0.50AC+0.19BC-0.64A^2 - 1.34B^2- C^2$.

Table 5. Analysis of variance for response surface methodology model

Source	Sum of Square	DF	MeanSquare	F-value	Pr > F	Significance
Modle	37.71	9	4.19	18.08	0.0005	significant
A: CH$_3$COONa	1.23	1	1.23	5.31	0.0546	
B: glucose	19.92	1	19.92	85.95	< 0.0001	**
C: potato	0.16	1	0.16	0.7	0.4290	
A.B	0.30	1	0.30	1.28	0.2957	
A.C	0.99	1	0.99	4.270	0.0776	
B.C	0.14	1	0.14	0.62	0.4576	
A^2	1.75	1	1.75	7.55	0.0286	*
B^2	7.57	1	7.57	32.68	0.0007	**
C^2	4.22	1	4.22	18.19	0.0037	**
Lack of fit	0.21	3	0.069	0.20	0.8942	not significant
Pure Error	1.41	4	0.35	-	-	
Residuals	1.62	7	0.23	-	-	
Cor Total	39.33	16	-	-	-	
R-Squared 0.9587			AdjR-Squared 0.9057			

Note: * represents significant; ** represents very significant.

According to analysis of variance, there are significant differences on the biomass between one degree term and quadratic term, in other words, there was no simple linear relationship between experimental factors and response. The model coefficients for each variable are also shown in Table 5. and F-value and P-value were employed to check the significance of each coefficient of the model. Correlation coefficient of the regression equation R^2(R-Squared) =95.87%, it showed that experimental data sufficiently were fitted to the model in the case of significant level α = 0.01 and 90.57% data response to the biomass depended on the selected three significant variables. Pr> F term represents the probability is greater than F, lack of fit was 0.8942, not significant, indicating that the equation was adequate for predicting biomass under all conditions, this proved that the model was selected appropriately.

The relationship between the response and experimental data of each variable can be demonstrated by three-dimensional response surface plots which represented the regression equation mentioned above (shown in Fig. 1-4). According to the analysis by the Design-expert 8.0.6, the optimal values of the three key variables for biomass of lipid fermentation of the endophytic fungus BS002 were crystalline sodium acetate 1.95 g/L, glucose 27.81 g/L, potato 246.47 g/L, respectively.

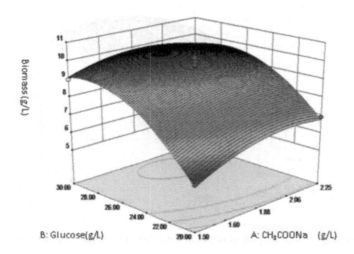

Fig. 1. Response surface figure of regression eqation of biomass VS CH₃COONa and glucose

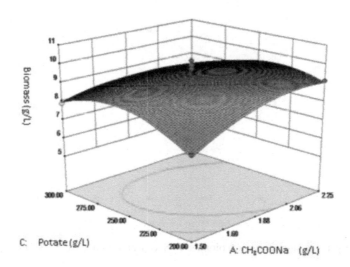

Fig. 2. Response surface figure of regression eqation of biomass VS CH₃COONa and potato

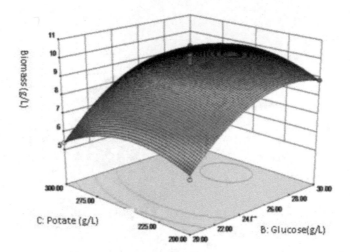

Fig. 3. Response surface figure of regression eqation of biomass VS potato and glucose

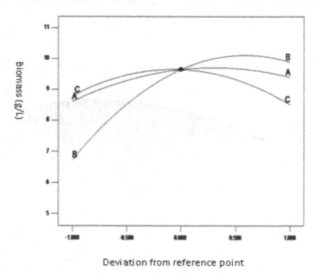

Deviation from reference point

Fig. 4. Binary regression curve of biomass VS concentration of CH_3COONa, potato and glucose

According to the response surface analysis charts, the biomass of endophytic fungus BS002 increased at first and then decreased with the increase in the value of any two parameters.

4   Conclusions

The test by RSM for fermentation condition optimization of endophytic fungus BS002 isolated from *Sophora flavescens* proved that cell dry weight 10.398 g / L on the base of optimized conditions was obtained, and it was increased by 29% compared with 8.062 g / L on the basic of medium, error was only 2.5%, compared to predicted values 10.1429 g / L. It was indicated that the optimal medium combination has important practical applications after the verification test.

Significant effects from liquid fermentation medium composition of endophytic fungus BS002 to cell dry weight were derived in the test. Mathematical model of three significant factors(CH_3COONa, potato, glucose) and biomass was established by RSM using the results of Plackett-Burman Design test. Regression effect of model was significant, it can predict the optimal conditions of the liquid fermentation. The optimal conditions are defined as follows: 100 mL in the container 250 mL, potato 246.47 g/L, glucose 27.81 g/L, CH_3COONa 1.95 g/L, concentration of fungus 10^7 cfu/ml, temperature 25 ℃, yeast extract 1.0 g/L, speed 150 rpm, fermentation for 4 d.

The applications of RSM are gradually widespread at home and abroad, RSM has been extensively applied in the optimization of medium composition, fermentation conditions and food manufacturing processes. It was reported that sulphuric acid-treated sugar cane bagasse hydrolysate can be efficiently used for the cell growth and lipid accumulation of T. fermentans, and it represented a 32.8% improvement in the lipid concentration and a 21.4% increase in the lipid coefficient by RSM[17]. The production of α-amylase by *Aspergillus oryzae* had about 20% increase by RSM for optimizing process parameters[21]. RSM applied in liquid fermentation conditions of endophytic fungus BS002, reduced the workload, and also obtained a better result. A certain foundation to go for separation and purification of the active substance and structure identification is laid.

References

1. Li, X., Yao, Y., Sun, G., et al.: Chemical Constituents from Marine-derived Fungus of Penicillium sp. Natural Product Research and Development 19, 804–806 (2007)
2. Regina, M., dos Santos, G., Edson, R.-F., et al.: Endophytic fungi from Melia azedarach. World Journal of Microbiology & Biotechnology 19, 767–770 (2003)
3. He, L., Liu, N., Wang, Y., et al.: Isolation an antimicrobial action of endophytic fungi from sophora flavescens and effects on microorganism circumstances in soil. Procedia Environmental Sciences 18, 264–270 (2013)
4. He, L., Ji, M.: Shenyang Agricultural Univercity, Shenyang (2012)
5. Kui, S.: Advances of Research on the Endophytic Fungus Resources of Medicine Plants. Science and Technology of Qinghai Agriculture and Forestry (1), 26–28 (2010)
6. Jianqiu, S., Liangdong, G., Wei, Z., et al.: Research Advances in the Diversities of Endophytic Fungi in Medicinal Plants and Their Bioactive Ingredients. Acta. Bot. Boreal. –Occident. Sin. 26(7), 1505–1519 (2006)

7. Lai, J., He, X., Jiang, Y., et al.: Preparative seperation and determination of matrine from the Chinese medicinal plant sophora flavescens Ait. by molecularly imprinted solid-phase extraction. Anal. Bioanal. Chem. 375(2), 264 (2003)
8. Hongli, Z., Yuejin, Z., Chongxue, H., et al.: Advances in the research on bioactivity of Sophora flavescens. Jour. of Northwest Sci-Tech Univ. of Agri. and For. 32(5), 31–36 (2004)
9. Xiaowei, Z., Jinrong, B., Ren, B.: Progress on the anti-tumor activities of matrine and oxymatrine. Chemical Reagents 32(1), 32–36 (2010)
10. Zhijie, C., Yongbin, H., Chang, S., et al.: Application of Placket-Burman Design for Determining Key Factors on Mycelium Growth and Extracellular Polysaccharides Excretion of Ganoderma lucidum. Food Science 26(12), 115–118 (2005)
11. Bi, J., Lu, L., Qin, Y., et al.: Screening of Main Component in the Medium impacting the Ability of the Zinc-richening Ganoderma lucidum with the Method of Plackett-Burman Design. Journal of Anhui Agri. Sci. 38(10), 5335–5337 (2010)
12. Guo, Z., Lu, J., Dai, D., et al.: Optimization of Biodegradation 0f Dibenzothiophene Using Response Surface Methodology. Transactions of Beijing Institute of Technology 32(1), 106–110 (2012)
13. Li, J., Zang, J., Xiao, L., et al.: Optimization of the Extraction Conditions of Celery Flavonoids by Response Surface M ethodolog. Journal of Anhui Agri. Sci. 40(5), 2687–2689 (2012)
14. Xia, B., Kang, J., Su, B., et al.: Response Surface Optimization of R. Palustris Fermentation. Food and Fermentation Technology 46(158), 29–35 (2010)
15. Chen, H., Yang, C., Wu, W., et al.: Optimization of the Fermentation Conditions of the Endophytic Fungus Hd3. Journal of Southwest China Normal University 36(1), 148–153 (2011)
16. Ambat, P., Ayyanna, C.: Optimizing Medium Constituents and Fermentation Conditions for Citric Acid Production from Palmyrajaggery Using Response Surface Methods. World J. Microbiol. & Biotechnol. (17), 331–335 (2001)
17. Huang, C., Wu, H., Li, R., et al.: Improving lipid production from bagasse hydrolysate with Trichosporon fermentans by response surface methodology. New Biotechnology 29(3), 373–378 (2012)
18. Jingwen, Z., Jilin, D., Ruiling, S.: The application of rye non-starch polysaccharides gel in sausage. China Brewing 31(8), 77–81 (2012)
19. Dai, W., Cheng, L., Tao, W.: Application of Response Surface Methodology in Optimization of Precursors for Taxol Production by Fusarium mairei K178. China Biotechnology 27(11), 66–72 (2007)
20. Qi, B., He, L., Liu, N., et al.: Fermentation Condition Optimization for Endophytic Fungi BS003 of Sophora Flavescens Using Response Surface Methodology. Acta Agriculturae Boreali-Occidentalis Sinica 21(9), 174–178 (2012)
21. Francis, F., Sabu, A., Madhavan Nampoothiri, K., et al.: Use of response surface methodology for optimizing process parameters for the production of ∝-amylase by Aspergillus oryzae. Biochemical Engineering Journal (15), 107–115 (2003)

Daily Sales Forecasting for Grapes
by Support Vector Machine[*]

Qian Wen[1,2], Weisong Mu[3], Li Sun[1], Su Hua[4], and Zhijian Zhou[1,**]

[1] College of Science, Applied Mathematics,
China Agricultural University, Beijing, China, 100083
[2] Department of Mathematics and Statistics, University of West Florida,
Pensacola, Florida, 32514
[3] College of Information and Electrical Engineering,
China Agricultural University, Beijing China 100083
[4] Department of Statistics, North Dakota State University, Fargo, North Dakota, 58102
qw2@students.uwf.edu, {slsally,zhijianzh}@163.com,
wsmu@cau.edu.cn, su.hua@my.ndsu.edu

Abstract. In this article, the quantity of grapes sold in one fruit shop of an interlocking fruit supermarket is forecasted by the method of support vector machine (SVM) based on deficient data. Since SVMs have a lot advantages such as great generalization performance and guarantying global minimum for given training data, it is believed that support vector regression will perform well for forecasting sales of grapes. In order to improve forecasting precision (FP), this article quantifies the factors affecting the sales forecast of grapes such as weather and weekend or weekday, results are suitable for real situations. In this article, we apply ε-SVR and LS SVR to forecast sales of three varieties of grapes. Moreover, the artificial neural network (ANN) and decision tree (DT) are used as contrast and numerical experiments show that forecasting systems with SVMs is better than ANN and DT to forecast the daily sales of grapes overall.

Keywords: support vector machine, artificial neural networks, grape sale forecasting, ε-SVR, LS-SVR.

1 Introduction

Grapes are special fruits that usually become ripe in summer, i.e., from July to September, and are very popular among fruit customers. Because people are more and more recognizing the nutritional value of grapes, the sales of grapes have also dramatically increased during summer. Unlike the large consumption of grape products in Europe and America, Chinese consumers prefer table grapes. However,

---

[*] This paper is supported by the China Agricultural Research System (CARS-30).
[**] Corresponding author.

D. Li and Y. Chen (Eds.): CCTA 2013, Part II, IFIP AICT 420, pp. 351–360, 2014.

grapes are difficult to store because they are perishable. For grape retailers , there are two causes for the loss of grapes: loss caused by storage because of a lack of refrigeration equipment and loss caused by customers who pick and excise grapes according to their preferences. Therefore it is very important for grape retailers to make the right decision in ordering because an insufficient quantity of grapes will not meet the customer demand and the shop owner will obtain less profit. On the other hand, too many grapes may result in a lack of freshness, thereby allowing the grapes be sold at discounted price, bringing a loss profit for retailers. As such, decision makers need an accurate method that is based on mathematics rather than on their experience, to determine the appropriate order quantity of grapes.

Sales forecasting is one of the major tasks in business administration. Precise forecasting of demands can not only decrease inventory costs but also improve the quality of customer service and gain competitive advantages. Recently, some techniques have been implemented to develop some models of forecasting demand for agricultural products with the aim of controlling inventory costs. However, for some fruits, especially grapes, the various factors involved (e.g. climate changes, holidays, and unfixed preference of consumers) are so complicated and changeable that forecast errors significantly influence inventory costs and profits (Roy and Samanta 2011). In this paper, a new algorithm with higher accuracy based on SVM is developed to forecast the demand for grapes, a method which has rarely been applied in such a field before. Because SVMs have greater generalization performance and can guarantee global minima for given training data, it is believed that support vector regression will perform well in forecasting grapes sales.

The rest of this article is organized as follows. The sales forecasting methods review is given in detail in Section 2, and models of SVM are presented in Section 3. Then the forecasting system framework based on the SVM is explored in Section 4. In Section 5, the proposed model is presented, and numerical examples are used to investigate the forecasting performance of the model. The conclusion, contributions of this article, limitations of the research, and some future research directions are provided in Section 6.

2   Selection of Forecasting Method

During the last few decades, many sales forecasting models such as time series, regression analysis, decision tree,   ANN and SVM have been developed in the field of perishable product. However not all these methods are suitable for grapes sales forecasting. Next, we will briefly introduce the traditional forecasting models and the SVM sales forecasting model.

2.1   Traditional Method for Forecasting

The traditional methods for forecasting models are mostly based on statistic methods. These methods range from the moving average and exponential smoothing to linear and nonlinear regression. Nonetheless all these models have deficiencies and cannot solve the problem of this article. ARMA model of time series is a method that uses the law of variation of the past variable to forecast future variation of the variable; however, this method cannot reflect what factors affect the quantity of sale. The regression analysis method is used to reflect the relationship between the quantity of sale and one or more independent variables, but this method is always based on a large number of data to solve the problem. As such it is not feasible to adopt regression analysis to forecast daily sales of grapes. Recently, ANN has received much attention in solving the problem of demand forecasting because of its competent performance in forecasting and pattern recognition. Many studies have attempted to apply the ANN model to time series forecasting. However, ANN models adopt the steepest descent algorithm to find optimal solutions, but they are unable to make sure that the error function of the neural networks converges to a global optimal solution. Moreover, a critical issue concerning neural networks is the over-fitting problem.

2.2   SVM for Forecasting

The SVM has recently been proposed as a new kind of learning network based on the statistical learning theories: the Huber robust regression theory and the Wolfe dual programming theory. SVM achieve good performances in terms of higher accuracy, better generalization and the global optimal solution (S.R.Gunn 1998; Vapnik 2000; Doumpos 2004). Originally, SVMs were developed for pattern recognition and classification problems(Cortes and Vapnik, 1995). Tang Hao (2007) used SVM in mechanical failure diagnosis and proposed the combination of principal component analysis (PCA) and SVM to improve the diagnosis rate dramatically. Wu Jiang (2007) applied SVM in computer-aided detection of cancer diseases and provided a reference of diagnosing cancer. In recent years, SVM has been used in regression problems, which making forecasting by SVM possible. Many scholars use SVM to forecast in various subjects. For example, Wu Qi (2008) forecasted the car sales by SVM based on the Gaussian loss function. Du Xiaofang (2011) used SVM, combined with fuzzy theory, to forecast the demand of perishable farm products.

Therefore, SVM is indeed an effective forecasting method as it needs only a small amount of data to forecast sales. This method applies particularly to the sales forecasting of grape, which lacks historical data. At present, there is no such research that applies the SVM method in forecasting the sales of grapes, thus we use SVMs to forecast the sales of grape in this paper.

3   SVMs Forecasting Model

It is well known that SVMs were developed by Cortes & Vapnik (1995) for binary classification and that they can also be applied in regression problems by introducing an alternative loss function. One character of SVMs is it is an algorithm that can only deal with linear problem. When the system is non-linear, the input vector x, is mapped into a high-dimensional feature space z, via a non-linear mapping, and then conducting linear regression in this space. The inner product of this mapping is called kernel function. In this article, the kernel function used is Radial basis function (RBF):

$$K(x,x_j) = \exp\left\{-\|x_j - x\|^2 / \delta^2\right\}.$$

The standard support vector regression model given by following equation:

$$y = f(x) = (w \cdot \phi(x)) + b \tag{3.1}$$

where $\phi(x)$ is in the high-dimensional feature space, which is non-linearly mapped from the input space x. The coefficients w and b are estimated by minimizing risk function $R(C)$:

$$\text{Minimizing } R(C) = \frac{1}{2}\|w\|^2 + C\frac{1}{n}\sum_{i=1}^{n} L_\varepsilon(d_i, y_i)$$

where constant $C > 0$ is penalty factor and $L_\varepsilon(d_i, y_i)$ is loss function.

3.1 ε-SVR Model

In regression, the quality of estimation is measured by the loss function. There are four possible loss functions that can be used: quadratic loss function, Laplacian loss function, Huber's loss function and ε-insensitive loss function. The ε-SVM model selects ε-insensitive loss function as its error measurement.

$$L_\varepsilon(d,y) = \begin{cases} 0 & |d-y| < \varepsilon \\ |d-y| - \varepsilon & otherwise \end{cases}$$

Based on ε-insensitive loss function, the decision function of ε-SVR model (S.R.Gunn, 1988) is

$$f(x) = \sum\left(\overline{\alpha_i} - \overline{\alpha_i^*}\right) K(x,x_i) + \overline{b}$$

$$\overline{b} = \begin{cases} y_j - \sum_{i=1}^{n}\left(\overline{\alpha_i} - \overline{\alpha_i^*}\right) K(x_i,x_j) + \varepsilon & \overline{\alpha_i} \in (0,C) \\ y_j - \sum_{i=1}^{n}\left(\overline{\alpha_i} - \overline{\alpha_i^*}\right) K(x_i,x_j) - \varepsilon & \overline{\alpha_i^*} \in (0,C) \end{cases}$$

Where $K(x_i, x_j)$ is RBF kernel function.

3.2   LS-SVR Model

The LS-SVR model selects quadratic loss function as its loss function. The formula of quadratic loss function is: $L_q(d_i, y_i) = \sum_{i=1}^{n}(d_i - y_i)^2$.

Combine the above loss function and RBF kernel function with the equation (3.1), we get the LS-SVR decision function (S.R.Gunn, 1988):

$$y = f(x) = \sum_{i=1}^{n}\overline{\alpha_i} K(x_i,x) + \overline{b}$$

where $\overline{b} = d_i - \dfrac{\overline{\alpha}}{C} - \sum\limits_{j=1}^{n} \overline{\alpha_j} K\left(x_j, x_i\right)$ and $K\left(x_i, x_j\right)$ is RBF kernel function.

4   Forecasting System Based on SVM

4.1   Selection of SVM Toolbox

For the moment, there are some toolboxes we can utilize such as LS-SVM toolbox and LIBSVM toolbox of MATLAB. Firstly, LIBSVM is a library for Support Vector Machines (SVMs) and developed by Chih-Jen Lin. This package has been actively developed by researchers since the year 2000 to help users to apply SVM easily. Also there has an easier edition at present for the user who does not know anything about SVM. This edition makes everything automatic--from data scaling to parameter selection (Chih-Chung Chang, 2011). Those are the reasons why we select LIBSVM tool in this paper. Secondly, The LS-SVM toolbox is mainly used with the commercial Matlab package. The Matlab toolbox is compiled and tested for different computer architectures including Linux and Windows. LS-SVM lab's interface for Matlab consists of a basic version for beginners (K. Pelckmans, 2003).

4.2   Forecasting Framework

The forecasting system framework is showed in the figure 1. At first, we need deal with data in two ways: One is data normalization processing (in order to avoid data overflow) including smooth the historical sale data (in order to eliminate singular values and noise). Another is to process dynamic information, such as weather data, week data, etc., which is corresponding to the historical sale data, as mentioned in Section 5.2. After that the set of date will separate into two parts, one is called training set and the other one is called testing set.

Subsequently, the training set inputted to SVM model is trained and learnt for adjusting the parameters to the optimal values. The future request is forecasted by the system after the machine completes learning. In addition, we obtain the best parameters C (penalty factor) and γ (a parameter of kernel function) by grid-search on C and γ with cross-validation. At last, forecasting is performed and the values

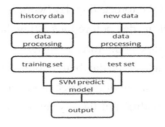

Fig. 1. Framework of forecasting

are obtained after test set is inputted to the trained SVM model. All these process will be done by the SVM toolboxes. What must be mentioned is we perform single-point forecast every time, in other word, there is only one value output in every iteration process. When the sale of a certain day is forecasted, we will put the real value of that day into the training set to renew the history data.

5   The Sale Forecasting of Grape

5.1   About Data

The data we used is obtained from a fruit supermarket called "Fu Man Jia". The data cover the time from the beginning of July 2011 up to the end of September in year 2012 since grape ripe on the large scale during this period. There are three kinds of grape sold in this market and they are XiaoMiFeng, JuFeng and MeiGuiXiang. We use all of them to test the efficiency of SVM forecasting model in this paper. The data of weather is collected from Website.

5.2   Index of Variables

Forecasting for sale of grape is a complicated procedure that involves multiply variables, and could be treated as regression function $y = f(x) = (w \cdot \phi(x)) + b$. The output value of the regression function is sale quantity y, and the input variable x contain many relevant factors, which control the sale, such as historical sales, weather information, holidays information, etc. The objective of this model is to find a mapping that has a high generalization performance from factors x to sale quantity y. According to historical sales, weather data, holiday's data, etc., we form 8 styles of training samples. Input variable of grape sale forecasting model are shown as:

$$X = \left(S_{d-1}, S_{d-7}, W_d, W_{d-1}, P_d, P_{d-1}, T_d, T_{d-1}\right)$$

where

S_{d-1}	Sales quantity at the day before the forecasting day
S_{d-7}	Sales quantity at the day 7 days before the forecasting day
W_d	Type of date at the forecasting day (workday or weekend)
W_{d-1}	Type of date at the day before forecasting day (workday or weekend)
P_d	Sale price of grape at the forecasting day
P_{d-1}	Sale price of grape at the day before forecasting day
T_d	The weather condition of the forecasting day
T_{d-1}	The weather condition of the day before forecasting day

Since the data's type of weather condition and holiday are not numerical value, we quantify them as follow:

(1) Quantified value of weather condition

Table 1. Quantified value of weather condition

Weather	value
sunny	1
cloudy	0.9
overcast	0.8
Light rain	0.7
moderate rain	0.5
Showery rain	0.4
downpour	0.2

(2) Quantified value of type of working day date

$$W_d = \begin{cases} 0 & monday, tuesday, wednesay, thursday, friday \\ 1 & saturday, sunday \end{cases}$$

5.3 Criteria of Forecasting System

In order to verify the validity of the prediction performance of SVM method, we use the day absolute error as statistical metrics as we only forecast one day's sale every time. Definition of criteria is illustrated in the following expression:

$$DAE = \left| \frac{d(i) - f(x_i)}{d(i)} \right| \times 100\%$$

Where $d(i)$ and $f(x_i)$ represent actual sales and the forecasting values respectively.

5.4 The Result of Forecasting and Analysis

The result is showed in the following figures and tables. The curves of figure 2, 3 and 4 show the comparison among real data, ε-SVR forecasting value, LS-SVR forecasting value and ANN and DT forecasting values. It is revealed from those figures that SVRs forecasting value are closer to real data than ANN and DT forecasting values. From table 2, 3 and 4, we find that the SVRs have smaller average relative error and maximum relative error. Even though we find the decision tree performs faster than other methods, that's not a decisive advantage with respect to forecasting grapes sale, since we just have small amount of data. As a conclusion, the forecasting systems of SVRs, though does not that satisfy the sale quantity, outperform the ANN and DT method.

All the points that have relative large forecasting error fall into following two categories. One is forecasting value is greater than the sale quantity. In this case, we found that there is no stock of grapes in the store at most of those situations, i.e., the demand quantity is greater than sale quantity. Therefore, with our method, we can not only satisfy the customer demand, but also increase profit of the store owner at those points. Another case is forecasting value is less than sale quantity. There are some special activities that need large amount of grapes may happen at those point. In this case, the customers always place order at least one day advance, our method will not bring loss to the store owner at some points. What is more, those situations rarely happen. Overall, the SVM methods we used are great ways to help store owner to gain higher profit.

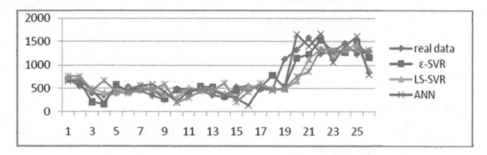

Fig. 2. The result of sale forecasting for XiaoMiFeng

Table 2. Comparison of real data and forecasting result for XiaoMiFeng

method	Average relative error	Maximum relative error	Time spent(s)
ε-SVR	0.2124	0.5395	50.32
LS-SVR	0.2229	0.5632	185.81
ANN	0.2890	0.7440	61.12
DR	0.2508	1.2559	5.02

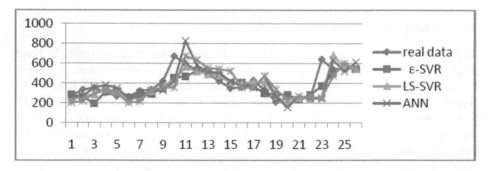

Fig. 3. The result of sale forecasting for JuFeng

Table 3. Comparison of real data and forecasting result for JuFeng

Method	Average relative error	Maximum relative error	Time spent(s)
ε-SVR	0.1407	0.4495	64.38
LS-SVR	0.1528	0.5938	238.27
ANN	0.1896	0.6223	70.30
DR	0.1467	0.6064	6.57

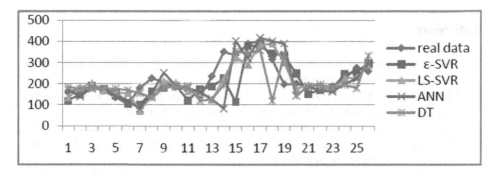

Fig. 4. The result of sale forecasting for Meiguixiang

Table 4. Comparison of real data and forecasting result for Meiguixiang

Method	Average relative error	Maximum relative error	Time spent (s)
ε-SVR	0.1926	0.6777	48.16
LS-SVR	0.1845	0.5957	196.05
ANN	0.2163	0.9715	50.34
DR	0.1725	0.7113	6.13

6   Conclusions

Forecasting is the foundation of fruit supermarket to make order plan and inventory control, while grapes sale has its own characteristics such as multi-dimension, small sample and nonlinearity. It is difficult for the decision maker to forecast the sale accurately by their experience. In this article, the ε-SVR and LS-SVR are used to forecast daily grapes sale, and the result is acceptable. Thus we provide an advanced intelligent forecasting technique for decision maker.

This article also has significant contribution in applications. For example, the forecasting technique we created can be applied in the management of fruit market more successfully. By applying this technique, the correct quantity of fruits with right quality in the appropriate time will be obtained and the shortages or over-stocking will be avoided properly.

However, there still have some limitations in our model. For example, we do not take into account the substituting fruit of the grape that may affects the grapes sale. Further study will focus on improving the algorithm accuracy while more practical factors are involved, so that more realistic sale forecasting result can be obtained in the future.

References

1. Smola, A.: Regression estimation with support vector learning machines. Master's thesis, Technische University at Munchen (1996)
2. Chakraborty, K., Mehrotra, K., Mohan, C.: Forecasting the behavior of multivariate time series using neural networks. Neural Networks 5(6), 961–970 (1992)
3. Chang, C.-C., Lin, C.-J.: LIBSVM: A library for support vector machines. ACM Transactions on Intelligent Systems and Technology 2, 27:1–27:27 (2011), http://www.csie.ntu.edu.tw/~cjlin/libsvm
4. Cortes, C., Vapnik, V.: Support vector networks. Machine Learning 20, 273–297 (1995)
5. Doumpos, M.: An Experimental Comparison of Some Efficient Approaches for Training Support Vector Machines. Operational Research 4(1), 45–56 (2004)
6. Tang, H., Qu, L.: Fault diagnosis of engine based on support vector machine. Journal of Xi'an Jiaotong University 9, 1124–1126 (2007) (in Chinese)
7. Wu, J., Dong, T.: SVM applied to modeling of cancer date. Science Technology and Engineering 20(7), 5363–5365 (2007)
8. Pelckmans, K., Suykens, J.A.K.: LS-SVMlab Toolbox User's Guide. Katholieke Universiteit Leuven. ESAT-SCD-SISTA Technical Report, 02-145 (2003)
9. Wu, Q., Yan, H.-S., Yang, H.-B.: A Forecasting Model Based Support Vector Machine and Particle Swarm Optimization. Power Electronics and Intelligent Transportation System (2008)
10. Roy, A., Samanta, G.P.: Inventory Model with Two Rates of Production for Deteriorating Items with Permissible Delay in Payment. International Journal of Systems Science 42, 1375–1386 (2011)
11. Gunn, S.R.: Support Vector Machines for Classification and Regression. ISIS Technical Report, University of Southampton, Department of Electronics and Computer Science (1998)
12. Vapnik, V.N.: The Nature of Statistical Learning Theory, 2nd edn. Springer, New York (2000)
13. Du, X.F., Leung, S.C.H.: Demand forecasting of perishable farm products using support vector machine. International Journal of Systems Science, 1–12 (2011)
14. Xu, X.-H., Zhang, H.: Forecasting Demand of Short Life Cycle Products by SVM. In: International Conference on Management Science & Engineering, vol. 9, pp. 10–12 (2008)

Research on Text Mining Based on Domain Ontology

Jiang Li-hua[1,2], Xie Neng-fu[1,2], and Zhang Hong-bin[3,*]

[1] Agricultural Information Institute of Chinese Academy of Agricultural Sciences,
Beijing, 100081
[2] Key Lab of Agricultural Information Service Technology of Ministry of Agriculture,
Beijing, 100081
[3] Institute of Agriculture Resources and Regional Planning of Chinese Academy
of Agricultural Sciences, Beijing, 100081

Abstract. This paper improves the traditional text mining technology which cannot understand the text semantics. The author discusses the text mining methods based on ontology and puts forward text mining model based on domain ontology. Ontology structure is built firstly and the "concept-concept" similarity matrix is introduced, then a conception vector space model based on domain ontology is used to take the place of traditional vector space model to represent the documents in order to realize text mining. Finally, the author does a case and draws some conclusions.

Keywords: Ontology, text mining, domain ontology, vector space model.

1 Introduction

Natural language is the main communication and expression thought tool in today's economic society. Although it has been studied for a long time, the understanding and using ability is still limited. The data mining technology based on statistics had matured and applied successfully in large scale relational database in the early nineteenth century. Naturally scholars had the idea of applying the technology of data mining to analyze the text block described by natural language and called it text mining or knowledge discovery in text. Different from the traditional natural language processing's focusing on understanding the words and sentences, the main goal of text mining is to find out the unknown and valuable knowledge or their relationship in large scale text sets. However, I found that most text mining lack of semantic considerations in application, only analyze grammatically, but not the content, so the results are always barely satisfied.

---

[*] Corresponding author.

D. Li and Y. Chen (Eds.): CCTA 2013, Part II, IFIP AICT 420, pp. 361–369, 2014.
© IFIP International Federation for Information Processing 2014

2   Text Mining Based on Ontology

Text mining, or knowledge discovery in text database, is the process of finding unknown, useful and understandable knowledge in large scale text database. The objects of text mining are semi-structured or unstructured. And they always contains multi-layer ambiguity, so a lot of difficulties of text mining are caused.

The traditional text mining method based on vector space model converts the text to word frequency vectors. The major defect of this method is neglecting the importance of semantic role leading to text mining results are unsatisfied. Therefore, semantic analysis and processing technology should be combine with text mining technology in order to develop more effective mining method to realize deep semantic level mining. Appling ontology to text mining provides theoretical support and a feasible approach to solve above problems.

At present, the representative semantic dictionaries applied common ontology are English WordNet and Chinese HowNet. There are many text mining methods based on WordNet and HowNet. But the text mining methods based on common ontology can not get a very good effect in partial field. Therefore a lot of research in recent years began to carry out the research of text Mining based on domain ontology. Bloehdom etc. put forward OnTology Based Text mining frame wOrk; The bag of words representation used for these clustering methods is often unsatisfactory as it ignores relationships between important terms that do not cooccur literally. In order to deal with the problem, Hotho etc. integrate core ontologies as background knowledge into the process of clustering text documents. Song etc. suggests an automated method for document classification using an ontology, which represses terminology information and vocabulary contained in Web documents by way of a hierarchical structure.

In our country, Knowledge Engineering Research Laboratory of computer science department in Tsinghua University has developed ontology data mining test platform based on semantic Web. And also there are some researchers who discussed applications of semantic processing technology in text mining. Xuling Zheng etc.proposed a corpus based method to automatically acquire semantic collocation rules from a Chinese phrase corpus, which was annotated with semantic knowledge according to HowNet（Zheng Xuling etc., 2007）. By establishing domain ontology as the way of knowledge organization, Guobing Zhou etc. introduced a novel information search model based on domain ontology in semantic context（Zou Guobing etc., 2009）. An Intelligent search method based on domain ontology for the global web information was proposed to solve the problem of low efficiency typical in traditional search engineers based on word matched technology by Hengmin Zhu（Zhu Hengmin etc., 2010）. In order to improve the depth and accuracy of text mining, a semantic text mining model based on domain ontology was proposed by Yufeng zhang etc.. And in this model, semantic role labeling was applied to semantic analysis so that the semantic relations can be extracted accurately (Zhang Yufeng etc., 2011).

Taken together, the research of semantic text mining based on domain ontology is still in research theory spread stage, but relative actively in foreign countries. But there is few whole text mining based on domain ontology solutions. And the research scope is only in foundation of shallow knowledge such as classification and clustering of text

(Bingham,2001; Montes-y-Gómez, 2001)but rarely in rich useful deep semantic knowledge such as semantic association foundation(Zelikovitz,2004) 、 topic tracking(Aurora, 2007) and trend analysis (Pui Cheong Fung, 2003)and so on.

3 Research on Key Technology of Text Mining Based on Domain Ontology

As an expert to guide the entire mining process, ontology is used to pre-process the text structure to realize semantic mining and improve mining effect.

3.1 Knowledge Representation Based on Domain Ontology

At present, most of the ontology system basic structure is similiar which are entity, conception, attribute and relation. Namely, the features and corresponding parameters of entities and conceptions are studied by certain rules. At the same time, the relationship of entity and cpnception is described. Agriculture ontology is choosen as the subject and domain knowledge organization in this paper. It can not only deal with the inner basic relation in agriculture subjection, but also more formal specific relationship. Formalized agriculture ontology is defined:

Agri_Onto=(Onto_Info,Agri_Concept, AgriCon_Relation, Axion)

Onto_Info is the basic information of ontology including name, creator, design time, midified time, aim, souce and so on; Agri_Concept is the set of agriculture knowledge conception; AgriCon_Relation is relationship set of conceptions including hierarchical relationship and non hierarchical relationship; Axion includes axiomatic set in ontology.

3.2 Conception Semantic Correlation

In domain ontology, there are words to present class and conception. And the words are not only serve as the bridge for class and conception, but also basic element. The relationship in ontology depend the words to connect with each other, so word set is the key of building agriculture ontology. In this paper, the class words and conception words in agriculture ontology are isolated to make up conception word set. T is conception sum in ontology. Matrix is used to construct conception correlation.

$$R = \begin{bmatrix} R(C_1,C1) & R(C_1,C_2) & ... & R(C_1,C_T) \\ R(C_2,C_1) & R(C_2,C_2) & ... & R(C_2,C_T) \\ .. & ... & ... & ... \\ R(C_T,C_1) & R(C_T,C_2) & ... & R(C_T,C_T) \end{bmatrix}$$

Formula(1)

In matrix, $R(C_1, C_T)$ is semantic correlation of C_1 and C_T. With respect to ontology conception correlation calculation, there are a lot of scholars to study the method. It is stated that there are always two methods: Information capacity and Conceptual distance. Conception semantic correlation method is adopted in this paper. Described as follows:

If conception C_i and C_j are synonymous relationship, the correlation of C_i and C_j is 1 and $R(C_i, C_j) = 1$; semantic correlation of no synonymous conceptions C_i and C_j is calculated by the following formula:

$$R(C_i, C_j) = \frac{(Dist\ (C_i, C_j) + \alpha) * \alpha * (d(C_i) + d(C_j))}{CE\ (C_i, C_j) * 2 * Dep * \max\ (|d(C_i) - d(C_j)|, 1)}$$

Formula(2)

Therein, $d(C_i)$ and $d(C_j)$ are their corresponding levels in the binary tree; $Dist(C_i, C_j)$ is the weight of all weighted edges in the shortest route from C_i to C_j;

; $CE(C_i, C_j)$ is the sun of edges in the shortest route from C_i to C_j; Dep is max depth of ontology tree; α is a controllable parameter, generally more than o or o. On this basis, semantic correlation matrix R can be built to represent all conceptions in agriculture ontology.

3.3   Calculation Documents Similarity

The key of automatic document clustering is to calculate similarity of documents. The most widely used method is cosine measure. Compare two documents

$$d_i = (t_{i1}, w_{i1}, t_{i2}, w_{i2}, ...t_{iu}, w_{iu})   \text{and}$$

$$d_j = (t_{j1}, w_{j1}, t_{j2}, w_{j2}, ...t_{jv}, w_{jv})$$

Included angle cosine is used to present the level of similarity of documents:

$$Sim(d_i, d_j) = \frac{(d_i, d_j)}{\|d_i\| * \|d_j\|} = \frac{(d_i, d_j)}{\sqrt{(d_i, d_i) * (d_j, d_j)}}$$

$$(d_i, d_j) = \sqrt{\sum_{m=1}^{u} \sum_{n=1}^{v} w_{im} * w_{jn} * Sim(t_{im}, t_{jn})}$$

$$\|d_i\| = \sqrt{(d_i, d_i)} = \sqrt{\sum_{m=1}^{u} \sum_{n=1}^{v} w_{im} * w_{jn} * Sim(t_{im}, t_{jn})}$$

Formula (3)

Therein, t_{im} is conception feature word; w_{im} is corresponding weight; $Sim(t_{im}, t_{jn})$ is semantic similarity of conception feature words, which can be got from formula (1).

4   Text Mining System Design Based on Domain Ontology

Combined with text mining and domain ontology, the text mining model based on domain ontology is put forward. The basic processing route is that: at first, "Conception - conception" correlation matrix of ontology is built. And the documents in which feature words are extracted are represented to space vector model based on conception. Then similarity between documents are calculated according to "Conception - conception" correlation matrix. At last, clustering analysis method is used to mine the deep knowledge.

4.1   System Frame

The text mining system frame mades up of six modules: text mining pretreatment, text feature extraction, text mining, ontology management, ontology reasoning, evaluation and output the modes.

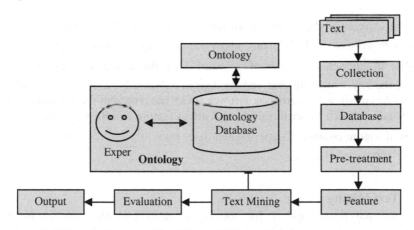

Fig. 1. System Structure

4.2   Structural Design

4.2.1   Text Data Pretreatment

The data source of text mining is unstructured text collections. They can are web pages, text documents, text files, word and excel documents, pdf files, E-mails and various forms of electronic documents. After the data resource acquired, they should be pre-treated. The process includes: data cleaning, such as denoising and incline deduplication; data selection: appropriate and related to field text data are selected; text segmentation: conception set in ontology is regarded as references to realize professional   vocabulary segmentation.

4.2.2   Text Feature Extraction

After data pretreatment, the text feature words must be extracted from the "clean" data. The process includes: (1) feature representation: The VSM is adopted to represent the documents. In the VSM, every document is presented to the following feature vector:

$$\mathsf{v}(D) = \left(t_1, w_1(d); t_2, w_2(d); ...; t_i, w_i(d); ...; t_n, w_n(d) \right)$$

Thereinto, t_i is key words; $w_i(d)$ is the weight of t_i in document d. $w_i(d)$ is often defined as the function of $\mathrm{tf}_i(d)$ that is the occurrence frequency of t_i in documentd, namely $w_i(d) = \psi(tf_i(d))$. The frequently-used ψ is the function TF-IDF:

$$\psi = tf_i(d) * \lg \frac{N}{n_i}$$

Thereinto, N is the sum of documents. n_i is the sum documents of t_i.

(2)Feature extraction: the disable word list and threshold λ_d are used to delete the unimportant words to reduce dimensions of document space. According to the function of TF-IDF to calculate the weight of key word t_i. If the weight is more than λ_d set in advance, the key words are reserved, or deleted. Obviously, a much larger threshold will filter too many words. And the left words can not represent the document content well. Instead, a much lower threshold will affect feature selection too little. Through analysis of a large number of experimental data. when $\lambda_d = 0.3$, a much better filter effect will come into being.

4.2.3   Text Mining

Clustering analysis algorithm for mining is used and the aim is to divide the objects set into multiple Classes made up of similar objects. Firstly, every document is represented to feature vector based on key words $\mathsf{v}(D) = \left(t_1, w_1(d); t_2, w_2(d); ...; t_i, w_i(d); ...; t_n, w_n(d) \right)$. Then match t_i with domain ontology conceptions, if so t_i is taken place by domain ontology conceptions. Otherwise, t_i is regarded as unknown word. After matching, the key words are taken place by domain ontology conceptions in order that text document is represented to conception set. Then clustering analysis is used to cluster large amount of documents to few meaning cluster quickly so that the hidden knowledge or mode is acquired.

4.2.4   Ontology Management

Ontology management is the key of the text mining model and provides semantic support for model. The main work of ontology management is building, storing, maintaining and optimizing the domain ontology. It provides a platform for users easy to build and maintain ontology database conveniently; to manage the ontology database in order to build, add, delete and modify the classes, relations and constraint rules in ontology; to find new conceptions or cases to extend ontology structure on the basis of text mining clustering algorithm.

4.2.5   Ontology Reasoning

The function of ontology reasoning is semantic reasoning clustering results and deleting redundant or useless cluster by domain ontology as background knowledge or priori knowledge. It can refine and generalize related knowledge to enhance effectiveness and feasibility of clustering results.

4.2.6   Evaluation and Output

The knowledge acquired from text mining may be inconsistent non-intuitive and not easy to understand. It is necessary to post process text knowledge. The main two indexes often used to evaluate text clustering effect are recall rate and accuracy rate to reflect completeness and correctness. The evaluation method which combined the two indexes is F measure.

$$\mathrm{Re}(i, j) = \frac{n_{ij}}{n_i}$$

$$\mathrm{Pr}(i, j) = \frac{n_{ij}}{n_j}$$
$$F(i, j) = \frac{2 * \mathrm{Pr}(i, j) * \mathrm{Re}(i, j)}{\mathrm{Pr}(i, j) + \mathrm{Re}(i, j)}$$

Therein, n_{ij} is the sum of class i in cluster j; n_i is the sum of documents of class i; n_j is the sum of documents in cluster j; n is the sum of documents.

4.3   Application Analysis

In order to verify the effectiveness of the system, 400 documents are selected from agricultural encyclopedia column of Chinese Agriculture Academy Science website as research objects which are 100 documents of farming, 100 documents of aquaculture, 100 documents of plant protection and 100 documents of veterinary medicine. After the 400 documents are pretreated, K-means based on space vector model and conception space vector model are used to realize cluster analysis. Clustering results are compared with accuracy, Recall and F measurement. The results are shown in the following table.

Table 1. Clustering results comparison

Arithmetic	Accuracy Rate	Recall Rate	F Measure
K-means Based on VSM	70.1	85.3	59.4
K-means Based on CVSM	84.2	93.8	89.9

As can be seen from the experimental results, space vector model based on domain ontology in which conceptions in domain ontology are instead of feature words so that correlation of feature words are reduced and dimensions of document vectors are decreased making better clustering results in accuracy, recall, F measure than K-means based on space vector model.

5   Conclusion

With the development of information technology and network resource, a flood of information is produced. To analyze the text content and potential valuable knowledge, this paper put forward text mining model based on domain ontology and introduced "conception-conception" correlation matrix and used space vector model based on domain ontology instead of space vector model to represent document and clustering algorithm to discovery knowledge.

References

1. Feldman, R., Dagan, I.: Knowledge discovery in textual databases (KDT). In: Proceedings of the First International Conference on Knowledge Discovery and Data Mining (KDD 1995), pp. 112–117. AAAI press, Montreal (1995)
2. Rosso, P., Ferretti, E., Jimenez, D., Vidal, V.: Text Categorization and Information Retrieval Using WordNet Senses. In: Proceedings of the Second Global Wordnet Conference GWC 2004, pp. 299–304 (2004)
3. Sedding, J., Kazakov, D.: WordNet-based Text Document Clustering. In: Proceedings of the Third Workshop on Robust Methods in Analysis of Natural Language Data (ROMAND), Geneva, pp. 104–113 (2004)
4. Ino, Y., Matsui, T., Ohwada, H.: Extracting Common Concepts from WordNet to Classify Documents. Artificial Intelligence and Applications, 656–661 (2005)
5. Shehata, S.: A Wordnet-based Semantic Model for Enhancing Text Clustering. In: 2009 IEEE International Conference on Data Mining Workshop, pp. 477–482 (2009)
6. Bloedorn, S., Cimiano, P., Hothon, A., Staab, S.: An Ontology-based Framework for Text Mining. LDV-Forum 20(1) (2005)
7. Hotho, A., Staab, S., Stumme, G.: Ontologies improve text document clustering. In: Third IEEE International Conference on Data Mining, ICDM 2003, pp. 541–544 (2003)

8. Song, M.-H., Lim, S.-Y., Park, S.-B., Kang, D.-J., Lee, S.-J.: Ontology-based automatic Classification of Web Pages. International Journal of Lateral Computing 1(1) (2005)
9. Zelikovitz, S.: Transductive LSI for Short Text Classification Problems. In: Proceedings of the 17th International FLAIRS Conference. AAAI Press, Miami (2004)
10. Aurora, P.-P., Rafael, B.-L., José, R.-S.: Topic discovery based on text mining techniques. Information Proceeding and Management (43), 752–768 (2007)
11. Fung, G.P.C., Yu, J.X., Lam, W.: Stock prediction: Integrating text mining approach using real-time news. In: Proceedings of the 2003 IEEE International Conference on Computational Intelligence for Financial Engineering, pp. 395–402 (2003)

Accuracy Loss Analysis in the Process of Cultivated Land Quality Data Gridding

Lingling Sang[1,3], Dehai Zhu[1,3,*], Chao Zhang[1,3], and Wenju Yun[2,3]

[1] China Agricultural University, Beijing, 100083, P.R. China
zhudehai@263.net
[2] Land Consolidation and Rehabilitation Center, Ministry of Land and Resources,
Beijing 100035, P.R. China
[3] Key Laboratory of Land Quality, Ministry of Land and Resources, Beijing 100035, P.R. China

Abstract. Spatial data gridding is one of the effective methods to solve the multi-source data fusion. In view of the current problems in the process of comprehensive analysis between the cultivated land quality data and other multi-source data, This paper, by adopting the method of the Rule of Maximum Area (RMA), converted the cultivated land quality data to the grid scale and analyzed the accuracy loss in the process of cultivated land quality data gridding in 6 grid scales (10M × 10M, 5M × 5M, 3M × 3M, 2M × 2M, 1M × 1M, 30S × 30S). Some conclusions have been reached. (1)The use of gridding methods will have assigned any analysis units to the specified data grid scales, and it provides a basis for spatial data integration, comprehensive analysis and spatial models construction;(2) Grid scale accuracy is higher, the original figure segmentation of cultivated land quality data is more serious, and grid results is more accurate, but grid computing time is increased step by step;(3) Through the study of the multistage of cultivated land quality data grid, the smaller the grid scale, the smaller the loss area of cultivated land quality, such as 10M × 10M gridding results lead to the most loss area, and Each grade area loss curve has a certain regularity;(4)From accuracy and computational efficiency, the most appropriate grid scale choice is 1M × 1M grid of 1:10000 cultivated land quality data of Daxing for the gridding processing.

Keywords: cultivated land, quality grade, gridding, accuracy analysis.

1   Introduction

Cultivated land quality grading results, namely agricultural land classification results, is under the unified national standard farming system, and are based on solar and temperature potential productivty of appointment crop.There includes cultivated land natural quality grade, utilization quality grade and economic quality composite grade.Cultivated land natural quality is assessed by the natural conditions of grading

---

* Corresponding author. College of Information and Electrical Engineering, China Agricultural University, 17 Qinghua East Road, 100083, P.R. China.

D. Li and Y. Chen (Eds.): CCTA 2013, Part II, IFIP AICT 420, pp. 370–380, 2014.
© IFIP International Federation for Information Processing 2014

units to calculate the theoretical yield of crop; and on the basis of cultivated land natural quality, there are assessed to get the grades of cultivated land utilization quality and the economic quality through the correcting step by step of the land use status and input and output level[1].The assessed results is mainly in view of the natural attribute, utilization, the social economic of cultivated land[2], and is quality evaluation in a specific time focusing on the capacity. However, the main parts of quality evaluation finally are the patches of cultivated land, and cannot achieve full coverage of the cultivated land quality assessment. In the process of practical application, the quantitative analysis which combined land quality data with land use data, soil data, environmental data, and other socioe conomic statistical data, must be the basis of administrative units to extract the relevant data. The work which the analytical methods are mostly correlation analysis is not only larger, but also is not conducive to developing mathematical models.

There are often different statistical units for model construction of multi-source data. This does not facilitate comparison and correlation analysis of multi-source data, and cannot fully play value of the existing data [3]. Spatial data gridding is one of the effective methods to solve this problem. It can not only improve the efficiency of management, but also brings about dynamic rule analysis under the support of GIS technology. The data are including the grid of the processed data and the derived results data. They could be formed on the space gradient between grid data which play the research of spatial differentiation , and formed in a grid-based data sequence which was based on the different units. Simultaneously, the gridded data have many advantages to match and integrate with multi-source data, especially they are suitable for spatial model construction to implement and express [3]. It is the basis work of a graphic rendering, scientific computing and space model implementation.

The study of spatial data gridding was carried out both at home and abroad. GIS data stored in the form of a polygon data, could also use the above method for grid transformation. There were different conversion methods between different data sources and grid data. The current gridding algorithms, most of which were regular (or irregular) the spatial distribution of point data. Spatial interpolation could achieve the conversion of point data from statistical units to the grid units. The traditional method was weighted average method to obtain the value of a grid point according to finding several nearest points in some rules spatially [4]. Interpolation methods could be used to deal with socio-economic data from the irregular grid to a regular grid [5]. In processing of meteorological data, the gridded results could better reflect the characteristics of the discrete data by comparing Kriging method and inverse distance interpolation method[6].Interpolation method of eight faceted search methods could ensure the accuracy of the grid interpolation for contour data[7].As well as, GIS spatial overlay between the layer files and the grid files could achieve data connection by the public fieled, and finally calculated attribute value of each grid cell using the weighted average method.In order to make remote sensing data can be matched with the grid scale data,we had to converse raster to vector data, and then used the method of grid processing.Of course, remote sensing data could also be obtained by resampling the grid cell size to match results. Moreover, Yang Cunjian and Cheng Jiehai discussed the accuracy of area loss during the conversion from the various land use types with different grid sizes [8,9,10].

Currently, there are no literature data to discuss the gridding methods and precision analysis for cultivated land quality. This paper chooses the Daxing district as a study area to analyze the accuracy of area loss during the conversion from 1:10000 cultivated land quality data to different grid scales.

2   Study Area and Data Sources

Daxing district, as the study area, is located in the south of Beijing, neighboring Tongzhou district in the east, west of Fangshan district across the Yongding River and Gu'an and Bazhou in Hebei Province in the south. The longitude is from 116 ° 13' to 116 ° 43 ', and the latitude is from 39 ° 26' to 39 ° 51'. There are 14 townships and 527 villages with a total area of 103 595.39 hectares, accounting for 40.18% which is the largest. There are not much available land area for development and utilization, the potential of land development is very limited.

The paper took 1:10000 cultivated land quality data in 2010 for the works, and the patches of existing cultivated land where the main object of the study. The nature quality of cultivated land in the study area was relatively homogeneous, of which the indicator differences in topsoil texture, profile configuration, salinization, organic matter content and other natural factors were not very significant. From the field investigation, most of the drainage condition indicators were 1-2 levels, and the irrigation rate indicators were" fully satisfied". According to the statistics of 1:10 000 cultivated land quality data in 2010 of Daxing district, cultivated land natural quality indexes were between 2 100 to 2 680, and the natural quality grade was from 11 to 14. Cultivated land utilization indexes were between 1 300 to 2 000, and the utilization quality grade was from 14 to 20, of which the area ratio of cultivated land of the maximum grade and the minimum grade was 4.9% and 6.9%. The grade distribution of cultivated land utilization index in Daxing district was put up a gradually decreasing tendency from the northeast to the southwest.

3   Gridding Method of Cultivated Land Quality Data

3.1   Selection of Grid Cell

The grid data is generated in the ArcGIS9.3.The methods are as follows:

(1) Defining the geographic coordinate system
Xi 'an 80 coordinate is one of the current commonly used coordinate system, so GCS_xian_1980 is defined as the geographic coordinate system in the study.
(2) Selecting the starting point and grid interval
The starting point is the intersection point with the equator and 0° longitude [11,12]. Firstly, the basic grid of the gird division of the China is based on the fixed longitude and latitude interval. Then the basic grid is divided into the next grid using the fixed longitude and latitude interval.

(3) First-level gridding results of the Chinese mainland
According to China's latitude and longitude range, the most Western longitude is 73.66E, the most eastern longitude is 135.08E, the most northern latitude is 53.52N and the most southern latitude is 4.00N. The latitudes from north and south are nearly 50 °, and the longitudes from west and east are nearly 60 °. The gridding results are as follows:

- ☞ There are divided into 64 zones (73°-136°) from the starting longitude73° along the longitude direction based on the longitude interval (1°);
- ☞ There are divided into 51 zones (4°-54°) from the starting latitude 4° along the latitude direction based on the latitude interval (1°);
- ☞ At this moment, the earth is divided into 64 × 51 grid cells, which constitutes the first-level subdivision units of the grid system, and the span of each grid is 1°×1°. Combined with the location of Chinese mainland, there are identified 1148 grids, each of which is 1°×1°grid of an area of approximately $0.739 \times 10^4 \sim 1.227 \times 10^4 km^2$.

(4) Determining the first-level grid of the study area
Using spatial overlay analysis in ArcGIS, we can pick out the first-level grid of the study area which are based on the above gridding results.

(5) Subdivision
The starting point is based on the first-level grid in the study area, then start to subdivide in latitude and longitude. There would establish 10M×10M, 5M×5M, 3M×3M, 2M×2M, 1M×1M, 30S×30S multi-level grids which preparing for gridding process and data analysis of cultivated land quality (Figure 1).

3.2   Selection of Gridding Method

Gridding is not the meaning of rasterization. Traditional rasterization can also be obtained by resampling the grid cell size to match results , and it is a process along with attribute information loss [13]. The reason is that the original grid exists in mixed types. This paper focuses on the different vector data which was gridded , and try to achieve the purposes of less attribute information loss.

According to the conversion of spatial data to the grid cell, the common methods are including the maximum value of the area method (Rule of Maximum Area, RMA) [14], the central attribute value (Rule of centric cell, RCC) [15] and the simple area weighted method. Among them, The RMA method is most commonly used that the value of a grid cell is the largest type of attribute values. If there are two or more dominant types, there can randomly select one of them as the output of the grid cell. The process is shown in Figure 3.

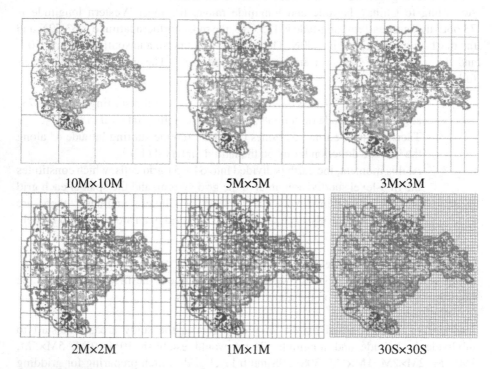

Fig. 1. Subdivision in study area (Note: M—minute ; S—second; 10M×10M—10 minute gird; 5M×5M—5 minute gird; 3M×3M—3minute gird; 2M×2M—2 minute gird; 1M×1M—1 minute gird; 30S×30S—30 second gird)

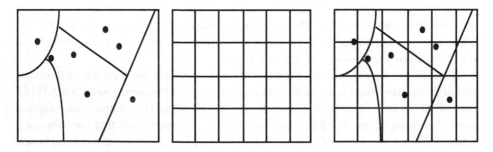

Fig. 2. Multi-source data into a regular grid

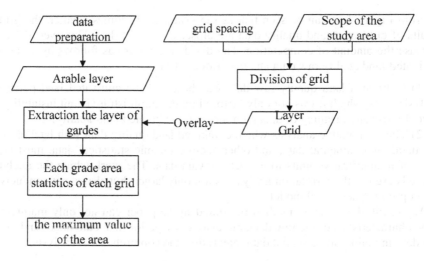

Fig. 3. Rule of maximum area

3.3   Accuracy Loss Analysis

In the research on gridding process and analysis of cultivated land quality data, firstly, selecting each grade area of cultivated land before the conversion as the reference data; then comparing the gridding data and the reference data of cultivated land quality data, there would achieve the patch number and area change results of cultivated land quality in different grid scales; finally, calculating such other area of precision loss with grid scale changes. Calculating formula is as follows:

$$E_i = A_{gi} - A_{bi} \tag{1}$$

$$L = \sum_{i=1}^{n} |E_i| \tag{2}$$

Where E_i is the area loss of grade i, the positive value is larger than the reference area, the negative value is smaller than the reference area; A_{gi} is the area of grade i after gridding; A_{bi} is the reference area of grade i; L is the overall area loss which is equal to the sum of the absolute value of area loss of each grade.

4   Results and Analysis

4.1   Changes of Spatial Distribution of Each Grade

Cultivated land quality data in Daxing district is processed by the above gridding method. In order to compare with the results, there are given 6 gridding results map with different grid scales in Figure 4.

From Figure 4, it can be seen that as the grid scale becomes smaller, the gridded results of cultivated land quality data are closer to the real data. However, there will increase the amount of computation. The disadvantages are as follows by comparing cultivated land quality raw data and the gridded data:

(1) Cultivated land quality raw data was based on the cultivated land patches as analysis units, which it was not only factored on the administrative unit instability, but it would lose any meaning of data monitoring for cultivated land quality;

(2) The quantitative analysis which combined land quality data with land use data, soil data, environmental data, and other socio-economic statistical data, must be the basis of administrative units to extract relevant data. The work which the analytical methods are mostly correlation analysis is not only larger, but also is not conducive to developing mathematical models.

Above all, the gridding method presented in the paper can not only maintain the basic characteristics of the raw data, but also it is superior than cultivated land quality raw data in multi-source spatial data integration and comprehensive analysis.

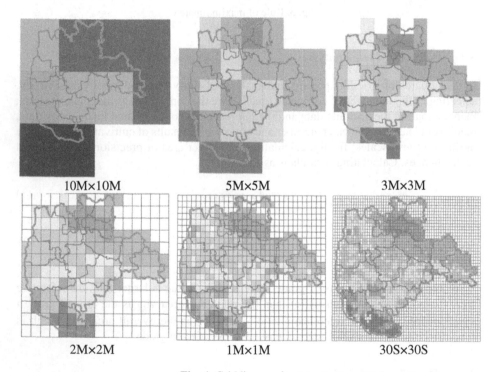

10M×10M   5M×5M   3M×3M

2M×2M   1M×1M   30S×30S

Fig. 4. Gridding results map

4.2   Analysis of Cultivated Land Quality Data

According to the statistics of 1:10000 cultivated land quality data in 2010 of Daxing district, there were 5191 patches of cultivated land. Through grid processing (10M × 10M, 5M × 5M, 3M × 3M, 2M × 2M, 1M × 1M, 30S × 30S), the original patches were divided and the grid scale cultivated land quality data information was shown in Table 1, where the 30S×30S grid division of the original data was most serious, eventually reached 10120 patches.

Table 1. Statistical analysis of cultivated land quality data based on grids

Grid	10M_Grid	05M_Grid	03M_Grid	02M_Grid	01M_Grid	30S_Grid
Number of patches	5388	5609	5915	6313	7432	10120
Area of the SD	439.47	155.55	65.63	32.08	10.02	3.85

In addition, there were some differences of each grade sequence integrity through grid processing of the original data in Table 2. The 10M × 10M gridding results were in loss of grades 15,18,20; The 5M ×5M gridding results were in loss of grades 15.The other girdding results were no loss of any grades, but their area loss was different.

Table 2. Grade sequence integrity after gridding (RMA)

	RMA_10M	RMA_5M	RMA_3M	RMA_2M	RMA_1M	RMA_30S
14	√	√	√	√	√	√
15	×	×	√	√	√	√
16	√	√	√	√	√	√
17	√	√	√	√	√	√
18	×	√	√	√	√	√
19	√	√	√	√	√	√
20	×	√	√	√	√	√

4.3   Accuracy Loss Analysis of Each Grade

Selecting tools from Arctoolbox\Data Management Tools\Generalization\Dissolve in ArcGIS, there can add up the cultivated land area of each grade after gridding, and compare with the reference data in Figure5 and Figure 6. Some conclusions have been reached.

(1) The smaller the grid scale , the less area loss of cultivated land quality;

(2) The 10M × 10M gridding results lead to the most area loss which is 280.01km^2 ; The 30S × 30S gridding results lead to the lest area loss which is 20.63km^2 ;

(3) Area loss curve of each grade has a certain regularity, which 10M × 10M, 5M × 5M, 3M × 3M, 2M × 2M gridding results were more area loss in the grades of 16, 17,18,19, and less area loss in the grades of 14,15;1M × 1M, 30S × 30S gridding results were less area loss in all the grades.

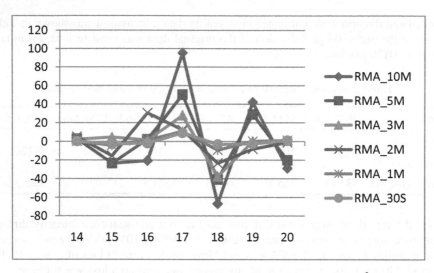

Fig. 5. Area loss of each grade in different girdding scales (km²)

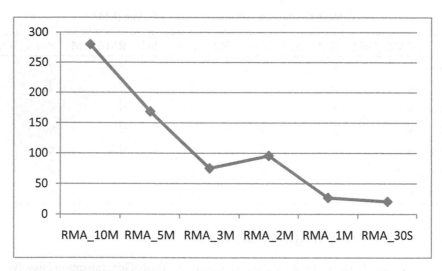

Fig. 6. The overall area loss of of each grade in different girdding scales (km²)

5   Conclusion and Discussion

This paper chose Daxing district as a study area and adopted the method of the Rule of Maximum Area (RMA) to analyze the accuracy of area loss during the conversion from 1:10000 cultivated land quality data to different grid scales and analyzed the accuracy loss in the process of cultivated land quality data gridding in 6 grid scales (10M × 10M, 5M × 5M, 3M × 3M, 2M × 2M, 1M × 1M, 30S × 30S). The foregoing analysis could be drawn :

(1) Grid scale accuracy was higher, the original figure segmentation of cultivated land quality data was more serious, and grid results were more accurate, but grid computing time was increased step by step;

(2) Through grid processing (10M × 10M, 5M × 5M, 3M × 3M, 2M × 2M, 1M × 1M, 30S × 30S) and studying the multistage of cultivated land quality data grid, the smaller the grid scale was, the smaller the area loss of cultivated land quality was, such as the 10M × 10M gridding results led to the most loss areas, and Each grade area loss curve had a certain regularity;

(3) The 1M × 1M and 30S × 30S girdding results were no loss of any grades which maintaining the grade sequence integrity after gridding. The area loss of cultivated land quality was less, and the number of patches by the 30S × 30S grid were 26.6% less than 1M × 1M grid. From accuracy and computational efficiency, the most appropriate grid scale was 1M ×1M grid of 1:10 000 cultivated land quality data of Daxing for the gridding processing.

The use of gridding methods would have assigned any analysis units to the specified data grid scales, and it provides a basis for spatial data integration, comprehensive analysis and spatial models construction. They can provide relevant researchers valuable references for spatial data processing and analysis. In addition, there are large differences in different scales and different research areas, such as the choice of the grid cell, the choice of the gridding methods and so on. It is to be carried out in-depth research in the future.

Acknowledgements. We are thankful that the study is supported by Chinese Universities Scientific Fund (2013BH034), and acknowledge Land Consolidation and Rehabilitation Center for providing essential raw data.

References

[1] GB/T 28407-2012.Regulation for gradation on agriculture land quality (2012)

[2] Lingling, S., Xiaopei, Z., Jianyu, Y., et al.: Construction and empirical analysis of green productivity evaluation system of agricultural land. Transactions of the CSAE 26(7), 235–239 (2010)

[3] Fan, Y., Shi, P., Gu, Z., et al.: A Method of Data Gridding from Administration Cell to Gridding Cell. Scientia Geographica Sinica 24(1), 105–108 (2004)

[4] Yan, B., Shunbao, L.: Implement Method of Vector Data Rasterization That without Attribute Information Loss: A Case Study of 1: 250 000 Land Cover Data of China. Journal of Geo-Information Science 12(3), 385–391 (2010)

[5] Li, D., Zhu, X., Gong, J.: From Digital Map to Spatial Information Multi-grid. In: IEEE International Geoscience and Remote Sensing Symposium, Anchorage, USA (2004)

[6] Li, D., Peng, M.: Transformation between urban spatial information irregular grid and regular grid. Geomatics and Information Science of Wuhan University 32, 95–101 (2007)

[7] Foster, I., Lman, C., Tuecke, S.: The Anatomy of the Grid: Enabling Scalable Virtual Organizations. International Journal Supercomputer Applications 15(3), 200–222 (2001)

[8] Cunjian, Y., Zengxiang, Z.: Models of accuracy loss during rasterizing land use vector data with multi2scale grid size. Geographical Research 20(4), 416–422 (2001)

[9] Jiehai, C., Fengyuan, W., Huanzhu, X.: Method for Spatial Data Scale Conversion. Geospatial Information 6(4), 13–15 (2008)

[10] Yang, C., Liu, J., Zhang, Z., et al.: Analysis of Accuracy Loss During Rasterizing Vector Data with Different Grid Size. Journal of Mountain Science 19(3), 258–264 (2001)

[11] FGDC-STD-011-2001 United States National Grid (2001)

[12] GB/T 12409-2009 Geographic grid (2009)

[13] Ma, T., Zhou, C., Xie, Y., et al.: A discrete square global grid system based on the parallels plane projection. International Journal of Geographical Information Science 23(10), 1297–1313 (2009)

[14] Sahr, K., Denis, W., Kimerling, J.: Geodesic Discrete Global Grid Systems. Cartography and Geographic Information Science 30(2), 121–134 (2003)

[15] Jing, Y., Xiaohuan, Y., Dong, J.: The grid scale effect analysis on town leveled population Statistical Data Spatialization. Journal of Geo-Information Science 12(1), 40–47 (2010)

Research and Application of Variable Rate Fertilizer Applicator System Based on a DC Motor

Honglei Jia[1], Xianzhen Feng[1], Jiangtao Qi[1,*], Xinhui Liu[2],
Chunxi Liu[3], Yongxi Yang[1], and Yang Li[1]

[1] Key Laboratory of Bionic Engineering, Ministry of Education,
Jilin University, Changchun 130022, China
[2] College of Mechanical Science and Engineering, Jilin University, Changchun 130022, China
[3] Jilin Province Academy of Agricultural Machinery, Changchun 130022, China
qijiangtao@jlu.edu.cn

Abstract. With the aim to simplify the electrical component structure of the present variable rate fertilizer applicator system, a kind of variable rate fertilizer applicator control system was developed. It is based on low voltage DC motor. The working voltage of this system is not higher than 12V DC, which can be supplied by the tractor's storage battery directly. With the encoder measuring the motor speed, it could response the running state of the motor in real time. A kind of control model based on PID algorithm was built. With this model, the influence of load variation on the motor speed could reduce. As a result, the control accuracy of this system could be improved. Field experiment has been conducted using this system. Experiment showed that when the travel speed was 3.60km/h, the maximum fertilizer rate could be 850 kg / hm^2. When the fertilizer rate was 200 ~ 600 kg/ hm^2, the mean error of this system was 1.71% while the maximum error was 2.56%.

Keywords: Variable rate technology(VRT), DC motor, PID control.

1   Introduction

Variable rate fertilizer technology is a kind of technology that variably input the fertilizer according to the actual need of the crops on the nutrients in the soil[1~4]. The benefit of improving agricultural production and ecological environment has been confirmed [5~7]. In order to master the core techniques in the variable rate fertilizer system, it is meaningful to develop our own variable rate fertilizer applicator system [8].

At present, many domestic research institutions and universities are doing research on the variable rate fertilizer applicator system. In ref.8 and ref.9, two kinds of variable rate fertilizer applicator control system based on CPLD and single-chip microcomputer were designed. The actuators of these systems are stepping motors [8~9]. In ref.10, ref.11 and ref.12, the actuators of the developed system are hydraulic

---

* Corresponding author.

D. Li and Y. Chen (Eds.): CCTA 2013, Part II, IFIP AICT 420, pp. 381–391, 2014.

motors[10~12]. The working voltage of the stepping motor and motor driver is higher than 12V DC, and it uses alternating current (e.g.110V AC). As for the system with hydraulic motor, the working voltage of electro-hydraulic proportional valve is also higher than 12V DC [14]. The storage battery voltage of the wheel tractor is 12V DC, in order to meet the requirements of the present system, inverter and transformer are needed. Thus, the electrical component structure of the system becomes complicated and the volume becomes huge, which also increases the cost.

In order to increase the system's control accuracy and simplify the electrical component structure, a kind of PID control system is developed in this paper. In this system, the DC motor is used as the actuator and an encoder is used to measure the motor speed. Besides, the working voltage of this system is not higher than 12V DC.

2   Experiments and Methods

2.1   The Composition of the Whole System

The composition of this system is shown in figure 1.

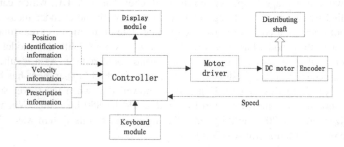

Fig. 1. Block diagram of the whole system

There are two kinds of working mode in this system: automatic mode and manual mode. In the automatic mode, the controller gains the position identification information, reads the corresponding grid prescription information stored in advance and gains the speed information from the speed sensor. With this information, the controller makes fertilization decision and generates pulse or analog voltage driving signal. The fertilization decision is made according to the fertilization decision-making formula, and the driving signal is transferred to the motor driver. The motor driver drives the DC motor to run. The encoder which is coaxially connected with the motor runs with the motor. At the same time, the controller gains the output information of the encoder and controls the motor running at a certain speed by applying the PID algorithm. Then the shaft of the fertilizer rotates driven by the DC motor through the chain. In the manual mode, the working process is the same as automatic mode except that the fertilizer rate information is manually inputted, not from the advanced stored grid prescription.

The grid position identification information can be acquired from the GPS devices [8~12] or by using the positioning technology based on speed sensor [14]. The Dead

Reckoning (DR) positioning method based on speed sensor is used as the positioning method in this system.

2.2   The Design of System Hardware

According to the function requirements of the system, the system mainly include: the single-chip microcomputer as the core controller, the speed acquisition module that gathers speed information, the keyboard module that used for inputting information, the display module that used for displaying the working state information of the machine, the motor driving module that used for driving the DC motor, the encoder used for gaining feedback of motor speed and the data recording module used for recording the working state of the fertilizer applicator.

2.2.1   The Selection of Motor Driver

Motor driver receives control signal from the controller, and outputs driving current to drive the motor. Its performance has direct influence on the stability and accuracy of the system. This system selects the motor driver produced by Beijing YongGuang gaote micro motor Co., LTD, the product model is YG8008-8EI. The driver has two kinds of control methods: pulse signal (+5V DC square-wave pulse) and analog voltage signal (-10V DC ~+10V DC). The working voltage is between +12V DC~+36V DC, which can be supplied by the tractor's storage battery directly.

2.2.2   The Selection of Controller

The controller has the following functions:① Generating pulse and analog driving signal. The signal is transferred to the motor driver to control the motor ② Receiving the conditioned speed sensor signal, used for computing the machinery speed and calculating the grid information ③ Receiving the input information from the keyboard module ④ Controlling the LCD display module to display the working state information of the machine ⑤Receiving the output of the encoder to measure the actual motor speed.

In order to meet the functions described above, the controller should have fast processing speed and rich timer/counter resources. This system selected a kind of 1T single-chip microcomputer, the product model is STC12C5A60S2. It has two programmable counter array (PCA) modules, which can be used to extend the timer resources to meet requirements of the system. At the same time, it is a kind of 1T controller. The processing speed can meet the system's requirement.

2.2.3   The Selection and Configuration of DC Motor and Encoder

The universal tillage machine 1GT-6 developed by College of Biological and Agricultural Engineering of Jilin University was used as the test prototype in this study. The rotational resistance of the shaft of the prototype was tested by using torque tester, the torque range is 2 N•m ~4 N•m. Considering the complexity of the operation environment of the agricultural machinery, taking great safety margin, the torque of the required motor should be more than 10 N•m. A kind of rare earth permanent magnet (REPM) DC motor produced by Beijing YongGuang gaote micro motor Co., LTD was used in this system, the product model is 130LYX-05F. Its maximum torque is 11 N•m, and its maximum speed is 300 r/min. According to the test, when the working voltage is 12V DC, the speed can be 100 r/min. If the travel speed is 3.60 km/h, the fertilizer rate can reach 850 kg/hm2, which meeting the requirements of this study.

The model of incremental encoder used by this system is PHB8-3600-G05L. It outputs 3 600 pulse per revolution and its working voltage is 5V DC. The encoder and the motor are coaxially connected by the encoder support. The encoder support is made of metal piece. As the encoder support is a bit flexible, the concentricity between the encoder and the DC motor can be ensured. The specific mechanism is shown in figure 2:

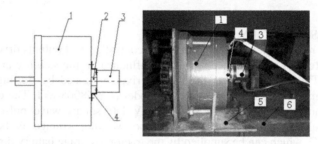

Fig. 2. Arrangement and setting of the DC motor and the encoder

1. Motor 2.Motor shaft 3.Encoder 4.Encoder support 5. Motor base 6. Machine beam

2.2.4   The Generation of the Analog Driving Voltage

The motor driver has two driving modes: pulse signal and analog voltage signal. The pulse driving mode is the prime selection, because the pulse signal can be generated by the controller directly. It is accurate and can also simplify the design of the system hardware. In order to guarantee the reliability of the prototype system, in this study, the analog driving circuit has also been designed. If something is wrong with the pulse driving mode, the operator can switch to analog signal driving mode quickly.

The principle diagram that produces analog driving voltage signal is shown in figure 3.

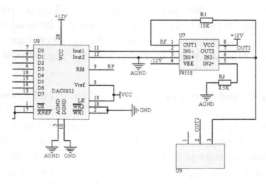

Fig. 3. Schematic of analog driving signal

The data bus D0-D7 of the D/A conversion chip DAC0832 is connected with the P0 port of the single-chip microcomputer. The data of D0-D7 controls the current of Iout1 and Iout2. F4558 is a kind of double operational amplifier chip, it converts Iout1 and Iout2 into the driving voltage signal. However, F4558 needs a negative power (-12V DC) supply. It would certainly increase the complexity of the hardware if voltage conversion chip is used. According to the datasheet of LCD module JM160128B, the 20th pin (VEE) of JM160128B can output -13.7V DC used to adjust the contrast of the LCD. The F4558 chip can use part of the voltage of this pin as the -12V DC power supply. In this way, we solve the power supply problem without increasing the cost of the hardware.

2.2.5 The Design of Speed Acquisition Module

Speed is one of the most critical factors for the variable rate fertilizer applicator system. The Hall sensor was used as the speed sensor in the speed acquisition module. Magnet slices installed on wheel hub of the rear wheel of the tractor, a total of 16 slices were used. The specific installation form is shown in figure 4.

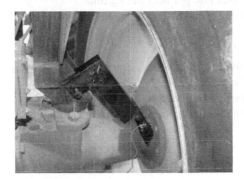

Fig. 4. Setting of speed sensor

2.2.6 The Design of Human-computer Interaction Module

In this system, grid number, fertilizer rate, travel speed, theory speed of the shaft of the fertilizer, and actual speed of the shaft of the fertilizer are the information needed to be displayed. The LCD module, whose product model is JM160128B, was used as the display module in this system.

The system selected 4×4 matrix keyboard to input information. There are 10 number buttons (0~9), 4 direction buttons (up, down, left and right), the "Confirm" button and the "Back" button.

2.3 The Design of System Software

The main function of the system software is counting the speed pulse, scanning and identifying the matrix keyboard, calculating pulse signal frequency used to control the motor, providing driving code for the LCD and recording the working state of the fertilizer applicator.

2.3.1   The Relationship between the Speed and Speed Pulse

The speed expression of the fertilizer is

$$v = \frac{C}{N_1} \times \pi \times D \times 3.6 \tag{1}$$

v —— Travel speed, km/h
C ——Pulse number per unit time, number/min
N1 ——Number of magnet slice, number
D ——Diameter of tractor rear wheel, m, D=0.57m

The fertilizer sowing amount of the feed is

$$q = \frac{10}{6} vBQ \times 10^{-3} \tag{2}$$

q ——Fertilizer sowing amount of every feed per minute, kg/min
B ——Row space of the fertilizer applicator, m, B=0.65m
Q ——Fertilizer rate per hectare, kg/hm2

The relationship between q and distributing shaft speed n is [15]

$$q = kn + b \tag{3}$$

n ——Distributing shaft speed, r/min
k ——Coefficient constant
b ——Coefficient constant

According to (2) and (3), the relationship between Q and n is

$$n = \left(\frac{10}{6} vBQ \times 10^{-3} - b \right) / k \tag{4}$$

2.3.2   The PID Algorithm to Control the Motor Speed

The PID control is a widely used control method in the control system [16]. It has the advantages of simple structure and good stability. Parameter setting is convenient and robust. In order to enhance the stability of the variable rate fertilizer applicator system and reduce the control error, the control model was established based on the PID algorithm as shown in figure 5, the differential expression of the mathematical model [17] is

$$Vo(t) = K_p \cdot e(t) + K_i \cdot \int_0^t e(t)dt + K_d \cdot \frac{de(t)}{dt} \tag{5}$$

e (t) =VO (t)-Vi (t)
K_p ——Proportional coefficient
K_d ——Differential coefficient
K_i ——Integral coefficient

The transfer function is

$$G(s) = \frac{Vo(s)}{Vi(s)} = K_p + K_d s + \frac{Ki}{s} \tag{6}$$

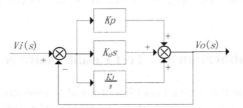

Fig. 5. Control mode of PID controller

2.4 Flowchart of System Software

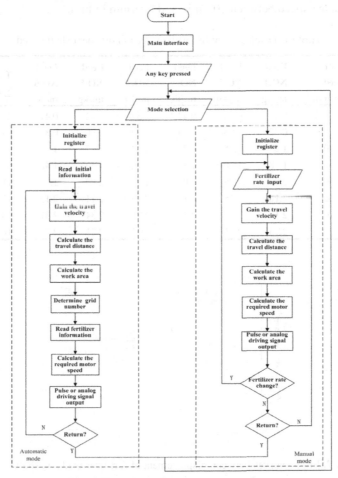

Fig. 6. Flowchart of the system procedures

The main interface will appear once the system is turned on. In the main interface, the system information and welcome message is displayed in this window. Only if the user presses one of the keys, the system will turn to the work mode selection interface. Then the software turn to "automatic mode" or "manual mode" according to the user's selection. The specific flowchart of the system software is shown in figure 6.

3   System Calibration and Test Result Analysis

In order to determine the constant coefficient k and b, calibration experiment should be carried out before the system starts to work[18]. The experiment used external force fertilizer, granular urea. The environment temperature is 10 °C. First of all, measuring the consistency of the fertilizer at the speed of n=45 r/min, and adjusting the feed to meet the consistency requirement. Then measuring the displacement of every feed at the speed between 10 r/min and 80 r/min in turn.

Table 1. Displacement of fertilizer feed in diffident shaft speed

Time t/s	Shaft speed n /r·min^{-1}	Feed NO.1 m_1/kg	Feed NO.2 m_2/kg	Feed NO.3 m_3/kg	Feed NO.4 m_4/kg	Feed NO.5 m_5/kg	Feed NO.6 m_6/kg	Total $\sum$/kg	Mean q/kg
60	10	0.41	0.44	0.42	0.39	0.42	0.41	2.48	0.41
60	20	0.82	0.85	0.84	0.80	0.90	0.84	5.04	0.84
60	30	1.22	1.25	1.26	1.18	1.27	1.22	7.40	1.23
60	40	1.48	1.50	1.49	1.46	1.52	1.48	8.94	1.49
60	50	2.00	2.05	2.04	1.90	2.03	1.97	11.99	2.00
60	60	2.12	2.15	2.15	2.10	2.11	2.13	12.78	2.13
60	70	2.10	2.31	2.40	2.31	2.34	2.37	13.83	2.31
60	80	2.76	2.75	2.80	2.75	2.81	2.74	16.68	2.78

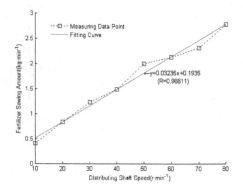

Fig. 7. Fitting curves between displacement of fertilizer feed and shaft speed

The linear fitting between q and n according to the experiment result is

$$q = 0.03235n + 0.1935 \tag{5}$$

Namely, k=0.03235, b=0.1935, with the equation (4), the final fertilization decision-making formula is

$$n = \left(\frac{10}{6} \times 0.65 vQ \times 10^{-3} - 0.1935 \right) / 0.03235 \tag{6}$$

In order to validate the performance of the system, in the autumn of 2012, field test was conducted in the test field of Changchun Academy of Agricultural Sciences, as shown in figure 8.

In order to test the fertilizer distributing performance of the prototype, the machine runs 20 meters at the speed of 3.60 km/h. The falling fertilizer was collected with plastic bags. In theory, the fertilizer sowing amount M is

$$\frac{M}{N_2 \times B \times 20} = \frac{Q}{10000} \tag{7}$$

N_2 —— number of fertilizer pipe, $N_2=4$
M —— theory fertilizer sowing amount, kg

Namely

$$M = 5.2 \times 10^{-3} Q \tag{8}$$

Table 2. Prototype test of distributing performance

Fertilizer rate Q/kg	Theory sowing amount M/kg	Actual sowing amount M'/kg	Relative error μ/%
200	1.04	1.05	0.96
300	1.56	1.58	1.28
400	2.08	2.05	1.44
500	2.60	2.54	2.31
600	3.12	3.04	2.56
Ave			1.71
Max			2.56

The field experiment shows that the maximum fertilizer rate can be 850 kg/hm2, when the travel speed is 3.60km/h. When the fertilizer rate is 200~600 kg/hm2, the mean error of this system is 1.71%, the maximum error is 2.56%. Its stable and reliable performance prove that it's a kind of feasible method for variable rate fertilizer applicator.

4   Conclusions

(1)With large torque DC motor as the actuator of variable rate fertilizer applicator system, the working voltage of the whole system is not higher than 12V DC, which can be supplied by the tractor's storage battery directly. The electrical component structure of the variable rate fertilizer applicator was greatly simplified.

Fig. 8. Field experiment

(2)With the incremental encoder being the motor speed detection device, the controller can monitor the motor speed in real-time. Establishing a control mode with the PID algorithm improves the control accuracy of the system.

(3)Applying this system to the field operation, it works stably. The maximum fertilizer rate can be 850 kg/hm2 when the travel speed is 3.60km/h. When the fertilizer rate is 200~600 kg/hm2, the mean error of this system is 1.71%, while the maximum error is 2.56%.

Acknowledgment. This study was supported by the Key Projects in the National Science & Technology Pillar Program of China during the 12th Five-Year Plan period (2011BAD20B09) and Scientific and Technological Development Projects in Jilin Province (No. 20125026).

References

1. Yang, C., Everitt, J.H., Bradford, J.M.: Comparisons of uniform and variable rate nitrogen and phosphorus fertilizer applications for grain sorghum. Transactions of the American Society of Agricultural Engineers 44(2), 201–209 (2001)
2. Zhang, S., Ma, C., Li, W., Xui, Y.: Experimental study on the influence of variable rate fertilization on maize yield and soil nutrients. Transactions of the Chinese Society of Agricultural Engineering 22(8), 64–67 (2006)
3. Wittry, D.J., Mallarino, A.P.: Comparison of uniform-and variable-rate phosphorus fertilization for corn-soybean rotations. Agronomy Journal 96(1), 26–33 (2004)

4. Saeys, W., Deblander, J., Ramon, H., Anthonis, J.: High-performance flow control for site-specific application of liquid manure. Biosystems Engineering 99(1), 22–34 (2008)
5. Zhijun, M., Chunjiang, Z., Hui, L., Wenqian, H., Weiqiang, F., Xiu, W.: Development and performance assessment of map-based variable rate granular application system. Journal of Jiangsu University: Natural Science Edition 30(4), 338–342 (2009)
6. Meng, Z., Zhao, C., Fu, W., Ji, Y., Wu, G.: Development and performance assessment of map-based variable rate granular application system. Journal of Jiangsu University: Natural Science Edition 30(4), 204–209 (2009)
7. Chengyu, Z., Haifeng, X.: Our country measuring soil fertilizer technology to increase revenue and reduce expenditure effect research—— Based on empirical analysis of jiangsu and jilin provinces. Agricultural Technical and Economic (2), 44–51 (2009)
8. Zhang, S., Qi, J., Liao, Z., Xu, Y.: Research and application of control system for variable rate fertilizer applicator based On CPLD. Transactions of the the Chinese Society of Agricultural Engineering 26(8), 200–204 (2010)
9. Xu, M., Chenglin, M., Guoqi, S., Jiang, Z.: Design of variable rate fertilizer applicator. Transactions of the Chinese Society for Agricultural Machinery 36(1), 50–53 (2005)
10. Xiu, W., Chunjiang, Z., Zhijun, M., Liping, C., Yuchun, P., Xuzhang, X.: Design and experiment of variable rate fertilizer applicator. Transactions of the Chinese Society of Agricultural Engineering 20(5), 114–117 (2004)
11. Chunying, L., Xi, W., Jun, Z., Zhimin, W., Weidong, Z.: Design of the velocity system of the valve-controlled motor on the variable rate fertilizer applicator. Journal o f Hei Long Jiang August First Land Relamation University 15(3), 47–50 (2003)
12. Baofu, H., Chunying, L., Xi, W., Aiping, L., Hongxia, Z.: Design of control system of variable rate fertilization. Journal of Heilongjiang Bayi Agricultural University 23(4), 68–71 (2011)
13. Linhuan, Z.: Study on hydraulic stepless speed control system of the variable rate fertilizer applicator. JiLin University, JiLin (2009)
14. Yu, Y., Zhang, S., Qi, J., Xu, Y., Wang, W.: Positioning Method of Variable Rate Applicators in Irregular Field. Transactions of the Chinese Society for Agricultural Machinery 42(2), 158–161 (2011)
15. Zhang, S., Ma, C., Du, Q., Nie, X., Wu, C., Han, Y.: Design of control system of variable rate fertilizer applicator in precision agriculture. Transactions of the Chinese Society of Agricultural Engineering 20(1), 113–116 (2004)
16. Yonghua, T.: New type of PID control and application. Machinery Industry Press, Beijing
17. Liang, C., Yi, S., Xi, W., Huai, B.F.: PID Control Strategy of the Variable Rate Fertilization Control System. Transactions of the Chinese Society for Agricultural Machinery 41(7), 157–162 (2010)
18. Wei, L., Zhang, X., Yuan, Y., Liu, Y., Li, Z.: Design and experiment of 2F-6-BP1 variable rate assorted fertilizer applicator. Transactions of the Chinese Society of Agricultural Engineering 28(4), 14–18 (2012)

Brief Probe into the Key Factors that Influence Beijing Agricultural Drought Vulnerability

Lingmiao Huang, Peiling Yang, and Shumei Ren

College of Water Resources & Civil Engineering, China Agricultural University,
Beijing 100083, China
huanglingmiao@cau.edu.cn, {yangpeiling,renshumei}@126.com

Abstract. Drought is a major disaster that Beijing agricultural systems faced with. The risk of drought disasters is a result of drought disaster together with vulnerability, the result of drought disaster only appeared in post-disaster, before the disaster occurred, we need to make some research on drought vulnerability that Beijing agricultural systems faced with, we choose VAM (vulnerability assessment method) that consist of three drought elements: exposure, sensitivity and adaptive capacity, based on two kinds of analytic hierarchy process method to determine the weight of each factors. try to verify the results of agricultural drought indicator system through the spatial distribution map of soil moisture, as well as explore the key factors that influence Beijing agricultural drought vulnerability.

Keywords: Vulnerability, Assessment, Indicator System, Analytic Hierarchy Process, Agricultural Drought, Soil Moisture.

1 Introduction

Vulnerability is context-specific and what makes one region or community vulnerable may be different from another community [1]. The main Vulnerability research bodies are diverse, such as ecosystem, the forests [2] and wetlands [3], etc. ;water resources system [4]; human-environment system [5], urban social living system [6], there are more widely agricultural system [7-11], etc.; current research also focus on vulnerability caused by climate change [6, 12-15]; as well as vulnerability studies on earthquake disasters [16], flood disasters [17-20] and caused by land use changes [21], etc.

The risk of natural disasters is a result of natural disasters together with vulnerability, the level of loss risk is directly related with social vulnerability, to a great extent, decided by the vulnerability of hazard bearing [22]. However, there are certain generic determinants of vulnerability including developmental factors that are likely to influence the vulnerability of a particular region or community even in diverse socioeconomic contexts [1]. Thus, one of the key features of vulnerability is its dynamic nature that may change as a result of changes in the biophysical as well as the socioeconomic characteristics of a particular region [23]. Hence, vulnerability assessments should be ongoing processes in order to highlight the spatial and temporal scales of vulnerability of a region [24].

D. Li and Y. Chen (Eds.): CCTA 2013, Part II, IFIP AICT 420, pp. 392–403, 2014.

This paper focuses on agricultural drought vulnerability which means the property or status which is sensitive to drought and easily threatens by drought, as well as caused loss [25]. The impact factors of drought vulnerability are multifaceted [26], meanwhile, vulnerability studies combined tightly with spatial scales [27]. Some scholars have tried a variety of methods to evaluate the overall vulnerability of agricultural drought; they appropriately selected the indicators mainly based on the main bodies and scales in evaluation and assessment [8, 9, and 28]. Currently, researches focused on choosing two or three factors from the three factors—— exposure, sensitivity and adaptability to determine the indicators [10, 29, and 30].

The aim of this paper is to develop and apply a quantitative approach to agricultural drought vulnerability assessment within Beijing to identify which of the country's regions and districts are most vulnerable to drought, also use a method to test the assessment results. To achieve this aim, the study objectives are:

1. To establish an indicator system for the assessment of agricultural drought based on the three elements of drought vulnerability;
2. To select 15 factors represent drought exposure, sensitivity and adaptive capacity, based on two kinds of analytic hierarchy process method to evaluate the exposure, adaptive capacity and sensitivity of Beijing's eleven regions and the districts within the most vulnerable regions;
3. To verify the results of agricultural drought indicator system through the spatial distribution map of soil moisture.
4. To explore the key factors that influence Beijing agricultural drought vulnerability.

2   Materials and Methods

2.1   Study Area

Beijing is located in the longitude of 115° 20'to 117° 30' and the latitude of 39° 25'to 41°. About 150km distance to the west of Bohai, located in the northwest edge of the North China Plain, the city's total area is 16410.54 km^2, mountain area account for 62%, and about 38% of the areas are plains.

The growth period of winter wheat in Beijing start from late September to mid-June, the rainfall of Beijing area is generally concentrated in the June to August, appear the phenomenon of precipitation and growth period of winter wheat water requirement dislocation, drought is a direct threat to stable and high yield of winter wheat, even has a serious impact on the socio-economic. It is urgent to carry out the study on Beijing drought vulnerability and reduce drought disasters.

2.2   Research Method

In this paper, using VAM vulnerability assessment methods [11], the three elements of vulnerability classified as exposure, sensitivity, and adaptive capacity. Exposure represent that system experience a degree of drought stress, it is related to the intensity, frequency and duration of the drought; sensitivity is the extent of agricultural system elements susceptible to the effects of drought; adaptive capacity is the behavior that stakeholders in the agricultural system taking to reduce the effects of drought in the pre-disaster and after disaster. The results of exposure, sensitivity, and adaptive capacity Co-expressed as drought vulnerability.

Using the formula [11]

$$V = f(E, S, A) \tag{1}$$

By the composite indicator method, the vulnerability results can be expressed as various indicators and the linear plus of their corresponding weights:

$$V = \sum_{i=1}^{n} I_i \times w_i + \sum_{l=1}^{m} I_l \times w_l + \sum_{r=1}^{s} I_r \times w_r \tag{2}$$

In the formula (1), V is drought vulnerability indicator of the evaluation unit; E is drought exposure indicator; S is drought sensitivity indicator; A is drought adaptability indicator. There are respectively multiplied by the evaluation indicator scores that belonging to the three elements and the corresponding weight of evaluation indicator. The I_i, I_l, I_r are the score of evaluation indicator i, l, r; W_i, W_l, W_r are the weight value of the evaluation indicator i, l, r.

2.2.1   Establishment of the Evaluation Indicator System

Starting from the three elements: exposure, sensitivity, and adaptive capacity to select representative sub-elements in order to refine the three elements.

Table 1. List of drought vulnerability indicators

Element	Evaluation indicators	The explain of evaluation indicators	Data sources	Time
Exposure (E)	Difference value between evaporation and rainfall/mm(+)	Reflects the degree of crop water demand is satisfied	China Meteorological Data Sharing Network	2010
	Average elevation/m(+)	Elevation is positively related to water abstraction difficulty during the drought period		
	Average slope/° (+)	The slope affect soil water retention capacity, increasing the severe drought period irrigation difficulty	International Scientific Database (90 meter resolution data)	
	The river network density/ (km/km²) (-)	Lack of precipitation, surface runoff is a major water sources of production and living		
Sensitivity (S)	Forest coverage /% (-)	Influence the climate of the area and the conservation of rainfall	Regional Statistical Yearbook of Beijing	2010
	The irrigation index (-)	The proportion of effective irrigation area in the total area of cultivated land		
	Winter wheat planting ratio (+)	The proportion of winter wheat acreage in the total cultivated area		

Table 1. (*continued*)

	Proportion of agricultural population(+)	The agricultural population groups are most sensitive to drought	
	Facilities agricultural area ratio(-)	The proportion of facilities agricultural area in the total cultivated area	
	Multiple cropping indicator (+)	Indicator of the degree of cultivated land use	
	Grain yield per unit area/ (t/hm^2)(-)	The lower, the more serious the impact of drought decrease yields, the more sensitive to drought	
Adaptive capacity (A)	Rural electricity consumption/10^4kwh(-)	Reflect the level of the rural economy, as well as the time and the frequency use of agricultural machinery	
	Net income of rural residents/Yuan(-)	Income determines the drought disaster recovery capabilities	Regional Statistical Yearbook of Beijing
	Arable land per capita/hm2(-)	The performance of the population effect and the pressure of land	2010
	Agricultural power in unit of arable land / (w/hm^2) (-)	Manifest the level of mechanization of agricultural production, but also reflect the size of the irrigation mechanical power during drought period	

Note: The table "+" and "-" represent the various evaluation and drought vulnerability positive or negative correlation (partial indicators refer to Chen Ping [29] indicator system)

2.2.2   Indicators Quantify and Weight Calculations

Analytic Hierarchy Process (AHP) is a method of evaluation and decision-making which is a combination of qualitative and quantitative analysis, quantitative analysis of the person's qualitative subjective judgment, making variety of heterogeneous data integration, it is a currently widely used to determine the weight. However, using AHP to build a judgment matrix will be a certain degree of subjectivity because the experts' judgment of the relative importance of the indicators varies different, while insufficient application of existing quantitative information is also an obvious inadequacy. Therefore, we can use two kinds of analytic hierarchy process method to calculate the weight of drought vulnerability indicators. Firstly, divide the evaluation behavior into three levels. Target layer is evaluation of vulnerabilities; criterion level is three drought vulnerability factors (exposure, sensitivity, adaptive capacity); lowest level is indicator layer which belonged to the upper evaluation indicators. Then, use the K-means algorithm for the discretization of each indicator data indicators, the data is divided into five levels. Finally, make assignments for each indicator positive correlation with vulnerability assigned as 2,4,6,8,10; negatively correlated with vulnerability assigned as 10,8,6,4,2.

2.2.3   Test for Vulnerability Indicator System

According to the daily moisture monitoring data from Beijing moisture sites in the year of 2009 to 2010, take the moisture values during the growth of winter wheat from September 2009 to June 2010, use the regression kriging interpolation method to draw soil moisture distribution ArcGIS map, verify the accuracy of the indicator system through the high and low values of the moisture distribution.

3   Results and Analysis

Making the weight of exposure, sensitivity and adaptive capacity in the criterion level equal. Using two kinds of analytic hierarchy process method: 0~2 Three Demarcation Method and 1~9 Demarcation Method to calculate the weight of the drought vulnerability indicators. Vulnerability evaluation results of all districts and counties in Beijing are shown in Figure1~3. Using K-means algorithm to clustering exposure, sensitivity and adaptive capacity, the cluster centers are shown in Table 2.

Table 2. The cluster centers of K-means algorithm

(0~2 Three Demarcation Method) Grade	1	2	3	4	5	(1~9 Demarcation Method) Grade	1	2	3	4	5
Exposure	1.47	2.02	2.28	2.32	2.44	Exposure	1.25	1.69	1.71	2.30	2.35
Sensitivity	1.27	1.68	1.74	2.06	2.55	Sensitivity	1.64	1.68	1.87	1.94	2.17
Adaptive Capacity	1.65	1.85	2.47	2.76	2.89	Adaptive Capacity	1.70	1.84	2.57	2.69	2.80

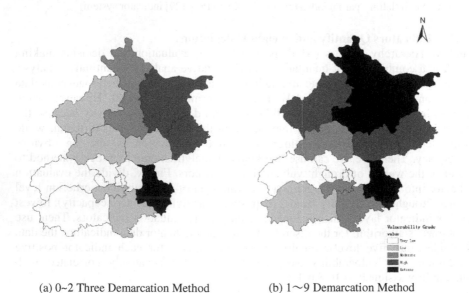

(a) 0~2 Three Demarcation Method   (b) 1~9 Demarcation Method

Fig. 1. Evaluation results of exposure

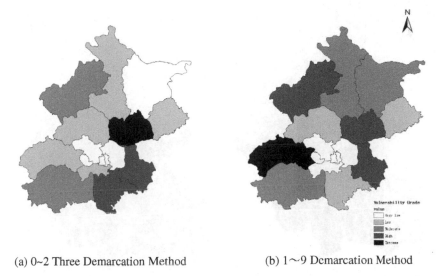

(a) 0~2 Three Demarcation Method   (b) 1～9 Demarcation Method

Fig. 2. Evaluation results of sensitivity

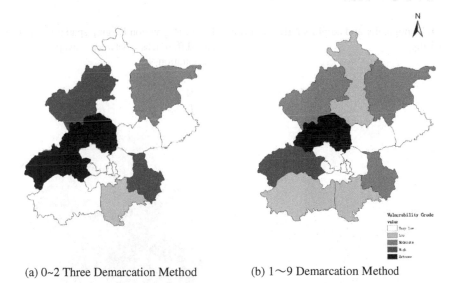

(a) 0~2 Three Demarcation Method   (b) 1～9 Demarcation Method

Fig. 3. Evaluation results of adaptive capacity

Beijing's topography and administrative divisions, the regression Kriging spatial interpolation on differences between evapotranspiration and precipitation are shown in Figure 4 and 5. Soil moisture site distribution of study area is shown in Figure 6, use the regression kriging interpolation method to deal with the moisture values during growth of winter wheat from September 2009 to June 2010 to draw soil moisture distribution ArcGIS map, shown as Figure 7, the level of vulnerability distribution shown in Figure 8, the distribution of soil moisture, superimposed soil moisture and adaptive capacity are shown in Figure 9and 10, respectively.

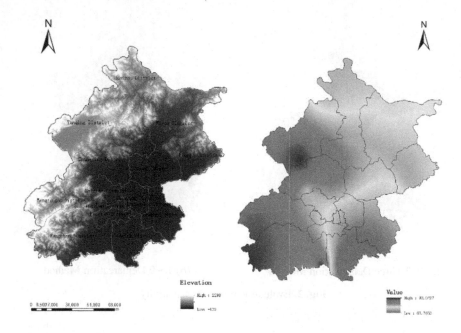

Fig. 4. Topography and administrative divisions of Beijing

Fig. 5. Regression Kriging spatial interpolation on differences between evaporative and precipitation

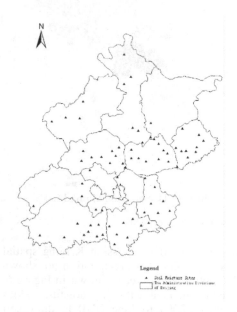

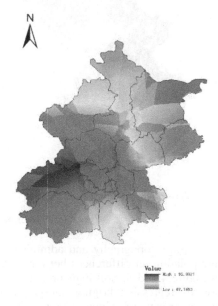

Fig. 6. Soil moisture sites in the study area

Fig. 7. Regression Kriging spatial interpolation on soil moisture

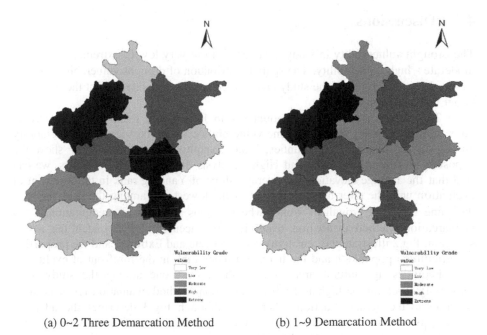

(a) 0~2 Three Demarcation Method   (b) 1~9 Demarcation Method

Fig. 8. The distribution of vulnerability grade

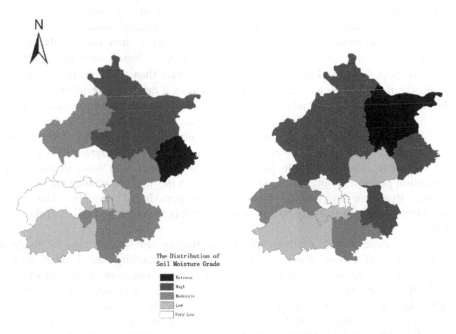

Fig. 9. The distribution of soil moisture grade

Fig. 10. The distribution of superposition of soil moisture and adaptive capacity

4   Discussions

The drought vulnerability in study area varied from very low to extreme, mainly are moderate - high vulnerability. The spatial distribution of drought vulnerability (Fig.8) shows the vulnerability of the study area can be roughly considered that the mountain larger than the plains.

1) The northern and western mountains in Beijing are mainly high - extreme vulnerability. Yanqing shows extreme vulnerability, the highest level of vulnerability among all the districts and counties. Take Yanqing for example, Fig.1 show the exposure in Yanqing are Low and High. Analysis the evaluation indicators, we can find that the average elevation and average slope of Yanqing rank in the forefront of evaluation unit, the river network density is the lowest one among the study area, at the same time, the growth of winter wheat serious shortage of precipitation, 1~9 Demarcation Method evaluation result is more accurate for it match the actual situation. Fig.2 illustrate the sensitivity are Moderate and Extreme, and the proportion of agricultural population and the irrigation index rank in the forefront of evaluation unit, Facilities agricultural area ratio is the lowest one among the study area, sensitivity level can be higher. 1~9 Demarcation Method evaluation result is more accurate to the actual situation. Therefore, although Fig.3 describes the adaptive capacity is moderate, still not enough to reduce vulnerability of agricultural systems caused by drought exposure and drought sensitivity.

From the view of three elements of drought vulnerability, the spatial distribution of exposure shows opposite trend. The spatial distribution of exposure is closely related to the terrain conditions and crop water supply and demand situation, the exposure element in northern mountains are significantly higher than that of the western mountains, mainly for the higher elevations, steeper slopes, also the differences between evaporative and precipitation is much larger than the western mountains, the northern mountains evaluation unit experiences a large degree of drought stress. The spatial distribution of adaptive capacity mainly related to agricultural power in unit of arable land and arable land per capita, adaptive capacity in the western mountains is higher than the northern mountains in this point. The common characteristics of the High-Extreme evaluation unit is that the agricultural power in unit of arable land and arable land per capita are both very low, indicating that population pressure acting on the land is bigger in the western mountains, as well as smaller irrigation mechanical power during periods of drought. The spatial distribution rule of sensitivity has no significant; it is the combined effect of the seven factors.

2) The eastern and southern plains in Beijing are mainly low - moderate vulnerability. Tongzhou shows extreme - high vulnerability, the highest level of vulnerability among all the districts and counties. Take Tongzhou District for example, Fig.1 shows the exposure in Tongzhou is extreme. Fig.2 represents the sensitivity is high, Fig.3 describe the adaptive capacity are extreme and moderate, respectively. Analysis the evaluation indicators, we can find that rural electricity consumption of Tongzhou ranks in the forefront of evaluation unit, agricultural power in unit of arable land is the lowest one among the study area, Therefore, we judge 1~9 Demarcation Method evaluation result is more accurate for it match the actual situation.

From the perspective of three elements of drought vulnerability, the spatial distribution rule of exposure, sensitivity and adaptive capacity is not obvious. In Shunyi, Tongzhou and Daxing these three areas, spatial distributions of exposure and sensitivity are more consistent, however the spatial distribution of adaptive capacity presents opposite trend, which is due to the income of rural residents is the main factor for adaptive capacity evaluation, in the southern evaluation unit, farmers generally have higher income level. When drought caused the loss of farmers, the rich farmers can easier and quicker recover from drought to pre-disaster levels. The interaction of three elements determine the vulnerability of drought in the region is mainly Low - Moderate vulnerability.

Fig.8 reflects the spatial distribution of exposure is high in northeast and low in southwest, apart from the western mountains, basically consistent with the distribution of soil moisture. The soil moisture during growth period of winter wheat, has taken irrigation conditions into account at this stage, considers the exposure and sensitivity of vulnerability, and does not involve the adaptive capacity. Thus, we can make the following assumptions: the distribution of vulnerability is the superposition of soil moisture and adaptive capacity.

Averaging and Superimposing soil moisture grade (Fig.9) and adaptive capacity grade to make the Fig.10, compared result with 1~9 Demarcation Method evaluation result, the superposition effect of the southern is better than the northern, in the southern except a small number of a grade difference, the rest are the same. However, in the north region except for a few results are the same, the rest several areas have one grade difference. That describes the northern region influence largely by the sensitivity factors. Therefore, regions or districts that its agricultural system elements susceptible to the effects of drought should consider the interaction of three elements of vulnerability.

5   Conclusions

In this study, make crop water demand and supply as an evaluation indicator of drought exposure, focus on the balance between water demand and supply while considering other factors, for example, natural, social and economic factors that influence the process of drought disaster. Try to use soil moisture to examine the results of the evaluation; the method is more accurate reflection of the potential losses of drought on different evaluation unit.

However, using AHP to build a judgment matrix will be a certain degree of subjectivity because the experts' judgment of the relative importance of the indicators varies different, while insufficient application of existing quantitative information is also an obvious inadequacy. When using the two kinds of analytic hierarchy process method to calculate the weight of drought vulnerability indicators, can avoid judgment matrix randomness which varies from person to person. This method is more objectivity. Compared the two results from 0~2 Three Demarcation Method and 1~9 Demarcation Method, the evaluation result of 1~9 Demarcation Method is more accurate for it match the actual situation.

Using K-means algorithm to making indicator data discretization, can avoid the disadvantages of equal-width interval method which is too simple and easily influenced by the amount of data. Indicator data discretization has directly impact on the assignment of the indicator data and the accuracy of the final vulnerability assessment. Therefore, a suitable discretization method is particularly important.

Drought vulnerability is the result of interaction of the three elements: exposure, sensitivity, and adaptive capacity. These three elements have its own rule in the spatial distribution. Single soil moisture distribution is an important factor that influences exposure, while having some impact on sensitivity, which lacks the element of adaptive capacity. In the development of drought-prevention policy, combining soil moisture and adaptability to determine drought risk can be used as a simple method for evaluation, however, areas or regions which affected largely by sensitivity should also take full account of the interaction of the three elements, and make it a starting point to develop the corresponding drought policy. Prevent or reduce disaster losses before the drought had occurred.

References

1. Brooks, N., Neil Adger, W., Mick Kelly, P.: The determinants of vulnerability and adaptive capacity at the national level and the implications for adaptation. Global Environmental Change 15(2), 151–163 (2005)
2. Lindner, M., Maroschek, M., Netherer, S., et al.: Climate change impacts, adaptive capacity, and vulnerability of European forest ecosystems. Forest Ecology and Management 259(4), 698–709 (2010)
3. Copeland, H.E., Tessman, S.A., Girvetz, E.H., et al.: A geospatial assessment on the distribution, condition, and vulnerability of Wyoming's wetlands. Ecological Indicators 10(4), 869–879 (2010)
4. Yu, C., Hao, Z.: Quantitative Assessment Research of Water Resources System Vulnerability of Shanxi Province: Bioinformatics and Biomedical Engineering. In: 3rd International Conference on ICBBE 2009. IEEE (2009)
5. Turner, B.L., Matson, P.A., McCarthy, J.J., et al.: Illustrating the coupled human–environment system for vulnerability analysis: three case studies. Proceedings of the National Academy of Sciences 100(14), 8080–8085 (2003)
6. Füssel, H.: Vulnerability: a generally applicable conceptual framework for climate change research. Global Environmental Change 17(2), 155–167 (2007)
7. Xu, W., Ren, X., Smith, A.: Remote sensing, crop yield estimation and agricultural vulnerability assessment: A case of Southern Alberta: Geoinformatics. In: 2011 19th International Conference on IEEE (2011)
8. Wilhelmi, O.V., Wilhite, D.A.: Assessing vulnerability to agricultural drought: a Nebraska case study. Natural Hazards 25(1), 37–58 (2002)
9. Wu, H., Wilhite, D.A.: An operational agricultural drought risk assessment model for Nebraska. Natural Hazards 33(1), 1–21 (2004)
10. Simelton, E., Fraser, E.D.G., Termansen, M., et al.: Typologies of crop-drought vulnerability: an empirical analysis of the socio-economic factors that influence the sensitivity and resilience to drought of three major food crops in China (1961-2001). Environmental Science & Policy 12(4), 438–452 (2009)
11. Fontaine, M.M., Steinemann, A.C.: Assessing vulnerability to natural hazards: impact-based method and application to drought in Washington State. Natural Hazards Review 10(1), 11–18 (2009)

12. Parry, M.L., Canziani, O.F., Palutikof, J.P., et al.: Climate change 2007: impacts, adaptation and vulnerability. Intergovernmental Panel on Climate Change (2007)
13. Polsky, C., Neff, R., Yarnal, B.: Building comparable global change vulnerability assessments: the vulnerability scoping diagram. Global Environmental Change 17(3), 472–485 (2007)
14. Metzger, M.J., Leemans, R., Schröter, D.: A multidisciplinary multi-scale framework for assessing vulnerabilities to global change. International Journal of Applied Earth Observation and Geoinformation 7(4), 253–267 (2005)
15. Turner, B.L., Kasperson, R.E., Matson, P.A., et al.: A framework for vulnerability analysis in sustainability science. Proceedings of the National Academy of Sciences 100(14), 8074–8079 (2003)
16. Rashed, T., Weeks, J.: Assessing vulnerability to earthquake hazards through spatial multicriteria analysis of urban areas. International Journal of Geographical Information Science 17(6), 547–576 (2003)
17. Shouyu, C., Yu, G.: Variable fuzzy sets and its application in comprehensive risk evaluation for flood-control engineering system. Fuzzy Optimization and Decision Making 5(2), 153–162 (2006)
18. Apel, H., Thieken, A.H., Merz, B., et al.: Flood risk assessment and associated uncertainty. Natural Hazards and Earth System Science 4(2), 295–308 (2004)
19. Zou, Q., Zhou, J., Zhou, C., et al.: Comprehensive flood risk assessment based on set pair analysis-variable fuzzy sets model and fuzzy AHP. Stochastic Environmental Research and Risk Assessment 27(2), 525–546 (2013)
20. Huang, D., Zhang, R., Huo, Z., et al.: An assessment of multidimensional flood vulnerability at the provincial scale in China based on the DEA method. Natural hazards 64(2), 1575–1586 (2012)
21. Metzger, M.J., Rounsevell, M., Acosta-Michlik, L., et al.: The vulnerability of ecosystem services to land use change. Agriculture, Ecosystems & Environment 114(1), 69–85 (2006)
22. Blaikie, P.M.: At risk: natural hazards, people's vulnerability, and disasters. Psychology Press (1994)
23. Adger, W.N., Kelly, P.M.: Social vulnerability to climate change and the architecture of entitlements. Mitigation and Adaptation Strategies for Global Change 4(3-4), 253–266 (1999)
24. Luers, A.L.: The surface of vulnerability: an analytical framework for examining environmental change. Global Environmental Change 15(3), 214–223 (2005)
25. Keyantash, J., Dracup, J.A.: The quantification of drought: an evaluation of drought indices. Bulletin of the American Meteorological Society 83(8), 1167–1180 (2002)
26. Wilhelmi, O.V., Wilhite, D.A.: Assessing vulnerability to agricultural drought: A Nebraska case study. Natural Hazards 25(1), 37–58 (2002)
27. Fekete, A., Damm, M., Birkmann, J.: Scales as a challenge for vulnerability assessment. Natural Hazards 55(3), 729–747 (2010)
28. Antwi-Agyei, P., Fraser, E.D., Dougill, A.J., et al.: Mapping the vulnerability of crop production to drought in Ghana using rainfall, yield and socioeconomic data. Applied Gegraphy 32(2), 324–334 (2012)
29. Ping, C., Xiaoling, C.: Evaluating drought vulnerability of agricultural system in Poyang Lake Ecological Economic Zone,China. Transactions of the CSAE 27(8), 8–13 (2011) (in Chinese with English abstract)
30. Huang, J., Liu, Y.: The assessment of regional vulnerability to natural hazards in China: World Automation Congress (WAC). IEEE (2012)

Chinese Web Content Extraction
Based on Naïve Bayes Model[*]

Wang Jinbo, Wang Lianzhi, Gao Wanlin, Yu Jian, and Cui Yuntao

College of Information and Electrical Engineering, China Agricultural University,
Beijing, 100083, China
wangcau@163.com, {ndjsj862,gaowlin,yj}@cau.edu.cn,
674853800@qq.com

Abstract. As the web content extraction becomes more and more difficult, this paper proposes a method that using Naive Bayes Model to train the block attributes eigenvalues of web page. Firstly, this method denoising the web page, represents it as a DOM tree and divides web page into blocks, then uses Naive Bayes Model to get the probability value of the statistical feature about web blocks. At last, it extracts theme blocks to compose content of web page. The test shows that the algorithm could extract content of web page accurately. The average accuracy has reached up to 96.2%.The method has been adopted to extract content for the off-portal search of Hunan Farmer Training Website, and the efficiency is well.

Keywords: Web Content Extraction, DOM Tree, Page Segmentation, Naive Bayes Model.

1   Introduction

Web content extraction is to extract the text which describe the page content; and it's also known as web theme block extraction [1]. It can be used for web data mining, classification, clustering, keyword extraction and the deep processing of web information. Web is semi-structured pages, so it contains a lot of advertising links, scripts, CSS styles, navigation and useless information. The main message is often hidden in the unrelated content or structure; and the noise makes it very difficult to extract page content. Therefore, how to quickly and accurately extract text content pages has been the focus of research at home and abroad [2].About web page text extraction, there is also a lot of research and methods. Now, three main web content extraction algorithms are as follows:

1. Wrapper-based approach, this method is to extract required information from specific web information sources and be expressed in a particular form. Wrapper-based approach can be accurate extracting and have high accuracy. But due to the complexity and irregular of web structure, a wrapper implementation generally for one website, it is difficult to meet for different web information extraction tasks [3].

2. Machine learning methods, by analyzing the structure of the page, and constantly generates new template and creates template library. Literature [4] takes machine learning methods for web thematic information extraction. Web page content

---

[*] Supported by The 12th Five-Year Plan science and technology support project (Project no.2012BAD35B02).

D. Li and Y. Chen (Eds.): CCTA 2013, Part II, IFIP AICT 420, pp. 404–413, 2014.

extraction based on templates has a relatively high degree of automation and is convenient for users. However, if you encounter a web page cannot find the corresponding template, the extraction will fail. As the template library continues to increase, the template library management will become increasingly complex [5].

3. Visualization layout classification method, a classical algorithm is VIPs put forward by Microsoft Asia research institute. It uses visual characteristics of the page structure excavation and makes full use of the web page background color, font color and size. However, due to the complexity of visual web, heuristic rules are so ambiguous that need to manually adjust the rules constantly. So how to ensure consistency of the rules is a difficulty [6].

The methods mentioned above all have some short comings and limitations. So this paper on the basis of predecessors' work and combining with the nature of html page in statistics and observation, according to the characteristics of the different features with different importance, it proposes an algorithm that uses Naïve Bayes Model [7] to train the block attributes eigenvalues of web page. After denoising the web page [8], divides web page into blocks and gets the statistical characteristics of the web block. The algorithm is easy to implement, without artificial participation and can extract web contents quickly and accurately.

2   Algorithm Framework

The algorithm is divided into training phase and testing phase. The training phase includes pretreatment of web pages and builds Naïve Bayes Model. The testing phase is based on the web pages pretreatment, using Naïve Bayes model which is built in training phrase to extract web content. Algorithm framework shows in figure 1.

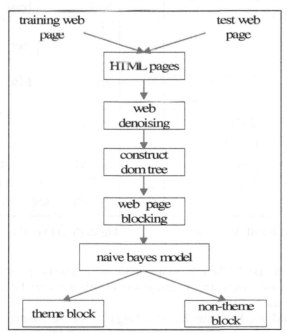

Fig. 1. Algorithm Framework

3   Web Page Pretreatment

Step one: Downloads news web pages respectively from Sohu, Netease, Sina, People's daily, Tencent. And each source downloads 200 web pages, then extracts web page manually, 500 as the training set, 500 as the testing set.

Step two: Denoising web pages and uses regular expression to delete CSS, scripts, comments on pages.

Table 1. Web Denoising Regex

noise type	regex
css styles	<[\\s]*?style[^>]*?>[\\s\\S]*?<[\\s]*?\V[\\s]*?style[\\s]*?>
script	<[\\s]*?script[^>]*?>[\\s\\S]*?<[\\s]*?\V[\\s]*?script[\\s]*?>
comments	<!—(.*?)-->

Step three: Resolve web page into a DOM tree [9]. Read the web page without noise into memory and use NekoHTML to modify the tags which is not regular, then resolves web page into a DOM tree [10]. The html in figure 2(a) corresponds to the DOM tree in figure 2(b) below:

Fig. 2. (a) HTML Web Page

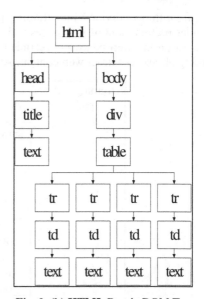

Fig. 2. (b) HTML Page's DOM Tree

Html document and DOM tree is a one-to-one relationship, and the DOM tree makes computer more convenient to process semi-structured html document, easier to block the web pages.

Step four: DOM tree blocking. By observing the website, finding that the text area is usually use tags such as table, td, tr, div to divide each block of text. So this article compares the above tags to DOM tree node properties, using the bottom-up approach to block the DOM tree. The block rules are as follows:

(1) Let DOM tree leaf node enter the queue.
(2) Scanning the DOM tree leaf node in turn, if the leaf node text is empty or is not block node , continue to scan the node's parent node until it encounter a block node whose text is not empty. Recording the node and compose it and its affiliated tags into blocks.
(3)Scanning the leaf node again, the same as (2), if it encounter a block node whose text is not empty. Recording the node and compose it and its affiliated tags into blocks. If the node and block node in (2) is sibling nodes, merges the block and block in (2).

4   Model Design

4.1   Naïve Bayes Model

Naive Bayes Model (Naive Bayesian Model, NBC) is the most widely used classification algorithm, it needs less estimated parameters, less sensitive to missing data, and its time complexity is low, classification is efficiency and stability.

Whether the web block is content or not is a Binary Classification Problem [11]. We use an n-dimensional feature vector $X = \{x_1, x_2, \ldots , x_n\}$ to represent a block, describing the n metrics about samples corresponding to the attributes $A_1, A_2, \ldots A_n$. $c_i \in C = \{c_1, c_2\}$ is a class variable, c_1 indicate that the page belongs to theme block, c_2 indicate that the page does not belong to theme block. To simplify the calculation, assume $x_1, x_2, \ldots , x_n$ are independent. That is, attribute values is independent between each other. It's why we call Naïve Bayes Naive [12]. A web block belonging to c_i classification's Naïve Bayesian formula shows as (1) [13]:

$$p(c_i \mid x_1, x_2, \ldots, x_n) = \frac{\prod_{j=1}^{n} p(x_j \mid c_i) * p(c_i)}{\prod_{j=1}^{n} p(x_j)} \tag{1}$$

4.2   Block Feature Extraction

In the paper, probability based statistical training webpage probability feature of each block is to determine the probability of block extraction test webpage of theme block. A lot of features affecting a block becoming subject block. By analyzing the structure of web pages, we can draw the conclusion that the following theme blocks [14] have several notable features:

(1)Hyperlinks are less, but navigation information blocks, advertising block contains a number of hyperlinks generally more.

(2)More text is in block; theme block is the region web information centralized, so the number of characters contained within the block is more. The noise block contains fewer amounts of characters.

(3)Theme block is used to describe the main content of a webpage, so it contains more punctuation, and noise block generally doesn't contain punctuation.

(4)<p> acts as paragraph mark, the theme contains a lot of information, and practical <p> labels often used to segment, while the noise block generally doesn't contain paragraph marks.

Based on the above characteristics , this article uses the number of characters within the block unlink , the ratio of the number of link characters and the total number of characters , the ratio of total number of punctuation and link characters , and the total number of <p> as Web page block's feature items.

Among this paper, *unlinktextsum* stands for the number of unlink characters, *linktextsum* stands for the number of link characters, *textsum* stands for the total number of text, and *puncsum* stands for the total number of punctuation.

Then the ratio of *linktextsum* to *textsum* named link shows as (2):

$$\text{link} = \frac{\text{linktextsum}}{\text{textsum}} \tag{2}$$

The ratio of *puncsum* to *linktextsum* named *punc* shows as (3):

$$\text{punc} = \frac{\text{puncsum}}{\text{linktextsum}} \tag{3}$$

4.3   Model Training

After generating DOM tree and blocking the training web pages [15], to work out the unlink characters number, ratio of unlink characters to character number, ratio of punctuation to link characters number, <p> tags number.

4.3.1   Unlink Characters Number

The more unlink characters in a block, the richer information the block contains, then it has a higher probability to be a theme block. Because hyperlinks are generally less than 20 characters, and blocks more than 100 characters are mostly theme block. In this article we will divide the number of block's unlink characters into 6 levels, that is, the number of unlink characters is less than 20, above 100, and four equal parts between 20 and 100. The unlinked character number scatters in an interval belonging to non-theme blocks and theme blocks' probability shows as (4), (5):

$$p(\text{linktextsum}_n \mid c_1) = \sum_{i=1}^{n} p(\text{linktextsum}_i \mid c_1) \qquad 1 \leq i \leq n \tag{4}$$

$$p(\text{linktextsum}_n \mid c_2) = \sum_{i=1}^{n} p(\text{linktextsum}_i \mid c_2) \qquad 1 \leq i \leq n \tag{5}$$

That is, each block *linktextsum*'s probability is the probability of *linktextsum* less than the block and the block's sum, and $n = \dfrac{linktextsum}{20} + 1$.

4.3.2   Link Characters and Character Number Ratios Probability

The lower link characters and character number ratios within the block, the higher probability a block to be a theme block. Navigation links blocks and advertising blocks of characters and the total number of characters ratio is generally greater than 50% and some even higher than 80%. So this article will divide link characters and character number ratios into 4 copies, that is, less than 10%, 10%-50%, 50%-80%, above 80%. The link characters and character number ratios scatters in an interval belonging to non-theme blocks and theme blocks' probability shows as (6), (7):

$$p(link_i \mid c_1) = \frac{linktextsum_i}{textsum_i} \qquad (6)$$

$$p(link_i \mid c_2) = \frac{linktextsum_i}{textsum_i} \qquad (7)$$

4.3.3   Punctuation and Link Characters Number Ratios Probability

Theme block contains much punctuation, but link text generally doesn't contain punctuation. The higher punctuation and link characters number ratios within the block, the higher probability a block to be a theme block. This article will divide ratios of punctuation number to link characters number into 3 copies, that is, less than 2%, 2%-10%, above 10%. The ratios of punctuation number to link characters number scatters in an interval belonging to non-theme blocks and theme blocks' probability shows as (8), (9):

$$p(punc_n \mid c_1) = \Sigma_{i=1}^{n} \; p(\frac{puncsum_i}{linksum_i} \mid c_1) \quad 1 \le i \le n \qquad (8)$$

$$p(punc_n \mid c_1) = \Sigma_{i=1}^{n} \; p(\frac{puncsum_i}{linksum_i} \mid c_2) \quad 1 \le i \le n \qquad (9)$$

That is, each block *punc*'s probability is smaller than the block *punc* and the block's *punc* probability sum.

4.3.4   <p> tags Number Probability

Web content containing much information, it often uses <p> tags for a paragraph replacement. So theme block contains many <p> tags. This article will divide <p> tags number into 3 levels, that is, 0 <p> tag, 0-3<p> tags, above 3 <p> tags. The <p> tags number scatters in an interval belonging to non-theme blocks and theme blocks' probability show as (10), (11):

$$p(\text{psum}_i \mid c_1) = \text{psum}_i \tag{10}$$

$$p(\text{psum}_i \mid c_2) = \text{psum}_i \tag{11}$$

4.3.5   Block Overall Probability

According to the formula (1) - (11):

The probability of a block to be a theme block shows as (12):

$$p(c_1 \mid unlinksum, link, punc, psum)$$
$$= \frac{p(unlinksum, link, punc, psum \mid c_1) * p(c_1)}{p(unlinksum, link, punc, psum)}) \tag{12}$$

According to the formula (1), (12):

$$p(c_1 \mid unlinksum, link, punc, psum)$$
$$= \frac{p(unlinksum \mid c_1) * p(link \mid c_1) * p(punc \mid c_1) * p(psum \mid c_1)}{p(unlinksum, link, punc, psum)} * p(c_1) \tag{13}$$

$p(c1)$ indicates the probability of a theme block in the training set, is a constant, and the denominator is also a constant.

Therefore, the probability of a block to be a theme block can be expressed as (14):

$$p(c_1 \mid unlinksum, link, punc, psum)$$
$$= p(unlinksum \mid c_1) * p(link \mid c_1) * (punc \mid c_1) * (psum \mid c_1) \tag{14}$$

Similarly, the probability of a block to be a non-theme block can be expressed as (15):

$$p(c_2 \mid unlinksum, link, punc, psum)$$
$$= p(unlinksum \mid c_2) * p(link \mid c_2) * (punc \mid c_2) * (psum \mid c_2) \tag{15}$$

If in a block $p(c1|unlinksum,link,punc,psum) >= p(c2|unlinksum,link,punc,psum)$, that is, the probability a block to be a theme block is bigger than the probability a block to be a non-theme block. Extract the block, put it into theme block queue, and output the block in queue.

5   Testing and Verification

In order to verify the effectiveness of the algorithm, we use java language to implement and test the proposed algorithm. Test procedure is as follows:

Download 100 pages respectively from *Sohu*, *Netease*, *Sina*, People's Daily and *Tencent*, totaling 500 web pages. These pages cover sports, entertainment, education, practical, financial and some other themes, almost all kinds of news.

Using the algorithm to extract text of the following web page from *Sina* Finance and Economics, the page to be extracted is shown in figure3:

The page URL is
http://finance.sina.com.cn/review/jcgc/20130606/182015723399.shtml .

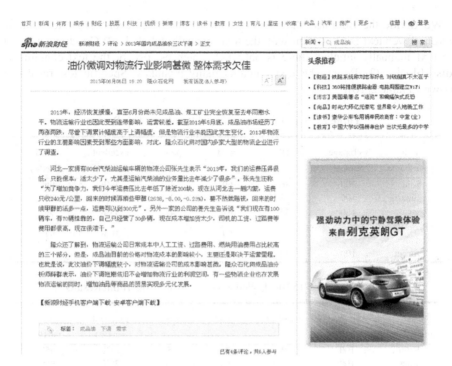

Fig. 3. The Original Page

Text extraction result is shown in Figure 4:

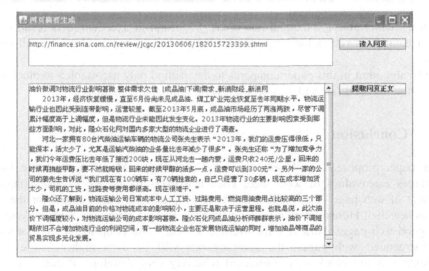

Fig. 4. Extraction Results

We divide the obtained theme information into three levels: (1) Excellent: the obtained web text is consistent with the text manually labeled. (2) Good: compared to the text manually labeled, there is only 1-2 sentences lost, or the text contains 1-2 noise blocks. (3) Poor: The text contains many mistakes. Specific test results are shown in table 2[16]:

Table 2. Algorithm Experimental Results in This Paper

web pagesource	web page number	excellent	good	poor	excellent rate(%)	good rate(%)
Sohu	100	36	60	4	36%	96%
Sina	100	38	59	3	38%	97%
Netease	100	35	61	4	35%	96%
People's daily	100	38	59	3	38%	97%
Tencent	100	32	63	5	32%	95%

Table 3. <table> to Block Web Page Extraction Algorithm Results

web pagesource	web page number	excellent	good	poor	excellent rate(%)	good rate(%)
Sohu	100	25	70	5	25%	95%
Sina	100	28	68	4	28%	96%
Netease	100	26	62	12	26%	88%
People's daily	100	30	66	4	30%	96%
Tencent	100	27	63	10	27%	90%

In the tables above, excellent rate is the proportion of excellent level result in all result data; good rate is the proportion of both excellent and good level in all result data.

The algorithm in this paper compares to the method only use <table> to block web page, both its good rate and excellent rates are significantly improved.

6   Conclusion

This paper proposed an algorithm using Naïve Bayes Model to train the block attributes eigenvalues of web page. Then it extracts theme blocks and composes content of web page. The method has been adopted to extract content for the off-portal search of Hunan Farmer Training Website, and the efficiency is well. Counting the good web pages extracted, the average accuracy rate is up to 96.2%. For some well-structured web pages, the accuracy rate will be even higher. An existing deficiency is the block tags considered relatively less, therefore, if consider more block tags, the accuracy of the system will also be enhanced. In future work we will do research for semi-structured web pages.

References

1. Li, X.: Harmonious man-machine environment, vol. 1, pp. 101–107. Tsinghua University press, Beijing (January 1, 2008)
2. Wu, Q., Chen, X., Tan, J.: Content Extraction Algorithm of HTML Pages Based on Optimized Weight. Journal of South China University of Technology (natural science edition) 39(4), 32–37 (2011)
3. Hsu, C.-H.: Initial Results on Wrapping Semi-structured Web Pages with Finite-State Transducers and Contextual Rules. In: Workshop on AI and Information Integration, in Conjunction with the 15'th National Conference on Artificial Intelligence (AAAI 1998), Madison, Wisconsin (July 1998)
4. Bar-Yossef, Z., Rajagopalan, S.: Template detection via data mining and its applications. In: 11th World Wide Web Conference (WWW 2002), Hawaii, USA (2002)
5. Yang, J., Li, Z.: DOM-based information extraction for WEB-pages topic. DOM-based Information Extraction for WEB-pages Topic 45(5), 1077–1080 (2008)
6. Deng, C., Yu, S.P., Wen, J.R., VIPS: A Vision-Based Page Segmentation [MSR-TR-2003-79] (2003)
7. Manning, C.D., Raghavan, P., Schutze, H.: Introduction to Information Retrieval, vol. 1, pp. 175–182. People's Posts and Telecommunications Press, Beijing (January 01, 2010)
8. Shoubin, D., Hua, Y.: The network information retrieval, vol. 1, pp. 93–99. Xi'an Electronic and Science University press, Xi'an (April 01, 2010)
9. MacDonald, M.: WPF Programming book, vol. 1, pp. 694–698. Tsinghua University Press, Beijing (June 1, 2011)
10. HTML resolve [EB/OL], http://litertiger.blog.163.com/blog/static/82453820069309334041O/
11. Borenstein, M.: Meta. Analysis: An Introduction. Science Press, Beijing (January 2013)
12. Zhao, Y., Xie, X., Xun, Y.: Application of Naive Bias classification. Electronic Production, 7 (2013)
13. Kupiec, J., Pedersen, J., Chen, F.: A Trainable Document Summarizer in Proceedings of the Eighth Annual International ACM SIGIR Conference on Research and Development in Information Retrieval, Seattle, Washington, pp. 68–73 (July 1995)
14. Wu, Q., Chen, X., Tan, J.: Webpage content extraction algorithm based on optimized weight. Journal of South China University of Technology (Natural Science Edition) (4) (2011)
15. Wang, C., Xu, J.: Webpage blocks and blocks of text extraction research based on the CURE algorithm. Microcomputer and its Application (12) (2012)
16. Guo, Y., Tang, H., Song, L., Wang, Y., Ding, G.: In: Source: Advances in Web Technologies and Applications - Proceedings of the 12th Asia-Pacific Web Conference, APWeb 2010, pp. 314–320 (2010)

Research on the Vegetable Trade Current Situation and Its Trade Competitiveness in China

Shasha Li[*]

Econometrics and Management College of China Agriculture University,
Haidian District Beijing China, 100083
susanlss2008@sina.com

Abstract. This paper analyzed the Current situation and status of China in the world vegetable trade, measured and analyzed main export varieties of vegetables trade competitiveness in China by using international market share(IMS) and trade competitiveness index(TCI). The conclusion is that: China's vegetable export scale expands gradually, frozen vegetables、 dehydrated vegetables、 and dried vegetables have significant comparative advantages and hold high stability in the international trade. the world economic recovery will provides the export of China's vegetable products with opportunities.

Keywords: Competitiveness, Varieties, Vegetable trade, The market share.

1   Introduction

Vegetable industry is an important part of agriculture, the development of vegetable industry in China have made great strides with the deepening of reform and opening up and accelerated industrialisation and urbanisation, vegetable planting area and output have formed into certain scale. Vegetable planting area is19.63958 million hectares in China in 2011 , accounting for 12.11% of total sown area of agricultural products; produced 679 million tonnes of vegetables with ￥1.26 trillion production value , accounting for 30 %of the total output value of agriculture. Overtaking proportion of grain output value for the first time, Vegetables has become the top agricultural products in China for the first time in 2011. At the same time, vegetables is China's export varieties with traditional comparative advantage among all the exported agriculture products as well, currently, China is the world's largest exporter of vegetables. The foreign trade scale of vegetable industry expands unceasingly since China's entry into WTO, China's vegetable exports have risen from 3.17 million tonnes to 7.72 million tonnes during 2001-2011, up 2.45 times, the average annual growth rate is 9.41%; Exports value, rising from $1.753 billion to $9.351 billion from 2001 to 2011, increased by 5.33 times, the average annual growth rate is19.15%. Export expansion speed higher than that of imports significantly.

---

[*] Shasha Li (1986-), Ph. D. student, research direction: agricultural economic theory and policy.

D. Li and Y. Chen (Eds.): CCTA 2013, Part II, IFIP AICT 420, pp. 414–422, 2014.
© IFIP International Federation for Information Processing 2014

An increasing number of the external environment uncertainty brings new difficulties and challenges to The development of vegetable trade in China with the continuous expansion of China's vegetable foreign trade scale, How to position the Chinese vegetable industry's position in international trade, How to making full use of the comparative advantages in China's agriculture[8], How to adjust the vegetable trade strategy in China duly to keep up with the trend of the vegetable trade patterns change, which will be of great significance to consolidate the vegetables exports great-power status and reinforce vegetable export international competitiveness in China.

2   Literature Review

There are lots of literatures research on China's vegetable trade, such as the potential impact of China's agricultural sector on world trade(Fuller,2001;Wu,2003;Huang ,2003)[9]-[11] ,which can be summed up in two aspects by making generalization and summary of existing research methods and ideas ,On the one hand, some scholars carried out the research by using the empirical analysis and normative analysis, Qi Zhang , Ming-yang Zhang (2013)adopted empirical analysis on the influence factors of Chinese vegetables export trade from the perspective of bilateral technical trade measures with trade gravity model ,pointed out that importer of per capita GDP, trade distance, importing countries domestic standards are important factors that affect Chinese vegetables export trade;[1]Da-xue Kan(2013) estimates the international market forces of China's vegetable industry by using the marginal cost model, results show that the international market power in China's vegetable industry appeared to descend after joining the WTO;[2]Yuan-yuan Hou,Li-li Wang(2011) built international competitiveness evaluation index system and used cluster analysis to compare the international competitiveness of 14 vegetable varieties and eight different kinds of vegetables in China, draw a conclusion that fresh vegetables, dehydrated vegetables have strong competitiveness in short term;[3]On the other hand, other scholars discussed China's vegetable trade issues from the following aspects such as Industrial chain, regional comparison and dynamic analysis and so on, Yong Tang,Jun Huang,Yue-yun Li(2006) measured the comparative advantage of China's vegetable production level by using the method of resource cost, pointed out that China's vegetable has a strong potential competitive advantage;[4]Hua Lin,Kai Wang(2010) analyzes Chinese vegetables export competitiveness against South Korea by using the revealed comparative advantage and export permeability index from regional comparative perspective, Results showed that the overall Chinese vegetables in Korea have absolute competitive advantage;[5]Feng-jie Pan,Yue-ying Mu(2011) draw conclusions that increasing trends of growing vegetables export quantity and amount, significant trend of diversification of vegetables export destination through analysis the changes in China's vegetable export trade from the perspective of dynamic.[6]

Synthesize existing literatures, the research on China's vegetable trade condition and trade competitiveness have carried out extensively. however, less research has been done on the competitiveness of the main vegetable varieties in China from global perspective, which remains to be improved. This paper will make further research and analysis on it. This article altogether is divided into five parts, besides the introduction and literature review, the third part analysis China's vegetable current trade status, the fourth part calculated and analysis the trade competitiveness of main vegetable varieties ; The fifth part is conclusions and policy recommendations.

3   China's Vegetable Trade Current Situation

The major characteristics of Chinese vegetables in foreign trade is export scale is far greater than imports(see Figure 1), the trade scale expanding year by year(see Table 1), trade surplus continues to rise steadily.

Looking from the trade growth process, vegetable imports volume basically remain at around 100000 tonnes during 2001-2007, exports volume keep increasing at 10% of the growth, trade surplus growth rate is fairly constant in the same slope as well. It is obvious to see impact of global financial crisis that outbreak in 2008 to Chinese vegetables in foreign trade, the growth rate of vegetable exports volume is only 0.32% in 2008, jumped to an all-time lowest record; vegetable trade surplus experienced negative growth for the first time with growth rate -1.31%;at the same time, vegetable imports volume is essential to maintain the original level. however, vegetables export trade began to gradually recover in China in 2009 as the economic stimulus policy Issued and implemented which promote the global economic recovery, exports volume and trade surplus year-on-year growth are1.92% and 20.65 respectively; increasing trend has shown in China's vegetable export trade in 2010, amplification of exports volume and trade surplus hit a record high, 190000 tonnes and 2.995 billion dollars respectively, year-on-year growth of 2.99%, 61.48%. Vegetable import was still keeping a steady growth in the year 2011, vegetable imports volume, exports volume and trade surplus rose by 4.55 million tonnes, 67000 tonnes and 7.814 billion dollars separately over the previous year.

Table 1. Import-export volume and import-export value in China during the year 2001-2011

Unit:10000 tonnes,100 million dollars

year	export volume	export value	import volume	import value
2001	317	17.52	10.02	3.25
2002	360	18.88	9.81	2.75
2003	432	21.99	9.51	1.75
2004	470	27.81	11.01	1.82
2005	520	33.02	10.5	1.65
2006	568	39.79	12.01	1.51
2007	622	42.16	10.5	1.25
2008	620	41.67	10.82	1.30
2009	636	49.96	9.01	1.25
2010	655	79.81	14.51	1.15
2011	772	93.51	16.51	1.11

Data sources: China trade foreign economic statistical yearbook and China customs database, calculated by the author

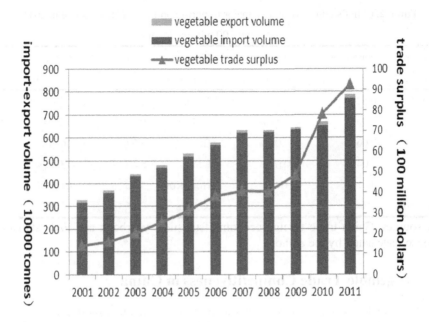

Fig. 1. Import-export volume and trade surplus in China during the year 2001-2011

Looking from major import-export market distribution of China's edible vegetable(see Table 2), The United States, the European Union, the association of southeast Asian nations (ASEAN) and Taiwan are major vegetable import destination, meanwhile, Japan, the European Union, America, ASEAN, Hong Kong and South Korea are a major vegetable output destination in China, China has claimed the position of No1 supplier of vegetable to Japan.[7]The association of southeast Asian nations (ASEAN) has become the top vegetable exporter of China in 2011, The edible vegetable bilateral trade value between China and ASEAN ascend to $1.457 billion, edible vegetable imports value and exports value accounted for 55.91% and 21.57%of edible vegetable total imports and exports value, trade vegetable varieties mainly include cassava, mung bean, red bean, etc; The USA is the second largest vegetable export country to China, vegetable imports value from the United States fell by 21.05% compared with the previous year, however, edible vegetable exports value to the United States increased by 13.84%, major trade varieties include dehydrated vegetables, frozen vegetables, pickled vegetables and so on. the European Union is the third largest vegetable export country to China, But the proportion of imports of edible vegetables from the European Union in China is only 6.5%, trade varieties mainly include dehydrated vegetables, frozen vegetables, etc. China import edible vegetables from Japan, Taiwan as well as export to Hong Kong, South Korea and other places , Trade products including melons, beans, carrots, Onions, tomatoes, etc. As a whole, China's vegetable import-export market distributed in Asia, Europe and the Americas, however, trade is mainly concentrated in Asia

Table 2. China's edible vegetable major import-export market trade value in 2011
Unit:100 million dollars

importing market	import value	Growth Year-on-year	Exporting market	import value	Growth Year-on-year
ASEAN	0.52	10.23%	Japan	14.05	17.97%
USA	0.30	-21.05%	ASEAN	9.65	5.72%
EU	0.06	100.00%	EU	8.71	16.76%
Taiwan	0.04	100.00%	USA	5.84	13.84%
Japan	0.01	0.00%	HongKong	4.23	64.59%
HongKong	--	--	Korean	2.26	15.35

Data sources: China trade foreign economic statistical yearbook and FAO statistical database, calculated by the author

4   Vegetable Trade Competitiveness in China

Seen from table 3,compared with world average price, the export price of major export vegetable varieties in China Showed the following characteristics:1. Export price of dehydrated vegetables, frozen vegetables, fresh vegetables, salted vegetables are lower than the world average level year in year out, Vinegar soak vegetables export price is slightly higher than the world;2. Vegetable prices rise and fall trends are almost consistent with the world average price;3.Among all the export vegetable varieties in China, fresh vegetables export price are far lower than the world average for the long years, next is pickled vegetable, difference in price between world price and dehydrated vegetables ,frozen vegetables is not very remarkable.

Overall ,judging China's export vegetables varieties trade competitiveness from the export prices, we may know that vegetable industry hold a significant export competitiveness on price, Among them, fresh vegetables have the most striking export competitiveness, followed by pickled vegetables, dehydrated vegetables, frozen vegetables, the vinegar soak vegetables export competitiveness is relatively weak.

Table 3. Varieties of vegetables export price and world export price in China during the year 2006-2010

year	Dehydrated vegetables		Frozen vegetables		Fresh vegetables		Vinegar soak vegetables		Pickled vegetables	
	China	World	China	World	China	World	China	World	China	World
2010	2.82	2.88	1.03	1.08	0.41	1.04	1.66	1.07	1.04	1.45
2009	2.03	2.51	0.92	1.11	0.33	0.89	1.44	1.05	0.96	1.50
2008	2.03	2.56	0.93	1.16	0.29	0.88	1.45	1.12	0.96	1.50
2007	2.24	2.50	0.86	1.05	0.25	0.92	1.23	0.99	0.96	1.38
2006	2.39	2.52	0.88	0.95	0.32	0.86	1.20	0.89	0.89	1.22

Data sources: FAO statistical database, calculated by the author

By analyzing the international market share changes of China's export vegetable varieties Frozen vegetables, dehydrated vegetables, fresh vegetables, vinegar soak vegetables pickled vegetables and dried vegetables(see Table 4),we can know that a majority of the vegetables varieties have high market share in international market; looking from the changes in proportion, the international market share of frozen vegetables always keep at 17%, holding long-term stability; Dehydrated vegetable international market share gradually upgraded as time goes on, showing strong international trade competitiveness; Fresh vegetable international market share fluctuated in a tight range, Keeping in 1% amplitude fluctuation on average; Vinegar soak vegetable international market share showing a trend of rising year by year, which reach the highest level in 2010, acting the potential of increasing trade competitive advantage; Pickled vegetable international market share remain above 14.34% throughout the year since 2001,which raised to 19% in the year 2010 for the first time, proving potential international trade competitiveness; Dried vegetable had the largest international market share among all the export vegetable varieties in China Maintaining above 65%, showing a rising trend, reflecting stable and strong trade competitiveness of the dried vegetable varieties in the international trade.

Table 4. International market share of export vegetable varieties in China during 2000-2010

Year	frozen vegetable	dehydrated vegetables	fresh vegetables	Vinegar soak vegetables	pickled vegetable	dried vegetables
2010	19.47%	51.47%	6.43%	5.11%	19.02%	66.65%
2009	16.28%	40.97%	5.75%	3.83%	16.22%	67.49%
2008	16.41%	41.03%	5.08%	2.14%	16.99%	68.63%
2007	17.29%	45.67%	4.77%	2.01%	19.83%	67.77%
2006	18.57%	47.83%	6.59%	2.42%	18.64%	69.34%
2005	16.86%	43.18%	6.69%	3.14%	18.18%	61.94%
2004	15.31%	38.15%	7.58%	3.11%	15.76%	57.05%
2003	13.95%	37.54%	9.79%	3.26%	14.34%	62.27%
2002	16.03%	39.53%	11.72%	2.26%	15.55%	62.08%
2001	19.45%	35.17%	10.43%	3.17%	15.59%	59.06%

Data sources: FAO statistical database, calculated by the author

From the perspective of international market share, the dried vegetables and dehydrated vegetables keep in 35% ~ 65% all the year round , frozen vegetables and pickled vegetables remain above 15% as well, illustrating China's these four types of exported vegetable varieties with strong export competitiveness. Looking at the vegetables trade competition index between China and its trade partners

In 2011(see Figure 5), the trade competition index of frozen vegetables, dehydrated vegetables, fresh vegetables of China against The association of southeast Asian nations (ASEAN) , EU, Japan South Korea, the United States and Australia are above 0.96, Indicates that China's frozen vegetables, dehydrated vegetables, fresh vegetables in the international market occupies a very strong competitive advantage; Vinegar soak vegetables showing significant trade competitiveness in China- Japan,

China-South Korea trade, holding strong competitiveness in China-USA, China-Australia trade, proving general trade competitiveness with trade competition index 0.48 and 0.5 separately among China-ASEAN, China-EU trade; Pickled vegetables trade competition index is above 0.96 in China-EU, China-Japan, China- South Korea trade, meanwhile, the value come to -0.10,0.26,0.71 separately in the trade relationship China-ASEAN, China-USA, China- Australia, the result show that China's Pickled vegetables has significant competitive advantage in China and the European Union, Japan, South Korea trade, hold strong competitive advantage comparing with Australia, take general competitiveness against the United States, stay in the weak competitive position in the trade with ASEAN.

Table 5. Trade competitiveness between China and its major vegetable trade partners in 2011

Varieties	USEAN	EU	Japan	Korean	USA	Australia
frozen	0.99	1.00	1.00	1.00	0.98	1.00
dehydrated	1.00	0.99	1.00	0.96	0.99	1.00
Fresh	1.00	1.00	1.00	1.00	0.97	1.00
Vinegar soak	0.51	0.48	1.00	1.00	0.87	0.97
Pickled	-0.10	0.96	0.99	0.96	0.26	0.71

Data sources: FAO statistical database, calculated by the author

5   Conclusion and Suggestion

(1) Main Conclusions

1. China' s vegetable foreign trade market is distributed in Asia, Europe, America, Oceania and Africa, seeing trade value and trade density, Chinese vegetables import and export market mainly concentrated in Asia, including Japan, South Korea north Korea and the ASEAN countries are China's major vegetables trade partners, The United States, the European Union is China's second and third largest vegetables trade partners respectively. The vegetable trade and cooperation between China and the world will further deepen as the world economic recovery and the rise of emerging economies.

2. seen the trends of China's vegetables import and export structure and trade surplus, the vegetable export scale and imports scale will continue to expand, besides, the trend of exports expanded faster than the growth rate of imports is very clear, the status of vegetable trade surplus will thus be sustained in the years to come.

3. From the result of comparison between China's export price and world export price, we know that the export price of dehydrated vegetables, frozen vegetables, fresh vegetables, salted vegetables are lower than the average level of the world, showing significant price advantage. Under the condition of regardless of non-price factor, production costs and export prices are the deciding factor for China's vegetables industry possess strong trade competitiveness.

4. From the result of international market share comparison, the international market share of China's dried vegetables, dehydrated vegetables keep more than 40%

all the year round, the proportion of pickled vegetables, frozen vegetables in the global total vegetables trade value remain above 15% for a long time within recent 10 years, to some extent, reflecting China's trade competitiveness of different vegetable varieties have differentiation.

5. The trade competitiveness index tell us that china's frozen vegetables, dehydrated vegetables, fresh vegetables has a significant competitive advantage in the vegetables trade with the association of southeast Asian nations (ASEAN) ,Japan, South Korea, the United States and Australia, Vinegar soak vegetables showed strong trade competitiveness in China-Japan, China-South Korean, China-Australia trade, Pickled vegetables stay in a significant trade competitive advantage position   in trading with EU, Japan, South Korean.

(2) Policy Suggestions

Based on the above analysis, this paper puts forward the following Suggestions

1. Optimize the structure of export vegetable varieties, improving the ability of exported vegetable varieties to earn foreign exchange . take the market as the guidance, In order to realize the resource allocation efficiency maximization as the goal, according to the comparative advantage of vegetables varieties such as resources endowment, policy orientation, system environment, market structure, etc,

adjusting and optimizing the structure of export vegetable varieties scientifically in accordance with different target market needs, to improve the overall capacity to earn foreign exchange of vegetables industry in China.

2. To enhance investment in agricultural scientific research, improve the sci-tech contents in vegetables, put both quantity and quality of vegetables into consideration. Taking science and technology as support, implementing Agricultural science and technology strategy , Increase investment in agricultural scientific research and agricultural technology talents cultivation, cultivating new high-yielding high-quality vegetable varieties, upgrading vegetable industry technology content and comprehensive competitiveness comprehensively, improving vegetable quality and benefits synchronous, to further consolidate and strengthen the strong trade competitive position of vegetables industry in China through the occupation of agricultural high-tech commanding heights to break the pattern of resources and environment constraints in vegetable production.

3. Build a vegetable production standardization system, improve the vegetable quality and safety coefficient. aiming at improving the vegetables varieties quality and safety, to build the national vegetable production standardization system by making goods regulations, technical standards to push the vegetable industry develop with the direction of standardization, industrialization, large-scale, to eliminate the vegetable quality and safety problems gradually, to circumvent trade risk such as green barrier, technical trade barrier and so on, to enhance the capability in shielding against risks and international export competitiveness of vegetable industry in China.

References

[1] Qi, Z., Zhang, M.-Y.: Bilateral technical trade measures on Chinese vegetables export trade impact analysis. Journal of International Trade (2013)

[2] Kan, D.-X.: Empirical study on China's vegetable industry in the international market forces. Jiangsu Agricultural Sciences (2013)

[3] Hou, Y.-Y., Wang, L.-l.: A comparative study of Chinese vegetables international competitiveness. Statistics and Decision (2011)

[4] Tang, Y., Tang, J.: Comparative advantage and export competitiveness analysis of Chinese vegetables. Journal of Agrotechnical Economics (2006)

[5] Lin, H., Wang, K.: Chinese vegetables export competitiveness and the expansion of trade space for Korea- setting United States as frame of reference. Journal of International Trade (2010)

[6] Pan, F.-J., Mu, Y.-Y.: analysis of China's vegetable export trade changes since entry WTO. China vegetable (2011)

[7] Mello. China is playing a growing role in global vegetable trade. AgExporter, Washington (2003)

[8] MOA. Agriculture in China, Ministry of Agriculture, the People's Republic of China (2004)

[9] Fuller, F., Beghin, J., De Cara, S., Fabiosa, J., Fang, C., Matthey, H.: China's accessionto the WTO: what is at stake for agricultural markets, Center for Agricultural and Rural Development. Iowa State University, Ames (June 2001)

[10] Wu, Z., Thomson, K.: Changes in Chinese competitiveness in major food products:implications for WTO membership. Journal of Chinese Economic and Business Studies (2003)

[11] Huang, J., Li, N., Rozelle, S.: Projections of food supply and demand and impacts ofgreen policies. In: Van T.F., Huang, J. (eds.) China's Food Economy in the Early, p. 21 (2004)

A Fault Data Capture Method
for Water Quality Monitoring Equipment
Based on Structural Pattern Recognition

Hao Yang[1,2,3,4,6], Daoliang Li[1,2,3,4,5,*], and Yong Liang[6]

[1] Key Laboratory of Agricultural Information Acquisition Technology,
Ministry of Agriculture, Beijing 100083, P.R. China
[2] Beijing Engineering and Technology Research Center for Internet of Things in Agriculture,
Beijing 100083, P.R. China
[3] China-EU Center for Information and Communication Technologies in Agriculture,
China Agricultural University, Beijing 100083, P.R. China
[4] Beijing Engineering Center for Advanced Sensors in Agriculture,
Beijing 100083, P.R. China
[5] College of Information and Electrical Engineering, China Agricultural University,
Beijing, 100083, P.R. China
[6] Institute of Informatics Science and Engineering, Shandong Agricultural University,
Tai'an, 271018, P.R. China
dliangl@cau.edu.cn

Abstract. To capture equipment fault in real time and automate fault diagnosis, a pattern recognition method, based on data eigenvector and TCP transport protocol, was proposed to capture Water Quality Monitoring equipment's fault information. Fault data eigenvector was designed after analyzing the equipment fault feature and capture strategy, structural pattern recognition strategy was confirmed and specific data frame was designed in response to the fault data eigenvector, by integrating the data frame design into the equipment's communication protocol, data related to different fault compiled into fault data frames by transmitters or communication module of equipment's different components, the remote sever captures equipment fault on transport via fault data frames according to the structural pattern recognition strategy. With 7 months of practical application in Taihu aquaculture project and research center of agricultural information technology, combining with historical fault data and contrast with artificial recognition result, the simulate experiment shows this method has higher response rate and process rate with a nice accurate.

Keywords: fault data capture, data eigenvector, transport layer, fault filter.

1   Introduction

Water quality parameters includes DO (dissolved oxygen), EC (electrical conductivity), SAL(solidity), WT (water temperature), PH and TUR (turbidity), all of

* Corresponding author.

D. Li and Y. Chen (Eds.): CCTA 2013, Part II, IFIP AICT 420, pp. 423–433, 2014.
© IFIP International Federation for Information Processing 2014

these parameters effect deeply to aquaculture management and breeding decision, so collecting water quality parameters is an extremely important work to aquaculture. The Water quality monitoring equipment in the paper is made to collect water quality parameters in time and offer water quality optimization services include SMS warning, oxygenation and other services. Equipment have been deployed in maricultural breeding bases and industrial aquaculture in Beijing, Tianjin, Hebei, Shandong, Jiangsu and Guangdong. Nevertheless, equipment fault prevents farmers from increasing more reliance on the equipment, researches must be done to develop fault diagnosis technology and form an integrated strategy.

ANNs requires large amount of sample data for training, as its process is pure mathematical procedure [1-13].There are many fault detection methods applied the support vector machine [14].

Fault detection should be an independent step in fault diagnosis process, fault diagnosis should base on fault data captured by fault detection step, and serious redundancy calculations exist in diagnosis process.

2   Data Eigenvector Analysis

2.1   Equipment Faults Description

Equipment faults are divided into five types due to component types. Namely sensor faults, transmitter faults, actuator faults, GPRS gateway faults and remote server faults, see fig1.

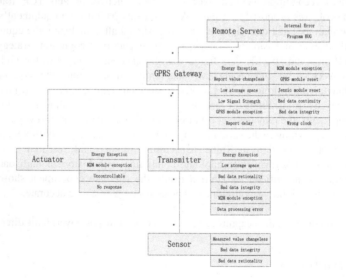

Fig. 1. Equipment fault tree

When combined with some high-class arithmetic like support vector machine to solve fault capture task in every node, long training time cost is not acceptable.

2.2 Data Eigenvector Related to different Equipment Faults

To an object represents a specific fault, recognition arithmetic need to judge whether it contains fault information according data eigenvector, for each fault type, there is a specific eigenvector and type record; these records are listed in Form 1.

Form1. Data type and eigenvector corresponding to different faults

Recognition target	Data type	Eigenvectors
Uncontrollable	Actuator state	Response data and executive result
No response	Actuator state	Response data
Energy Exception	M2M state	Power supply mode & nominal voltage & measured value
Low storage space	Device information	Residual capacity calculated value
Jennic module exception	M2M state	Jennic module reset times
M2M module exception	M2M state	M2M signal value and emission frequency
Data process error	M2M state	Conversion formula
Wrong clock	GPRS state	GPRS gateway clock
GPRS module exception	GPRS state	State code & SIM card state & heart beat interval
GPRS module exception	GPRS state	Reset times
Low Signal strength	GPRS state/Device state	Signal emission frequency
Internal error	Message from server	Error code
Program BUG	None	Server crash
Bad data integrity	Record	Actual record count divide theoretical record count & sensor state
Bad data continuity	Record	Actual record interval time
Report delay	Record	Report time minus collected time
Report value changeless	Record	value & sensor collect voltage value
Bad data rationality	Record	Value change rate & value

3 Method and Capture Strategy

3.1 Data Transfer Process on Transport Layer

Data frame format was proposed in the following passage. Form.2 is the transmitting data frame format.

Form2. Transmitting Data Frame Format

Length	2 bytes	1 bytes	1 bytes	1 bytes	2 bytes	M bytes	N bytes	2 bytes	2 bytes
Content	Start segment	Frame type	Frame length	Frame code	Control code	Address segment	Data segment	CRC check	Terminal
Description			Head				Eigenvector data frame		Tail

The eigenvector data frames and fault filter introduced in following passages are based on the data frame format and this protocol.

3.2   Define Eigenvector Data Frames and Frame Filter

The design of eigenvector data frame is shown in form 3 below. Eigenvector type is corresponding to the frame code; form 4 is the detail design of eigenvector data frame model.

Form3. Eigenvector data frame format

	Type	GPRS gateway clock/Report time	Eigenvector type	Eigenvector data
Length	2 bytes	6 bytes	1 bytes	N

Form4. Detail Design of Eigenvector Data Frame Model

Eigenvector source	Frame code	Type code	Eigenvetor type	Description
GPRS gateway state	0x01	Link heartbeat:[0x01][0x00];	Device heart beat interval time	Hexadecimal value
			Reporting cycle	Hexadecimal value
		State heartbeat:	Acquisition cycle	Hexadecimal value
		long connection: [0x02][0x01]	Storage cycle	Hexadecimal value
		Short connection:[0x02][0x02]	Signal strength	Hexadecimal value
		no connection: [0x02][0x03]	Not reported data count	Hexadecimal value
			Gateway voltage	Hexadecimal value

M2M state	0x02	Power on: [0x01][0x00]	Energy type	0x00:battery
		Pause: [0x02][0x00]		0x01:electric supply
				0x02:solar pannel
		Power off:[0x03][0x00]	Device reset time	Hexadecimal value
			Not reported data count	Hexadecimal value
			Deivce voltage	Hexadecimal value
Record	0x03	Channel count:Generated by sending device, equals the amount of channels belong to the device	Channel number	Hexadecimal value
			channel type	DO:0x00;
				EC:0x01;
				WT:0x02;
				PH:0x03;
			channel value	Hexadecimal value
Actuator state	0x04	Power on: [0x01][0x00]	Channels corresponding to control action	Hexadecimal value
		Pause: [0x02][0x00]	Control trigger	0x00:unknown
		Power off: [0x03][0x00]		0x01:timing
				0x02:automatic
				0x03:remote control
				0x04:manual operation
				0x05 exception protect
				0x06:Illeagal control
				0x07:SMS control
		No action: [0x04][0x00]	Control type	0x00:power on
				0x01:pause
				0x02:power off
				0x03:no action
Actuator state	0x04		State feedback type	0x00:poer on
				0x01:pause
				0x02:power off
				0x03:running
			State feedback channels	Hexadecimal value
			Original channel value	Hexadecimal value

A frame filter is designed here to filter eigenvector information from data frames with a high rate. The filter design is showed in Form 5.

Form5. Data filter design

Process	Byte number	Mask code	Description
1	1	0xFF	Extract the first byte, whose low three bits maybe the high three bits of the frame length.
2	4	0xFF	Extract the length of this frame
3	5	0xFF	Extract the frame code.
4	6	0xF0	Address type
4	6	0x0F	Address type
4	7	0xF0	Address type
4	7	0x0F	Address type
5	8	0xF0	Address description
6	9~9+M	0xF...F	Device address
7	9+M~11+M	0xFFFF	Data type
8	11+M~17+M	0xF...F	Report time/ GPRS Gateway clock
9	17+M~18+M	0xFF	Eigenvector type
10	18+M~18+M+N	0xF...F	Eigenvector value

3.3   A Method to Capture Fault by Filtering Data Frame

Form 6 is the recognize rule.To explain the process more clearly and detail, the filtering process is shown in the following fig2. Fig3 introduces recognition strategy respectively for GPRS gateway fault, because of space limit, device fault, record fault and actuator fault recognition strategy is not introduced here.

Form6. Recognize rule

Data type	Recognize parameter	Recognize result
GPRS Gateway state	No GPRS connection	GPRS offline
GPRS Gateway state	GPRS signal strength<16	Communication error: low signal strength
GPRS Gateway state	SIM card service stopped or Reset times in 24 hours>1 or Heart beat interval time>10minutes	GPRS module exception
GPRS Gateway state	Reporting cycle> heart beat interval time Device voltage<5 GPRS clock time difference>5 minutes	Bad data integrity Low device voltage Wrong GPRS clock
Device state	Device power off	Device shut down
Device state	Device voltage as the high byte and energy type as the low byte	<0x0502, solar pannel exception
Device state	Device voltage as the high byte and energy type as the low byte	<0x0501,electric supply exception
Device state	Device voltage as the high byte and energy type as the low byte	<0x0500,battery exception
Device state	Device reset time>1 in 24 hours	M2M module exception
Device state	Signal strength<50	Communication error:Low signal strength
Device state	Report time difference>5minutes	Report delay
Device state	Not reported data count>200 in 24 hours	Bad data integrity
Record	Report data value changeless	M2M communication exception
Record	Report data value out the threshold	Bad data rationality
Record	Unnormal report data value change rate	Bad data rationality
Record	Report time difference>5minutes	Report delay
Record	(Data count/heart beat count)/(Heart beat interval time/data collection cycle)	>1,GPRS connection fault
Record	(Data count/heart beat count)/(Heart beat interval time/data collection cycle)	<1,Bad data integrity
Actuator state	Actuator state code equals [0x02 0x00] or [0x03 0x00]	Uncontrollable
Actuator state	Actuator state code equals [0x04 0x00]	No response
Actuator state	Report delay	Report time difference>5minutes

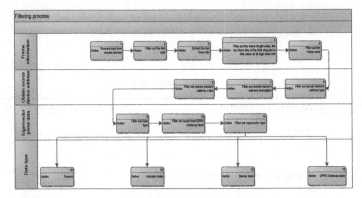

Fig. 2. Filtering process

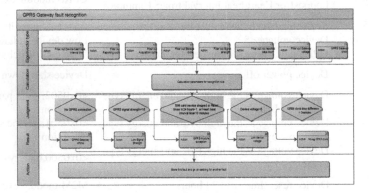

Fig. 3. GPRS Gateway Fault Recognition Strategy

4   Simulation Experiment and Discussion

4.1   Simulation Strategy and Build Simulation LAN

To verify the strategy's processing rate and accuracy, different personal computers with simulating IP addresses are regarded as the GPRS gateways to report collected data to remote server. The experiment LAN structure is shown in fig4.

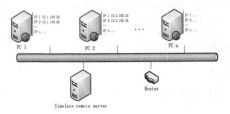

Fig. 4. Experiment LAN Structure

The router simulate the GPRS service, simulate remote server is another PC, simulate remote server receive the data frame via TCP/IP protocol and telnet service, then decode and filter the data frames with programs based on the fault recognition strategy in chapter 3.3. The program are written in java and based on apache mina framework.

4.2   Experiment Result and Evaluation

1290 records of fault are found from working log of operation and maintenance team is used in the experiment. The program was divided into two executable jar file, namely Simulation Server and Simulation Gateway. Fig 5 to 6 shows simulation experiment process and result.

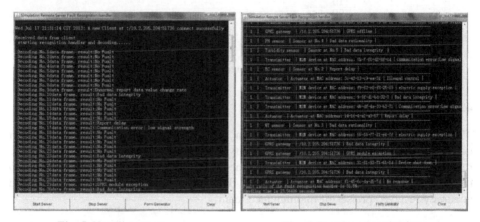

Fig. 5. Handling process   **Fig. 6.** Result and evaluation

The first experiment is to get the accuracy of the recognition method under different actual fault record ratio, and total records' count is fixed to 200. The experiment result is shown in fig7. The second experiment is about the accuracy under different total record count, the result is shown in fig8.

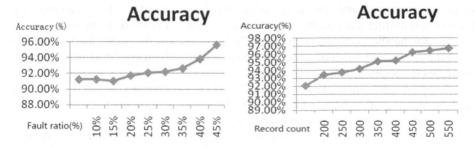

Fig. 7. Accuracy under Different Actual Fault Record Ratio

Fig. 8. Accuracy under Different Total Record Count

The last experiment is to test the time-cost under different fault record count, as the strategy time-cost is mainly on recognized fault records' handling, so this experiment change the fault record count and take the average fault records' handling time as the time-cost evaluation parameter, total record count is fixed to 400. The average fault records' handling time equals total handling time divides 400. The result proves the handling time is proportional to fault record count, as shown in fig .22.

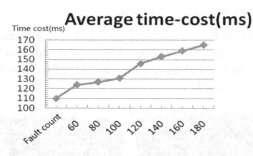

Fig. 9. Average time-cost

4.3   Application Effect

The recognition strategy was applied in the Water Quality Monitoring system developed by China Agricultural University, the program has been deployed on the remote server of the system since December, 2012, by now it recognized 246586 records, and is still running on the server. Its high efficient, low time-cost and nice accuracy received praises of maintenance team.

5   Conclusions

The proposed pattern recognition method, based on data eigenvector and TCP transport protocol analyses the Water Quality Monitoring equipment comprehensively. The author designed the data transport protocol based on transmit rule, based on this transport protocol, the data filter is proposed. According to actual diagnosis experiment and knowledge, added threshold to the data filter, as a result, the filter is able to filter out data that beyond the threshold. A novel recognition method is proposed to handle fault data with high efficiency and low time-cost, the method is a combination of four strategies, the strategy is specifically, the method manage these strategies to cooperation more efficient.

After 7 months' application, program based on the proposed method performed well on Water Quality Monitoring system. Work logs and praises from maintenance team prove the method feasible and efficient.

Nevertheless, because of the shortage of eigenvector analysis and filter design, recognition misses and errors exist, the method has potential to be improved better, further research should focus on the reduction of the recognition misses and errors. Moreover, relative diagnosis algorithm is necessary to be designed.

Acknowledgement. Financial support from state science and technology support program (2011BAD21B01). Data, diagnosis knowledge and device support from device maintenance team of Research Center of Agricultural information technology. China Agricultural University is gratefully acknowledged.

References

[1] Wu, K., Choudhury, D., Matsumoto, H.: Wireless Power Transmission, Technology, and Applications. In: Proceedings of The IEEE, vol. 101(6SI), pp. 1271–1275 (2013)

[2] Xiaodan, P., et al.: Uplink transmission in the W-band (75-110 GHz) for hybrid optical fiber-wireless access networks. Microwave and Optical Technology Letters 55(5), 1033–1036 (2013)

[3] Lakshmanan, S., Sivakumar, R.: Proteus: Multiflow Diversity Routing for Wireless Networks with Cooperative Transmissions. IEEE Transactions on Mobile Computing 12(6), 1146–1159 (2013)

[4] Cheng, S.T., et al.: Hierarchical Distributed Source Coding Scheme and Optimal Transmission Scheduling for Wireless Sensor Networks. Wireless Personal Communications 70(2), 847–868 (2013)

[5] Tahir, Y.H., et al.: Transmission of Visual Data over Wireless Fading Channel in Real-Time Systems Based on Superposition Coding Scheme. Arabian Journal for Science and Engineering 38(6), 1459–1469 (2013)

[6] Chunlai, Y., et al.: Study on Wireless Energy and Data Transmission for Long-Range Projectile. IEEE Transactions on Plasma Science 41(5), 1370–1375 (2013)

[7] Refaat, S.S., et al.: ANN-based for detection, diagnosis the bearing fault for three phase induction motors using current signal. In: 2013 IEEE International Conference on Industrial Technology (ICIT 2013), pp. 253–258 (2013)

[8] Talebi, H.A., Khorasani, K.: A Neural Network-Based Multiplicative Actuator Fault Detection and Isolation of Nonlinear Systems. IEEE Transactions on Control Systems Technology 21(3), 842–851 (2013)

[9] Jettanasen, C., Ngaopitakkul, A.: A novel probabilistic neural network-based algorithm for classifying internal fault in transformer windings. IEEJ Transactions on Electrical and Electronic Engineering 8(2), 123–131 (2013)

[10] Prieto, M.D., et al.: Bearing Fault Detection by a Novel Condition-Monitoring Scheme Based on Statistical-Time Features and Neural Networks. IEEE Transactions on Industrial Electronics 60(8), 3398–3407 (2013)

[11] Li, B., et al.: Applying the fuzzy lattice neurocomputing (FLN) classifier model to gear fault diagnosis. Neural Computing & Applications 22(3-4SI), 627–636 (2013)

[12] Liu, H., et al.: Adaptive neuro-fuzzy inference system based faulty sensor monitoring of indoor air quality in a subway station. Korean Journal of Chemical Engineering 30(3), 528–539 (2013)

[13] Hashemi, M., Safizadeh, M.S.: Design of a fuzzy model based on vibration signal analysis to auto-detect the gear faults. Industrial Lubrication and Tribology 65(3), 194–201 (2013)

[14] Keskes, H., Braham, A., Lachiri, Z.: Broken rotor bar diagnosis in induction machines through stationary wavelet packet transform and multiclass wavelet SVM. Electric Power Systems Research 97, 151–157 (2013)

The Analysis of County Science and Technology Worker Internet Usage and Its Influence Factors

Chen Huiping[1], Tian Zhihong[2,*], Wang Yubin[2,*], and Han Xue[3]

[1] Agricultural Information Institute of Chinese Academy of Agricultural Sciences,
Beijing, 100081, China
[2] College of Economics and Management, China Agricultural University, Beijing 100083, China
wyb@cau.edu.cn
[3] Department of Agricultural and Consumer Economics, College of Agricultural,
Consumer and Environmental Sciences, University of Illinois at Urbana-Champaign,
Champaign, IL 61820, USA

Abstract. The county science and technology workers are critical components of Chinese science and technology workers, playing an important role in science and technology application and popularization. Improving the skills for Internet use of Chinese county science and technology workers can promote the level of science and technology in our counties and towns and expand the way of transformation of advance scientific and technological achievements. This paper analyzes county science and technology worker Internet use and the determinants based on the Second National Survey data of county science and technology workers. The results show that, among all Chinese science and technology workers in counties, there are only 51% were proficient in Internet use while computer configuration, education, age, length of service are the crucial factors affecting their proficiency . Hence, three recommendations are proposed to improve the skills for internet use of county science and technology workers.

Keywords: county, science and technology workers, internet use, determinants.

1 Introduction

Science and technology workers are those who are engaged in generation, development, dissemination and application of systematic scientific and technical knowledge, with corresponding scientific and technological work as their occupation in modern society. From the perspective of occupation, science and technology include scientific researchers, engineering technicians, agricultural technicians, health technicians and natural science teachers. Science and technology workers in our country are characterized by its large size and significant impact. With the increasing impact of science and technology innovation on the development of China, science and technology workers are also paid more and more attention by the public and also different talents projects are launched endlessly.

---

*Corresponding author.

D. Li and Y. Chen (Eds.): CCTA 2013, Part II, IFIP AICT 420, pp. 434–443, 2014.
© IFIP International Federation for Information Processing 2014

County science and technology workers are the main component of Chinese science and technology workers, the main force in application and promotion of science and technology, and the important power of promoting the progress of science and technology, the development of regional economy and the construction of new socialist countryside. Counties consist of more than 90% of China's land area and more than 80% of China's population.

With the advancement of science and technology, computers and internets become popular. As to science and technology workers, under the background of knowledge "explosion" and information "flood", computer and Internet provide a vast information space and dramatically reduce science and technology workers' workload. The powerful search feature of the Internet facilitates retrieving literature, and greatly improves the efficiency of scientific research.

Currently, research on science and technology workers are all based on census data or survey data, involving all aspects of science and technology workers. According to the "Second National Survey of scientists " data, science and technology workers' job satisfaction, views on academic misconduct, and status of participation in science activities were analyzed by Zhang Jian [1], Zhao Yandong [2], Xue Shu [3] et al. Based on questionnaire data, Lu Genshu [4] summarized the characteristics of social responsibilities of scientists, He Guangxi [5] analyzed the public evaluation and attitudes to science and technology workers.

In the existing literature, Internet use and behavior have been studied mainly through descriptive statistics, nonparametric statistics and Logistic regression and other methods. Cao Rongrui [6] analyzed the purpose, behavior, and obsession status of Shanghai college students' Internet use via the questionnaire data. The determinants of Beijing residents' Internet use were studied by Xu Meng [7] using hierarchical linear model. Research about determinants, consequence variables, and intervening or regulatory mechanism on Internet use were summarized by Wang Jinliang [8].Previous studies indicate that socio-demographic factors, family factors, regional factors, psychological factors, and other factors can affect personal Internet use behavior significantly.

There are also many related research abroad. Education, gender, age, income, place of residence, broadband applications, and so on are consider as the most important factors of internet use in the study of Sciadas (2002) [9], Singh (2004) [10], Whitacre and Mills (2006) [11]. Larry McKeown (2007) [12]analyzed the factors associated with internet use in rural and small town of Canada. In 2005, only 58% of residents living in rural and small town areas accessed the Internet, well below the national average. Individuals that are older, those with lower levels of education and those living in households with lower incomes were less likely to have used the Internet. Research has got deeply with the population of internet. Ohbyung Kwon(2010) [13] make an empirical analysis on the factors affecting social network service use, individual characteristics and psychological factors have been considered in.

There are many studies on science and technology workers but research about Internet use has not been studied. Although there are many studies on Internet use, the observations in those studies use the Internet for both work and entertainment. A few observations use the Internet mainly for entertainment, especially the college students in the study of Internet use behavior. As for science and technology workers, their purpose of using the Internet is to facilitate work and to improve productivity. Therefore, the study about Internet use and determinants of county science and technology workers is very necessary.

2   Data Source and Methodology

2.1   Data Source

Data used in this study are the research data from the "Second National Survey of county science and technology workers". In order to understand the difficulties in work and life of county science and technology workers, the China Association for Science implemented and led the Second National Survey of work conditions of county science and technology workers in 2011, which involves 206 Counties of 31 provinces (autonomous regions and municipalities) in total. 12360 copies of questionnaires were issued and 18913 valid copies were returned.

In this study, the county agricultural science and technology workers are divided into 5 groups, such as agro-technicians, health workers, engineering technicians, science teaching staff and student village officials.

Agro-technicians: refers to those who are engaged in agriculture technology in companies or institutions, including senior agronomists (veterinarians, livestock engineers), agronomists (veterinarians, livestock engineers), assistant agronomists (veterinarians, livestock engineers), technicians and technical staffs without professional titles.

Health technicians: refers to those who are engaged in medical work in companies or institutions, including licensed doctors, assistant practicing doctors, registered nurses, pharmacists, inspection technicians, imaging technicians, health supervisors and doctors (pharmacists, nurses, technicians) on probation and other health professionals and technical personnel without professional titles.

Engineering technicians: refers to those who are engaged in engineering technology in companies or institutions, including senior engineers, engineers, assistant engineers, technicians and other technical personnel without professional titles.

Natural Science teachers: refers to those who teach mathematics, physics, chemistry, biology, information technology, labor and technical and physical geography for primary education and secondary education (junior and senior); teachers who are engaged in science, engineering, agriculture, medicine and other disciplines in educational institutions for vocational education (secondary, tertiary), higher education.

Student village officials: refers to the fresh or previous college graduates, who served as the village party branch secretary, deputy secretary, committee director and assistant, or other positions of the village "two committees" in the rural areas (including community), or who engaged in secretarial, finance and accounting for rural credit cooperatives, science parks and other public service institutions.

2.2   Analytical Method

The analyses of Internet use of county scientific and technical workers mainly uses comparative analysis and proportion analysis. The analysis of determinants is conducted by constructing Probit model and processing data via Stata 12.0.

According to existing research, socio-demographic factors affect Internet use behavior significantly. Based on the previous studies and the data obtained in this survey, variables such as gender, age, length of service, education degree, job title, industry, nature of the enterprise/institution are considered as major factors affecting Internet use of county science and technology workers. Internet use is the dependent variable which is classified into two types, skilled and unskilled. Value is assigned to other variables. Table 1 shows the setting of the Probit model of every variable.

Table 1. Probit model variable setting and definition

Variable Name	Value	Definition
dependent variable: Internet use	0, 1	Skilled = 1, unskilled = 0, no = 0
Explanatory variables:		
worker to computer ratio	0, 1	"one worker to one computer" = 1, one worker to multiple computers"= 1, no computer = 0
gender	0, 1	Male = 1, Female = 0
age	1-6	Before 1950 =1, 1950-1959= 2, 1960-1969 =3, 1970-1979 =4, 1980-1989 =5, After 1990 =6
length of service	1-7	Less than 3 years =1, .3~5 years=2, 6~10 years =3, 11~20 years =4, 21~30 years =5, 31~40 years =6, more than 41 years=7
education degree	1-6	Junior high school and below = 1, senior high school / Higher vocational education/ technical secondary school= 2, junior college = 3, , undergraduate=4, master = 5, Dr.= 6
Job title	1-5	senior = 1, sub-senior = 2, intermediate = 3, primary = 4, no title = 5
Industry	1-5	agricultural technicians = 1, engineering technicians= 2, health technicians = 3, natural sciences teachers= 4, student village officials= 5
nature of the enterprise/institution	1-7	government offices at low level = 1, fiscal funding institutions = 2, non-fiscal funding institutions = 3, state-owned enterprises= 4, private enterprises = 5, private non-enterprise= 6, social groups = 7

3   Empirical Analyses

3.1   Basic Characteristics of Science and Technology Workers

In the observations, 18,913 county science and technology workers, males and women accounted for 56% and 44%, respectively. As for age, in all observations, 31~50 age people accounted for 66%; 18~30 age people accounted for 27%; only 7% of people are over the age of 50. Corresponding to age, 34% of respondents' length of service is from 11 to 20 years; 26% of the respondents' length of service is from 21 to 30 years; and11% is from 6 to 10 years and under 3 years. It is observed that most of science and technology workers engaged in the same profession from the beginning of their career, since the proportion of 30 to 50 years old worker and the proportion of the workers with10 to 30 years length of service is similar.

From view of the education background, 49% of the respondents are with undergraduate degree; 36% of the respondents are with junior college degree; only 2% of the respondents are with graduate or higher degree. In terms of the job title, most of the titles of respondents are intermediate and primary. The proportion of these two titles were 36% and 28%, respectively and the proportion of sub-senior and senior were 10% and 4%, respectively.

According to the industry, the proportion of agricultural science, health technicians, natural sciences teaching staff, engineering technicians and student village officials among the respondents were 26%, 24%, 23%, 21% and 5%, respectively. Except the student village officials, the proportion of science and technology workers in other industry personnel is closer.

In the survey, the proportion of institutions is 79%, ranking top; the proportion of government departments is 12%, ranking second; the third is private enterprise, accounting for 5%; private non-enterprise, state-owned enterprises, community groups, and other units account for 4%.

3.2   Internet Use of Science and Technology Workers

Equipped with computer and Internet is precondition for using the Internet. Although computers and the Internet have been popularized in urban areas, some of science and technology workers in some rural areas cannot use the Internet due to the absence of computers and the Internet. Survey data indicate that 90% of science and technology workers are equipped with computers in the units, in which 88% have access to the Internet, that is 88% of the science and technology workers may use computer to surf the Internet.

There are 63% of the respondents that share computes with others. Their work condition is poor. There are only 27% of respondents that are equipped with their own computers. For the access to the Internet, there are only 55% of respondents whose Internet use is unrestricted. In this case, among the investigated county science and technology workers, those who can access the Internet freely accounted for 49.5%.

In spite of poor work conditions and infrastructure, among all the investigated county science and technology workers, 51% of them believe that they are proficient in sending and receiving e-mail and searching information and data; only 7% are unable to access the Internet. Even though the popularity of the Internet technique of county science and technology workers is quite low, it will be improved with the further promotion of Internet technology.

With regard to the person using the Internet skillfully, the proficiency is significantly different in gender. The proportion of female county science and technology workers who can use the Internet skillfully is higher than that of male. Female choosing "quite skilled" account for 51% of all female; male choosing "quite skilled" account for 50% of all male, while the proportion of male choosing "no" is 8%, 2% higher than that of female.

The proportion of sciences and technology workers who can use the Internet skillfully in different industries varies. Specific conditions are shown in Table 2. We can observe that Internet use of engineering technicians and natural sciences teacher is better than that of agricultural technicians and health technicians, which is related to the characteristics of their industries. The proportion of persons using the Internet expertly among the student village officials is very high, mainly because the student village officials are mostly the generation after 80s and familiar with the Internet in school.

Table 2. The computer configuration and Internet use of county science and technology workers in different industries

Industry	Agricultural technicians	Engineering technicians	Health technicians	Science teachers	Student village officials	Average
skilled user proportion	47%	57%	37%	54%	85%	51%
unskilled user proportion	9%	5%	11%	6%	2%	7%

3.3   Analysis of Determinants

The Probit model was estimated via maximum likelihood estimation. The results are shown in Table 3. As we can see, influence of every variable on the dependent variable is significant except nature of the enterprise/institution. If one is male and equipped with computer, younger and with high degree and short work years, the more likely to access internet.

Whether computers are equipped in enterprises/institutions is the most influential factor of county science and technology worker proficiency in Internet use. Increasing 1% of the worker to computer ratio can increase the Internet utilization rate 0.67%. The financial difficulty and the fact that leaders overlook the importance of equipping computer result in that science and technology workers have no access to computers.

Hence, the possibility of using the Internet proficiently dramatically reduced. Increasing the worker to computer ratio is the most important factor of promoting Internet use of scientific and technology workers.

Age is also an important factor affecting the Internet skill of county science and technology workers. The younger one has the higher possibility of using the Internet skillfully. The number of observation born before 1950 and born in 1950-1959 is small. In addition, behavior of two age groups is similar, so the simulated result is not significant. Because the rise of computer and internet is in the late 1990s and technology is changing rapidly, many technology workers, especially mid-aged, learn new things slowly. Also, since training of Internet technology in enterprises/institutuions and society is limited, some county science and technology workers are not proficient in using the Internet or unable to access the Internet by the limitation of their own internet skills, even equipped with computers and the Internet.

The effect of work age and real age are similar, but directions are opposite. The longer one works, the lower possibility one is able to use the Internet. The one with long length of service learns new technology about the Internet slowly and inactively. Probably because the number of observation worked under 3 years and worked between 3-5 years is small and their behavior is similar, the results of two variables above are insignificant.

Education degree affects the Internet use of county science and technology workers markedly. The higher one's degree is, the higher possibility of one's ability to use the Internet will be. Among 6 dummy variables of education degree, the variables of undergraduate and master degree have significant differences from junior high school and below and the other 3 variables are adverse. Generally, those who are able to obtain higher degree are those with strong desire of knowledge and ability of learning. If this person has experience in using computers and the Internet after having a job, his desire of knowledge will be significantly stronger and speed of learning the Internet will be faster than the others. If the person has experience in using computer and the Internet in school, the higher his degree, the higher possibility of using the Internet skillfully will br. Requirement for Internet skills in school education is increasing in degree.

Table 3. Probit model estimation results

variable name	Coefficient	Std. Err.	P>z
PC configuration-1	0.6796	0.0500	0.0000
Gender-1	0.0991	0.0288	0.0010
Age			
2	0.1691	0.2808	0.5470
3	0.5740	0.2765	0.0380
4	1.0022	0.2784	0.0000
5	1.3393	0.2831	0.0000
6	1.2809	0.3452	0.0000

Table 3. (*continued*)

Length of service			
2	-0.1081	0.0711	0.1290
3	-0.1922	0.0698	0.0060
4	-0.3618	0.0754	0.0000
5	-0.3164	0.0839	0.0000
6	-0.4838	0.1002	0.0000
7	-0.5645	0.1330	0.0000
Education degree			
2	-0.1453	0.1723	0.3990
3	0.1892	0.1679	0.2600
4	0.6238	0.1681	0.0000
5	1.2816	0.2093	0.0000
6	0.4764	0.4919	0.3330
Job title			
2	0.0830	0.0436	0.0570
3	-0.3513	0.0420	0.0000
4	-0.1130	0.0407	0.0060
5	0.0822	0.1495	0.5820
6	0.2551	0.0945	0.0070
Industry			
2	-0.0022	0.0791	0.9780
3	-0.1924	0.0720	0.0080
4	-0.3123	0.0766	0.0000
5	-0.3149	0.0815	0.0000
Nature of the enterprise/institution			
2	-0.0816	0.0536	0.1280
3	-0.1674	0.0663	0.0120
4	-0.2436	0.0955	0.0110
5	0.0959	0.0825	0.2450
6	0.2896	0.1950	0.1370
7	-0.0576	0.1707	0.7360
constant	-1.2825	0.3404	
Log likelihood = -5582.7732		Prob > chi2 = 0.0000	
LR chi2(33) = 2167.90		Pseudo R2 = 0.1626	

4   Conclusion and Recommendations

As an important mean of communication and research, Internet technology is an important skill which technical personnel need to have when science and technology becomes more and more important and changes rapidly. However, among the numerous county science and technology workers who make significant contributions, there is still a portion of them who are unable to use the Internet or to know little about Internet use and 51% of them think they can use the Internet skillfully. To improve the Internet skill of China's county science and technology workers is to enhance the level of science and technology in county and to advance the expanding way of achievements transformation. So, it is necessary to progressively enhance the Internet skills of county scientific and technical workers by effective means.

Based on the analysis of determinants, the following suggestions are offered:

(1) To better the work conditions for county science and technology workers, to increase worker to computers ratio, to achieve everyone has an access to a computer or one people per computer if possible. As long as appropriate hardware is sufficiently provided, science and technology workers will have the motivation and possibility of improving their Internet use skills.

(2) To strengthen on-the-job training of county science and technology workers, to train them regularly and especially to train middle-aged and older ones about Internet use skills. Since the Internet emerges in a short time and develops rapidly, many people know about the Internet after having their first job. Mutual learning between colleagues is inconvenient but Internet related training organized by enterprises/institutions can work more effectively.

(3) To improve the compensation incentive mechanism in order to encourage science and technology workers to make progress and to make more contribution. For some science and technology workers with longer length of services, who have made certain achievements, got some job titles and have little learning motivation, we can stimulate their learning enthusiasm so as to improve their Internet use skills by such methods as setting job grading, examining regularly, grading division results, and awarding the staff who got great achievement and so on.

Acknowledgments. This study is supported by China Association for Science and Technology research project "The Second Survey of County S&T Workers in China "(KXDC201002). Thanks all participants of the project.

References

[1] Zhang, J., Zhang, Z.: The empirical analysis on working satisfaction of S&T professionals——based on the data of Tianjin. Forum on Science and Technology in China (3), 112–116, 123 (2010)
[2] Zhao, Y., Deng, D.: Chinese science and technology personnel's views on academic misconduct: A survey result of 30,000 science and technology personnel. Science Research Management (8), 90–97 (2012)

[3] Xue, S., He, G., Zhao, Y.: Level and obstacle of S&T Professionals'participation in popularization of science in China-—based on the data of the general survey of Chinese S&T Professionals. Forum on Science and Technology in China (1), 126–130 (2012)

[4] Lu, G., Sun, H., Li, K., et al.: Analysis of basic characteristics of social responsibility consciousness of Chinese scientific and technical workers. Journal of Xi 'An Jiaotong University (Social Science) 35(5), 110–119 (2012)

[5] He, G., Wang, F.: Public image of S&T professionals and its cognitive basis in China. China Soft Science (7), 83–93 (2009)

[6] Cao, R., Jiang, L., Liao, S., et al.: Report on Internet use of college students in ShangHai. Newsman (04), 58–63 (2012)

[7] Xu, M.: Determinants of Internet Adoption in Beijing's Residents-——a HLM Analysis. Library and Information Service 56(14), 77–81, 46 (2012)

[8] Wang, J., Su, Z.: Current directions on Internet use: antecedents, outcome variables, and influencing mechanism. Journal of Southwest University (Social Science Edition) 20(3), 82–88 (2012)

[9] Sciadas, G.: Unveiling the digital divide. Connectedness Series(7), Statistics Canada (2002)

[10] Singh, V.: Factors Associated with Household Internet Use in Canada, 1998-2000. Rural and Small Town Canada Analysis Bulletin, Statistics Canada (2004)

[11] Whitacre, B., Mills, B.: A need for speed? Rural Internet connectivity and the noaccess / dial-up / high-speed decision. American Agricultural Economics Association, Long Beach (2006)

[12] McKeown, L., Noce, A., Czerny, P.: Factors Associated with Internet Use: Does Rurality Matter? Rural and Small Town Canada Analysis Bullet, Statistics Canada (2007)

[13] Kwon, O., Wen, Y.: An empirical study of the factors affecting social network service use. Computers in Human Behavior 26(2), 254–263 (2010)

A Smart Multi-parameter Sensor with Online Monitoring for the Aquaculture in China

Fa Peng[1,2], Jinxing Wang[2], Shuangxi Liu[2], Daoliang Li[1,3,*], Dan Xu[3], and Yang Wang[4]

[1] Beijing Engineering and Technology Research Center for Internet of Things in Agriculture, Beijing 100083, China
[2] College of Mechanical and Electronic Engineering, Shangdong Agricultural University, Taian271000, China
[3] College of Information and Electrical Engineering, China Agricultural University, Beijing 100083, China
[4] College of Engineering, China Agricultural University, Beijing 100083, China
dliangl@cau.edu.cn

Abstract. PH, DO,ORP, EC and water-level are important parameters of the aquaculture monitoring. But the high cost of foreign sensors and high-energy consumption of Chinese sensors make it impossible for wide use in China. This paper uses MCU STM8L152 to realize the ultralow power design. With simple hardware structure design, the cost of the multi-parameter sensor can be reduced .The experiment data of the multi-parameter sensor contrasting with the results obtained by Hach multi-parmeter meter, indicates that the sensor is reliable to monitor the water quality with low cost, high efficiency and good precision.

Keywords: Multi-parameter, aquaculture in China, Ultralow power, Low cost.

1   Introduction

From the 1970s, the portable water quality analyzer has been appeared in Japan; the water quality on-line monitoring started to be used in 1980s [1-2]. However, the first water quality monitoring system was used in China in 1988, which was in inadequate due to various reasons: (i) most parts of the system were imported from the oversea; (ii) it was very costly for the maintenance in proper operation; (iii) it didn't adapt to the aquaculture in China.

With the transformation of Chinese agriculture from the traditional way to the modern way, the monitoring of water quality attracts more and more attention of Chinese researchers. For example, Sun has designed the online multi-parameter water quality analyzer; Zhang has finished the research on multi-parameter microsensor array [3-4]. All these things make a great progress of Chinese sensor technology. However, foreign corporations still take control of the sensor-market in China at present; for ion electrode of sensors is very mature and the foreign sensors have higher precision and

---

[*] Corresponding author.

D. Li and Y. Chen (Eds.): CCTA 2013, Part II, IFIP AICT 420, pp. 444–452, 2014.
© IFIP International Federation for Information Processing 2014

stability than Chinese sensors. But the high cost made it impossible that foreign sensors or analyzers are used in China widely.

The on-line monitoring has advantages over traditional monitoring approaches such as sampling followed by laboratory analysis; the on-line monitoring can collect the data anytime and help to know the dynamic information of some element real-time. But data collection and management, energy efficiency still exist hindering the long-term using of on-line monitoring [5].A large variety of faults can impact on data quality, including sensors affected by aging, biofouling or leaking of internal solutions, or simply sensors with bad connections to the data collection device. What's more, sensor networks should be focused on in order to realize the commercialization of multi-parameter sensor.

This paper aims to present a multi-parameter sensor with online monitoring for the aquaculture in China, which makes the real-time monitoring come true with low cost, high efficiency and good precision. The multi-parameter sensor is based on advanced microsensor, microprocessor and smart data acquisition and transfer technology. As a result, pH, dissolved oxygen (DO), electrical conductivity (EC), oxidation-reduction potential (ORP) and water level can be measured.

2   System Design

2.1   Detection Principle

There are many ways to get the information of water quality; for multi-parameter sensor, ion-selective electrodes and other physical electrodes are applied. So, only the detection principles about ion-selective electrodes and other physical electrodes are introduced here.

Nernst equation is a mathematical description of an ideal pH or ORP electrode behavior [6-8]. It's an important connection between the electric potential difference and the density of the active material in electrochemical system:

$$E = E^0 - \frac{RT}{nF} \ln \frac{a_{i1}}{a_{i2}} \tag{1}$$

Where E is the single electrode reduction potential; E^0 is the standard electrode r eduction potential; T is the absolute temperature (K); R is the universal gas con stant: $R=8.31 \ J \cdot mol^{-1} \cdot K^{-1}$; n is the number of moles of electrons transferred in the half- reaction; F is the Faraday constant, the number of coulombs per mole of el ectrons: $F= 9.6467 \times 10^4 \ C \cdot mol^{-1}$; a_i is the chemical activity for the relevant speci es, where a_{i1} is the reductant and a_{i2} is the oxidant. $a_i = C_i f_i$, where C_i is the con centration of species i; f_i is the activity coefficient of species i, when C_i is less th an 10^{-3} mol·L^{-1}, $f_i \approx 1$.

At room temperature (25°C), RT/F can be replaced by a constant, then we get Eq.(2):

$$E = E^0 + \frac{0.05916}{n} \lg \frac{a_{i2}}{a_{i1}} \tag{2}$$

Clark electrodes are widely used to monitoring the dissolved oxygen. When an appropriate polarization voltage is applied on the two electrodes, the dissolved oxygen can come through the polymer film and participate in the reaction at the cathode:

Oxidation at the anode: $4Ag+4Cl^-{\rightarrow}4AgCl+4e^-$

Reduction at the cathode: $O_2+2H_2O+4e^-{\rightarrow}4OH^-$

Overall reaction: $4Ag+O_2+2H_2O+4Cl^-{\rightarrow}4AgCl+4OH^-$

Based on the Faraday's law Eq.(3), the diffusion current is proportional to the oxygen concentration at a certain temperature when the electrode is selected. That means once we get the measured current, the oxygen concentration can be known [9].

$$i = K \times N \times F \times A \times C_s \times P_m / L \tag{3}$$

Where K is the constant; N is the number of moles of losing electrons in the reaction; F is the Faraday's constant; P_m is the permeability coefficient of the film; L is the thickness of the film; A is the cathode area; C_s is the oxygen partial pressure(dissolved oxygen concentration) in the sample[10-11].

The measurement of electrode conductivity is based on the principle of electrolytic conduction. When electrodes are immersed in the solution with appropriate voltage, positive and negative ions in solution will move along the direction of electric field, which makes the solution become the conductor. The electric capability of conductor can be demonstrated by conductance or conductivity. The relationship between conductance (G) and conductivity (σ) is expressed by Eq.(4):

$$G = 1/R = s/(\rho \cdot l) = \sigma \cdot s/l = \kappa^{-1} \cdot \sigma \tag{4}$$

Where R is the conductor resistance (Ω); l is the length of the conductor (cm); s is the cross sectional area of the conductor (cm^2); ρ is the resistivity of the conductor $(\Omega \cdot cm)$; κ is the constant relating to conductivity cell.

Conductance is proportional to conductivity; in order to monitoring the conductivity of acquaculture water, we need to design the circuit and get the information of conductance.

As we know, the deeper of water a probe locates at, the larger pressure it suffers. Piezoresistive pressure probe is based on Pascal's principle [12]. When there is no pressure on the probe, there's no output voltage. The output voltage increases with pressure if the probe is immersed in water.

2.2   Hardware Structure Design

In Fig.1, analog signals of water quality can be collected by detecting probes. However, chemical probe is sensitive to temperature changing and interfering ions can reduce the accuracy. Meanwhile, the limited range of microprocessor is exceeded by the analog signals if detecting probes are connected to the microprocessor directly. So we need the signal conditioning circuit to solve these problems. Not only should the signal conditioning circuit remove noise, but it can adjust the amplitude of analog signals. The signal conditioning circuit consists of operational amplifiers and some filter circuits.

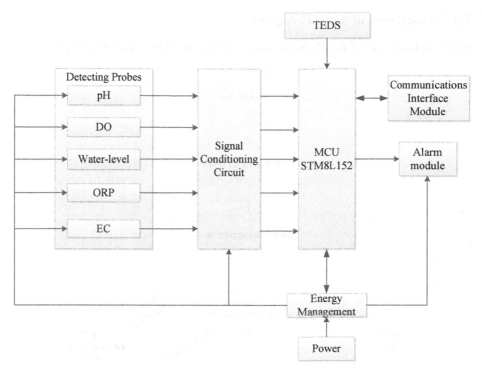

Fig. 1. Schematic diagram of multi-parameter sensor

In order to reduce the energy consumption of the multi-parameter sensor, ST's ultra low power MCU STM8L152 is chosen to be a core processor. STM8L152 is integrated with 64 Kbytes of high-density embedded Flash program and 4 Kbytes of RAM. Its ultralow power consumption is $1\mu A$ in active-halt mode. STM8L152 can operate either from 1.8 to 3.6 V (down to 1.65 V at power-down) or from 1.65 to 3.6 V[13]. All these features make the microprocessor suitable for ultralow power sensor. What's more, analog to digital conversion chip is embedded in STM8L152, which makes it possible of converting analog signals from signal conditioning module into digital signals. What's more, up to 28 analog channels are very helpful for following study. With energy management module, microprocessor can control the power supply for each probe, preventing the noise by disturbances.

The spreadsheets (TEDS) in the flash of STM8L152, which store the channel information and calibration parameters, are designed to realize self-identification, self-diagnosis and self-calibration of the multi-parameter sensor. The communication interface module can be RS485 bus or GPRS, which is wildly used to communicate with PC monitoring plat. When monitoring values exceed the warning values, alarm signal is emitted by the alarm module. Moreover, the energy management module can provide stable voltage for detecting probes, signal conditioning circuit and alarm module. Rechargeable batteries and solar energy as the power are alternative energy methods.

2.3   Description of System Program

Software flow chart of the Multi-parameter sensor's system is shown as Fig.2.

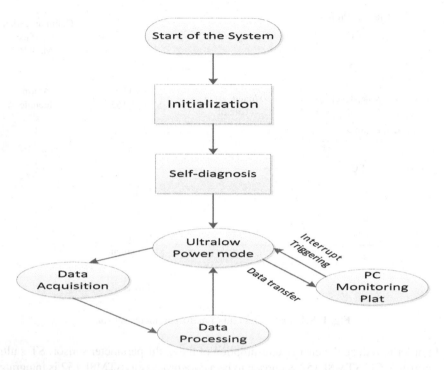

Fig. 2. Multi-parameter sensor's system program

The Multi-parameter sensor' system program is composed of the main program and interrupt service program. The purpose of the main program is to finish the initialization after system power on. Whether the supply voltage holds in a proper range can be tested by the self- diagnosis program of the main program. Meanwhile, the main program also can diagnose the nature of detecting probes and other faults of multi-parameter sensor. After the initialization and self-diagnosis, the device can operate normally. Then power supply of peripheral devices is stopped by the microprocessor and all the idle ports are placed on a high or low level simultaneously, which means the system enters into interrupt service program (ultralow power mode).

Ultralow power mode is designed to improve the energy efficiency. The focus of ultralow power mode is on interrupt triggering caused by PC monitoring plat. When collecting the data, the microprocessor executes the instruction of data acquisition sent by PC monitoring plat via either RS485 bus or GPRS. Then peripheral devices get the power supply; with controlling the power supply for each probe, microprocessor can acquire analog signals. After converting analog signals into digital signals, the data will go through compensation calculation and calibration calculation so as to get actual values. If actual values exceed the warning values, alarm signal is emitted by the alarm module. When finish the data acquisition, the system is back to ultralow power mode. What's more, system parameters and sensor calibration can be realized through serial ports.

3   Results and Discussion

3.1   Preparation of the Experiment

The standard solutions of pH 4.00, 6.86 and 9.18 were prepared for verify the multi-parameter sensor's reliability and accuracy. So the electronic analytical balance of Denver TP114 was purchased from America. The potassium hydrogen phthalate ($KHC_8H_4O_4$) of 10.12 g was dissolved in distilled water and diluted to 1000ml with a 1000 ml volumetric flask, when the buffer at pH 4.00 was made. Before use, the distilled water boiled for 15~30 min in order to remove the carbon dioxide in the water. For the buffer of pH 6.86, the potassium dihydrogen phosphate (KH_2PO_4) of 3.39 g and anhydrous di-sodium hydrogen phosphate (Na_2HPO_4) of 3.35 g were dissolved in distilled water and diluted to 1000ml. The borax of 3.80 g is dissolved in distilled water and diluted to 1000ml in order to get the buffer at pH 9.18.

High-low temperature test chamber of Surui RGDJ-100, high precision water bath of Jinghong DKB-501S and Hach HQ40d multi-parmeter meter were purchased for the experiment. The conductivity of standard EC solution depends on the mess of KCL, so precious weighing is needed. With Denver TP114, standard EC solutions of different conductivity were prepared.

Table 1. Required mess of KCl for standard EC solution (20°C)

Standard EC (mS/cm)	15	25	40	60
KCL(g)	5.0442	8.588	14.0072	21.3669

The standard solution of hydroquinone is a very reliable material for verifying the accuracy of ORP measurement. Three portions of 10 g hydroquinone were respectively dissolved in the buffer of 1000 ml at pH 4.00, pH 6.86 and pH 9.18; meanwhile, there should be some solid hydroquinone guaranteed the saturation condition of the solution. Precise ruler was used as the criteria for the measurement of water level.

3.2   Characterization and Validation of the Multi-parameter Sensor

The multi-parameter sensor and Hach HQ40d multi-parmeter meter were put into high precision water bath of Jinghong DKB-501S. The amount of DO was measured per five degree centigrade increase. From Table 2, the accuracy of the multi-parameter sensor with DO probe is less than 1%, which means the multi-parameter sensor can be applied to monitor the amount of DO in the field.

Table 2. Detection accuracy of the multi-parameter sensor with DO probe

Temperature (°C)	Reference value (mg/l)	Measured value(mg/l)	Relative error(%)
5	9.97	9.88	-0.90%
10	9.16	9.12	-0.44%
15	8.88	8.86	-0.23%
20	8.50	8.49	-0.12%
25	6.93	6.95	0.29%

Table 3. Detection accuracy of EC in different temperatures

Temperature (°C)	Reference value (mS/cm)	Measured value (mS/cm)	Reference value (mS/cm)	Measured value (mS/cm)
5	15	15.32	40	39.82
15	15	15.16	40	40.51
25	15	15.28	40	40.06
35	15	15.22	40	40.72
5	25	25.43	60	61.27
15	25	25.21	60	60.62
25	25	25.20	60	60.31
35	25	25.10	60	60.43

Table 4. Detection accuracy of pH in different temperatures

Temperature (°C)	Standard solution	Measured value	Relative error
10	4.00	4.03	0.03
20	4.00	4.05	0.05
30	4.02	4.04	0.02
10	6.85	6.85	0
20	6.88	6.87	-0.01
30	6.85	6.87	0.02
10	9.33	9.22	-0.11
20	9.23	9.27	0.04
30	9.14	9.31	0.17

Table 5. Detection accuracy of the water-level

Reference value (m)	Measured value(m)	Relative error(%)
7.67	7.62	-0.65%
5.68	5.63	-0.88%
7.49	7.42	-0.93%
6.50	6.54	0.62%
6.93	6.95	0.29%
7.82	7.87	0.64%
8.54	8.49	-0.59%

Table 6. Detection accuracy of the ORP

pH	Reference value (mV)	Measured value(mV)	Relative error(mV)
4.02	218.5	219.0	0.5
6.90	59.0	58.3	0.7
9.21	-76.5	-76.1	-0.4

At the same way, the multi-parameter sensor and standard EC solutions in DKB-501S had shown the detection of EC. The values measured by the multi-parameter sensor were found to be in a good agreement with standard solutions. Measurement error is less than 1mS/cm as shown in Table 3. DKB-501S can guarantee the precision of the measurement thanks to isolating solutions from the atmosphere.

In Table 4, Standard solution means the standard solutions of pH 4.00, 6.86 and 9.18. Besides, the experiment was performed in high-low temperature test chamber of Surui RGDJ-100. When the temperature increases every time, one hour is needed to get the stable and reliable data. At the room temperature, the maximum error of the multi-parameter sensor with pH probe is 0.05. Relative error of the multi-parameter sensor with water-level probe is less than 1% and relative error of the multi-parameter sensor with ORP probe is less than 1mV, from Table 5 and 6.

All these results demonstrated the reliability, accuracy and sensitivity of the multi-parameter sensor for online monitoring with pH, DO, EC, ORP and water level, which means compensation calculation and calibration calculation are effective to get actual values. So the multi-parameter sensor we designed can be used for online monitoring the aquaculture in China with acceptable errors.

4   Conclusion

It has been demonstrated that a multi-parameter sensor with online monitoring for the aquaculture in China is designed. The multi-parameter sensor makes the real-time monitoring come true with low cost, high efficiency and good precision. Detection principles of water quality are introduced in this paper.

In particular, STM8L152 is chosen to be a core processor embedding analog to digital conversion chip, which simplifies the system's design. What's more, active-halt mode of the chip is helpful the ultralow power design. In addition, compensation calculation and calibration calculation guarantee the accuracy of data. With energy management module, microprocessor can control the power supply for each probe, preventing the noise by disturbances. With simple hardware structure, the cost of the multi-parameter sensor is very low, which means the multi-parameter sensor can be used in China widely.

Acknowledgement. This research is founded by "Special Fund for Agro-scientific Research in the Public Interest" (201203017), Key Program for International S&T Cooperation Projects of China （2013DFA11320） , Key Projects in the National Science & Technology Pillar Program during the Twelfth Five-year Plan Period (2012BAD35B03).

References

[1] W.P.: Introduction to on-line aquaculture water quality monitoring system. Fishery Modernization (4), 17–19 (2006) (in Chinese)

[2] Lee, P.G.: A Review of Automated Control Systems for Aquaculture and Design Criteria for Their Implementation. Aquacultural Engineering 14, 205–227 (1995)

[3] Sun, M.J., Li, X.Y., Li, X.Y.: Design and Application of Online Multi-parameter Water Quality Analyzer Based on Cygnal Microcomputer C8051F020. Control and Instruments in Chemical Industry 36(6), 56–58 (2009) (in Chinese)

[4] Zhang, H.H., Liu, B., Jin, Q.H., Liu, Y.S.: Research and fabrication of multi-parameter microsensor array for water quality monitoring based on integrated reference electrode. Transducer and Microsystem Technologies 30(10), 99–101 (2011) (in Chinese)

[5] Philip, W., et al.: Environmental sensor networks in ecological research. New Phytologist (182), 589–607 (2009)

[6] Bard, A.J., Faulkner, L.R.: Electrochemical Methods: Fundamentals and Applications, 2nd edn. Wiley, New York (2001)

[7] Trinidad, P., de Leon, C., Walsh, F.C.: The use of electrolyte redox potential to monitor the Ce(IV)/Ce(III) couple. Journal of Environmental Management 88(4), 1417–1425 (2008)

[8] David, J.M.: ORP Measurements in Water and Wastewater. Ultrapure Water Journal 25(5), 26–31 (1993)

[9] Liu, Q., Zou, Y.Q., Xing, H.Y.: Measuring Instrument of Dissolved Oxygen Based on MSP430. Instrument Technique and Sensor (9), 33–35 (2009) (in Chinese)

[10] Milka, T.N., Nikolova, V., Petrov, V.: New real-time analytical applications of electrochemical quartz crystal microbalance: Stoichiometry and phase composition monitoring of electrodeposited thin chalcogenide films. Analytica Chimica Acta 573, 34–40 (2006)

[11] Liu, X.W., et al.: Nano-structured manganese oxide as a cathodic catalyst for enhanced oxygen reduction in a microbial fuel cell fed with a synthetic wastewater. Water Research 44(18), 5298–5305 (2010)

[12] Liu, L., Deng, S.J., Zhang, J.: Research of Technologies of Water Level Detection for Water Supply and Drainage System. Industry and Mine Automation 37(12), 21–24 (2011) (in Chinese)

[13] ST. STM8L15X microcontroller family (2010), http://www.st.com

An Intelligent Ammonia Sensor
for Livestock Breeding Monitoring

Yang Wang[1], Zetian Fu[1,2,*], Lingxian Zhang[2], Xinxing Li[2], Dan Xu[2], Lihua Zeng[2],
Juncheng Ma[2], and Fa Peng[3]

[1] College of Engineering, China Agricultural University, Beijing 100083, China
[2] College of Information and Electrical Engineering,
China Agricultural University, Beijing 100083, China
[3] College of Mechanical and Electronic Engineering,
Shandong Agricultural University, Taian271000, China
fzt@cau.edu.cn

Abstract. Ammonia concentration is the major parameter to evaluate livestock breeding farms atmosphere quality and it also is regarded as the key indicator to describe the production of livestock breeding farms. Based on the oxidation characteristics of ammonia, this paper presented a new intelligent detecting instrument, the intelligent ammonia sensor, for the measurement of ammonia concentration, which used the microcontroller STM8L152 as the key control module. However, the TEDS module, which is useful to self-identification, self-diagnosis and self-calibration of ammonia sensor, is in the flash of the STM8L152 and only if the STM8L152 conveys the warning signal to the alarm module, will the alarm module starts the worker loop.

Keywords: ammonia concentration, livestock breeding farms, intelligent sensor, STM8L152.

1 Introduction

Ammonia is a valuable chemical product and material due to its numerous applications in the environmental protection, clinical diagnosis, industrial processes, food processing, and power plants [1-5]. However, it is a kind of colorless gas with a special odor which is very harmful to human body and livestock. The investigation results show that ammonia can result in serious economic losses by irritating the livestock's respiratory system, which may cause abortion or neonatal malformations such as birth defects, retinitis, and brain damage [6]. In livestock breeding farms, a large amount of ammonia is released through ammonification, a series of metabolic activities that decompose organic nitrogen like manure from livestock [7]. It is performed by bacteria and fungi. The released ammonium ions and gaseous ammonia will be converted to nitrite and nitrate by bacteria [8]. The worldwide ammonia

---

* Corresponding author.

D. Li and Y. Chen (Eds.): CCTA 2013, Part II, IFIP AICT 420, pp. 453–460, 2014.

emission resulting from livestock is approximated to be 20–35 Tg/year [9]. Therefore, an effective method for monitoring ammonia is badly needed for the livestock breeding farms environ-mental measurements and control. The numerous efforts have been paid to developing ammonia sensors and many scientific papers have been written concerning ammonia sensors for different domain using several sensing principles. However, there is no adequate research on intelligent ammonia sensor for livestock breeding farms in China, because of following reasons: (I) farmers don't pay attention to the ammonia; (ii) the high cost of ammonia sensor makes it hard to afford; (iii) it does not adapt to the Chinese livestock breeding farms.

With the development of technology and Precision Agriculture, more and more Chinese scientists and research workers pay attention to monitoring ammonia for livestock breeding farms. Among the various sensors, electrochemical sensors have many advantages, such as miniaturization, low cost, rapid response, the inherent high sensitivity, and selectivity [10]. Besides, on-line monitoring has advantages over traditional monitoring approaches on sampling followed by laboratory analysis, because on-line monitoring can collect data anytime and help to know the dynamic information of some element in real-time. But data collection and management, energy efficiency still exist hindering the long-term using of on-line monitoring [11]. Nowadays, foreign corporations still take control of the sensor-market in China. Foreign ammonia sensors are very successful and they are of higher precision and stability than Chinese sensors. But high cost made it impossible for the foreign sensors to be widely used in China. Therefore, electrochemical ammonia sensors which can be used in on-line monitoring for livestock breeding farms are worth developing.

This paper aims to present an intelligent ammonia senor for livestock breeding. With the conditioning circuit, the noise of the ammonia can be reduced, which precision can be guaranteed. The low cost of design makes the long-term monitoring come true. By using microprocessor and advanced transfer technology, the volume of the ammonia sensor can realize mini-type. As a result, NH_3 concentration and the temperature can be measured.

2 System Design

2.1 Detection Principle

There are two ways, current mode and potentiometric mode, to get the information of ammonia concentration. For intelligent sensors, current mode is applied for lower cost. So, only the detection principles about current mode are introduced here. Ammonia sensors of current mode are based on the principle of potentiostatic polarization method [12] which means response current is measured by excitation voltage .The working principle, structure design and critical fabricating technology of ammonia sensor with three-electrode structure are introduced as follow.

The sensor consists of three electrodes – sensing electrode, negative electrode and reference electrode which are separated by a layer of electrolyte thin film and connected by an external low impedance circuit. The oxidizing reaction of ammonia is on the surface of sensing electrode and the reaction generate an internal current between sensing electrode and negative electrode. The current value is related to ammonia concentration so that the ammonia concentration can be detected while connecting a load resistance to external circuit. The function of reference electrode is keeping electromotive force stable. Besides, no current flows through reference electrode so that each voltage of electrode could be remained stable. In conclusion, three electrodes is made for the intelligent ammonia sensor for livestock breeding farms in China, which was due to various advantage: (I) the intelligent ammonia sensor has a greater measuring range; (ii) it is no different of polarization; (iii) the output could be keep linear. When an appropriate polarization voltage is applied on the sensing electrode and negative electrode, the ammonia can come through the polymer film and participate in the reaction at the negative electrode:

Oxidation at the sensing electrode: $12NH_3+I_2+6H_2O \rightarrow 2IO_3^-+12NH_4^++12e^-$

Reduction at the negative electrode: $3O_2+6H_2O+12e^- \rightarrow 12OH^-$

Overall reaction: $12NH_3+I_2+12H_2O+3O_2 \rightarrow 2IO_3^-+12NH_4^++12OH^-$

Based on the Faraday's law Eq. (1), the diffusion current is proportional to the ammonia concentration at a certain temperature when the electrode is selected. That means once we get the measured current, the ammonia concentration can be known [13].

$$i=K \times N \times F \times A \times C_s \times P_m / L \qquad (1)$$

In this equation, K represents a constant, while N represents the amount of electrons in the reaction, F represents the Faraday's constant, A represents the cathode area, C_s represents ammonia partial pressure in the sample, P_m represents the permeability coefficient of film, and L represents thickness of the film [14-15].

Eq. (2) is a mathematical description of an ideal ammonia concentration or ORP electrode behavior. It's an important connection between the electric potential difference and the density of the active material in the electrochemical system:

$$S = K\log_e 1/(1-C) \qquad (2)$$

Where : S is output signal, K is the constant, and C is ammonia concentration. Eq. (2) shows that the signals are nonlinear. In most cases, the bias can be ignored and it can be compensated if needed.

In order to monitoring the ammonia concentration of livestock breeding farm, we need to design the circuit and get the information.

2.2 Hardware Implementation

2.2.1 System Design

In Fig.1, the ammonia sensor consists of 10 modules and their relations and functions are explained as following.

First of all, the power modules supply electricity to the entire system. Then NH3 concentration can be collected by NH3 detecting probes while TEMP detecting probe is collecting temperature to achieve temperature compensation of NH3 concentration. As NH3 probe need steady current source and TEMP probe need AC current source, excitation signal source consists of 2 modules which provide different sources for two probes. The signal conditioning circuit, which consist of V-I Converter-amplifying circuit and filtering-amplifying circuit, can remove noise and adjust the amplitude of analog signals. The core processor of sensor is MCU STM8L152 with 64 Kbytes of high-density embedded Flash program and 4 Kbytes of RAM. Moreover, the MCU controls the normal operation of the rest modules.

The TEDS module, which is useful to self-identification, self-diagnosis and self-calibration of ammonia sensor, is in the flash of the MCU. Communication interface module, which is important to communicate with ammonia sensor, is also controlled by the MCU. Only if the MCU convey to alarm module the warning signal will it start the worker loop. Energy management module, which is controlled by the MCU, is designed to make the voltage of the power module stable.

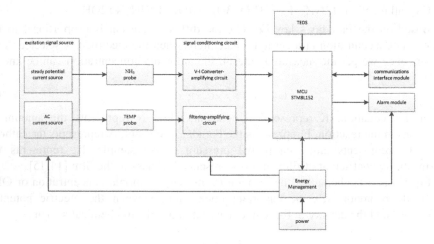

Fig. 1. Schematic diagram of intelligent ammonia sensor

2.2.2   Circuit Design

The circuit of intelligent ammonia sensor is designed into Integrated Circuit. The ammonia probe adopts the techniques of three-electrode. A very important point of the circuit is the inverse relationship between the circuit noise and response time. In order to get the best precision of the sensor, the optimal between the circuits with loop voltage is less than 10mV. What's more, fast response speed can be guaranteed with least resistance of the circuit. The reference electrode and the induced electrode should be shorted to make sure the sensor is in the state of preparation for work; so the field-effect transistor is used for connecting reference electrode to the ground.

2.3   Description of Software

Intelligent ammonia sensor has modes of working: waiting mode of ultralow power can realize the energy saving; data acquisition mode can get the precise ammonia concentration. With the channel information and calibration parameters in the ferroelectric memory, the style of probe, the serial number and data-structure can be identified. What's more, the self-diagnosis program can also judge whether the intelligent ammonia sensor works normally. Waiting mode of ultralow power aims to improve the system efficiency; the data acquisition need to be triggered by the external signal. In data acquisition mode of STM8L152, the embedded analog-to-digital conversion is good for the simplification of system design. Single conversion mode is adapted to converter the analog data into digital data. In order to guarantee the conversion precision, the average value of 100 conversions is regarded as the measurement result.

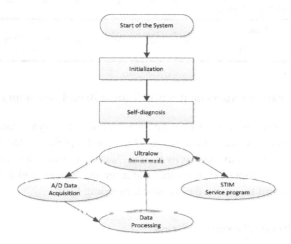

Fig. 2. Intelligent ammonia sensor's system program

System task is triggered by the serial port to receive external trigger signal occurs. And the signal is collected by following pathways: the host computer via RS485 bus, using the microcontroller's serial port to send data acquisition command, the sensor probe under the control of the microcontroller upload the collected data to the microcontroller and the resulting actual measured value of each parameter. If the measured value exceeds the preset alarm value, the alarm means generates an alarm signal.   Additionally, the system can also accept serial signals, setting parameters, sensor calibration.

3   Result and Discussion

3.1   Preparation of the Experiment

The gas of ppm N 5, 10, 25, 50, 75 and 100 were prepared for verify the sensor's reliability and accuracy. The text space is a seal box of constant volume. At the same pressure, the concentration of ammonia can be controlled by changing the discharging time. The required discharging time for standard NH3 concentration results as shown in Table 1. Two pairs of Chinese brand ammonia sensors were prepared for comparative experiment.

Table 1. Required discharging time for standard NH_3 concentration (20℃)

Standard NH_3 concentration (ppm)	0	5	10	25	50	80	100
Time (second)	0	8	16	40	80	128	160

3.2   Comparative Experiment with Chinese Brand Ammonia Sensors

At the completely same qualification, we conducted a systematic controlled trials comparing various performance. As can be seen in table 2，the experiment proved that the intelligent ammonia sensor had less reaction time and recovery time. Moreover, the experiment showed that the ammonia sensor had a better accuracy and it can be competent to survey the livestock breeding farms.

3.3   The Effects of External Stress

A sudden pressure change on the process of sensor detection generates an instant response, and then the peak signal falls immediately. To solve this problem, a restrictor will be placed in front end of the sensor. The experiment showed that the sensor could conquer the effects of external stress completely. And the specific data is simple so that it can be shown from table 2.

Table 2. Experiment of standard NH_3 concentration (20℃)

Standard NH_3 concentration (ppm)	0	5	10	25	50	80	110
Experiment value (mV)	5	63	146	295	591	947	1106
Comparative value 1 (ppm)	0	7.88	11.88	26.25	49.9	74.68	108.13
Comparative value 2 (ppm)	0.08	3.3	7.6	21.03	58.13	73.31	91.25

3.4 Calibration of the Ammonia Sensor

The demarcation curve can be obtained from Table 2, shown as Fig. 3. Least square m ethod is used for curve fitting in this paper. The fitting formula is Eq.3.

$$y=0.083(x-5) \tag{3}$$

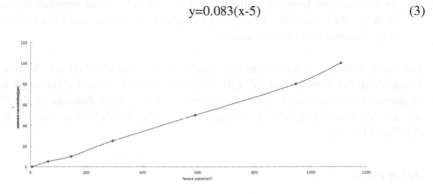

Fig. 3. Output character of intelligent ammonia sensor on 20°C

3.5 Characterization and Validation of the Intelligent Ammonia Sensor

From table 3 we can get the conclusion that the accuracy of the intelligent ammonia sensor is less than 5%, which means the intelligent ammonia sensor can be applied to monitor the ammonia concentration of livestock breeding farms because the accuracy of the intelligent ammonia sensor is satisfied for the livestock breeding. So compensation calculation and calibration calculation are effective to get actual values.

Table 3. Characterization and Validation of the intelligent ammonia sensor (20°C)

Standard NH$_3$ concentration (ppm)	0	5	10	25	50	80	100
sensor value (ppm)	0	6	9	27	48	83	104

4 Conclusion

In this article, we carefully studied the detection methods for measuring the ammonia concentration, designed a low-power, low noise, high speed, intelligent ammonia measuring device.

(1) The system chose STM8L152 microcontroller as the core of the detection control unit, its rich pin function and the built in functional unit simplify the system's design, which improved stability and anti-jamming capability of the system.

(2) The system choose single-chip PWM as output control, and other functional units selected low-power devices for designing, which effectively reduced the system power consumption. The overall average power consumption of the system is only 0.03W.

(3) As can be seen from the experimental data of the system repeatability tests and linearity tests, the system has good linear consistency no matter how low or high the concentrations of ammonia is.

Acknowledgement. This research is founded by "Special Fund for Agro-scientific Research in the Public Interest" (201203017), Key Program for International S&T Cooperation Projects of China（2013DFA11320）, Key Projects in the National Science & Technology Pillar Program during the Twelfth Five-year Plan Period (2012BAD35B03).

References

[1] Timmer, B., Olthuis, W., van den Berg, A.: Sensors and Actuators B:Chem. 107(2), 666–667 (2005)

[2] de Mishima, B.A.L., Lescano, D., Holgado, T.M., et al.: J. Electroanal. Chem. 506(2), 127–135 (2001)

[3] de Vooys, A.C.A., Koper, M.T.M., van Santen, R.A., et al.: Electrochim. Acta 43, 395–400 (1988)

[4] de Vooys, A.C.A., Mrozek, M.F., Koper, M.T.M., et al.: Electrochem. Comm. 3(6), 293–299 (2001)

[5] Ji, X.B., Banks, C.E., Compton, R.G.: Analyst 130(10), 1345–1360 (2005)

[6] Wang, Y.D., Wu, X.H., Su, Q., et al.: Solid-State Electronics 45(2), 347–348 (2001)

[7] Campbell, N.A., Reece, J.B.: Biology, pp. 15–30. Pearson Education Inc. (2002)

[8] Kowalchuk, G.A., Stephen, J.R.: Ammonia-oxidizing bacteria: amodel for molecular microbial ecology. Ann. Rev. Microbiol. 55, 485–529 (2001)

[9] Warneck, P.: Chemistry of the Natural Atmosphere, pp. 101–130. Academic Press Inc. (1998)

[10] Wallgren, K., Sotiropoulos, S.: Sensors and Actuators B: Chem. 60(2/3), 174 (1999); Yagodina, O.V., Nikolskaya, E.B.: Analytica Chim. Acta 385(1-3), 137 (1999)

[11] Philip, W., et al.: Environmental sensor networks in ecological research. New Phytologist (182), 589–607 (2009)

[12] Xu, J., Hua, K., Wang, Y.: Research of ammonia sensor. Journal of Zhengzhou Institute of Light Industry (Natural Science Edition) 19(4), 39–42 (2004)

[13] Liu, Q., Zou, Y.Q., Xing, H.Y.: Measuring Instrument of Dissolved Oxygen Based on MSP430. Instrument Technique and Sensor (9), 33–35 (2009)

[14] Milka, T.N., Nikolova, V., Petrov, V.: New real-time analytical applications of electrochemical quartz crystal microbalance: Stoichiometry and phase composition monitoring of electrodeposited thin chalcogenide films. Analytica Chimica Acta 573, 34–40 (2006)

[15] Liu, X.W., et al.: Nano-structured manganese oxide as a cathodic catalyst for enhanced oxygen reduction in a microbial fuel cell fed with a synthetic wastewater. Water Research 44(18), 5298–5305 (2010)

Research on the Knowledge Based Parameterized CAD System of Wheat and Rice Combine Chassis

Xingzhen Xu[1], Shuangxi Liu[2], Weishi Cao[1], Peng Fa[1], Xianxi Liu[2], and Jinxing Wang[2,*]

[1] Shandong Provincial Key Laboratory of Horticultural Machineries and Equipments, Shandong Agricultural University, Taian 271018, China
[2] College of Mechanical and Electronic Engineering, Shandong Agricultural University, Taian 271018, China

Abstract. In this paper, through using Pro/Toolkit to do secondary development of Pro/E, it completes the parametric modeling of every key parts of rice and wheat combined harvester chassis by using the object-oriented technology, and combines with Microsoft Access database to store the relevant knowledge. Meanwhile, it collects lots of related knowledge about rice and wheat combine harvester chassis from various channels and establishes a knowledge base of key components of the chassis after sorting them which can achieve the purposes of rapid designing chassis.

Keywords: Knowledge base, Agricultural machinery chassis, Parametric modeling, Secondary development.

1   Introduction

In recent years, product development gradually transformed from the data intensive to knowledge intensive which will make the CAD system develop towards digitization, integration and intelligent direction. [1] As a creative and complex activity, product design needs a lot of knowledge, a large number of historical data query, parts typical structure, process plan, evaluation results of manufacture and test, calculation parameters and performance test parameters. The knowledge is scattered storage in each designers' minds, drawer and archives which can not be merged into geometric model in traditional CAD system. So it is impossible to realize the reuse of knowledge-based resources that reduces design efficiency greatly. [2]

Knowledge-Based Engineering (KBE) technology is a new intelligent design method towards modern design requirements. [3] With the technology of parametric design based on knowledge we can greatly improve the design speed, improve product design quality reduce the research cost and shorten the development period. The knowledge base is both the foundation of applying all kinds of design knowledge and also the key technology which can realize the knowledge based CAD system. [4]

---

[*] Corresponding author.

D. Li and Y. Chen (Eds.): CCTA 2013, Part II, IFIP AICT 420, pp. 461–468, 2014.

The knowledge base supports rapid design, knowledge reuse and sharing in the process of product development. By using the object-oriented technology to realize the parametric modeling of every key parts of rice and wheat combined harvester chassis, using Pro/Toolkit to realize the secondary development of Pro/E and combining with Microsoft Access database to store the relevant knowledge, this paper designs the knowledge based parameterized CAD system of wheat and rice combine chassis.

2   The General Structure of Chassis Parameterized CAD System

2.1   The Main Idea of Parametric Design

In the process of the traditional design, geometric model of parts is fixed size, so it is complex to modify the shape of parts that even a small detail modifications need re-drawing. The product structure is determined in the process of parametric design. It designs different specifications of the products according to different structure parameters which are determinate by specific conditions and specific parameters. Its basic task is to replace the formal parameters of original graphics with the one concrete structure parameters that accord with the graphics we should design. Its concrete structure parameter is in correlation with specific products.

The basic process of the program parameters design includes that: creating original graphics, determining the graphic parameters, determining the relation between the original graphics parameters and the specific structure parameters by professional knowledge, completing the design drawings and related documents. The whole process requires a database and database management system to storage and management the various kinds of data and graphics. By using the technology of parametric design can easily modify graphics. And designers can also inherit the experience and knowledge of past design. Then designers can concentrate on creative concept and overall design to give full play to creativity and improve the efficiency of design.

2.2   System Synthesis

The system consists of three modules, as shown in Figure 1. They are the parametric module of Pro/E, the parametric module based on knowledge, database management module of Access database. The Pro/E module includes the creation of three-dimensional model and the implementation of model parameter-driven. The parametric module based on knowledge includes the parametric design verification module, product examples and specific model, the knowledge of parameters variation and parts constraints. The database management module of Access database includes system authority management, model base management and database management.

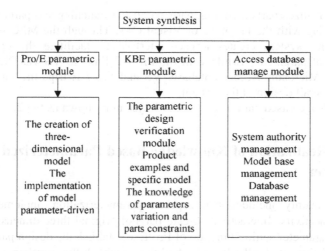

Fig. 1. Parameterized CAD system of wheat and rice combine chassis

2.3   The System Overall Design Concept

Create three-dimensional models by interactive mode and build design parameter under Pro/E. Then retrieve the design parameters of models and provide editing functions of parameters and regeneration functions of the new three-dimensional model according to the new design parameter.

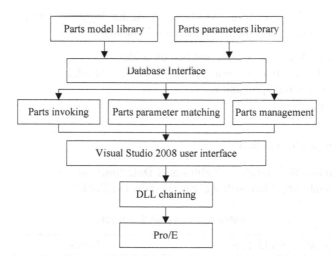

Fig. 2. The knowledge based parameterized CAD system

By using Microsoft Access as the database, parts model library and parts parameters library which store two-dimensional models, three-dimensional models and parts related knowledge, parameter respectively are established in this system. The database connects to Visual Studio 2008(VS) by the corresponding database interface and realizes parts invoking, parts parameter matching and parts management by programming with the language of Visual C++. Through the MFC programming under VS, this system creates a user interface to facilitate the operation and management of parts. Meanwhile with the Pro / E links to VS via DLL mode, we complete the Microsoft Access database, VS and Pro / E tripartite interaction and achieve the overall design of the system.

The knowledge based parameterized CAD system is shown as Fig.2.

3   The Realization of Knowledge Based Parameterized CAD System

The system mainly includes two aspects: parameterization implementation and database connectivity. Interactively create parts of original three-dimensional model, and use the parameter setting function of Pro/E to establish the design parameters and dimension relations under the Pro / E. And then establish the correlative forms in the Microsoft Access relational database. Using Visual C++ to map a Crecordset class object for interactive, then retrieve the design parameters of the model by the Pro/Toolkit application, finally input parameters and regenerate the model according to the data source object. The Access database and Pro/E is connected via VS2008 platform.

3.1   The Realization of Parametric System

Using Pro/Toolkit procedure development technology to carry on secondary development in Pro/E, this method consists of four steps:

1. Write source files (Source code and resource files)
2. Compile and link, create the executable file (DLL or EXE)
3. Register Pro/Toolkit application
4. Run the application

3.2   Database Connection

3.2.1   Establish the Database Tables and Data Sources
Design a data table by Microsoft Access, as shown in Table1.

Table 1. The model file directory

Field Name	Field Type	Notes
Number	AutoNumber	As the primary key, documents identification
File Name	text	File name, including the extension operator
Path	text	The storage path of documents

After design data tables, create a new data source DB in Administrative Tools under the menu of Control Panel, and connect to the database created. The process is shown in Figure 3.

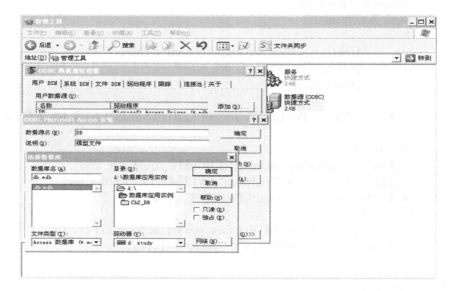

Fig. 3. The connection process of data source

3.2.2   Create MFC Program and Design Dialog

Build a new MFC DLL program under VS, complete the basic setup and add Pro/Toolkit header files and initialization function user_initialize () and termination functions user_terminate () in the item master file.

Add dialog resources in the menu of resources under VS. And add the appropriate button controls in the dialog box, complete ID and attribute settings. After finish designing dialog interface, double-click interface to generate CDBDlg dialog class.

3.2.3   Set Up the Data Table Class

In this paper, we connect databases by setting up the data table class using ODBC. Using CRecordset class to bind the data table, click [Add Class] under the menu [Project] in VS, select the MFC ODBC user and then choose the added data source DB, define the name of data class and complete the work of adding class.

3.2.4   The Key Program Code of Connection

Through the MFC programming, database is connected to Pro/E and the dialog function is completed, the key part of the code is as follows:

```
BOOL CDBDlg::OnInitDialog()
{
      CDialog::OnInitDialog();
      //connect to database
      if(!m_Set.Open())
      {
            AfxMessageBox("Database connection failed ! ");
            SendMessage(WM_CLOSE,0,0);
            return FALSE;
      }
      m_List.SetExtendedStyle(LVS_EX_FULLROWSELECT|LVS_EX_GRIDL
INES);
      //insert column
      m_List.InsertColumn(0,"Number",LVCFMT_LEFT,50);
      m_List.InsertColumn(1,"FileName",LVCFMT_LEFT,80);
      m_List.InsertColumn(2,"Path",LVCFMT_LEFT,150);
   ShowList();
      return TRUE;   // return TRUE unless you set the focus to a control
      // unusual:  OCX Property page should return FALSE
```

3.3   Application Example

3.3.1   The System Interface

This paper designs the knowledge based parameterized CAD system of wheat and rice combine chassis and realize the Microsoft Access database, VS2008 and Pro/E tripartite interaction. A screenshot of system interface is shown in Figure 4.

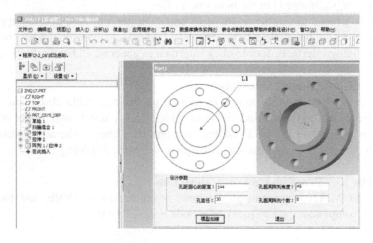

Fig. 4. The system interface

3.3.2   The Knowledge Database

This paper collects lots of related knowledge about rice and wheat combine harvester chassis from various channels and establishes a knowledge base of key components of the chassis after sorting it. The knowledge database consists of case library, rule library, parameter library and material library. Part of the knowledge of wheat and rice combine chassis is shown in figure 5.

参数ID	参数名	参数代号	参数单位	参数说明
1001	发动机额定功率	P	kW	此值为整机所提供给传动系的设计要求，有些功率用马力作单位，1马力=
1002	发动机的额定转速	ne	r/min	(null)
1003	最大行走速度	Vm	km/h	联合收割机一般前进速度范围为V=1～20kM/h、倒挡的速度范围是V倒=2.
1004	最大负重	G	N	整机最大质量*重力加速度g
1005	最大爬坡度	α	度（°）	一般要求车辆爬坡度30%，即20°左右。
1006	驱动轮动力半径	R	m	驱动轮动力半径，依据《拖拉机设计手册》，可按轮胎新胎外半径乘0.93
1007	最大收获率	ξmax	hm^2/h	收获效率指纯小时生产率（hm^2/h） 1hm^2=10000m^2,1公顷为15亩

公式ID	公式名称	公式结论代号	公式体	公式结论单位	说明
1001	最小作业速度	Vmin	ξmin*1000...	km/h	ξ_min是纯工作小时生产率最小值，单位是hm^2/h
1002	最大作业速度	Vmax	ξmax*1000...	km/h	ξ_max是纯工作小时生产率最大值，单位是hm^2/h
1003	行走底盘传...	Ped	G*Vmax*f/...	kW	本公式计算行走底盘所需的最大功率，G*f得到整机
1004	行走底盘传...	Ted	9550*Ped/ne	N·M	确定了行走底盘传动系的最大功率和相应的转速，ξ
1005	总的最大传...	imaxx1	G*R*(f*co...1		最低速时，传动系传到驱动轮上的扭矩应能够驱动
1006	总的最大传...	imaxx2	120*ne*π*...1		传动系总的最大传动比应能够使车辆速度降到联合
1007	总的最大传...	imaxs1	G*0.8*φ*R...1		根据驱动轮与地面的附着情况，确定最大传动比的值

Fig. 5. Combine working parameters and engine parameters

4   Conclusions

Based on above studies, we develop a rice and wheat combine harvester chassis parametric knowledge base. The implementation of the system adopts four basic modules: the user interface module using VC++6.0 to develop with VS2008 as a platform, 3D model library module created by Pro/E, Pro/Toolkit program modules to achieve Pro/E system menu loading and parametric, knowledge base module in which the Microsoft Access as the database.

Acknowledgment. Funds for this research were provided by the National Science and Technology Plan Projects, the major projects foster Agricultural machinery professional chassis digital design and complex piece of lean manufacturing (2011BAD20B01).

References

1. Chi, W., Gang, G., Fu-an, T., Binhui, Y., Guo, W.: Knowledge-based Construction of Knowledge Base of Parametric Design System for Steering Gears. Mechanical Engineering & Automation 48(3), 1–7 (2008)
2. Li, Z.: Chen Zhiying The application of Knowledge based Engineering on CAD. Mechanical Manufacture 45(6), 1–3 (2007)
3. Zhi, L., Xianlong, J., Huaiyu, J., Xiaowei, Z.: The Knowledge Representation and Reuse in Product Design. Journal of Shanghai Jiaotong University 40(7), 1184–1186 (2006)

4. Zhongtu, L.: Research on the Key Issues of Knowledge based CAD System, pp. 142–271. Huazhong University of Science & Technology Wuhan (2005)
5. Hao, Z., Wenhua, Z., Peng, C.: Research and design of mechanical standardized parts based on knowledge engineering. Manufacturing Automation 33(10), 71–78 (2011)
6. Jinju, C., Deyu, W., Lijuan, X.: MA Chong MA Chong Mid -ship Section Structural Design and Optimization Based on Knowledge Based Engineering. Journal of Shanghai Jiaotong University 46(3), 368–373 (2012)
7. Kitakura, Y., Nakajima, H., Yamamoto, K., et al.: Remodeling the combine harvesterfor the adaptive use in the harvesting buckwheat in early stage. Bulletin of the Fukui Agricultural Experiment Station 45, 24–34 (2008)
8. Perng, D.A., Chang, C.F.: A new feature-based design system with dynamic editing. Computers & Industrial Engineering 32(2), 383–397 (1997)
9. Angele, J., Fensel, D., Landes, D., Studer, R.: Developing Knowledge-Based Systems with MIKE. Journal of Automated Software Engineering 5(4), 389–418 (1998)
10. Kitamura, Y., Mizoguchi, R.: Ontology-based description of functional design knowledge and its use in a functional way server. Expert Systems with Application 24(2), 153–166 (2003)
11. Cristiano, J.J., Liker, J.K., White, C.C.: Customer-driven product development through quality function deployment in the U.S. and Japan. Journal of Product Innovation Management 17, 286–308 (2000)
12. Calkins, D.: Learning all about knowledge based engineering. Intelligence (1996)
13. Zhiyong, X., Mingzhong, Y.: Research on Integration of product innovation design method based on knowledge engineering. Journal of Wuhan University of Technology (Information & Management Engineering) 29(6), 110–113, 147 (2007)
14. Shangshang, Z., Yunhe, P., Shijian, L., Yueting, Z.: Research on Product Innovative Design Technology Based on Knowledge. China Mechanical Engineering 13(4), 337–340 (2002)

An Intelligent Search Engine
for Agricultural Disease Prescription

Weijian Ni, Mei Liu, Qingtian Zeng[*], and Tong Liu

Shandong University of Science and Technology Qingdao,
Shandong Province, 266510 P.R. China
niweijian@gmail.com

Abstract. Generic search engines have played a significant role in helping people locate their needed information on the web. However, they don't perform as desired on domain-specific queries. In this paper, we focus on the domain of agriculture and develop a novel search engine specifically for agricultural disease prescription retrieval. In order to improve the performance of search for prescription documents, we exploit the domain-specific characteristics embedded in agricultural disease prescription, and propose a domain-specific query expansion approach as well as a BM25-based structural retrieval function. An intelligent search engine for agricultural disease prescription is then implemented based on the proposed retrieval model. User interfaces of the developed search engine are demonstrated.

Keywords: Search Engine, Agricultural Disease Prescription, Query Expansion, Information Retrieval.

1   Introduction

A document of agricultural disease prescription generally covers multifaceted information about some agricultural disease, including symptom, transmission route, control method and etc. Given an appropriate piece of agricultural disease prescription, farmers are likely to know about the treatment plans, causes and other aspects of the agricultural disease he was encountering. Due to massive amounts of agricultural disease prescriptions on the web, search engine becomes an indispensable tool for farmers when looking for the right treatment plan.

However, providing farmers with agricultural disease prescriptions of practical application is not a trivial task. As an example, a farmer may expect to know about *control methods* of an agricultural disease he encounters during plant cultivation, by typing a keyword query describing the *symptom* to search engines; nevertheless, a document containing extended yet repeated information about the input symptom is more likely to be returned than that containing information about control method. That is partly because most traditional information retrieval models are based on keyword-match between user's query and candidate documents. However, when searching for

---

[*] Corresponding author.

D. Li and Y. Chen (Eds.): CCTA 2013, Part II, IFIP AICT 420, pp. 469–477, 2014.
© IFIP International Federation for Information Processing 2014

agricultural disease prescription documents, user's intent is not only embedded in the literal denotation of the query, but also dependent on the semantics associated with the initial query which is hard to be captured in the keyword-match retrieval schema.

Table 1. An example of agricultural disease prescription document

芹菜病毒病（Celery Virus Disease）
一、症状（I. Symptom）
全株染病。初叶片皱缩，呈现浓、淡绿色斑驳或黄色斑块，表现为明显的黄斑花叶，严重时，全株叶片皱缩不长或黄化、矮缩……
二、病原（II. Etiology）
由黄瓜花叶病毒（CMV）和芹菜花叶病毒（CeMV）侵染引起。两种病原引起的叶症状相似。芹菜花叶病毒（CeMV）粒体线形，寄主范围窄，主要侵染菊科、藜科、茄科中几种植物，病毒汁液稀释限点100～1000倍，钝化……
三、传播途径（III. Transmission route）
CMV和CeMV田间主要通过蚜虫传播，或通过人工操作接触磨擦传毒……
四、发病条件（IV. Epidemic factor）
栽培管理条件差，干旱、蚜虫多发病重……
五、防治方法（V. Control method）
主要采取防蚜、避蚜措施进行防治。其次是加强水肥管理，提高植株抗病能力，以减轻为害。其他方法参见番茄病毒病……

Compared with plain-text documents, an essential difference of agricultural disease prescription documents is that the passages therein typically play a clear role in describing different aspects of the corresponding agricultural disease, such as symptom and control method. The roles may be either provided explicitly by document authors just as the example shown in Table 1, or implied in the content of text. In other words, agricultural disease prescription document is essentially a type of structured document in which passages with their roles make up of the fields or components. Obviously, the performance of prescription retrieval would be promoted if the document structure embedded in agricultural disease prescriptions was exploited in a principled way.

Based on the above observation, we propose a novel structural retrieval model and further develop an intelligent search engine for agricultural disease prescriptions. Specifically, we propose a field probabilistic model and a field associative model to formulate the structure information embedded in prescription documents, and incorporate the document structure into the prescription retrieval process through domain-specific query expansion and structured BM25 retrieval function. The domain-specific search engine is then implemented using Apache Lucene toolkit [1].

2   Related Work

2.1   Information Retrieval

Information Retrieval (IR) has gained substantial research attentions due to massive amount of data emerging, especially on the web [2]. Traditional IR models such as vector space model, language model, probabilistic model and learning-to-rank [3], have become one of core functions of modern search engines. A key to IR models is to score a document against a user's query through evaluating how the document fits user's information need.

If documents have structure more than plain-text, it would be hard to incorporate the structure information into traditional IR models because they are mostly developed for plain-text documents. Therefore, a number of retrieval models scoring documents through considering structure evidences, referred to as Structural Retrieval, have been proposed. A basic idea of structural retrieval is to combine the scores calculated on each components within the document using traditional score functions [4,5]. More recently, structural retrieval models have found wide applications in job-resume matching [6], question answering [7] and etc. However, to our knowledge, no work exists on adopting structural retrieval model for the structured agricultural disease prescription documents.

2.2   Agriculture-specific Search Engine

According to the statistics, by the end of 2009, there had been more than 30,000 agricultural web sites on the Internet, which cover heterogeneous agricultural information about market trends, agricultural technologies and etc. Since generic search engines usually don't incorporate characteristics of agriculture domain, it is necessary to develop agriculture-specific search engines to help farmer acquiring their interested information from the Internet. Key techniques of agricultural IR have been widely studied recently. As example, Huang [8] proposed an agriculture deep web entry discovery algorithm to acquire agriculture-specific deep web resources; Zhou [9] focused on filtering the agriculture unrelated topic pages based on URL and content during web pages crawling.

Several Chinese agriculture-specific search engines have been put into production, e.g. AgriSou [10], Sounong [11] and Agr365 [12]. There are also numbers of world wide agriculture-specific search engines: AgNIC (Agriculture Network Information Collaborative) [13] is a knowledge discovery system and collaborative platform for agricultural resource interchanging and searching supported by U.S. national agricultural library; Agriscape [14] is an online directory on agriculture related information and provides a simple query interface.

Most of the existing agriculture-specific search engines aim to retrieve various types of agricultural information such as farm price and agricultural news, which is so heterogeneous that embodies little common domain-specific characteristics; while we aim to develop an agriculture-specific search engine especially for one type of agricultural information, i.e., agricultural disease prescription. Furthermore, the agriculture-specific search engine in the paper is developed based on a novel structural retrieval model instead of tradition IR models.

3   Architecture of Prescription Search Engine

Figure 1 shows the architecture diagram of the developed agricultural disease prescription search engine based on structural retrieval. There are mainly two components: structure information model and structural retrieval function. The former aims to derive the structure information embedded in prescription documents that would contribute to retrieval accuracy; the latter then provides users with prescription documents with high relevance w.r.t. users' queries.

In the following section, we will present the proposed structural retrieval model in details. Formally, let $C = \{D_1, \cdots, D_n\}$ be a prescription collection composed of n documents and assume every passages in a prescription document are tagged with one (or possibly a few) field f from a given field set $F = \{f_1, \cdots, f_k\}$.

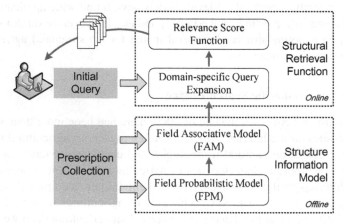

Fig. 1. Architecture diagram of the developed search engine

4   Structural Retrieval Model for Prescription Search

4.1   Structure Information Model

Field Probabilistic Model
Due to the different semantics of prescription fields, the affinities of a term with different fields are not necessarily always equal. For example, the terms *yellow* and *patches* would give a stronger hint on the *symptom* than other aspects of a disease, so it is necessary to pay emphasis on the *symptom* field when searching for the query term *yellow patches*. Therefore, we propose a Field Probabilistic Model (FPM) to infer field affinity for each term.

A FPM is a multinomial distribution $p(f|w)$ that attempts to capture the probability of a field f when seeing a given term w. By Bayes' rule, the posterior probability

$p(f|w)$ can be computed by combining the prior probability of each field and the conditional term likelihood. That is,

$$p(f|w) = \frac{p(w|f)p(f)}{p(w)} = \frac{p(w|f)p(f)}{\sum_{f' \in F} p(w|f')p(f')} \tag{1}$$

where $p(f)$ is the prior probability of field f observed in prescription documents and $p(w|f)$ is the probability of seeing a term w in a given field f. We estimate the both probabilities by collection statistics of a given prescription collection C.

Field Associative Model

In agricultural disease prescription documents, there are often potential associations among field terms. Taking agricultural diseases with the symptom *oozing black liquid* as an example, if the control methods containing the terms *quarantine* and *spraying suspension agent* were found to be more effectual than that containing the term *removing weeds* in related prescription documents, the documents with the term *oozing black liquid* in the *symptom* field, as well as the term *quarantine* or *spraying suspension agent* in the *control method* field, would be preferred for the query *oozing black liquid*. Therefore, we propose a Field Associative Model (FAM) to discover these strong inter-term associations.

A FAM is a set of association rules [15] each of which is an implication of the form $R: A \Rightarrow B$, where both the antecedent and consequent are sets of terms. The utility of each rule is measured by two metrics, namely *support* and *confidence*.

Support of a rule is defined as in a given prescription collection the percentage of documents where every terms in the rule are seen. It can be simply computed as the term set probability: $sup(A \Rightarrow B) = p(A \cup B)$. *Confidence* of a rule is the probability of seeing terms in consequent in the prescription set where the terms in antecedent appear, i.e., $conf(A \Rightarrow B) = p(B|A)$.

Intuitively, high *support* of a rule implies that the rule covers a considerable part of the prescription collection, while high *confidence* implies that the terms in consequent are highly likely to appear if the terms in antecedent are seen. Therefore, in FAM, only the rules whose *support* and *confidence* satisfies minimum thresholds could be considered meaningful. Practically, we view a prescription document with its composed terms as a piece of transaction data, and employ FP-growth [16], a frequent-itemsets mining algorithm with high time and space efficiency, to mine FAM-rule candidate from the given prescription collection.

4.2   Structural Retrieval Function

Based on the above structure information models, we propose a prescription retrieval mechanism to provide users with proper prescription documents. In particular, we first propose a domain-specific query expansion approach to capture user's actual query intent, and then propose a score function that calculate relevance of prescription documents w.r.t. the expanded user's query.

Domain-specific Query Expansion

Instead of employing traditional query expansion approach in generic IR, we leverage the derived structure information about agricultural disease prescriptions to uncover possible query intents behind query terms. The idea is to view FAM-rules as a prescription knowledge repository composed of informational associations among term sets, and expand the query terms in antecedent of a FAM-rule by the terms in its corresponding consequent.

Formally, given a set of FAM-rules $\mathcal{R} = \{(R_1: A_1 \Rightarrow B_1), \cdots, (R_m: A_m \Rightarrow B_m)\}$ and an initial user's query composed of t terms: $Q = \{q_1, \cdots, q_t\}$, the expanded user query takes the following form:

$$Q' = Q \cup Q_E \tag{2}$$

where Q_E is the set of expanded query terms and

$$Q_E = \{q' \mid \forall R_i \in \mathcal{R}, \exists q_j \in Q, q_j \in A_i \wedge q' \in B_i\} \tag{3}$$

Note that the terms q in the expanded user query Q' are assigned with different weights w_q according to the expanding confidence. Generally, for the terms $q \in Q, w_q = 1$; whereas for the terms $q \in Q_E$,

$$w_q = \frac{\sum_{R \in \mathcal{R}_q} conf(R)}{|\mathcal{R}_q|} \tag{4}$$

where $\mathcal{R}_q$ is the set of FAM-rules according to which the term q is expanded and $conf(R)$ is *confidence* of a given FAM-rule R.

As mentioned above, the field affinity of each term varies among prescription fields, which is depicted in FPM, we further structurally expand user's query using FPM to derive the underlying structure of users' search intent.

Formally, given a set of prescription fields $F = \{f_1, \cdots, f_k\}$ and a query $Q' = \{q'_1, \cdots, q'_s\}$ expanded using FAM, the query further expanded using FPM can be written as:

$$Q'' = \begin{matrix} & \begin{matrix} f_1 & \cdots & f_k \end{matrix} \\ \begin{matrix} q'_1 \\ \vdots \\ q'_s \end{matrix} & \begin{pmatrix} w_{1,1} & \cdots & w_{1,k} \\ \vdots & \ddots & \vdots \\ w_{s,1} & \cdots & w_{s,k} \end{pmatrix} \end{matrix} \tag{5}$$

where each element $w_{i,j}$ in the matrix indicates query intensity of the expanded query term q'_i on the field f_j, and $w_{i,j} = w_{t_i} \cdot p(f_j|t_i)$.

Prescription Score Function

After expanded using FAM and FPM, user's query will be structurally represented as a matrix. It makes most traditional IR models difficult to be leveraged directly. In the paper, we extend the traditional BM25 model [17] to deal with the structured query in a natural way. The traditional score function of BM25 model is shown as follows:

$$score(D, Q) = \sum_{q \in Q} \frac{(k_1 + 1) \cdot tf(q, D)}{k_1 \cdot \left(1 - b + b \cdot \frac{|D|}{avgdl}\right) + tf(q, D)} \cdot \log \frac{N - df(q) + 0.5}{df(q) + 0.5} \quad (6)$$

where tf and df are the term frequency function and the document frequency function, respectively. k_1 and b are free parameters.

However, for structured documents and queries, it would not be adequate to calculate term frequency by simply counting occurrence of a term in a document. Thus we adapt the score function in BM25 model by revising the term frequency function. The idea is that an occurrence of some query term in a field is weighted by the field weight of that term. Assume that a prescription document is segmented according to prescription fields, i.e., $D = \{D_{f_1}, \cdots, D_{f_k}\}$, and each field of document D_{f_j} is viewed as unstructured text, term frequency of a term q'_i ($1 \leq i \leq s$) in the structured query Q'' is calculated as:

$$tf_{struct}(q'_i, D) = \sum_{j=1,\cdots,k} tf(q'_i, D) \cdot w_{ij} \quad (7)$$

Then, we substitute Equation 7 into Equation 6 and get the relevance score function for agricultural disease prescription retrieval.

5   Implementation

A search engine for agricultural disease prescription was developed based on the proposed structural retrieval model. The indexing and retrieval function are implemented using Apache Lucene toolkit [1]. To construct prescription database for the search engine, we crawled web pages about agricultural diseases from several agricultural websites. For the pages in each individual website, a set of rules were manually designed to extract the text of prescription and to segment the text according to predefined prescription fields in {*Symptom, Etiology, Pathogenesis, Infection_Way, Control_Method*}. In total, there are 7903 prescription documents indexed in the developed search engine.

Figure 2 shows the query interface of the developed search engine. The user can type his query in the top box on the view. Our search engine also supports facet search and each of the facets is a prescription field. Below the query box, the user can find the retrieved prescription documents ranked by the relevance score. Unlike generic search engines that return users with excerpts of relevant webpages, each result of the developed search engine lists excerpts of every prescription fields and a photo of that disease (if any exist). When the user clicks on the titles of retrieved results, the details of the corresponding agricultural disease prescription will be displayed in a pop-up window as show in Figure 3. All the information about the selected agricultural disease is organized by the prescription fields. A navigation menu on the top-right side of the view is designed to help user browse details of the prescription document. We can roughly say that, through using our system, it would be much easier for farmers to locate their needed information in agricultural disease prescriptions.

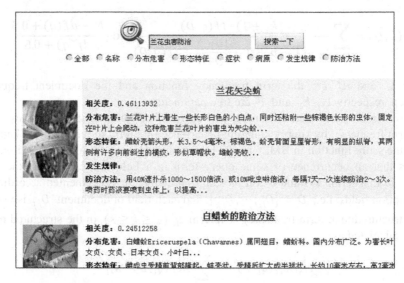

Fig. 2. Query interface of prescription search engine

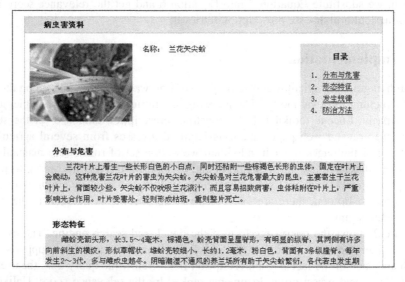

Fig. 3. Detail display interface of prescription search engine

6   Conclusion and Further Work

In this paper, we developed a domain-specific search engine for agricultural disease prescription. To exploit the domain characteristic of agricultural disease prescription, we based the search engine on a novel structural retrieval model. The proposed retrieval model includes modeling structure information embedded in prescription documents, structurally expanding user's query and a structural retrieval function. We constructed a real-world prescription collection and implemented the search engine using Apache Lucene toolkit. The search results are structurally organized in the user interface to facilitate user's information needs.

As a primary effort in domain-specific search engine for agricultural disease prescription, there is still much room for improvement. One direct and effective way to improve retrieval result is to enlarge the indexed prescription collection. In this way, more potential relevant prescription would be included in retrieval results. Besides, instead of BM25, we will leverage other recently proposed IR models, e.g. learning to rank, in this IR task.

Acknowledgments. This paper is supported partly by National Natural Science Foundation of China(No. 61170079 and 61202152), Excellent Young Scientist Foundation of Shandong Province (No. BS2012DX030), Higher Educational Science and Technology Program of Shandong Province (No. J12LN45), Postdoctoral Science Foundation of China (No. 2012M521363), National Statistical Science Foundation of China (No. 2012LY001), Special Fund for Agro-scientific Research in the Public Interest (No. 201303107), Special Fund for Fast Sharing of Science Paper in Net Era by CSTD (No. 2012107) and Sci. & Tech. Development Fund of Qingdao(13-1-4-153-jch).

References

1. Apache Lucene (2013), http://lucene.apache.org/
2. Manning, C.D., Raghavan, P., Schütze, H.: Introduction toInformation Retrieval. Cambridge University Press (2008)
3. Liu, T.-Y.: Learning to Rank for Information Retrieval. Springer (2011)
4. Wilkinson, R.: Effective retrieval of structured documents. In: Proceedings of SIGIR, pp. 311–317 (1994)
5. Ogilvie, P., Callan, J.: Combining document representations for known-itemsearch. In: Proceedings of SIGIR, pp. 143–150 (2003)
6. Yi, X., Allan, J., Croft, W.B.: Matching resumes and jobs based on relevance models. In: Proceedings of SIGIR, pp. 809–810 (2007)
7. Zhao, L., Callan, J.: Effective and Efficient Structured Retrieval. In: Proceedingsof CIKM, pp. 1573–1576 (2009)
8. Huang, H.: Complex Adaptive Agriculture Vertical Search Model and its Implementation. Dissertation: University of Science and Technology of China (2010)
9. Zhou, P.: Research on key techniques of agricultural search engine. MS Thesis: Capital Normal University, China (2009)
10. AgriSou (2013), http://www.agrisou.com/
11. Sounong (2013), http://www.sounong.net/
12. Agr365 (2013), http://so.ag365.com/
13. AgNIC (2013), http://www.agnic.org/
14. Agriscape (2013), http://www.agriscape.com/
15. Han, J., Kamber, M., Pei, J.: Data Mining: Concepts and Techniques, 3rd edn. Morgan Kaufmann, Massachusetts (2011)
16. Han, J., Pei, J., Yin, Y.: Mining frequent patterns without candidategeneration. In: Proceedings of SIGMOD, pp. 1–12 (2000)
17. Robertson, S.E., Walker, S., Hancock-Beaulieu, M.: Okapi atTREC-7. In: Proceedings of TREC, pp. 199–210 (1998)

Design of Animal Myocardial Contractile Force Detection System Based on Tissue Engineering

Guiqing Xi[1], Ke Han[2], Ming Zhao[2], Caojun Huang[1], and Feng Tan[1]

[1] College of Information and Technology Heilongjiang Bayi Agricultural University Daqing 163319, China
[2] Harbin University of Commerce, Haerbin 150028, China
xiguiqing@163.com

Abstract. This paper uses the sensor convert animal myocardial contractile force into a voltage signal, which is collect by a MCU acquisition, transferred to the computer, using the PC visualization language VB compiled a data acquisition and processing platform, collected data can be stored into a computer, and real-time curve of the data changes is plotted at any time. Through testing the tissue engineered cardiac tissue strips micro contractility, it was proved that the platform operation is of stable performance, accurate data acquisition, processing method effective, the system can also be applied in other similar signal detection and acquisition and other fields.

Keywords: tensile testing, data acquisition, tissue engineering, real-time plotting.

1   Introduction

Tissue engineering is a life science emerging in the 1980s, and the structure of tissue-engineered cardiac muscle tissue is an important aspect of tissue engineering. Now after mix-culture of the collagen and myocardial cells can obtain a consistent beat, the beating frequency coincide that of the natural growth of myocardial tissue. But myocardial beat generation of contractile force is also an important indicator tested in tissue-engineered myocardial performance, however the accurate detection of the contractile force method is still not mature enough.

The computer acquisition of the real-time analog signal has a relatively mature technology. The data collection and transmission mainly use the A/D data acquisition card and the existing communication protocols (such as RS232/485)[1].So how to utilize the existing high-level language to design data operation and processing platform with many functions is a key issue.

This article starts from the device of contractile force detection, and then designs the cantilever stretch detect structure. Using MCU collects force signals from the sensor and transmits to the computer. Based on the implementation of data acquisition, use VB language to design the myocardial contraction operation platform which is facilitate to the operation of the data acquisition and the processing function[2].

D. Li and Y. Chen (Eds.): CCTA 2013, Part II, IFIP AICT 420, pp. 478–485, 2014.

2   Experiments and Methods Myocardial Contractile Force Detecting Device

Simple mechanical contraction force stretching machinery should has the advantages of simple structure, good specimen tensile coaxial, convenient assembly and disassembly, and preload adjustment method is reliable, good experimental repeatability etc...According to the above requirements, this paper designs such cantilevered contractile tension detecting device as the Fig.1.

As shown in Fig.1, contractile force detection device is composed of several parts including the pedestal / bracket, force sensor, a pre-load nut and support cantilever beam etc...what's more the tension sensor with temperature compensation is the core part of the device, and the strain gauges were embedded in silicone adhesive and is isolated from the air, avoiding the external corrosion, all the other parts are made of stainless steel, thus the whole device can be placed in water, it is necessary for the measurement of myocardial strip contraction force, because tissue engineered cardiac tissue strips stays in culture liquid and if left exposed in the air, myocardial cells will die soon[3]. Near the bottom of the bracket in somewhere can be folded into 90 degrees so that make sure the myocardial strip in horizontal position, which enables the detection of myocardial placed in the culture dish containing a culture liquid container.

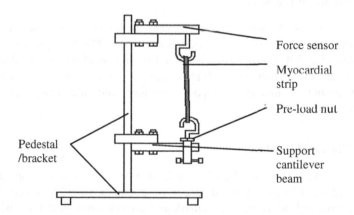

Fig. 1. Schematic of contractile force detecting device

Using the coaxial positioning principle make the sensor and the support cantilever beam are kept in the same plane, which are respectively provided with a hook, the hook through the special structure and maintain strict with the axis, it ensures that the stretching of the strict coaxiality, reduces the test principle error[4]. Through the pre-loaded nut to be measured myocardial strip preload in the device, then the weak beats of the myocardial strip can be detected by the sensor.

3   Computer Data Acquisition System

Force sensor detects the contractility signal through the sensor converted into a weak electrical signal, through the A/D converter for converting to the digital signal, using SCM acquisition, transmit to the computer ; for storing, printing, modification and analysis processing by the data acquisition and processing platform. Throughout the course of the basic frame diagram shown in Fig.2.

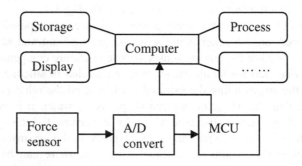

Fig. 2. Frame diagram of data acquisition and processing

3.1   The Choice of the Sensor

As a result of myocardial contractile force is very small so it needs to select the micro force measuring sensor, through the experiment choose LDWX-1 type micro tension sensor, it adopts the foil strain gauge affixed to the alloy steel sphere, is subjected to a tensile force, has the advantages of high accuracy, good stability, low temperature drift, good output symmetry, compact structure and measuring micro force characteristics.

3.2   The Choice of the ADC

ADC choose AD574, it is launched by the United States of America analog digital company (Analog), 12 bit high speed successive comparative A/D converter, a built-in bipolar circuit hybrid integrated remarkable conversion monolithic, with fewer external components, low power consumption, high precision, and has the advantages of automatic zero adjustment and automatic polarity conversion function, only need small external RC-pieces can constitute a complete A / D converter[5].

3.3   The Choice of the MCU

The MCU mainly completes the data collecting, processing, transmitting and receiving orders and other functions; it is the core of the system. As the entire core of the system controller is real-time at work, in order to meet the design requirements, which requires its power consumption must be very low, the speed should be quick and reliable performance. So the controller chooses low power MSP430 microcontroller, with its own unique strengths to reducing the chip's supply voltage and a flexible and controllable operating clock and other aspects have its one's own knack in.MSP430 series single chip power supply voltage ranges 1.8-3.6V. Thus could let the clocking at 1MHz operating conditions, chip current will be in about 200 ~ 400uA, clock shutdown mode with the lowest power only 0.1uA, other properties include the speed, reliability and other aspects are also consistent with the system requirements.

4 Host Computer Data Acquisition and Processing Platform Design

4.1 The Main Function of the System and the Basic Structure of the Procedure

Based on the VB language development of testing platform to realize the files reading and writing, data acquisition control and preliminary processing, acquisition of tension values in real-time curve drawing in the plane coordinate system; image recording, image redraw, coordinate axis size transform functions.

This platform program is designed according to several function module structure, simple flow chart as shown in Fig.3, which has a more detail description of the processing function module of the change curve of the image to redraw, the other modules are marked position in the process ,their specific design will be introduced step-by-step.

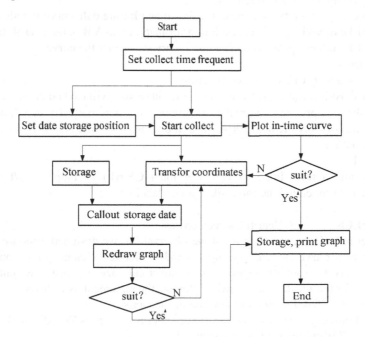

Fig. 3. System program

4.2   The Data Acquisition Module Program

VB program via the serial port control chip utilize the data sent from MCU, storage, and sent out acquisition command etc; before reading the signal, initialize acquisition channel as follows:

Mscomm1.Commport=2   Select COM2

Mscomm1.Settings="9600,N,8,2"   // Set the communication parameters

Mscomm1.Inputlen=0   //Read into all of the characters from buffer for receiving

Mscomm1.OutbufferSize=256   // Set the send buffer size

Mscomm1.InbufferSize=512   / / Sets the receive buffer size

Mscomm1.PortOpen=True   Open COM2

In order to eliminate the error caused by the interference signal in the acquisition process, we designed a continuous acquisition sampling five times for each signal, get the mean value of four significant figures, and so that we can t reduce the interference error to some extent[6].

4.3   Curve of Real-Time Plotting Module

Curve of real-time plotting module is the main function of this platform, including real-time display of curves and data collection, tracking and calculation of the value, the time interval between the peak of the window display and mouse click coordinates real-time display and other functions.

The following procedures achieve the real-time plotting data curve function:

Picture1.DrawStyle = 0' //Set resolution for mapping as VB default resolution

Picture1.Line ((m - pinl), yali)-(m, data), curvec   //Plot the curve

yali = data

Call draw_axis // Call draw axis function

If the acquisition time is set very long , the curve size will exceed computer screen, then we take the approach that make the screen curve automatically saved, then clear the screen and display the next screen data ,the achieve specific statement as follows:

If m >= x0 Then

k = k + 1

SavePicture Picture1.Image, Label5.Caption + CStr(k) + ".bmp"   //Full screen to save the picture, called the input file name + label

m = 0

Picture1.Cls   // Clear the screen redraw

In order to facilitate the data real-time observation, we designed a mouse click on any point of a curve, it will pop up a window marked change point coordinates, allowing users to roughly estimate the numerical size, in order to change the coordinates for the ideal figure drawing. Realize the statement as follows:

t = Format(x, "fixed   // Get the value of time axis

press = Format(y, "fixed")' //Get the value of pressure axis Coordinate display

Show'   // Window displays the value [7].

4.4   Coordinate Transformation and Redraw Graphics Module

The transformation of coordinates for adapting to the different size tension measurement value mapping, in the measurement of tissue engineered cardiac tissue strips that tension measurement value very small sample, we need a coordinate scale which is relatively concentrated, and so we use section method for drawing or redrawing curve.

Sensor measuring range is 0 ~ 5N, then we will divide coordinate axis into 3 types. one measuring range is more than 2N, its coordinates minimum scale is 0.5;a force measurement in the range of 0.2 ~ 2N, its scale is 0.1;the other is a force measurement range less than 0.2N, the corresponding coordinate minimum scale is 0.01.Start is the system default first coordinate mode[8]. In order to allow users to pay attention to selection of coordinate system, the system doesn't draw the coordinates, but in need of drawing graphics that it prompts the user to choose the coordinate system.

The main program calls the appropriate code to change the current drawing coordinate system, so as to redraw graphics, it can also change graphics that have been drawn .The following figure is a representation of the function of the specific effect. From the Fig.4 ~ 6 comparison, we can see in Fig.5, expressing data peak interval and familiar change trend more clearly, so using this method solves the appropriate display of graphics[9].

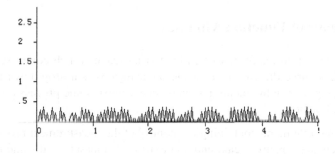

Fig. 4. Coordinate drawn in the range more than 2N data curves

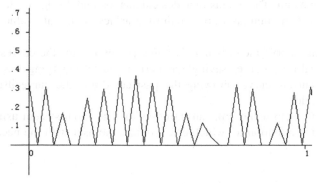

Fig. 5. Coordinate drawn in the range 0.2 ~ 2N data curves (abscissa elongated 5 times)

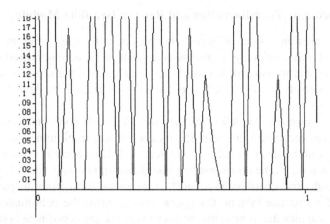

Fig. 6. Coordinate drawn in the range less than 0.2N data curves (abscissa elongated 5 times)

Of course, the graphics processing not only these features, but also including the display of the curve of the past, amplification, modify and so on, so the operation of the graphics module actually also has some of the features that of image viewing software.

5   Meaus of Function Modules

The data acquisition platform written by VB language, which comprises a very eye-catching data curve drawing area, while other regions is a drop-down menu and the corresponding icon in the menu, the menu can achieve some process control and data processing function.

(1)File menu: File menu containing the control of the measurement process, including measurement start, pause break, end and exit of the measure and some useful control functions; also including the initialization of measurement, setting sampling time and frequency of sampling.

(2)Processing menu: The menus includes calculation and display of data values of the maximum and minimum, as well as redraw graphics coordinate value input interface and so on.

(3)Curve menu: curve menu is mainly some processing for the curves, including the display of real-time curve, saving the current curve, enlarging the current curve, printing the current curve, showing the past curve and the curve itself and set the background color.

(4)Help menu: Help menu links an introduction text about system using method and matters needing attention to, which can provides the user with the necessary tips and help.

6   Conclusion

This paper introduces the application of VB and microcontroller used in the development of tissue engineering cardiac tissue contraction force measuring system. After the test results showing that the system has good stability, wide measuring range, data processing and high drawing function. This system has good portability, and can be applied to many similar systems. It provides a convenient operation, accurate data system in some experimental data testing for tissue engineering.

Acknowledgment. Funds for this research was provided by Science and technology research Projects by Educational Commission of Heilongjiang   Province of China (12531154).

References

1. Fang., W., et al.: Cardiac excitation contraction coupling in mammalian development in postnatal changes. Progress in Physiological Sciences 44(3), 227–232 (2013)
2. Lüs, L.S., He, W., et al.: Bioreactor cultivation enhances NTEB formation and differentiation of NTES cells into cardiomyocytes. Cloning and Stem Cells (2008)
3. Jian-sheng, W., Jian-meng, W.: Biomedical Signal Detection and Processing. Time Education (9), 6–7 (2011)
4. Lo, L., et al.: Design of Weak Signal Detection Based on MSP430G2452. Industrial Control Computer 26(5), 15–16 (2013)
5. Jianhua, W., et al.: Development of steel strand tension measurement sensor. Transducer and Microsystem Technologies 29(5), 83–86 (2010)
6. Guiyun, X., et al.: Development of high speed data acquisition system based on Vibration Analysis. Experimental Technology and Management 30(7), 47–50 (2013)
7. Kai, X., Jian, Z.: MSP430 series single-chip microcomputer system project design and practice, pp. 3–6. Mechanical Industry Press, Beijing (2009)
8. Yongqiang, S.: The realization way of the data acquisition in the VB environment. China Auto Industry (2004)
9. Yang, L.: Design of pressure sensor data acquisition system PC software based on VB. Mechanical Engineer (12), 56–58 (2012)
10. ZHanming, L., et al.: Design of data acquisition system based on VB and Advantech data acquisition card. Computer and Modernization (7), 236–238 (2012)

Analysis of Airflow Field of Toss Device of Yellow Corn Forage Harvester

Yan Huang[1], Manquan Zhao[1,*] and Hantao Liu[2]

[1] Mechanical and Electrical Engineering College of Inner Mongolia Agricultural University,
Hohhot 010018, China
[2] Food Science and Engineering College of Inner Mongolia Agricultural University,
Hohhot 010018, China
{wuxinglaozu,nmgzhaomq}@163.com, 28715369@qq.com

Abstract. Use a very important tool called ICEM CFD integrated in ANSYS Workbench to mesh the model of toss device and then apply the software FLUENT to simulate numerically and analyze the velocity distribution and pressure distribution, based on the RNG k-epsilon model. Numerical simulation results showed that air flow field and pressure distribution of toss device were asymmetry. Maximum wind speed of fan exit was 42.8 m/s, which met the actual needs. In a word, the design of toss device was reasonable. But there was the existence of the secondary flow in toss cylinder, making some gas couldn't flow smoothly, thus affecting delivery efficiency. Through analysis of the stress field in the blower, the results showed that static pressure of windward side blade increased from roots to ends and flow channels between fan blades in different positions showed different static pressure characteristics. The maximum static pressure of windward side blade was 878 pa, the minimum static pressure of lee side blade just passing the export of fan was -950 pa, the negative pressure meaning suction. The conclusions provided a reference for structural optimization and performance improvement of the toss device of Yellow Corn Forage Harvester .

Keywords: Forage harvester, ANSYS FLUENT, Air flow field, Numerical simulation.

1 Introduction

Crop straw resources in China were very rich, containing large amounts of corn straws[1-3]. 9HS-170 type Yellow Corn Forage Harvester was united harvest machine which was mainly used for sequential feeding, chopped, propulsion and toss loading corn stalk .

The working subassembly contained roll device, toss device, pressure device, as well as transmission system and propulsion system etc. Toss device was the core part of the machine. Its working principle was that the material delivered by the screw

---

* Corresponding author.

D. Li and Y. Chen (Eds.): CCTA 2013, Part II, IFIP AICT 420, pp. 486–491, 2014.

conveyor were threw along toss cylinder from exit under the action of high speed rotation blade. The main existing problems of toss device were big power consumption and low efficiency. In order to reduce the power consumption of toss device and improve the efficiency of toss. Scholars both at home and abroad did a number of studies[4-8].

In recent years, with the development of computational fluid dynamics, CFD technology replaced the classical fluid mechanics of some approximate calculation method and graphic method. Facing fluid flow within the fluid machinery problems of fan and pump, in the past mainly by means of the basic theoretical analysis and lots of physical model experiment, now mostly adopt the way of CFD. CFD technology has now reached the level of analyzing and solving complex problems such as 3 D viscous turbulent flow and vortex motion [9-11].

The paper used software of Soildworks to establish the three-dimensional model and used software of ICEM CFD to mesh toss devices by unstructured tetrahedral and evaluate the quality of the grid. Finally, use ANSYS FLUENT 12.1 to simulate numerically its internal flow field, so as to understand the distribution law of flow field and provide theoretical basis for optimization design personnel, making up for the deficiency of the traditional experiment.

2   The Mathematical Model

The choice of turbulence model had a great influence on the result when analyzing the air flow field of toss device. General numerical calculation methods were broadly divided into the following three categories: Direct Numerical Simulation, Reynolds Averaged Navier-Stokes, Large Eddy Simulation. Among two equation model K-epsilon Model included Standard k-epsilon Model, RNG k-epsilon Model, Realizable k-epsilon Model. Standard k-epsilon Model was stable and relatively accurate ,which was widely used in engineering application. RNG k - epsilon model was similar to the standard k-epsilon (SKE) model in the form. It made improvements in the following areas compared with the standard k – model [12]:

(1) RNG model added a condition to the epsilon equation, which effectively improved the accuracy.
(2) The model took the turbulent vortex into account and improved the accuracy in this respect.
(3) RNG theory for turbulent Prandtl number provided an analytical formula, however , Standard k - epsilon model used a constant that the user provided.
(4) Standard k - epsilon model was a model of high Reynolds number, RNG theory provided an analytical formula considering low Reynolds number flow viscosity.

These features made RNG k - epsilon model had higher reliability and accuracy than Standard k - epsilon model.

The paper selected RNG k-epsilon model, which had better performance in simulating complex flow such as jet impact, separated flow, secondary flow, rotating flow in order to understand movement regularity of air flow field of the toss device better.

3   Multiple Reference Frame and the Boundary Conditions

Multiple reference frame (MRF) model was used to solve periodic rotation in the case of a transient when analyzing air flow field of toss device[13]. Otherwise, set rotating area parameters in the software of Fluent, including motion type, which was the moving reference frame and the rotation axis position measured in the ICEM CFD. The other important conditions such as the rotation speed was 1270 rpm as well as defining the interface between static domain and rotating domain, the entrance boundary condition was pressure inlet, the exit boundary condition was pressure outlet, at the same time ,setting the wall of fan blades was movement. Its rotating speed was zero relative to the speed of the rotating field.

4   Physical Model and Numerical Calculation Method

Create and assemble components such as fan and shell and toss cylinder, etc. based on structure size of toss device and then handle and simplify model appropriately. The simple structure model of toss device was shown in Fig.1. After that the model was imported into the ICEM CFD in order to be meshed by unstructured tetrahedral grid. At the same time, set the static field and rotating field and the location of the inlet and outlet. The calculation method was Semi-Implicit Method for Pressure-Linked Equations (SIMPLE), interpolation method of the convection item was the First - Order Upwind, which was easy to converge[13].

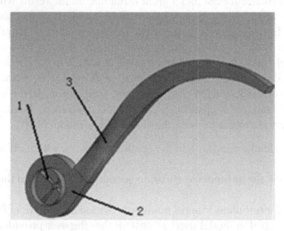

1.Blade   2. Fan shell   3.Toss cylinder

Fig. 1. 3D entity model of toss device

5   Numerical Simulation and Analysis

After 607 times of iterative computation, the results converged and its convergence precision was 1e-3.

As can be seen from the velocity vector plot (Fig.2), the wind speed near the end of the blade was between 35.7~42.8 m/s ,close to the theoretical calculation value of 40 m/s, indicating the model was correct and the data was reliability. In addition, when the blower working normally, the speed of the middle area was smaller than the edge and the maximum speed existed in the area between blade and shell was 42.8m/s, meeting the design requirement. Furthermore, as shown in Fig.3, there was gas reflux in the middle part of toss cylinder, effecting the toss efficiency. Optimizing personnel may appropriately change related parameters of toss device, making the effect of negative impact minimize as well as reducing noise.

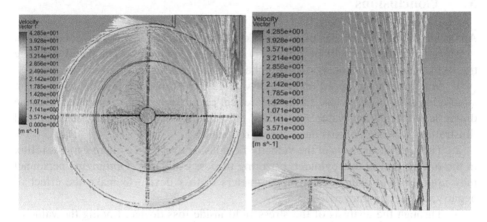

Fig. 2. Velocity vector plot of the blower   **Fig. 3.** Velocity vector plot of toss cylinder

As shown in Fig.4 and Fig.5, static pressure of windward side of blade increased from roots to ends , the maximum static pressure at the end of the windward side of blade surface was 878 pa. The minimum static pressure of lee side blade just passing the export of fan was -950 pa, the negative pressure meaning suction.

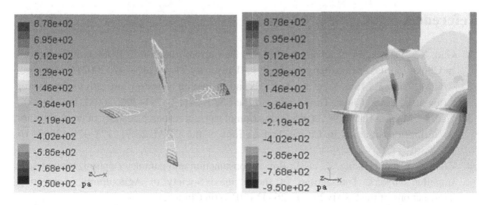

Fig. 4. Static pressure contour of blade   **Fig. 5.** Static pressure contour inside the blower

Flow channels between fan blades in different positions showed different static pressure characteristics. Static pressure of flow channels that were gradually far from the exit decreased and static pressure of flow channels that were gradually close to the exit increased. The distribution was stronger asymmetry. According to pressure distribution of blade, maximum stress concentrated at root and ends of blade ,so the structural at root and ends of blade needed to be strengthened in order to improve the service life.

6   Conclusions

Airflow Field of Toss Device of Yellow Corn Forage Harvester was very complicated.
    More information could be gotten by numerical simulation, compared with the traditional experimental analysis, using the software of ANSYS FLUENT :
    1   Maximum wind speed of fan exit was 42.8 m/s ,meeting design requirement, simulation results were close to the theoretical calculation results compared with the theoretical calculation value, verifying the validity of the model selection and illustrating the design of toss device were rationality.
    2   Through the analysis of the velocity field inside toss device, there was gas reflux in the middle part of toss cylinder, effecting the toss efficiency. Optimizing personnel may appropriately change relative parameters of toss device to make the effect of negative impact minimize.
    3   Through the analysis of the stress field inside toss device, finding the value of static pressure was asymmetry, illustrating that pressure distribution was associated with the geometric size of fan exit and blade shape.

Acknowledgements. Fund origin of this project: The Inner Mongolia Autonomous Region Technology Innovation projects （20101734）, The Inner Mongolia Agricultural University Technology Innovation Team projects (NDPYTD2010-8).

References

1. Gao, X., Ma, W., Ma, C., et al.: Analysis on the Current Status of Utilization of Crop Straw in China. Journal of Huazhong Agricultural University 21(3), 242–247 (2002) (in Chinese)
2. Cui, M., Zhao, L., Tian, Y., et al.: Analysis and evaluation on energy utilization of main crop straw resources in China. Transactions of the CSAE 24(12), 291–296 (2008) (in Chinese)
3. Gao, L., Ma, L., Zhang, W., et al.: Estimation of nutrient resource quantity of crop straw and its utilization situation in China. Transactions of the CSAE 25(7), 173–179 (2009) (in Chinese)
4. Zhai, Z., Gao, B., Yang, Z., et al.: Power consumption and parameter optimization of stalk impeller blowers. Transactions of the Chinese Society of Agricultural Engineering. Transactions of the CSAE 29(10), 26–33 (2013) (in Chinese)
5. Chancellor, W.J.: Influence of particle movement on energy losses in an impeller blower. Agricultural Engineering 41(2), 92–94 (1960)

6. Shinners, K.J., Koegel, R.G., Pritzl, P.J.: An upward cutting cut-and-throw forage harvester to reduce machine energy requirements. Transactions of the ASAE 34(6), 2287–2290 (1991)
7. Zhai, Z., Wang, C.: Numerical Simulation and Optimization for Air Flow in an Impeller Blower. Transactions of the Chinese Society for Agricultural Machinery 39(6), 84–87 (2008) (in Chinese)
8. Song, J., Liu, Y., Ma, W., et al.: Internal flow field simulation and structure parameters optimization of centrifugal fan of snow remover. Transactions of the Chinese Society of Agricultural Engineering. Transactions of the CSAE 1(3), 83–87 (2011) (in Chinese)
9. Li, G., Wang, Y., Lü, X., et al.: Numerical simulation of three-dimensional flow field in centrifugal pump with deviated short splitter vanes. Transactions of the CSAE 27(7), 151–155 (2011) (in Chinese)
10. Li, X., Yuan, S., Pan, Z., et al.: Numerical simulation of whole flow field for centrifugal pump with structured gird. Transactions of the Chinese Society for Agricultural Machinery 44(7), 50–54 (2013) (in Chinese)
11. Huang, S., Wang, G.: Analysis of flow field asymmetry and force on centrifugal pump by 3-D numerical simulation. Transactions of the Chinese Society for Agricultural Machinery 37(10), 66–69 (2006) (in Chinese)
12. Wu, G., Song, T., Zhang, Y.: FLUENT introductory and proficient in case, pp. 38–39. Publishing House of Electronics Industry, Beijing (2012) (in Chinese)
13. Wei, L., Weidong, S., Xiaoping, J., et al.: Numerical calculation and experimental study of axial force on multistage centrifugal pump. Transactions of the CSAE 28(23), 52–59 (2012) (in Chinese)

A GPRS-Based Low Energy Consumption Remote Terminal Unit for Aquaculture Water Quality Monitoring

Dan Xu[1], Daoliang Li[1,*], Biaoqing Fei[1], Yang Wang[2], and Fa Peng[3]

[1] College of Information and Electrical Engineering,
China Agricultural University, Beijing 100083, China
[2] College of Engineering, China Agricultural University, Beijing 100083, China
[3] College of Mechanical and Electronic Engineering,
Shangdong Agricultural University, Taian 271000, China
dliangl@cau.edu.cn

Abstract. The monitoring of water quality parameters such as DO, pH, salinity and temperature are necessary for the health of seafood such as sea cucumber. However, traditional monitoring system is based on cable data acquisition that has many disadvantages. Nowadays, GPRS is the most commonly accepted way for wireless transmission. Based on it, a type of low energy consumption RTU is developed and applied. In this paper, details of the design are introduced. In hardware design of this type of RTU, STM8L152 is selected in the MCU module to accomplish the function of ultralow power consumption, and solar battery is designed to solve the problem of power supply. In software design of the RTU, the sleep/online mode conversion is programed to reduce the energy consumption. It is comparatively low-priced and can detect necessary parameters for aquaculture. Performances of the RTU are tested in experimental stations and compared with two advanced water quality analyzers. Results show that it shows almost the same variation characteristics as those of HACH and YSI.

Keywords: RTU, GPRS, aquaculture, low energy consumption.

1 Introduction

Aquaculture is a fast growing food-producing part in the world. [1] In China, there are lots of problems in traditional aquaculture, such as pollution and consumption. [2] In modern aquaculture management, it's most important to monitor water quality reliably and control water environment in time. It's useful in enhancing the fish concentration and growth rate, at the same time reducing the occurrence of fish diseases. Dioxide oxygen (DO), pH, salinity and temperature are most important factors in sea water quality for aquaculture. In order to improve the level of aquaculture, these parameters must be precisely monitored. [3]

Traditional monitoring system is based on cable data acquisition that has more and more disadvantages. [4] In the process of wiring cable technology, there are a series

D. Li and Y. Chen (Eds.): CCTA 2013, Part II, IFIP AICT 420, pp. 492–503, 2014.

of restrictions, such as high temperature, high pressure, high altitude and high risks. First, it is difficult for the installation of data acquisition devices, which will increase the maintenance workload. Second, quantities of cables will have a bad impact on visual appeal and increase the cost. Third, cables face the challenges of aging and the risk of bitten by rodents and other animals, which leads to a rise of fault rate. To tackle these issues, wireless data acquisition is adopted.

Within the past decade, large numbers of water quality monitoring instruments have been commercialized. Some integrated with remote real-time water quality monitoring systems have been developed and deployed by scientists, governmental agencies and industries throughout the world in modern aquaculture. [5] Nowadays two wireless ways are mainly introduced, wireless sensor network (WSN) and General Packet Radio Service (GPRS). [6] Wireless sensor network is a kind of wireless network without infrastructure. [7] It receives and sends messages through wireless and self-organization multi-hop routing. [8] It is a relatively new wireless technology that has sharp technical problems that is instability. On contrast, GPRS is a relatively mature technology that has been adopted widely in daily life, such as mobile office, mobile commerce, mobile information service, mobile internet and multimedia business. [9]

GPRS is a new type of data transmission technology based on Global System of Mobile Communication (GSM). It adopts packet switching mode, which occupy wireless resources only in the process of sending and receiving data. In theory, it can reach transmission data rate as high as 171.2 Kbit/s. Except for the advantage in speed, it is also always on-line. That means users can keep contact with the net at any time. [10]

Based on the advantages of GPRS, a new type of remote terminal unit (RTU) is developed and applied in the experiment stations. It detects DO, pH, salinity and temperature at the same time.

In this paper, details of design of this type of RTU are introduced. And tests of working performances about its detecting parameters are done and analyzed. Finally we reach a conclusion about application of this RTU.

2   Materials and Methods

2.1   Selection of Monitoring Sites

To test the performances of this type of RTU, two experimental stations in the city of Weihai in Shandong Province are established. One is in Wendeng Ocean and Fishery Bureau, another is in Shandong Xunshan Aquatic Product Group Corporation.

The city of Weihai (N 36°41'~37°35', E 121°11'~122°42') locates in the eastern most part of Shandong peninsula. With three sides (east, south, north) facing the sea and a coastline of 985.9 km, it's also sees the Korean peninsula and the Japanese islands across the sea. [11] It is rich in marine resources. And the market driven stimulus keeps mariculture a rapid growth year by year. [12] However, the science and innovation level is still low. Extensive pattern remains on the development and utilization of marine resources. Many problems still exist, such as the great need of professionals, the lack of high-tech research results, and the low level and slow pace in the industrialization of the project.

Based on those advantages and places need improvement, high-tech research results are integrated into this area and two experimental stations in the city of Weihai are built. The two experimental stations aim mainly at raising sea cucumber and abalone. Sea cucumber and abalone are both traditional Chinese seafood with high edible and medicinal value. In recent years, due to the worldwide over-development of sea cucumber resources and the sharp decline of its population, sea cucumber artificial breeding is springing up. [13] However, these precious sea creatures also demands strict living conditions. DO, PH, salinity and temperature have to be precisely detected and controlled in a defined range. For example, sea cucumber is cold temperature zone species. Its living water temperature is -1.5~30℃. It takes in less food and be in the process of half dormancy when the water temperature is below 3°C. It reaches the highest food consumption when the water temperature is 10~15°C. Then food consumption falls sharply when it is 10~15°C. Then it goes into aestivation when it is over 20°C. So, the most moderate water temperature is 3~20°C. And it stops growing when it is below 2°C and over 23°C. [14] Thus, large numbers of RTUs are installed which can detect these parameters in the two experimental stations.

From July 25th to July 28th in the year of 2013, performances of the RTUs installed in the two experimental stations are tested. Large numbers of experiments are done and useful data is collected for future analysis. Lots of problems are found out. Most problems are solved, while some problems are leaving unresolved but already reported for further study.

2.2   Hardware Design

This type of RTU consists mainly of 4 modules. They are sensor module, MCU module, GPRS module and power module. Their relations are shown in figure 1.

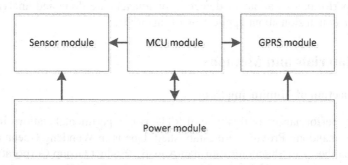

Fig. 1. Relations between different modules

The sensor module can detect necessary parameters of water quality. It consists of DO sensor, pH sensor, salinity sensor and temperature sensor. The MCU module processes the data collected by sensors, then sends the data to GPRS module. The GPRS module sends the processed data to the monitoring center. The power module is controlled by the MCU module and supply power to the other three modules.

After PCB plate-making and circuit debugging, hardware performs well. Then details about the system are shown as follows.

2.2.1 Sensor Module

The sensor module is important part of the RTU. It detects on-line necessary water quality parameters, including DO, pH, salinity and temperature. The temperature sensor is attached to DO sensor electrode. So there are altogether three electrodes. Sensors send data about water quality to MCU through the signal wires, at the same time suspending in the water depending on signal wires.

4-core water-proof cables are selected as signal wires. They connect sensors and MCU. They have three functions. The first function is to supply power to the sensors. The second function is to send data collected by sensors to MCU. The third function is to act as suspension wire to suspend sensors in the water. The lengths of signal wires are adjustable so that sensors can detect water quality information in different depths according to different application requirements.

2.2.2 MCU Module

The MCU module is core part of the RTU. It has four functions. First, it controls the power module that supplies power to each module. Second, it controls sensors to collect data. Third, it controls the GPRS module to send the data wirelessly. Last, it processes all the data. To cut down the cost and energy consumption, STM8L152 is selected in the MCU module. [15] It is an ultra-low power consumption MCU chip of ST Corporation. And it has the following advantages. [16]

First, it has a wide working voltage range: from 1.8 V to 3.6 V or form 1.65 V to 3.6 V. (Its minimum working voltage is 1.65 V in the mode of power-down).

Second, it has five ultralow energy consumption modes: low power running mode (5.1 μA), low power waiting mode (3.0 μA), real time clock operation suspend mode (1.2 μA), self-wakeup suspend mode (0.91 μA) and SRAM content retaining suspend mode (350 nA). They play a key role in cutting down the energy consumption of the whole system.

Third, its operating temperature range is from -40℃ to 85℃, which enables it working under different severe environments.

Forth, it has rich internal functions. It integrates SPI function internally which makes it convenient for the operation of wireless communication. And it integrates EEPROM and LCD function to facilitate the expansion of the function.

Fifth, compared with MSP430 series MCU, STM8L series MCU has lower prices and higher cost performance.

2.2.3 GPRS Module

The core chip of the GPRS module is SIM900. [17] It is an ultra compact and reliable wireless module presented by the corporation of SIMCom. It is a complete Quad-band GSM/GPRS module in a SMT type and designed with a very powerful single-chip processor integrating AMR926EJ-S core, allowing you to benefit from small dimensions and cost-effective solutions.

Featuring an industry-standard interface, the SIM900 delivers GSM/GPRS 850/900/1800/1900MHz performance for voice, SMS, Data, and Fax in a small form factor and with low power consumption. With a tiny configuration of 24mm×24mm×3 mm, SIM900 can fit almost all the space requirements in your M2M applications, especially for slim and compact demands of design.

2.2.4 Power Module

The power module consists of solar panel, charging control module and storage battery. The charging control module converts solar energy collected by the solar panel directly into electrical energy and stores it in the storage battery. The storage battery supplies power to the whole system.

The core chip of the power module is CN3063. [18] CN3603 is a single-cell lithium battery charge management chip that can be used in the solar power supply. Thermal modulation circuit can control the chip temperature in a safe range when in face of high power consumption of the device or high ambient temperature. Charging current is set through an external resistor. When the input voltage is powered down, CN3063 enters the mode of low-power sleep automatically. At this time the current consumption of the battery is less than 3 μA. It also involves other functions like lockout of low voltage input, automatic recharge, battery temperature monitoring and indicators of charging states and end of charging states.

2.2.5 Operating Principle

In the RTU system, the input end of charging control module connects with the solar panel. The output end of charging control module connects with the storage battery. The storage battery connects with the MCU power input end. The power output end of MCU connects with charging control module, sensors and GPRS module. The charging control module, storage battery, MCU and GPRS module are placed in the circuit box. Among them, MCU module and GPRS module are integrated in the main circuit board. Antenna is settled on the main circuit board to send data. There are four sensor interfaces on the bottom of circuit. Signal wires of sensors connect with circuits in the circuit box through these interfaces. RS485 is adopted as the communication interface. There are also fixing holes on the bottom of circuit box. Ribbons can pass through these fixing holes to tie the RTU to the support rod above the water surface. Solar panel is fixed on top of the circuit box to collect solar energy. The 4-core waterproof cables act as signal wires. They connect to sensors to one end, and connect to MCU in the circuit box through waterproof interfaces. The lengths of signal wires are adjustable so that sensors can be placed in different level in the water.

2.3   Software Design

2.3.1   Energy Consumption Reducing Software

This system works in two modes, they are online and sleep. It is necessary for reducing the energy consumption. When the system is not on the process of collecting or processing data, it is set on sleep mode. When the water quality parameters need to be detected, it turns to online mode automatically to implement the data collecting, processing and sending. The intervals of sleep/online mode changing are previously integrated in the MCU. The intervals of online detecting can be the same. Every few hours, intervals of detecting can be changed according to values of parameters. For example, DO can be detected more intensely at the time before down when it is low.

2.3.2   Operating Principle

The system sets a sampling interval through programming. Sensors collect data once among each interval. The MCU supplies power to sensors only on the process of data sampling, while cuts off the power at other time, so as to cut down the energy consumption. The data collected by sensors is sent to MCU through signal wires for processing. The processed data is sent through GPRS to the monitoring center.

The flow chart of main program is shown in figure 2.

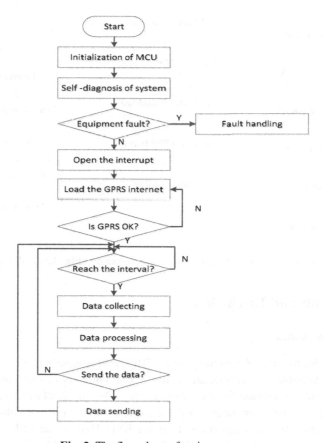

Fig. 2. The flow chart of main program

Charts of subprograms are shown in figure 3.

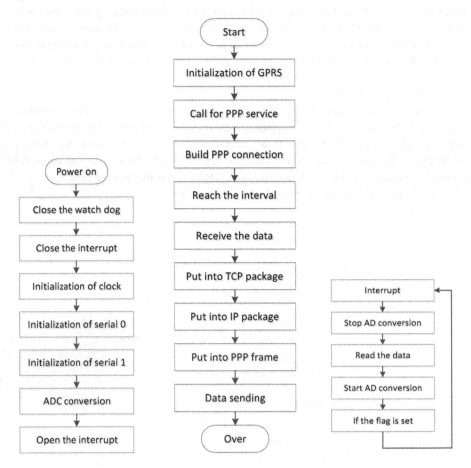

Fig. 3.1 Initialization   **Fig. 3.2** GPRS data transmission   **Fig. 3.3** Analog data acquisition

3   Results and Discussion

3.1   Experiments

From July 25th to July 28th in the year of 2013, performances of the RTUs installed in the two experimental stations are tested. Large numbers of experiments are done and useful data is collected for the future analysis. From 12:30 to 17:15 of each day, DO, pH, salinity and temperature in these experiment stations are collected by three different instruments for comparison. They are RTU, HACH and YSI.

HACH helps water resource professionals generate reliable data throughout the entire cycle of water, from measuring precipitation to monitoring estuaries and ground water, and everything in between. Hydrolab multi-parameter water quality instruments are built using the industry's leading sensor technology. [19]

YSI has been used for many years in facilities that process wastewater generated by metal finishing plants, but recently it has become prominent in municipal wastewater treatment plants. [20] The probe contains a sensor that measures electrical charges from particles, called ions, and these charges are converted to millivolts (mV) that can be either negatively or positively charged. And like all sampling measurements taken by operators, they are snapshots in time that can indicate process efficiency and identify treatment problems before they affect effluent quality. When using continuous monitoring and control instrumentation, this snapshot can become a real-time indicator.

HACH and YSI are all advanced water quality monitoring instruments produced in America that can detect precise water quality parameters respectively. To have a contrast with the performances of RTU, useful data is collected from time to time in different test points by these three instruments.

3.2   Results

DO values collected by these instruments from time to time are shown in figure 4. (Data is collected in July 28th)

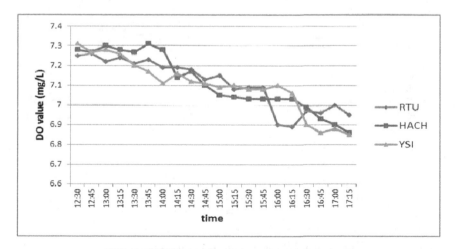

Fig. 4. DO values collected by three instruments

PH values collected by these instruments from time to time are shown in figure 5. (Data is collected in July 27th)

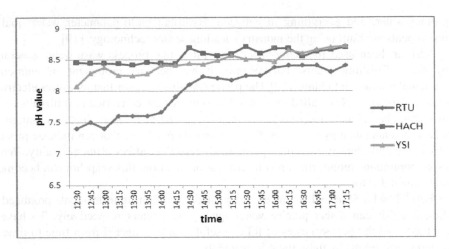

Fig. 5. PH values collected by three instruments

Salinity values collected by these instruments from time to time are shown in figure 6. (Data is collected in July 26th)

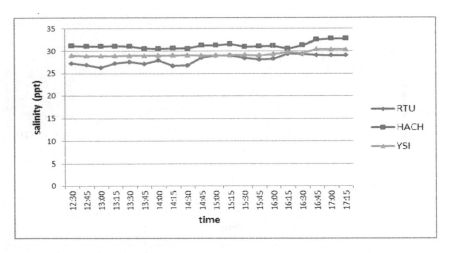

Fig. 6. Salinity values collected by three instruments

Temperature values collected by these instruments from time to time are shown in figure 7. (Data is collected in July 25th)

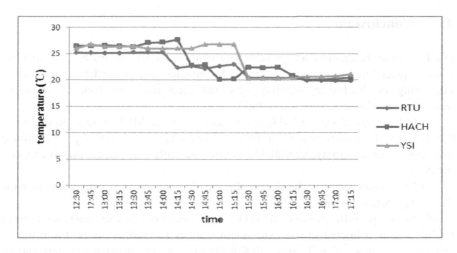

Fig. 7. Temperature values collected by three instruments

As these figures show, parameters collected by RTUs are almost the same precision as the other two famous instruments. It means that this type of RTU is in good performances.

Abnormal data may be caused by different environmental problems. Such as the instability of sensors, failure of power, errors in operation, and so on.

3.3   Discussion

Data collected by HACH and YSI shows the variation characteristics of these four parameters. From 12:30 to 17:15 of each day, DO, pH, salinity and temperature show different variation characteristics. DO decreases over time during this period. While pH increases over time during this period. Salinity has low fluctuations but it increases as the sea water evaporates and it reaches its peak at about 17:00. Temperature reaches its peak at about 14:00 and then it decreases as time goes. Data collected by RTU shows almost the same variation characteristics as those of HACH and YSI. Though not stable sometimes, it shows comparative good performances on the whole.

Though RTUs in most test points perform well, data collected in some test points of these experiment stations is abnormal. But in fact it has nothing to do with the RTU itself. It is sensors that are not stable. For example, some DO values detected by sensors are normal in the first couple of days. But they turn abnormal after about 14 or 15 days. So the DO sensors will have to be cleaned in about 10 days or so to gain accurate values. It is very inconvenient. So the urgent affairs should be the independent development of more stable and more accurate sensors.

4   Conclusion

(1) This study researches into a new type of RTU based on GPRS. It is applied in some aquaculture experiment stations. It provides a new technical solution for improving the backward breeding way and aquaculture environment monitoring method of the Chinese aquaculture field.

(2) In hardware design, STM8L152 is selected in the MCU module to accomplish the function of ultralow power consumption, and solar battery is designed to solve the problem of power supply. STM8L152 is comparatively low-priced, thus it cuts down the cost.

(3) In software design, the sleep/online mode conversion is programed to reduce the energy consumption.

(4) To test performances of this RTU, large numbers of experiments are done and useful data is collected in the experiment stations for further study. From the data analysis, we can see that it shows almost the same water quality parameters variation characteristics as those of HACH and YSI. Though not stable sometimes, it shows comparative good performances on the whole.

Acknowledgements. This work was supported by the National Agricultural Science and technology achievement transformation project (2012GB2E000330), the state of 12th five-year science and technology support projects (2012BAD35B03) and (2011BAD21B01), and the programs "Agro-scientific Research in the Public Interest" (201203017).

References

1. Hongbin, L., Guoxiang, H., Dakui, F., Bangding, X., Lirong, S., Yongding, L.: Prediction and elucidation of the population dynamics of Microcystis spp. in Lake Dianchi (China) by means of artificial neural networks. Ecological Informatics, 184–192 (2007)
2. Marcella, B., Wolf, A.: Planning for sustainable aquaculture. Tilapia Farming in the United States, China and Honduras, 1–39 (2013)
3. Liu, A., Junsheng, G., Meiyan, W.: Effects of Nanomaterials on Water Quality of Aquiculture. In: 2013 Third International Conference on Intelligent System Design and Engineering Applications, pp. 688–691 (2013)
4. Mingyou, Q., Lei, H., Wensheng, G.: Design and Realization of a Spot Data Acquisition System Based on ZigBee. Telecommunication Engineering, 34–38 (2008)
5. Ma, D., Ding, Q., Li, Z., Li, D., Wei, Y.: Prototype of an Aquacultural Information System Based on Internet of Things E-Nose. Intelligent Automation & Soft Computing, 569–579 (2012)
6. Yuwen, S., Mingxia, S., Yingjun, X., Mingzhou, L., Longshen, L., Xiaoli, K.: Design on the System of Collection, Storage and Release for Field Information. Advanced Materials Research, 347–353 (2011)
7. Jingtao, W., Changbao, M., Yongbo, L.: Design of Wireless Sensor Network Node for Water Quality Monitoring. Computer Measurement & Control, 2575–2578 (2009)

8. Zhiguo, D., Deqin, X., Yunhua, Z., Guozhen, O.: Design of Water Quality Monitoring Wireless Sensor Network System Based on Wireless Sensor. Computing Engineering and Design, 55 (2008)
9. Xiaoyan, Z., Yong, X., Tingbo, J., Jinhua, W., Hongwu, H.: Research on the Power-Frequency Electric Field Measuring System Based on GPRS. Applied Mechanics and Materials, 246–247 (2012)
10. Jianxiang, L., Chongguang, F., Tao, L., Haibo, L.: Design of Electric Vehicle Data Acquisition and Transmission System Based on GPRS and Zigbee. In: Proceedings of the 2012 Second International Conference on Electric Information and Control Engineering, pp. 542–545 (2012)
11. Song, P.: Thought and Suggestions about Development of Blue Economy in the City of Weihai. Northern Economy, 70–72 (2011)
12. Sun, Z., Shuchao, L.: "Blue and Yellow" Strategic Perspective Transformation and Development of Mariculture. Shandong Social Sciences, 144–148 (2012)
13. Min, S., Ji, C., Daoliang, L.: Water temperature prediction in sea cucumber aquaculture ponds by RBF neural network model. In: 2012 International Conference on Systems and Informatics, pp. 1154–1159 (2012)
14. Ji, C., Li, D., Du, S., Wei, Y., Tai, H.: A Wireless Sensor Network Based Water Temperature Stratification Monitoring System for Aquaculture of Sea Cucumber. Sensor Letters, 1094–1100 (2011)
15. Li, S.: Study on Vehicle Detection Technology Based on AMR Sensor. Journal of Hubei University of Education, 48–50 (2012)
16. Zheng, G., Xu, Z., Qiu, Y.: Design of a Low Energy Consumption Wireless Sensor. Fujian Computer, 39–40 (2012)
17. Jay, M., Shah, S.: Simplified Secure Wireless Railway for Public Transport. In: 2013 Fifth International Conference on Computational Intelligence, Communication Systems and Networks, pp. 77–82 (2013)
18. Ianmin, H., Gao, Y.: Greenhouse wireless sensor network monitoring system design based on solar energy. In: International Conference on Challenges in Environmental Science and Computer Engineering, pp. 475–479 (2010)
19. Chen, G.: Detecting total phosphate in water with the instrument HACH. Industrial Water Treatment, 62–63 (2010)
20. Kaishan, S., Lin, L., Tedesco, L., Clercin, N., Hall, B., Shuai, L., Kun, S., Dawei, L., Ying, S.: Remote estimation of phycocyanin (PC) for inland waters coupled with YSI PC fluorescence probe. Environmental Science and Pollution Research, 5330–5340 (2013)

The Model for the Agricultural Informationalization Benefit Analysis

Lifeng Shen, Xiaoqing Yuan, and Daoliang Li[*]

College of Information and Electrical Engineering, China Agricultural University, PRC.100083
P.O. Box 121, College of Information and Electrical Engineering, China
Agricultural University, 17 Tsinghua East Road, Beijing, 100083, P.R. China
dliang1@cau.edu.cn

Abstract. Based on the assessment of Agricultural Informationalization level, this paper proposed a model for the Agricultural Informationalization benefit analysis whit the Cobb-Douglas production function, to reveal the connection between agri-information index and the agricultural output, and also we made an empirical analysis of Shandong province. It is proved that the Agricultural Informationalization and the farming population both have a direct and positive impact on the development of rural economy in Shandong Province, especially, the development of the Agricultural Informationalization level has a significant influence on the increase of the rural economy in Shandong province , while the impact brought by agricultural fixed asset investment remains limited. By 1% increase of Agricultural Informationalization level, the rural economy increases by 0.565%, which indicates that Agricultural Informationalization is quite beneficial for the development of Shandong's agricultural economy.

Keywords: Agricultural Informationalization, benefit analysis, Cobb–Douglas production function.

Informationalization has become an increasingly major drive force of economic development and a significant indicator amidst the assessment of the comprehensive national power and the international competitiveness for a nation or region. Agricultural Informationalization is not merely the fundament and a significant component of national economic Informationalization, but the important means of balancing the urban and rural development and stimulating the agricultural economy. [1] Agricultural Informationalization is going to be the significant symbol of agriculture in the 21st century, and it is an inexorable trend in the contemporary modern agricultural development. [2]

With lately increasing attentions from the local governments, the Informationalization level in agriculture-oriented rural areas has been continuously improved, and the contribution of the developing Agricultural Informationalization has made to the agricultural productivity has been remarkably valued in these places. A number of domestic researchers [3-11] have studied the economic expansion in

---

[*] Corresponding author.

D. Li and Y. Chen (Eds.): CCTA 2013, Part II, IFIP AICT 420, pp. 504–512, 2014.

informationalising regions and its connection to the gross output from economics perspectives, while most of them focused on the analysis of how the general informationalising process has motivated the economic development, rather than any empirical research on the function mechanism and inner relationship.with benefit analysis. This research attempted to examine problems that exist within agri-informationalising, to target any links or stages where no benefit has been made, and also to provide the Agricultural Informationalization investment which has not currently made profits with confidence and scientific suggestions in their decision-making. Moreover, it helps local governments find out their advantages and disadvantages in rural Agricultural Informationalization, to provide the appropriate strategic positioning of Agricultural Informationalization.

This paper established a model to reveal the relationships of Agricultural Informationalization index and agricultural output with Cobb-Douglas production function, and empirically analyzed the Agricultural Informationalization of Shandong Province. It attempted to provide the nation and local governments a significant reference in terms of targeted Agricultural Informationalization investment and Agricultural Informationalization policy making, so that the benefits of Informationalization amidst the construction of socialist new countryside can be better deployed. Hence this research will be remarkably influential for its social and economic meanings.

1   Model Building for Agricultural Informationalization Benefits Analysis

1.1   Measurement Model

Cobb-Douglas production function for the assessment of Agricultural Informationalization benefits has been widely adopted by both domestic and foreign scholars in empirical researches. Based on the literature review and practical development of Agricultural Informationalization, this paper attempted to assess the Agricultural Informationalization benefits with Cobb-Douglas production function.

For calculation and analyzing the benefit of Agricultural Informationalization, this research modified the Cobb-Douglas production function based on the new growth theory inaugurated by Paul M. Romer. Romer articulated that, the development of science and technology should also be included in the Cobb-Douglas production function, besides capital and labour [12]. Therefore, this research believes that information, the maximum return of technological developmen t, can be regarded to replace technological advance and become the third factor to affect Cobb-Douglas production function in terms of input. Cobb-Douglas production function can be modified as follow [13]:

$$Y=AK^{\alpha}L^{\beta}I^{\gamma} \tag{1}$$

The log-linear model of formula (1) is:

$$log(Y)=log(A)+\alpha log(K)+\beta log(L)+\gamma log(I) \tag{2}$$

where Y, K, L, I, respectively represent the total value of agricultural output, capital output, labor output, and the Agricultural Informationalization index, and α, β, γ respectively indicate the changes of capital output, labor and Informationalization; A is a constant that implies other factors that may affect the agricultural benefits.

1.2   Model Testing

With SPSS17.0 and the Enter method, Y', K', L', I' were all included for the linear regression analysis. The regression result was shown in *Table 1*, based on which the regression model can be built. By categorizing the regression model, a total of three variables, including agricultural fixed asset investment, the farming population and the general agri-informational index, are revealed, as well as the other five variables of sub-indexes and the total value of agricultural output.

Table 1. The result of regression model (general Agricultural Informationalization index)

Items	Constants	K'	L'	I'
Standardised coefficients		a	j	g
B	b	h	w	m
(T-statistic)	t	c	f	i
R-squared		$r2$		
F- statistic		f		
Durbin-Watson statistic		d		

The above regression model reveals both the selected samples' and the overall goodness of fit, and allows F-test to verify the linearity degree of the model and T-test to verify all variables with their explanatory capabilities to economic improvement. If the goodness of fit, the linearity degree and resolution are all high, and variables K, L, I and constants get through the test, it can be generally concluded that the model has been successfully verified, and it has a good fineness to the reality. The model is presented as follow by *Formula 3*:

$$Y' = b + hK' + wL' + mI' \qquad (3)$$

Formula 3 can be transformed into:

$$Y_t = e^b K_t^h L_t^w I_t^m \qquad (4)$$

Similarly, five more regression model can be established for the other five indexes, including Agricultural Informationalization infrastructure (F), the Informationalization of agricultural production (P), the Informationalization of agricultural operation(C), the Informationalization of agricultural management (M), the Informationalization of agricultural services(S),which would not repeat here.

2   The Agricultural Informationalization Benefit Analysis of Shandong Province

2.1   The Assessment of Agricultural Informationalization in Shandong Province

A total of 5 major indexes, including Agricultural Informationalization index, agricultural infrastructure informationalization index, agricultural production informationalization index, agricultural operation informationalization index, agricultural management informationalization index, and agricultural services informationalization index, should be included to establish of Agricultural Informationalization assessment system, based on which a comprehensive index model for the assessment of Agricultural Informationalization can be built for the study of assessment in Shandong Province from the year 2003 to 2011. The detailed process has been fully developed in my doctoral dissertation; therefore only the result was displayed in this paper, while no explanation will be given to the model building and calculating processes here.

Table 2. The Agricultural Informationalization index of Shandong Province (2003-2011)

	2003	2004	2005	2006	2007	2008	2009	2010	2011
Agricultural Informationalization infrastructure (I)	0.047	0.067	0.090	0.110	0.128	0.155	0.178	0.201	0.231
the Informationalization of agricultural production (F)	0.083	0.109	0.142	0.193	0.239	0.295	0.358	0.421	0.495
the Informationalization of agricultural production (P)	0.003	0.006	0.010	0.015	0.013	0.024	0.029	0.035	0.040
the Informationalization of agricultural operation (C)	0.010	0.017	0.024	0.033	0.039	0.045	0.050	0.057	0.064
the Informationalization of agricultural management (M)	0.020	0.044	0.067	0.083	0.102	0.133	0.158	0.191	0.254
the Informationalization of agricultural services (S)	0.140	0.187	0.242	0.269	0.302	0.345	0.377	0.400	0.415

2.2   Regression Model

The Agricultural Informationalization index in the model (I) and the five major indexes to show the Agricultural Informationalization situation in Shandong Province were all well displayed in *Table 2*. By referring to *Shandong Province Statistic Yearbook [Shandong sheng tongji nianjian]* and *China's Rural Statistic Yearbook [Zhongguo nongcun tongji nianjian]*, the statistic data of agricultural gross output value in the period from 2003 to 2011 (Y), agricultural fixed asset investment (K), farming population (L) could be seen. The gross output value of agriculture, fixed asset investment and farming population in the year of 2012 have not been published yet. The detailed data were displayed in *Table 3*.

Table 3. Data of economic growth in Shandong Province

	Gross output values of agriculture (Y/100 million yuan)	Agricultural fixed asset investments (K/100 million yuan)	Farming populations (L/10 thousan people)
2003	2902.5	296.0	2638.3
2004	3453.9	116.4	2542.1
2005	3741.8	1491.6	2350.3
2006	4058.6	1186.2	2328.0
2007	4766.2	1141.3	2265.2
2008	5613.0	1304.0	2313.5
2009	6003.1	1586.7	2297.4
2010	6650.9	1823.0	2273.1
2011	7409.8	2470.4	2211.6

As time goes by, comparability of data in different years weakens. For a considerable comparability of the data, the price index has to be invited to eliminate the impacts brought by price fluctuation. In 2003, for the base period, the influence of price fluctuation on gross output values of agriculture and agricultural fixed asset investments had been excluded, so the data in model remained consistency. The organized data were shown in *Table 4*.

Table 4. Gross output values of agriculture and agricultural fixed asset investments excluding the price impacts

	Gross output values of agriculture (Y/100 million yuan)	General indexof rural residents consumptio n price (the base: 100)	Gross values excluding price impact (Y'/100 million yuan)	Agricultural fixed asset investments (K/100 million yuan)	Price index of fixed asset investme nts (the base: 100)	Fixed asset investments excluding prince impact (K'/ 100 million yuan)	Farming populatio n (L/10 thousand people)
2003	2902.5	——	2902.5	296.0	——	296.0	2638.3
2004	3453.9	104.6	3302.0	116.4	107.4	108.4	2542.1
2005	3741.8	102.4	3493.4	1491.6	102.9	1349.7	2350.3
2006	4058.6	101.0	3751.7	1186.2	101.8	1054.4	2328.0
2007	4766.2	105.3	4184.0	1141.3	104.0	975.4	2265.2
2008	5613.0	106.2	4639.7	1304.0	107.7	1034.8	2313.5
2009	6003.1	100.1	4957.2	1586.7	96.9	1299.4	2297.4
2010	6650.9	103.5	5306.4	1823.0	103.6	1441.1	2273.1
2011	7409.8	105.9	5582.5	2470.4	106.8	1828.5	2211.6

The data needed in the model of *Formula 3*, after taking the logarithms of gross agricultural output values (*Y*), agricultural fixed asset investments (*K*), farming populations (*L*) and national Agricultural Informationalization index, were organized as below in *Table 5*.

Taking logarithms on Gross output values of agriculture(*Y*), agricultural fixed asset investments *(K)*, farming populations *(L)* and national Agricultural Informationalization index ,the log transformation of the data could be rewritten in table 5.

Table 5. Data of agri-economic development model for Shandong Province

	Ln(Y')	Ln(K')	Ln(L')	Ln(I)	Ln(F)	Ln(P)	Ln(C)	Ln(M)	Ln(S)
2003	7.9733	5.6904	7.8779	-3.0576	-2.4889	-5.8091	-4.6565	-3.8947	-1.9661
2004	8.1023	4.6856	7.8407	-2.7031	-2.2164	-5.1160	-4.1044	-3.1304	-1.6766
2005	8.1586	7.2076	7.7623	-2.4079	-1.9519	-4.6052	-3.7508	-2.6971	-1.4188
2006	8.2300	6.9607	7.7528	-2.2073	-1.6451	-4.1997	-3.4173	-2.4865	-1.3130
2007	8.3390	6.8829	7.7254	-2.0557	-1.4313	-4.3428	-3.2416	-2.2813	-1.1973
2008	8.4424	6.9420	7.7465	-1.8643	-1.2208	-3.7297	-3.0989	-2.0155	-1.0642
2009	8.5086	7.1697	7.7395	-1.7260	-1.0272	-3.5405	-2.9868	-1.8483	-0.9755
2010	8.5767	7.2731	7.7289	-1.6045	-0.8651	-3.3524	-2.8735	-1.6581	-0.9163
2011	8.6274	7.5113	7.7015	-1.4653	-0.7032	-3.2189	-2.7442	-1.3704	-0.8795

To assess the Agricultural Informationalization benefits from different angles, the Agricultural Informationalization index and the five major informationalization indexes were taken as variables, which means $Ln(I)$, $Ln(F)$, $Ln(P)$, $Ln(C)$, $Ln(M)$ and $Ln(S)$ were taken into the calculation of $I't$ for the analysis of regression model. Due to the similar calculating processes, this article exemplified the model of the general Agricultural Informationalization index, while the other five indexes' models could be adapted to the same procedure.

With SPSS17.0 and the Enter method, the linear regression results of $Ln(Y')$, $Ln(K')$, $Ln(L')$ and $Ln(I)$ from *Table 5* were displayed in *Table 6* as follows. By building a regression model for the regression results and organizing the regression model, the three variables, including agricultural fixed asset investments, farming populations and the general Agricultural Informationalization index, along with the total output value of agriculture, finally were used for the equation simulating the economy in Shandong Province.

Table 6. Regression results of Agricultural Informationalization index

Items	Constants	K'	L'	I'
Standardised coefficients		-0.011	0.352	1.323
B	-1.229	-0.003	1.389	0.565
(T-statistic)	(-0.333)	(-0.147)	(2.931)	(13.826)
R-squared		0.994		
F- statistic		265.205		
Durbin-Watson statistic		2.595		

As shown in *Table 7*, the value of R^2 was 0.994, which indicated the relatively satisfying fit of selected samples was good; the value of F was 265.205, showing that the considerable linear degree of the whole model was fine; D-W was 2.595, indicating the data series was no first-order autocorrelation; numbers below T item showed that the variables in the model were accountable to explain their respective influence on economic growth. Based on all verified results, the following model was established.

$$Y' = -1.229 - 0.030K'_t + 1.389L'_t + 0.565I'_t \qquad (5)$$

The model could also be transformed into:

$$Y_t = e^{-1.229} K_t^{-0.030} L_t^{1.389} I_t^{0.565} \qquad (6)$$

Similarly, models for agricultural infrastructure informationalization index (*F*), agricultural production informationalization index (*P*), agricultural operation informationalization index (*C*), agricultural management informationalization index (*M*) and agricultural services informationalization index (*S*) could be respectively established.

2.3   Interpretation of Results

From the regression model, we found that the standardized coefficients of the agricultural fixed asset investment, farming population and Agricultural Informationalization index were -0.011, 0.352 and 1.323 in the Agricultural Informationalization index model. It indicated that the Agricultural Informationalization, especially the farming population had profounder influence on the rural economic growth in Shandong Province, just in contrast with the agricultural fixed asset investment .From the other 5 regression index models, the results show that the agricultural fixed asset investment had less effect than Informationalization and farming population, that indicates Agricultural Informationalization and farming population played more important role than agricultural fixed asset investment in promoting the rural economic growth in Shandong Province.

From another point of view, we found that the output elasticity of of K', L' and I' were respectively -0.011, 0.352 and 1.323 in the Agricultural Informationalization index model. According to the concept of output elasticity, each 1% increase in Agricultural Informationalization leads to Shandong's agricultural economic growth by 0.565%, which means the agricultural economic growth in Shandong can enormously benefits from Agricultural Informationalization, while agricultural fixed asset investment improvement bought negative impacts on agricultural productive benefits growth in Shandong Province. Also, the results of other 5 indexes reveal the similar laws that Agricultural Informationalization and farming population had positively promote the rural economy development in Shandong Province.

In summary, these seven models above have revealed similar situation that Agricultural Informationalization has positively accelerated the agricultural economic growth significantly, which says that there is a preliminary success of Agricultural Informationalization development in Shandong Province. On the contrary, the increase of fixed asset investments has not been conducive to the agricultural economic growth, probably because of the lack of legitimate plan in advance of some fixed asset investments, or the absence of strict surveillance over the use of funds which causes a lower utilization of funds and leads to the inadequate capital input in agricultural economic development, so the pull effect of capital input on agriculture economic growth is not obvious.

3   Conclusion

With more inputs from the local governments, the general Agricultural Informationalization level in China is constantly improving, and more and more attention has been paid to the agricultural output benefits, which brought by informationalization development. However, the previous researches and practices have not clarified the relationship between Agricultural Informationalization and agricultural output, which only made the benefit of developing Agricultural Informationalization blurred for local governments. But, the benefit analysis of Agricultural Informationalization can effectively dissect the function mechanism and inner relationsship of how Agricultural Informationalization contributes to agricultural economic growth. Based on that, this paper established the Agricultural Informationalization benefits analytical model to interpret the contribution of Agricultural Informationalization to agricultural economy in Shandong Province. The results show that the problem how Agricultural Informationalization accelerates the agricultural economy can be well solved by this approach, and so this paper laid the foundation for the future construction of Agricultural Informationalization.

Acknowledgements. This paper was supported by Shandong Province Self-innovation Projects (2012CX90204).

References

1. Jianguo, W.: Study on the significance and development of agricultural informatization. Anhui Agricultural Science Bulletin 12(8), 181 (2006)
2. Tingting, H., Dehua, L.: The Measure and Influence Factors Analysis of Agricultural Informationization Level in China. Information Science 26(4), 565–571 (2008)

3. Charles, J.: Inform action resources and economic productivity. Information Economics and Policy (1), 67–70 (2007)
4. Dewan, Kraemer: Information Technology and Productivity: Evidence from Country - evel Data. Management Science (4), 53–56 (2008)
5. Gill, G., et al.: Dumagan and Isaac Turk. Economy-Wide and Industry-Level Impact of Information Technology (4), 51–55 (2006)
6. Weill, P.: The relationship between investment in information technology and firm performance: a study of the valve manufacturing sector. Information Systems Research (4), 92–94 (2004)
7. Yosri, A.: The relationship between information technology expenditures and revenue contributing factors in large corporations. Doctoral Dissertation, Walden University (3), 38 (2007)
8. Zhigang, Z.: Empirical study on relationship between informatization and regional economic development. Commercial Times (12), 125–126 (2010)
9. Jin, X.: Analysis of Impacts of Regional Informatization on Economic Growth. Statistical Research 27(5), 74–80 (2010)
10. Shaolin, W., Weiping, L.: Influence of informatization on economic growth – A case in Guangdong province. Zhishi Jingji (5), 83–84 (2010)
11. Mingyuan, M., Xiangyang, Q.: Research on Beijing Rural Informatization Performance Evaluation System. Chinese Agricultllml Sciellce Bulletin 27(30), 285–289 (2011)
12. Romer, P.M.: The origins of endogenous growth. Journal of Economic Perspectives 8, 3–22 (1994)
13. Li, W.: Empirical Analysis of Information Contribute to Shandong Rural Economic Growth. Northwest Agriculture and Forestry University (2010)

Strategic Optimal Path and Developmental Environment on Photovoltaic Industry in China Based on an AHP-SWOT Hybrid Model

Yiding Zhang[1,2] and Songyi Dian[2]

[1] Wu Yuzhang Honors College, Sichuan University, Chengdu 610225, China
[2] School of Electrical Engineering and Information Technology,
Sichuan University, Chengdu 610065, China
yiding_zhang@126.com, scudiansy@scu.edu.cn

Abstract. Utilizing the qualitative analysis of SWOT and quantitative method of AHP, this paper presented the AHP-SWOT hybrid model of the photovoltaic industry in China in order to calculate the influence of the strengths, weaknesses, opportunities & threats (SWOT factors and SWOT sub-factors), and the SO strategy, WO strategy, ST countermeasure and WT countermeasure on the development of the photovoltaic industry in China. The results show that the influence of the threats are the most, the weaknesses and opportunities second, and the strengths the least among four SWOT factors of the development of the photovoltaic industry in China. Thereby, among 12 SWOT sub-factors, complete industrial chains are the most important factor in the strengths factors, lack of core technology is of the most significance in the weaknesses factors, latent necessity of the photovoltaic industry development as a result of global energy crisis is the most valuable opportunities factor, and external trade environment deteriorating due to international trade protection is the most threats factor. Among four combination path of the development strategy, the WT countermeasure possesses the most valuable positive influences on the development of the photovoltaic industry in China, which is the optimal path for the strategy alternatives.

Keywords: photovoltaic industry, development strategies, strategic path, AHP, SWOT, China.

1   Introduction

After a period of the photovoltaic industry development, China has become one of the world's largest photovoltaic manufacturing countries. However, the capacity of photovoltaic industry in China is given priority to with export, more than 90% of which were obliged to depend on the international market. At present, the export of Chinese photovoltaic industry was impacted because of the global financial crisis and the European debt crisis, and the domestic markets were not completely open so that the development of Chinese photovoltaic industry has been involved in the bottleneck

D. Li and Y. Chen (Eds.): CCTA 2013, Part II, IFIP AICT 420, pp. 513–522, 2014.

period [1]. China-Eu photovoltaic conflict finally had been resolved through a price undertaking, it also is a wake-up call for Chinese industry development.

In order to find out the problems that photovoltaic industry in China were facing and explore its development strategies, domestic scholars have done a lot of qualitative analysis and statistical description work on photovoltaic industry, which mainly focused on status definition[1], development status, international competitiveness[2], industrial cluster[3], operating and collaborative performance[4, 5], and development strategies. From the perspective of development strategies, however, the quantitative optimization really matters to help to choose the strategic path of photovoltaic industry in China and promote its sustainable development. This paper combined SWOT method with AHP method, which took full advantage of the combination of quantitative and qualitative analysis [6-11]. By using this hybrid method, this paper qualitatively analyzed the strengths, weaknesses, opportunities, and threats (SWOT) of the photovoltaic industry development in China, quantitatively evaluated sub-factors of SWOT and optimized the strategic path of photovoltaic industry development in China.

2   SWOT Analysis of Photovoltaic Industry Development in China

2.1   SWOT Factor Analysis of Photovoltaic Industry Development in China

The photovoltaic industry in China has faced both restrictions and opportunities since 2012. The photovoltaic industry development in China is a complex system affected by national economy and social development. The basic elements of SWOT method consist of strengths, weaknesses, opportunities, and threats. Strengths and weaknesses are internal factors while opportunities and threats are external factors. Utilizing the SWOT method, the factors should be divided into strengths, weaknesses, opportunities, and threats according to the affection on the photovoltaic industry development in China [6, 8, 11].

SWOT analysis matrix was constructed, and contained 12 SWOT factors affecting the photovoltaic industry development in China (Fig.1). Among 12 SWOT factors, there were three SWOT sub-factors for strengths, weaknesses, opportunities, and threats, respectively [8, 9]. In SWOT sub-factors from strengths, there were three factors for Photovoltaic industry is a newly developing resource industry, Complete industrial chains, and Obvious strengths of cluster development in photovoltaic industry; from weaknesses, three factors for The photovoltaic industry chain mainly amassed on the mediate part of low value, Relatively high cost of producing photovoltaic production component, and Lack of core technology; from opportunities, three factors for Government support in photovoltaic industry, Latent necessity of the photovoltaic industry development as a result of global energy crisis, and Juncture of integration in photovoltaic industry; from threats, three factors for The photovoltaic industry chain mainly amassed on the mediate part of low value, Relatively high cost of producing photovoltaic production component, and Lack of core technology.

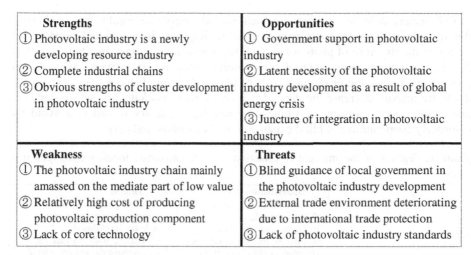

Strengths	Opportunities
① Photovoltaic industry is a newly developing resource industry ② Complete industrial chains ③ Obvious strengths of cluster development in photovoltaic industry	① Government support in photovoltaic industry ② Latent necessity of the photovoltaic industry development as a result of global energy crisis ③ Juncture of integration in photovoltaic industry
Weakness	**Threats**
① The photovoltaic industry chain mainly amassed on the mediate part of low value ② Relatively high cost of producing photovoltaic production component ③ Lack of core technology	① Blind guidance of local government in the photovoltaic industry development ② External trade environment deteriorating due to international trade protection ③ Lack of photovoltaic industry standards

Fig. 1. SWOT analysis matrix of the developmental strategy for the photovoltaic industry in China

2.2 Strategy Combination and Path Analysis of the Photovoltaic Industry Development in China

Development strategies could be obtained by combining and adjusting the factors of strengths, weaknesses, opportunities and threats according to the effects on photovoltaic industry system in China. The factors of different strategies could be combined according to the effects, as has the different influence on the photovoltaic industry system in China. According to analysis on the inner strengths, weaknesses and the external opportunities and threats of the photovoltaic industry in China, two strategies of SO strategy, WO strategy and two countermeasures of ST countermeasure, WT countermeasure could be formed by matching the four SWOT factors[8, 9].

The four paths are of different characteristics (Table 1). On basic strategy, SO means aggressive attack, which should take advantages, and seize opportunities. On tactics, strategic measures is to greatly develop the photovoltaic energy industry, further perfect the photovoltaic industry chains, adjust and optimize the government's supporting policies, promote the development of photovoltaic industry, and complete the top design, exert the advantage of cluster development in photovoltaic industry.

ST means corresponding defense, which should take advantages and avoid threats. On tactics, strategic measures is to enhance inter-regional cooperation, reasonably plan area distribution in photovoltaic industry, develop the market in China, set up standards in photovoltaic industry and carry out market access rules, and speed up the integration of photovoltaic industry chains.

WO means gradual advance, which should seize opportunities and change weaknesses. On tactics, strategic measures is to adjust and optimize the industrial structure, integrate the photovoltaic industry, train more qualified staff, increase investment in scientific research, make breakthrough in technology and master core skills in photovoltaic industry, reduce the production cost of photovoltaic products, and strengthen government's supporting force, improve the top design, positively develop high-end industry chain junction of high additional values.

WT means defense or retreat, which should overcome weaknesses and avoid threats. On tactics, strategic measures is to improve inter-regional cooperation, optimize the structure of photovoltaic industry chain and regional distribution, reduce the reliance upon raw materials on foreign, increase investment, master core technology, reduce the production costs of photovoltaic products, actively explore the domestic market, decrease the dependence on overseas markets, and set up standards and market access system of Chinese photovoltaic industry in order to avoid the disorderly cooperation and blind expansion of photovoltaic industry.

Table 1. Features of the strategic combination of the photovoltaic industry development in China

External enviroment Internal factors condition factors	Opportunities (O) ① Government support in photovoltaic industry ② Latent necessity of the photovoltaic industry development as a result of global energy crisis ③ Juncture of integration in photovoltaic industry	Threats (T) ① Blind guidance of local government in the photovoltaic industry development ② External trade environment deteriorating due to international trade protection ③ Lack of photovoltaic industry standards
Strengths (S)	**SO strategy: aggressive attack (take advantages, seize opportunities)**	**ST strategy: corresponding defense (take advantages, avoid threats)**
①Photovoltaic industry is a newly developing resource industry ②Complete industrial chains ③Obvious strengths of cluster development in photovoltaic industry	① Greatly develop the photovoltaic energy industry ② Further perfects the photovoltaic industry chains ③ Adjust and optimize the government's supporting policies, promote the development of photovoltaic industry ④ Complete the top design, exert the advantage of cluster development in photovoltaic industry	① Enhance inter-regional cooperation, reasonably plan area distribution in photovoltaic industry ② Develop the market in China ③ Set up standards in photovoltaic industry and carry out market access rules ④ Speed up the integration of photovoltaic industry chains
Weaknesses (W)	**WO strategy: gradual advance (seize opportunities, change weaknesses)**	**WT strategy: defense or retreat (overcome weaknesses, avoid threats)**
① The photovoltaic industry chain mainly amassed on the mediate part of low value ② Relatively high cost of producing photovoltaic production component ③ Lack of core technology	① Adjust and optimize the industrial structure, integrate the photovoltaic industry ② Train more qualified staff, increase investment in scientific research, make breakthrough in technology and master core skills in photovoltaic industry, reduce the production cost of photovoltaic products ③ Strengthen government's supporting force, improve the top design, positively develop high-end industry chain junction of high additional values	① Improve inter-regional cooperation, optimize the structure of photovoltaic industry chain and regional distribution, reduce the reliance upon raw materials on foreign ② Increase investment, master core technology, reduce the production costs of photovoltaic products, actively explore the domestic market, decrease the dependence on overseas markets ③ Set up standards and market access system of Chinese photovoltaic industry in order to avoid the disorderly cooperation and blind expansion of photovoltaic industry

3   Construction of the SWOT-AHP Model of Photovoltaic Industry Development in China

3.1   The SWOT-AHP Model of Photovoltaic Industry Development in China

Based on the goal of sustainable development of photovoltaic industry in China, the SWOT-AHP model of photovoltaic industry development in China was constructed within the theoretical logic framework of the analysis on system factors and factor weight, decision making and strategy choice (Fig. 1)[8, 9].

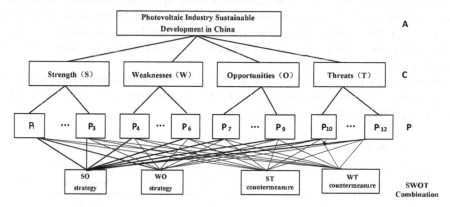

Fig. 2. The SWOT-AHP model of photovoltaic industry development in China

According to the general goal of the photovoltaic industry development in China, a bottom-up hierarchy was constructed. The top level was the objective level which is the sustainable development of the photovoltaic industry. The basic elements of SWOT analysis were middle constraining elements, which constituted the criteria level. The bottom level was the strategy combination of the photovoltaic industry development in China, which was the alternatives level [8, 9].

3.2   The Comparison Matrix by Layers

Between layer A and C showed in Fig.1, the comparison matrix of A-C layer can be established as $A = (c_{ij})_{n \times n}$, whose elements are evaluated using a 9-point scale. The paired comparison of element i with element j in the C layer is placed in the position of C_{ij} of the comparison matrix of A-C layer [6, 7].

The comparison matrix is a square matrix as $A = (c_{ij})_{n \times n}$, and:

$$c_{ij} > 0; \quad c_{ij} = 1/c_{ji}; \quad c_{ii} = c_{jj} = 1 \tag{1}$$

Similarly, the comparison matrix of C-P layers can be obtained in the same method used in forming the comparison matrix of A-C layers. Data on paired comparison matrices were collected from reviewers. Scores of c_{ij} were estimated on the average by inviting the field experts. The participating decision makers provided paired comparisons for each level of the hierarchy in order to obtain the weight factors of each element on that level, and with respect to one element in the next higher level [6, 7].

3.3   Overall Rank of the Hierarchical Level

Single ranks of the hierarchical level are used to calculate the importance of the elements of layer k+1 to layer k (meaning each element of C to A and P to C in this paper), and the elements of every layer are ranked according to the relative score of the paired comparisons [11].

The consistency index (CI) was calculated based upon the maximum eigenvalue. The consistency ratio (CR) is calculated by dividing the CI by the ratio index (RI). The CR has to be lower than 0.1, otherwise the matrix will be considered inconsistent. If the matrix is inconsistent, the eigenvector generated from this matrix will be rejected [11].

The overall rank of the hierarchical level is the important ranking of the elements of the strategy combination (SO, WO, ST, WT) to the general objective of A. The method algorithm is shown as Table 2.

Table 2. Calculation method for overall ranking of the hierarchical level

Items	C_1 W_{C_1}	C_2 W_{C_2}	C_3 W_{C_3}	C_4 W_{C_4}	Single ranks of A–P $\sum_{i=1}^{4} W_{C_i} W_{P_j}$ (j=1 2 ..., n)
P_1	$W_{P_1^1}$	$W_{P_1^2}$	$W_{P_1^3}$	$W_{P_1^4}$	$\sum_{i=1}^{4} W_{C_i} W_{P_1^i}$
P_2	$W_{P_2^1}$	$W_{P_2^2}$	$W_{P_2^3}$	$W_{P_2^4}$	$\sum_{i=1}^{4} W_{C_i} W_{P_2^i}$
$\vdots$	$\vdots$	$\vdots$	$\vdots$	$\vdots$	$\vdots$
P_n	$W_{P_n^1}$	$W_{P_n^2}$	$W_{P_n^3}$	$W_{P_n^4}$	$\sum_{i=1}^{4} W_{C_i} W_{P_n^i}$
Σ	1.00	1.00	1.00	1.00	1.00

Overall rank of the hierarchical level		
SO		$\left(\sum_{i=1}^{4} W_{C_i} W_{P_j} \ (j = 1\ 2\ ...,\ n) \right) * W_{SO}$
WO		$\left(\sum_{i=1}^{4} W_{C_i} W_{P_j} \ (j = 1\ 2\ ...,\ n) \right) * W_{Wo}$
ST		$\left(\sum_{i=1}^{4} W_{C_i} W_{P_j} \ (j = 1\ 2\ ...,\ n) \right) * W_{sT}$
WT		$\left(\sum_{i=1}^{4} W_{C_i} W_{P_j} \ (j = 1\ 2\ ...,\ n) \right) * W_{WT}$

Notes: In the SWOT-AHP model of photovoltaic industry development in China, n represented the numerical value of SWOT Sub-factors in index alternatives, for 12 in this paper.

4   Calculation Results of the SWOT-AHP Model of Photovoltaic Industry Development in China

Invite the experts, scholars and managers participating in the survey, and filling the strategic survey questionnaires of photovoltaic industry development in China in order to relative importance of the SWOT elements.

4.1   Single Ranks of the Hierarchical Level and Consistency Test

(1)A-C Single ranks
According to the calculating method above, the weights of the middle level to the top level can be figured out, the results was shown in Table 3:

Table 3. The comparison matrix of A-C layer

A	C_1	C_2	C_3	C_4	Weights of the middle level
C_1	1.0000	0.4727	0.7091	0.3406	0.1340
C_2	2.1154	1.0000	1.5000	0.7205	0.2835
C_3	1.4102	0.6667	1.0000	0.4803	0.1890
C_4	2.9360	1.3880	2.0820	1.0000	0.3935

Consistency test: λ_{max} =4.000123, CI=(4.000123-4)/3 = 0.000041, RI = 0.882 (Obtained by looking up tables, the same below）, CR= CI / RI =0.000041/0.882= 0.000046 <0.1(Pass the consistency test)。
(2) C-P Single ranks
The weights of indexes in the alternative level to indexes in the criteria level were shown in Table 4. The CRs were 0.0003, 0.0001, 0.0001and 0.0001, which meant that the results all passed consistency test.
(3) P-SWOT Single ranks
The weights of indexes in the alternative level to indexes in the bottom level were shown in Table 4. The CRs were 0.000077, 0.000179, 0.000009, 0.000188, 0.000109, 0.000069, 0.000041, 0.000267, 0.000240, 0.000338, 0.000017 and 0.000098, which meant that the results all passed consistency test.

4.2   Overall Rank of the Hierarchical Level

The weights of four strategies in bottom level to the sustainable development of the photovoltaic industry were shown in Table 4.

Table 4. The overall ranking results of the hierarchical level

SWOT factors	A-C weights	CR	SWOT sub-factors	C-P weights	A-P weights	P-SWOT Weights				
						SO	WO	ST	WT	CR
S	0.1340	0.0003	P_1	0.3316	0.0444	0.3886	0.2976	0.1984	0.1154	0.011505
			P_2	0.4538	0.0608	0.3400	0.2330	0.2913	0.1356	0.003067
			P_3	0.2145	0.0287	0.3515	0.2182	0.2500	0.1803	0.031795
W	0.2835	0.0001	P_4	0.2501	0.0709	0.1157	0.2975	0.1704	0.4163	0.053029
			P_5	0.3262	0.0925	0.0664	0.3357	0.1923	0.4056	0.052427
			P_6	0.4237	0.1201	0.0624	0.3757	0.1205	0.4414	0.076473
O	0.1890	0.0001	P_7	0.3100	0.0586	0.3697	0.2848	0.2118	0.1337	0.033883
			P_8	0.5195	0.0982	0.5064	0.3157	0.1230	0.0549	0.038559
			P_9	0.1705	0.0322	0.3978	0.3099	0.2195	0.0727	0.040398
T	0.3935	0.0001	P_{10}	0.3100	0.1220	0.0803	0.1732	0.3283	0.4181	0.069052
			P_{11}	0.5195	0.2044	0.1326	0.2286	0.2723	0.3665	0.081185
			P_{12}	0.1705	0.0671	0.1212	0.1806	0.3176	0.3806	0.054057
A-SWOT Weights						0.1991	0.2686	0.2266	0.3056	

5   Strategic Path Selection of Photovoltaic Industry Development in China

5.1   Effects Analysis of the SWOT Factors

The analysis results of the SWOT factors showed that the rank order of four SWOT factors was threats >weaknesses>opportunities>strengths based on the degree of influence. The results stated that the threats and weaknesses faced by photovoltaic industry development in China were obvious. The threats had great influence on photovoltaic industry development in China while strengths had the little influence.

In the strengths group of SWOT, the rank order of the factors was $P_2> P_1>P_3$, the greatest advantage was complete industrial chains. The second greatest advantage was Photovoltaic industry for a newly developing resource industry. The least advantage was obvious strengths of cluster development in photovoltaic industry.

In the weaknesses group, the rank order of the factors was $P_6>P_5>P_4$, the biggest weaknesses was Lack of core technology. The second biggest weakness was relatively high cost of producing photovoltaic production component. The third biggest weakness was the photovoltaic industry chain mainly amassed on the mediate part of low value.

In the opportunities group, the rank order of the factors was $P_8>P_7>P_9$, the biggest opportunity was latent necessity of the photovoltaic industry development as a result of global energy crisis. The second biggest opportunity was government support in photovoltaic industry. Juncture of integration in photovoltaic industry was at the last, which had the smallest influence.

In the threat group, the rank order of the factors was $P_{11}>P_{10}>P_{12}$, the biggest threat was external trade environment deteriorating due to international trade protection. The second biggest threat was lack of photovoltaic industry standards. Blind guidance of local government in the photovoltaic industry development was at the last, which had the smallest influence.

5.2   Selection of Strategic Path Combination

The results in Table 4 showed that the weights of the four combinations (SO strategy, WO strategy, ST countermeasure and WT countermeasure) was 0.1991, 0.2686, 0.2266, 0.3056, which meant that the rank order was WT countermeasure>WO strategy > ST countermeasure >SO strategy. The results showed WT countermeasure or WO strategy were the better choices than anyone of SO strategy and ST countermeasure, and the WT countermeasure was the best choice.

6   Conclusions

According to the SWOT-AHP quantitative analysis of photovoltaic industry development in China, WT countermeasure and WO strategy were the better choice, and the WT countermeasure was the best choice. The basic strategy is to choose the WT countermeasure as the strategic optimal path in order to overcome weaknesses and avoid threats. On tactics, strategic measures is to strengthen the regional cooperation, optimize the structure of photovoltaic industry chain and regional distribution, reduce the dependence of raw materials on foreign; to increase investment, master the core technology, reduce the production costs of photovoltaic products, actively explore the domestic market; to set up the standards and market access system of Chinese photovoltaic industry in order to avoid the disorderly competition and blind expansion of photovoltaic industry.

References

1. Yanxia, G., Yanyun, G.: The present situation analysis and development countermeasures of solar photovoltaic power generation. Shanxi Architecture 39(3), 199–201 (2013) (in Chinese)
2. Jing, F.: Analysis on the Status of international competitiveness and Development Approaches of China's PV industry. Journal of Hebei University (Philosophy and Social Science) (2), 1–12 (2013) (in Chinese)
3. Chong, T.: Research on the Photovoltaic Industry Development Paths of Liaoning Province Based on Industrial Cluster Theory—Taking Jinzhou City as a Case. Science and Technology Management Research (20), 137–139, 147 (2013) (in Chinese)

4. Yanfang, L., Yazheng, L.: A Study on the Operating Performance of PV Industry Listed Companies—On the basis of 23 PV industry listed companies. Science Technology and Industry 12(10), 77–80 (2012) (in Chinese)
5. Zhengnan, L., Chunqi, L., Guodong, W.: Research on Evaluation Index System of Collaborative Performance of Photovoltaic Industrial Chain. Science & Technology and Economy (1), 106–110 (2013) (in Chinese)
6. Duchelle Amy, E., Guariguata Manuel, R., Giuliano, L., Antonio, A.M., Andrea, C., Tadeu, M.: Evaluating the opportunities and limitations to multiple use of Brazil nuts and timber in Western Amazonia. Forest Ecology and Management 268, 39–48 (2012)
7. Lingxian, Z., Zetian, F., Xiaoshuan, Z.: Optimization on Structure of China Agricultural Domestic Support based on Increasing Farmers' Income. Systems Engineering Theory & Practice (4), 9–18 (2007)
8. Seungbum, L., Patrick, W.: SWOT and AHP hybrid model for sport marketing outsourcing using a case of intercollegiate sport. Sport Management Review 14(4), 361–369 (2011)
9. Miika, K., Pekka, L., Mikko, K., Jyrki, K.: Making use of MCDS methods in SWOT analysis—Lessons learnt in strategic natural resources management. Forest Policy and Economics 20, 1–9 (2012)
10. Zavadskas, E.K., Turskis, Z., Tamosaitiene, J.: Selection of construction enterprises management strategy based on the SWOT and multi-criteria analysis. Archives of Civil and Mechanical Engineering 11(4), 1063–1082 (2011)
11. Aguarón, J., Moreno-Jiménez, J.M.: The geometric consistency index: Approximated thresholds. European Journal of Operational Research 147, 137–145 (2003)

Design of the Unmanned Area Fetching Trolley

Xuelun Hu, Licai Zhang, Yaoguang Wei[*], and Yingyi Chen

College of Information and Electrical Engineering,
China Agricultural University, Beijing 100083, China
{982537081,804306590}@qq.com, weiyaoguang@gmail.com,
chyingyi@126.com

Abstract. This paper discusses the project which is one of the Beijing City College Students' Scientific Research and Entrepreneurial Action Plan Project named "unmanned area fetching trolley", The project used infrared sensor to measure the distance between the car and the obstacle, The MCU would analyze and process this signal, and generate two different PWM signals respectively to drive motors when the distance is less than a certain value, so as to realize avoiding obstacle automatically. After Hall sensor detected the PWM wave corresponding to the actual speed of the car, the PID module would compare the PWM wave of the actual speed and the PWM wave which is generated by MCU, so as to determine whether the output is desired and achieve closed-loop control, thus can make the control of speed is more precise. The mechanical arm is composed of five steering, the upper monitor control it through the wireless module, each instruction can make the rotation angle of each actuator accurate to 1 degree, so the mechanical arm is flexible and precise. The project ultimately achieved the desired effect that avoid obstacle automatically, upper monitor control accurately. After expansion and improving, it can be used for automatic weeding, automatic cleaning, automatic patrol, automatic sowing, automatic harvesting and so on.

Keywords: avoid obstacle automatically, mechanical arm, fetching trolley.

1   Introduction

Nowadays, automation has entered and played an important role in all trades and professions, such as industry, agriculture, transportation, national defense and other aspects. Automation also makes significant contribution in the development of the national economy. It is the degree of automation, which represents the level of the industry's development.

Agriculture is the foundation of national economy, agriculture machinery automation is the center of agricultural modernization. It embodies the latest achievements of modern science and technology, and has become the necessary condition of increasing production, improving labor productivity, and reducing the heavy manual labor. In general, realizing automation in agricultural production has the following

---

[*] Corresponding author.

D. Li and Y. Chen (Eds.): CCTA 2013, Part II, IFIP AICT 420, pp. 523–533, 2014.
© IFIP International Federation for Information Processing 2014

significance: improving labor efficiency; shorten the production cycle; improving economic efficiency; reducing the labor intensity; and emboding a country's level of science and technology[1].

An trolley which can avoid obstacle automatically is able to apply to industry and agriculture, such as automatic weeding, automatic cleaning, automatic patrol, automatic sowing, automatic harvesting and so on. This paper discusses the unmanned area fetching trolley's implementation amply, and introduces the required modules and their functions, even analyses the application domain of this trolley.

This unmanned area fetching trolley can get into the place which is dangerous, inconvenient or harmful for people, such as high temperature,radiant and so on. The trolley's walking distance can coverage all-round of the region,so as to fetching the objects. Nowadays, the self-navigation, Global Position System (GPS) and the machine vision can achieve this function, they are high precision, applicable to a wide range, easy to use relatively, but also have the limitation of expensive. This unmanned area fetching trolley equiped with the infrared obstacle avoidance and the wireless module to realize the low cost and wireless intelligent control.

2   The Implementation of the Project

2.1   Requirement Analysis and the Terms of Settlement

The main idea of this project is to make a trolley which is capable of walking automatically, avoiding obstacle automatically, fetching objects by mechanical hand accurately.

The implementation of this unmanned area fetching trolley needs to solve two major problems which are the automatic obstacle avoidance and the wireless remote control. Through analysis and calculation, we select the infrared sensor which is low cost and a wireless control module which is high precision finally[2].

2.2   Content and Technique of the Trolley

2.2.1   Content of the Research

The car is equipped with infrared sensor, the signal which access from the sensor is the input of the MCU(STC12C5A60S2). The MCU would analyze and process this signal, and generate two different PWM signals respectively to drive motors when the distance is less than a certain value, so as to realize avoiding obstacle automatically. After Hall sensor detected the PWM wave corresponding to the actual speed of the car, the PID module would compare the PWM wave of the actual speed and the PWM wave which is generated by MCU, so as to determine whether the output is desired and achieve closed-loop control, thus can make the control of speed is more precise. Furthermore, this trolley can be used in multiple aspects, such as automatic weeding machine, automatic vacuum cleaner, automatic patrol car and so on. In this article, we equipped it with mechanical arm to achieve the innovation[3].

2.2.2   Technique of the Research

(1) The module diagram of the trolley as shown in figure 1, from the graph we can find out the relation between each module in this trolley.

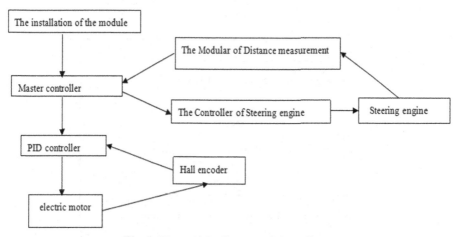

Fig. 1. The module diagram of the trolley

(2) MCU(STC12C5A60S2)

STC12C5A60S2 is a MCU which has single clock / machine cycle(1T) produced by STC, it is a new generation of 8051 single chip microcomputer which has high speed, low power consumption and strong anti-interference. It integrates MAX810 special reset circuit, 2 channel of PWM, 8 channel of high speed 10 bit A/D converter, it mainly used for motor control and strong interference situation. The pin diagram of STC12C5A60S2 as shown in figure 2.

	PDIP-40	
CLKOUT2/ADC0/P1. 0 — 1		40 — VCC
ADC1/P1. 1 — 2		39 — P0. 0/AD0
RxD2/ECI/ADC2/P1. 2 — 3		38 — P0. 1/AD1
TxD2/CCP0/ADC3/P1. 3 — 4		37 — P0. 2/AD2
SS/CCP1/ADC4/P1. 4 — 5		36 — P0. 3/AD3
MOSI/ADC5/P1. 5 — 6		35 — P0. 4/AD4
MISO/ADC6/P1. 6 — 7		34 — P0. 5/AD5
SCLK/ADC7/P1. 7 — 8		33 — P0. 6/AD6
P4. 7/RST — 9		32 — P0. 7/AD7
RxD/P3. 0 — 10		31 — EX_LVD/P4. 6/RST2
TxD/P3. 1 — 11		30 — ALE/P4. 5
INT0/P3. 2 — 12		29 — NA/P4. 4
INT1/P3. 3 — 13		28 — P2. 7/A15
CLKOUT0/T0/P3. 4 — 14		27 — P2. 6/A14
CLKOUT1/T1/P3. 5 — 15		26 — P2. 5/A13
WR/P3. — 16		25 — P2. 4/A12
RD/P36 — 17		24 — P2. 3/A11
XTAL2 — 18		23 — P2. 2/A10
XTAL1 — 19		22 — P2. 1/A9
Gnd — 20		21 — P2. 0/A8

Fig. 2. The pin diagram of STC12C5A60S2

The MCU received the infrared sensor's input signal though P1.0 (ADC) and communicated with PID module by serial port P3.0 and P3.1, thus the MCU can produce two different PWM waveforms to control two motor's speed, so as to make the trolley rotate to different directions[4]. Flow chart as shown in figure 3.

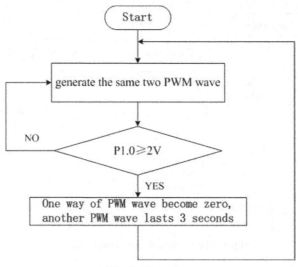

Fig. 3. Flow chart

(3) Infrared sensor (GP2Y0A21YK0F)

The light wave is the electromagnetic wave which wavelength between 10~106nm. The Infrared wavelength range is between 780~106nm[5]. The wavelength can be determined by the follow formula :

$$\lambda_0 = \frac{hc}{A} = \frac{1.239}{A} \tag{2-1}$$

Among the formula, C is the speed of light, H is the Planck constant.

GP2Y0A21YK0F is a distance measuring sensor unit, composed of an integrated combination of PSD(position sensitive detector), IRED (infrared emitting diode) and signal processing circuit[6]. The block diagram of GP2Y0A21YK0F as shown in figure 4.

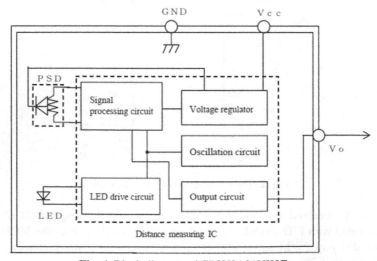

Fig. 4. Block diagram of GP2Y0A21YK0F

The barrier of different distance can make infrared sensor produce different voltage output. The closer obstacle stay, the higher voltage export. Output is connected with the SCM's P1.0.Measurement range of the sensor is 3cm to 80cm, Example of distance measuring characteristics as shown in figure 5.

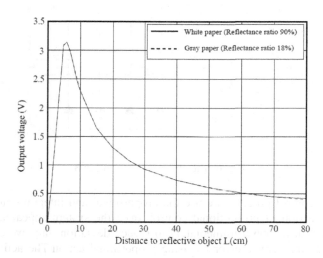

Fig. 5. Example of distance measuring characteristics(output)

(4) Speedometer

It is the Hall sensor that form the module of Speedometer .Each motor is equipped with two Hall sensors, the sensor is conducted when the magnet on the motor rotate over the Hall sensor, then the corresponding port output a low level. The sensor generate a pulse when each times the Magnet over the sensor. So the number of the output pulse represents the speed of the motor. As the magnet pass two Hall sensors successively, the waveform of the two sensor will have skewing. From the skewing we can know the direction of rotation of the motor.

(5) Motor driver (L298) and PID controller (SCM STC12C5410AD)

PID controller communicate with MCU. According to the Output of the MCU, the controller can generate different PWM waveform and deliver the PWM waveform which is amplifying by the motor drive to two motor. In addition, the PID controller is linked to the Speedometer, then delivering the actual speed of the motor to PID controller, furthermore, it would determine whether the output is desired or not. Thus we achieved the closed-loop control, so the speed control is more precise.

L298 is a monolithic motor driver, which has integrated high voltage, high current, and dual full bridge. It is designed to connect standard TTL logic level and driving inductive loads (such as relays, coils, DC and stepper motor)[7]-[8]. The external circuit of L298 as shown in figure 6.

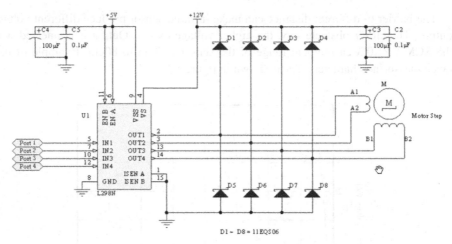

Fig. 6. The external circuit of L298

In the PID control process, we use the proportional and integral control. The following is the action of proportional control, once the system appeared deviation, it can adjust control response immediately to reduce deviation, thus we can accelerate the adjustment and reduce error by using proportional action.The action of integral control is to make the system to eliminate the steady state error and improve indiscrimination degree. The integral control is in progress until the output is no error. In our project, the KP(proportional coefficient) is 1.8, KI(Integral time constant) is 0.67. The system structure diagram as shown in figure 7.

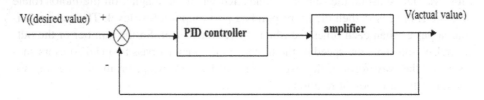

Fig. 7. System structure diagram

2.3   Content and Technique of the Mechanical Arm

2.3.1   Research Content
The trolley is equiped with the mechanical arm which is composed of five steering, the upper monitor control mechanical arm through the wireless module[9], each instruction can make the rotation angle of each actuator accurate to 1 degree, so the mechanical arm is flexible and precise.

2.3.2   Research Technique
(1) Servo Controller(STC10F08XE)
Servo Controller is connected with upper monitor through the serial port and it can control 16 ways steering simultaneously by decoding the upper monitor's instructions

so as to produce different PWM waveform to control the rotation of each steering engine[10]. So the circuit is simplified largely. Hereon,we use five of the ways to control five steering. The control flow chart as shown in figure 8.

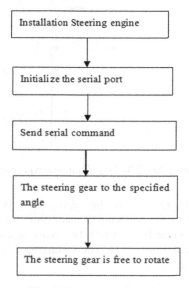

Fig. 8. The control flow chart

(2) Wireless module (APC220-43)

APC220-43 is a wireless data transmission module which highly integrated half duplex micro power, it embedded MCU and high performance RF chip ADF7020-1. Its anti-interference ability and the sensitivity is very high. A part of the Wireless module is connected with the controller's serial port, while the other part is connected with the upper monitor, thus it can realize wireless communications. The control range is up to 15 meters. Wiring diagram of APC220-43 and upper monitor as shown in figure 9, wiring diagram of APC220-43 and terminal equipment (servo controller) as shown in figure 10.

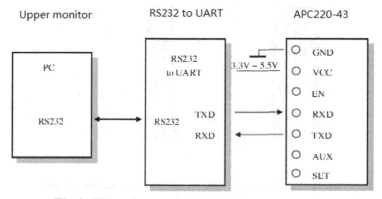

Fig. 9. Wiring diagram of APC220-43 and upper monitor

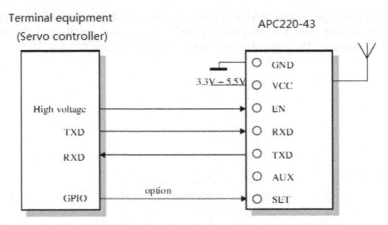

Fig. 10. Wiring diagram of APC220-43 and terminal equipment (servo controller)

The servo control can be precisely to the degree through the instruction, so as to achieve the precise control of the mechanical hand, thus we can make five servos of the manipulator act at the same time. These two reasons are the greatest advantage relative to the remote manual control[11]. The schematic diagram and number of mechanical arm as shown in figure 11.

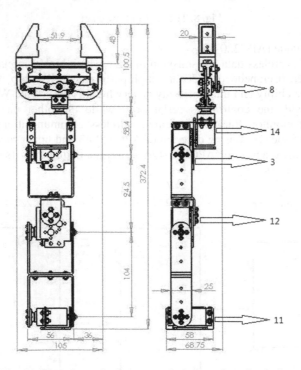

Fig. 11. The schematic diagram and number of mechanical arm

Through experiment and calculation, we can figure out the rotation angle of each servo manipulator arm of stretch, fetch and reset[12]. Instruction format is as follows:

D<time> #<ch>A<degree>...#<ch>A<degree>T<step>!   (2-2)

<time> represents the waiting time to run this command, it's unit is ms.

<ch> represents the number of steering engine which is under control.

<degree> represents the angle of the steering engine, the range of it is from 0 to 180 degrees.

<step> represents the number of step of the steering engine to complete this action, the time of completing each step is 20ms, the number of steps multiply 20ms equal to the finish time.

"!"represents the command come to end.

Three action commands are as follows:

Action	Instruction	The meaning after decoding
stretch	D1500 #08A10#14A90#03A160 #12A60#11A60T50!	This instruction is executed after 1.5 seconds, the steering engine number of 8 turn to 10 degrees, the steering engine number of 14 turn to 90 degrees, the steering engine number of 03 turn to 160 degrees, the steering engine number of 12 turn to 60 degrees, the steering engine number of 11 turn to 60 degrees, a second after the execution of this instruction. The time of complete this instruction is 1s.
fetch	D1500 #08A90#14A90#03A160 #12A60#11A60T50!	This instruction is executed after 1.5 seconds, the steering engine number of 8 turn to 90 degrees, the steering engine number of 14 turn to 90 degrees, the steering engine number of 03 turn to 160 degrees, the steering engine number of 12 turn to 60 degrees, the steering engine number of 11 turn to 60 degrees, a second after the execution of this instruction. The time of complete this instruction is 1s.
reset	D1500 #08A90#14A90#03A90# 12A90#11A90T50!	This instruction is executed after 1.5 seconds, the steering engine number of 8 turn to 90 degrees, the steering engine number of 14 turn to 90 degrees, the steering engine number of 03 turn to 90 degrees, the steering engine number of 12 turn to 90 degrees, the steering engine number of 11 turn to 90 degrees, a second after the execution of this instruction. The time of complete this instruction is 1s.

3   Analysis of the Results

3.1   Innovation of the Project

(1)The domination of the trolley which can avoid obstacle automatically achieved the closed-loop, so the domination is more precise.

(2)Using upper monitor to send commands so as to make the control more precise and more accurate, the mechanical hand separate itself from the telecontroller.

(3)When the trolley which can avoid obstacle automatically encountered every obstacles, it can turn left 30 degrees and then move on ,each encounter obstacles will make the trolley stay for a while, so as to fetching objects.

3.2   The Expandable Portion of the Project

(1)The cameras extract the unmanned area actual situation to accomplish fetching things.

(2) Adjusting the steering when the trolley encountered obstacles let the trolley cover more place of the unmanned area.

3.3   Results of the Project

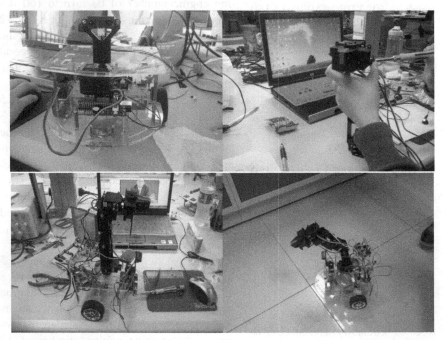

4   Conclusion

The "unmanned area fetching trolley" is implemented by multiple modules, each module own the fixed work mode and implementation effect, so it is convenient for us to debugging. Available pins of the SCM can be used, and the compatibility of the program is good. In large and medium-sized machines, the trolley can work stable when you change the drive module and the drive motor to correct type. Avoiding obstacle automatically also can be applied to many aspects. After expansion and improving, it can be used for automatic weeding, automatic cleaning, automatic patrol, automatic sowing, automatic harvesting and so on[13]. So the project have certain degree of applicability and expansibility.

Acknowledgements. This work was supported by the National AgriculturalScience and technology achievement transformation project (2012GB2E000330), theFundamental Research Funds for the Central Universities（2013QJ053）.

References

1. Jie, Z.: Characteristics and development of mechanical automation technology. City Construction Theory Research (Electronic Edition) (33) (2012)
2. Jing, C.: The design of control system of intelligent car based on STC12C5A08S2. Yinshan Academic Journal (Natural Science Edition) 25(4), 40–43 (2011)
3. Fanxin, Y., Xiaoming, L., Xianlun, H., et al.: The production of Wireless remote control infrared obstacle avoidance car. Electronic Engineering of Yantai Nanshan University (36), 52–53 (2011)
4. Xuejian, W., Long, C.: Design and implementation of intelligent obstacle avoidance car base on Single Chip Micyoco. Network and the Technology (1), 148–150 (2012)
5. Zhicong, Z.: Discussion on technology of infrared obstacle avoidance robot. Chinese Collective Economy (10), 168–169 (2008)
6. Yongchao, H., Wei, W., Sensen, D., et al.: Infrared obstacle avoidance of mobile robot and monocular vision tracking of. Mechanical & Electrical Engineering 23(6), 60–62 (2006)
7. Guodong, T., Yunguo, G.: Singlechip control of a stepping motor based on L297/L298 chip. Micro Computer Information 22(12-1), 134–136 (2006)
8. Guangqi, X., Min, Y., Yinfeng, W., et al.: MCU drive control of stepping motor. Journal of XiangNan University 32(5), 37–41 (2011)
9. Bo, X., Shuangxi, G.: Wireless remote visual control of robot. Wireless Internet Technology (9), 141 (2012)
10. Yan, W., Qinghua, Y., Guanjun, B., et al.: Design and test of mechanical arm joint picking optimization. Journal of Agricultural Machinery 42(7), 191–195 (2011)
11. Hongxing, G., Lenian, Z., Jianhong, T., et al.: Research on Motion Control of Manipulator Based on Microprocessor. Mechanical Manufacturing and Automation (5), 150–152 (2010)
12. Yong, C., Jiaqiang, Z., Weibin, G.: Kinematics analysis and motion control for a weeding robotic arm. Journal of Agricultural Machinery 38(8), 105–108 (2007)
13. Qingbo, Z.: Study on avoidance techniques and avoidance control of fruit picking robot. Jiangsu University (2008)

Intelligent Ammonia-Nitrogen Sensor
Which Based on Ammonia Electrode

Fan Zhang, Yaoguang Wei[*], Yingyi Chen, and Chunhong Liu

College of Information and Electrical Engineering, China Agricultural University,
Beijing 100083, China

Abstract. To solve the problems that the traditional ammonia detection methods were complex, not easy to maintenance and difficult to realize quick measurement in situ, an intelligent ammonia sensor has been designed in this paper. The intelligent ammonia sensor integrates ammonia electrode, pH electrode and Ammonium ion electrode together to realize the In situ detection of ammonia. Because the output signal of ammonia electrode is weak and easy to be disturbed by external interference, a low-pass filter circuit has been designed, this kind of circuit have a good effect. The test results have shown that the sensor is easy operation, low cost and no pollution.

Keywords: aquaculture Ammonia nitrogen Ammonia sensitive electrode Sensor on-line monitoring.

1   Introduction

Our country is a big agricultural country. Aquatic production has been 15 consecutive years ranked first in the world. However, most farmers rely on their own experience, with color, smell, water taste or observed other aquatic animal's abnormal behavior to evaluation aquaculture water quality. For aquaculture, dissolved oxygen, ammonia nitrogen, PH is one of the important water quality parameters need to be monitoring[1]. Ammonia existing in water has certain toxicity for aquaculture products. It influents the quality of aquatic products, restrict the sustainable development of aquaculture, especially with the promotion of high density factory farming technology, ammonia pollution control demand is increasingly prominent. Ammonia harm the aquatic organisms mainly refers to the dangers of non-ionic ammonia, after non-ionic ammonia entering aquatic organisms[2] . It has a significant impact on enzyme hydrolysis reaction and membrane stability. Demonstrated difficulty in breathing, not feeding, decreased immunity, convulsions, coma and other phenomena, affect the growth and reproduction of aquatic organisms, even lead to aquatic organisms decimated, even causing loss to the economy[3].

There are different ways to determine the content of ammonia nitrogen in the water[4][5]. The existing method of ammonia nitrogen determination has some

---

[*] Corresponding author.

D. Li and Y. Chen (Eds.): CCTA 2013, Part II, IFIP AICT 420, pp. 534–543, 2014.
© IFIP International Federation for Information Processing 2014

shortcomings[6]. Such as titration's sensitivity is not high enough. Spectrophotometry needs a large amount of chemical reagent and complicated steps[7]. Ammonium ion electrode method is easily affected by other monovalent cations. Optical fluorescence technology is not mature. Spectrometry instruments are expensive. These methods are difficult to meet the needs of the scene in situ detection with high frequency. With ammonia sensitive probe we can realize quick measurement in situ. The sensor has some good features, such as simple operation, low cost, pollution-free, and don't need to pretreatment the water. Based on the above analysis, an compound ammonia sensors which integrated ammonia sensitive probe and PH probe has been designed in this paper. A low cost ammonia nitrogen on-line monitoring method will be the main research content, and find a solution of electrochemical intelligent ammonia sensor.

2   Measurement Principle

Ammonia nitrogen content in the water is in the form of free ammonia NH_3 and ammonium ion NH_4^+ chemical combination of the existence of the amount of nitrogen[8]. It is an important index of water pollution. When free ammonia NH_3 reaches a certain concentration is harmful to aquatic organisms. For example, it will be able to cause toxic effects on some kind of fish when free ammonia over 0.2mg/L .The solubility of ammonia in water at different temperatures and PH is different, when the PH content is high, it will have a higher proportion of free ammonia, on the contrary[9], a higher proportion of the ammonium ions. Under a certain condition, the ammonia and the ammonium ion has the following balance equation: $NH_3 + H_2O \leftrightarrow NH_4^+ + OH^-$

In this study, Ammonia electrode 9512HPBNWP was elected to measure ammonia content in water. The ammonia electrode is a composite electrode, PH glass electrode as indicator electrode, silver - silver chloride electrode as the reference electrode. Put the electrodes inside a plastic sleeve which containing 0.1mol/L ammonium chloride liquid-filled, and equipped with gas-sensitive film. Add ionic strength to the aqueous sample solution, PH may be raised to 11 or more, and ammonium salts are converted to ammonia, because of diffusion, ammonia gas will pass through the membrane (water and other ions can't pass the gas membrane). After ammonia gas into the inner filling, will present the following balance: $NH_3 + H_2O = NH^{4+} + OH^-$

Ammonia leads the balance equation shift to the right, then the value of PH increased with the entry of ammonia. Finally PH glass electrode measured value changes. At constant ionic strength, temperature, nature and electrode parameters, the measured electromotive force and the ammonia concentration in water samples meet the Nernst equation[10]. We can determine the nitrogen content of the sample from the measured potential value. Finally draw a standard curve by measured voltage signal to determine the concentrations of the unknown samples. Fig.1 shows measuring device of ion selective electrode.

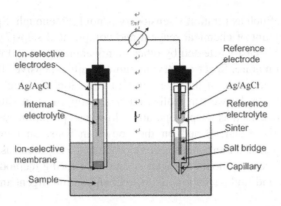

Fig. 1. Measuring device of ion selective electrode

As content of free ammonia in the solution changes, formed the overall balance:

$$E = E^0 - \frac{RT}{nF}\ln\frac{\alpha_{i1}}{\alpha_{i2}} = E^0 + 2.303\frac{RT}{nF}\lg\frac{\alpha_{i1}}{\alpha_{i2}} \tag{1}$$

In the formula above, E is a single electrode potential; E^0 is the potential difference between standard electrodes; T is the absolute temperature; R is the gas constant and equal to 8.31J / (mol × K); n is the transferred charge moles under E^0; F is the Faraday constant equal to 96467C.

Nernst equation is an important formula which linking the potential difference of the chemical system and electro-active substance activity (concentration) together. It is also an important theoretical basis for the electrochemical analysis method[11].

Under the action of strong alkaline solution Ammonium ions is converted to the dissolved ammonia. Since ammonia through the semipermeable membrane into the internal electrolyte, free ammonia and H^+ in the electrolyte thin are combined to form the ammonium ion. Assuming the electrolyte is not a PH buffer solution, as a characteristic of alkaline ammonium action will then increase the PH value of the test solution, as a characteristic of alkaline ammonium the PH value of the test solution will increase. Using 0.1mol/L NH4Cl as buffer solution, due to the concentration of NH_4^+ is significantly higher than reaction of NH_4^+ therefore consider that NH_4^+ concentration is a constant. The change of PH value is determined by the change of concentration of NH_4^+. Therefore, there are following equations:

$$a_{H^+} = K\frac{a_{NH_4^+}}{a_{NH_3}} = k'\!\Big/\!P_{NH_3} \tag{2}$$

$$K = a_{H^+} + P_{NH_3}\!\Big/\!a_{NH_4^+} \tag{3}$$

$$E = E^0 - \frac{2.303\,RT}{F}\lg P_{NH_3} \tag{4}$$

P_{NH} is the partial pressure of ammonia in the sample or in the thin layer, According to Henry's law P_{NH3} =K[NH3], K is Henry's law constant, so the battery voltage and ammonium ion concentration in the sample under Nernst relations:

$$E = E^0 - \frac{2.303RT}{F} \lg NH_3$$

(5)

The formula expressed that potential difference between two electrodes and the measured molar concentration of ammonia solution has a logarithmic relationship. Therefore, it is only need to test potential difference between the two electrodes and the temperature. Then the concentration of ammonia in the solution can be measured by calculating.

3 Circuit Design of Ammonia Sensor

3.1 Overall Framework

Smart sensors include ammonia-sensitive probe ammonia, ammonium ion probe, pH and temperature probe, signal conditioning modules, TEDS memory, microcontrollers MSP430, bus interface module, power management modules, etc. As shown above. A free ammonia signal will be obtained by Ammonia-sensitive probe, a ammonium ion signal will be got through the ammonium ion probe, PH signal got by the pH probe. Signals are transferred to MSP430 MCU A/D input port through the transmission circuit. And then the concentration of free ammonia, the concentration of ion ammonium, value of pH, water temperature and nitrogen content are calculated by the microcontroller. Finally, the bus interface module output variables.

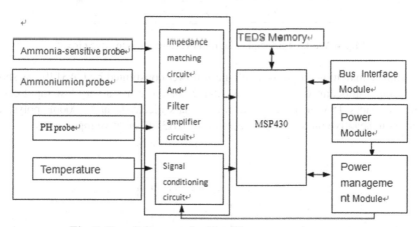

Fig. 2. Overall framework of Intelligent ammonia sensors

3.2 Power Modules

For glass electrode's characteristics that output positive voltage in acid solution and output negative voltage in alkaline solution, a negative 3.3V power supply module was designed.

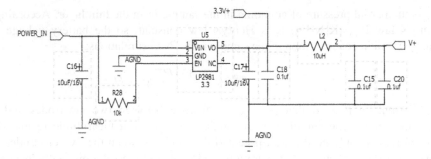

Fig. 3. +3.3V schematic circuit diagram

3.3   Singal Transmission Modules

Since ammonia electrode and pH electrode output impedance are particularly high, so
the first stage of the amplifier circuit must use high input impedance op amp to match
impedance. In addition, during the test it is easy to see that the electrode probe output
signal susceptible to interference by 50Hz signal, so the low-pass filter circuit is added
to the signal conditioning module.

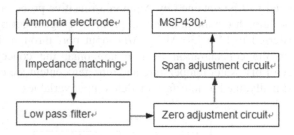

Fig. 4. Working principle diagram of Ammonia sensitive circuit

Impedance matching circuit is a voltage follower constituted by CA3140. As shown,
CA3140 input impedance is as large as 1.5TΩ, therefore has a very low input current.
Connect resistors 15M to positive output of glass electrode and 1000pF polystyrene
capacitors to negative output. Then access to the CA3140 positive input port.

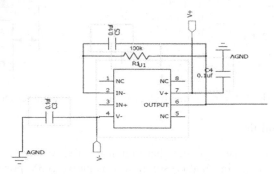

Fig. 5. Impedance matching circuit

Low-pass filter circuit is a Π-type RC filter constituted by a TLC27L4, and it is also a bidirectional integration filter. We can adjust the RC value to precisely control the time constant. In this experiment, the low-pass filter cutoff frequency is 50Hz. When the external interference is greater than 50Hz, the interference signal will be attenuated less than 45db. This circuit has a good filtering effect.

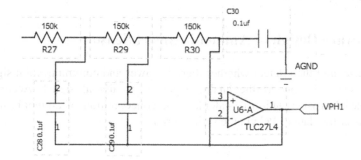

Fig. 6. Low-pass filter circuit

3.4   Temperature Compensation

Temperature compensation circuit is used to ensure the circuit working properly and stability in a certain temperature range. Some devices have difference of the positive temperature coefficient and the negative temperature coefficient such as transistors, diodes, resistors. When temperature rises, positive temperature coefficient devices' effect will increase, and the negative coefficients devices have opposite effect.

Since platinum RTD has good stability, on line analyzers always use it to automatically compensate for temperature. The working principle of platinum PRD is transform the changes of temperature into the changes of resistance. The following figure is Pt100 platinum resistance's resistance changes with temperature. in the range of 0-100℃.

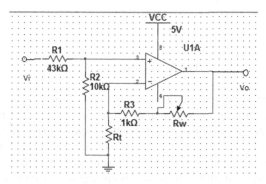

Fig. 7. Temperature compensation circuit

Operational amplifiers and platinum thermal resistance may constitute automatic temperature compensation circuit. In this circuit A is the integrated operational amplifier, Rt is a platinum RTD, Vi is the output signal of the transmitter. When the input signal Vi is not the maximum value, the output signal V0 is always constant in the range of 0-100℃. So this kind of circuit realized the purpose of automatic temperature compensation

4   Software Design of Ammonia Smart Sensor

First, initialize, and then detect whether there is power and communication signals. If a signal is entered, convert the analog signal is to digital signal. According to predetermined Nernst equation to calculate the concentration of ammonia nitrogen. Finally transfer the data to PC via the serial port[12].

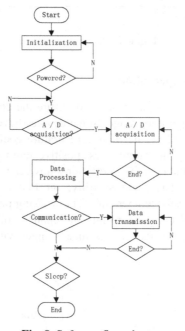

Fig. 8. Software flow chart

The digital filter is an important foundation for digital signal processing in the signal filtering, monitoring and parameter estimation process. It is the most widely used of a new system. Digital filter is complete signal filtering functions, using finite precision arithmetic to achieve discrete-time linear time-invariant systems. The inputs are a group of analog signal sampling and quantization coding of the digital, the output is another set of digital which had been digital conversion. Digital filter has high stability, high accuracy, flexibility and other prominent features.

IIR digital filter system function can be written in the form of closed functions Which using recursive structure, namely the structure with a feedback loop. IIR filter structures are usually composed by the basic operations such as delay, multiplied by the coefficient and adding.

An N-order IIR filter system function can be expressed as:

$$H(z) = \frac{\sum_{k=0}^{M} b_k z^{-k}}{1 + \sum_{k=1}^{N} a_k z^{-k}} \tag{6}$$

$$y(n) = \sum_{k=0}^{M} b_k x(n-k) - \sum_{k=1}^{N} y(n-k) \tag{7}$$

5   Experimental Data Analysis

Firstly, checking electrode slope, Obtaining the slope value provides the best means for checking electrode operation. Slope is defined as the change in milivolts observed with every tenfold change in concentration. This experiment used 9512HPBNWP ammonia electrode. When the solution temperature is in the range of 20°C to 25°C, the slope should be between -54mV to -60mV. If the slope is not in this range, which means that ammonia electrode is not working normally, troubleshoot need to be done firstly.

Analyze the factors that affect ammonia electrode.

In the conditions of a temperature of 25 °C, using standard ammonia solution to configure sample solution of 0.01mg/L, 0.1mg/L, 1mg/L and 10mg/L. added Ionic strength agent to adjust the PH. When PH is equal 7,8,9,10,11, it can measured values of ammonia-sensitive transmitter output of different concentrations of the sample solution. From the measured data can be seen, when the PH is greater than 11, ammonia nitrogen in the solution can be completely converted to free ammonia overflow and accurately measuring the amount of free ammonia.

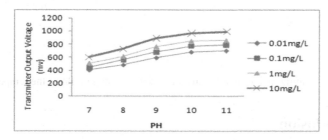

Fig. 9. Curve of ammonia electrode transmitter output voltage follows the PH value

Using standard ammonia solution to configure sample solution of 0.01mg/L , 0.1mg/L, 1mg/L, 10mg/L. Measured ammonia electrode transmitter output value once every 10 °C In the range of 10°C to 50°C. The measurement results as shown below. When the PH value is greater than 11, it can be seen that the effect of temperature on the concentration of free ammonia is not obvious.

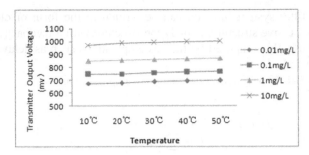

Fig. 10. Curve of ammonia electrode transmitter output voltage follows the temperature

Measure the effect of temperature and PH on concentration of free ammonia with a fixed concentration of a sample solution. Recorded ammonia electrode transmitter output value once every 10 ℃ In the range of 10℃ to 50℃. While adjust the pH by the ion strength agents, making pH of the sample solution is stable at 7, 8, 9, 10, 11, 12. By experiment, obtain a curve of ammonia electrode transmitter output voltage follows the temperature and PH in solution of 10mg/L. From the figure can get the conclusion that the same with the above two curves. When the PH is greater than 11, ammonia nitrogen in the solution can be completely converted to free ammonia overflow and accurately measuring the amount of free ammonia. The effect of temperature on the concentration of free ammonia is not obvious from 10℃ to 50℃, Therefore, in order to simplify processing, the temperature may be regarded as constant.

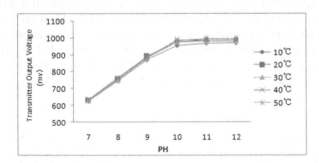

Fig. 11. Curve of ammonia electrode transmitter output voltage follows the temperature and PH

6   Conclusion

According to the actual needs of aquaculture water quality monitoring, to solve the problems that the traditional ammonia detection methods were complex, not easy to maintenance and difficult to realize quick measurement in situ, Ammonia electrode and PH electrode was chosen as the sensing probe. Using both hardware and software filtering to solve the problems that Ion selective electrode output signal is weak and susceptible to be disturbed. A smart ammonia sensor is developed. According to the

Nernst equation and characteristics of interactions among free ammonia, ammonium ions, PH and temperature, an algorithm and a relationship between ammonia concentration and voltage signal from ammonia-sensitive probe can be got. This kind of smart sensor achieved rapid detection of ammonia concentration and has many advantages such as need no pretreatment of water samples, simple operation, low cost and no pollution.

Acknowledgment. This work was supported by the National Agricultural Science and technology achievement transformation project (2012GB2E000330), the Fundamental Research Funds for the Central Universities (2013QJ053.

References

[1] Palani, S., Liong, S.Y., Tkalich: An ANN application for water quality forecasting. Marine Pollution Bulletin 56(9), 1586–1597 (2008)

[2] Dutot, A.L., Rynkiewicz, J., Steiner, F.: A 24-h forecast of ozone peaks and exceedance levels using neural classifiers and weather predictions. Environmental Modelling & Software 22(9), 1261–1269 (2007)

[3] Palani, S., Liong, S.Y., Tkalich: An ANN application for water quality forecasting. Marine Pollution Bulletin 56(9), 1586–1597 (2008)

[4] Rodriguez, M.J., Sérodes, J.: Assessing empirical linear and non-linear modelling of residual chlorine in urban drinking water systems. Environmental Modelling and Software 14(1), 93–102 (1998)

[5] Alp, M., Cigizoglu, H.K.: Suspended sediment load simulation by two artificial neural network methods using hydrometeorological data. Environmental Modelling & Software 22(1), 2–13 (2007)

[6] Waich, K., Mayr, T., Klimant, I.: Fluorescence sensors for trace monitoring of dissolved ammonia. Talanta 77(1), 66–72 (2008)

[7] Yang, M.-Z., Dai, C.-L., Wu, C.-C.: A Zinc Oxide Nanorod Ammonia Microsensor Integrated with a Readout Circuit on-a-Chip. Sensors (2011)

[8] Temple-Boyer, P., Hajji, B., Alay, J.L., Morante, J.R., Martinez, A.: Properties of SiOxNy films deposited by LPCVD from SiH4/N2O/NH3 raseous mixture. Sensors and Actuators A: Physical 74 (2009)

[9] Sharma, A.L., Kumar, K., Deep, A.: Nanostructured polyaniline films on silicon for sensitive sensing of ammonia. Sensors and Actuators A: Physical 198

[10] Huang, J., Wang, J., Gu, C., Yu, K., Meng, F., Liu, J.: A novel highly sensitive gas ionization sensor for ammonia detection. Sensors and Actuators A: Physical 150

[11] Waich, K., Mayr, T., Klimant, I.: Fluorescence sensors for trace monitoring of dissolved ammonia. Talanta 77(1), 66–72 (2008)

[12] Li, J., Pilkington, N.T.: Embedded architecture description language. Journal of Systems andSoftware (2009)

Dissolved Oxygen Prediction Model
Which Based on Fuzzy Neural Network

Liu Yalin[1,2,3], Wei Yaoguang[1,2,3,*], and Chen Yingyi[1,2,3]

[1] Key Laboratory of Agricultural Information Acquisition Technology, Ministry of Agriculture,
Beijing 100083, P.R. China
[2] Beijing Engineering and Technology Research Center for Internet of Things in Agriculture,
Beijing 100083, P.R. China
[3] China-EU Center for Information and Communication Technologies in Agriculture, China
Agricultural University, Beijing 100083, P.R. China
Beijing Engineering Center for Advanced Sensors in Agriculture,
Beijing 100083, P.R. China

Abstract. In crab ponds, dissolved oxygen is the foundation for pond cultivation's survival. The changes of dissolved oxygen content are influenced by multiple factors. Higher levels of dissolved oxygen content are crucial to maintaining healthy growth of crab breeding. Affected by physic-chemical process of aquatic water, the changes of dissolved oxygen content have a large lag. In order to solve the problem of dissolved oxygen forecast, the prediction model which based on fuzzy neural network has been proposed in this paper. It integrated the characteristic of learning fuzzy logic and neural networks optimized performance to realize the dissolved oxygen prediction. The prediction results have shown it more suitable for dissolved oxygen prediction than grey neural network method. The prediction accuracy can meet the need of dissolved control.

Keywords: Dissolved oxygen(DO), Fuzzy neural network, Prediction.

1 Introduction

Dissolved oxygen is the most important factor for fishes healthy growth. Pond cultivation must maintain a certain level of dissolved oxygen to make the fish health growth. Meanwhile, dissolved oxygen is playing a dominent role in adjusting the substances of oxidative decomposition in water. High levels of dissolved oxygen content in water can suppress and mitigate the toxic effects of ammonia and hydrogen sulfide and other substances on fish. Currently, farmers monitor the donamic changes of dissolved oxygen mainly based on the observation of biological activities. Such "after-control" methods always lead to a negative effect on the growth of cultured organisms. Therefore, how to grasp the dynamic change laws of dissolved oxygen in pond water, forecasting the situation of low DO content and take acts to keep the DO content stable in the pond is the urgent problems to be solved.

---

* Corresponding author.

D. Li and Y. Chen (Eds.): CCTA 2013, Part II, IFIP AICT 420, pp. 544–551, 2014.

The DO content is not only influenced by the effects of water physical and chemical properties, such as water temperature, PH, salinity, electrical conductivity, but also influenced by the atmosphere environments factors. The interaction between these factors is complicated. It cannot take precise mathematical model to describe these nonlinear relations. In recent years, domestic and international scholars have put forward lots of methods for dissolved oxygen forecasting, such as time series analysis, neural networks, and statistical analysis etc.

Time series analysis is a quantitative analysis method, Time series analysis comprises methods for analyzing time series data in order to extract meaningful statistics and other characteristics of the data. Time series forecasting is the use of a model to predict future values based on previously observed values. Simple or fully formed statistical models to describe the likely outcome of the time series in the immediate future, given knowledge of the most recent outcomes. Due to the fact that the water quality changes are affected by multiple factors, there exists ramdomness. Hnece, time series analysis has a certain limits when conducting dissolved oxygen forecasts. Jiao Ruifeng, etc. Proposes a prediction model which based on grey relational analysis and Monte Carlo method to forecast the reservoir water quality. Lu Qi adopts gray neural network model which based on Grey theory and neural network theory to forecast the lake permanganate index. The disadvantage is that NNs can learn the dependency valid in a certain period only. The error of prediction cannot be generally estimated. The advantages are that neural network has a very strong nonlinear fitting capability which maps arbitrarily complex nonlinear relationships. It's also easy to learn and convenient for the realization of computer. In addition, it has strong robustness, memory capacity, nonlinear mapping ability and self-learning ability. However, it also has problems, for example, premature convergence etc.

Focused on the issue of current methods, the prediction model which based on fuzzy neural network has been proposed in this paper. With the combination of learning, imagination, adaptation and fuzzy information processing, fuzzy neural network can improve the overall learning and expression ability of the system.

2   Fuzzy Neural Network Prediction Algorithm

2.1   Fuzzy Neural Network

A fuzzy neural network or neuro-fuzzy system is a learning machine that finds the parameters of a fuzzy system (i.e., fuzzy sets, fuzzy rules) by exploiting approximation techniques from neural networks. Fuzzy neural network is not only has the traits of general neural network, but also has some special characteristics. For example, due to the calculation method of fuzzy mathematics, fuzzy neural nework makes some processing units easier, further quickening the speed of information processing. And because it uses the fuzzy operating mechanism, it strengthens the fault tolerance of the system. But the most important is that fuzzy neural network enlarges the scope of system information processing, enabling the system to process deterministic and non-deterministic information at the same time, which fortifies the flexibility of system information processing to a great extent.

Both neural networks and fuzzy systems have some things in common. They can be used for solving a problem (e.g. pattern recognition, regression or density estimation) if there does not exist any mathematical model of the given problem. They solely do have certain disadvantages and advantages which almost completely disappear by combining both concepts.

A fuzzy system demands linguistic rules instead of learning examples as prior knowledge. Furthermore the input and output variables have to be described linguistically. If the knowledge is incomplete, wrong or contradictory, then the fuzzy system must be tuned. Since there is not any formal approach for it, the tuning is performed in a heuristic way. This is usually very time consuming and error-prone.

The basic idea of fuzzy neural network is that it tries to integrate the fuzzy system representation, self-adaptation of neural network and knowledge adjustment and discovery together into a system. Fuzzy neural network discovers and adjusts the membership functions of the fuzzy subsets self-adaptingly, and classifies language values self-organzingly. The learning ability of neural network is brought into play, and the system knowledge is moderately adjusted, thereby leading to a higher intelligence and adaptability of fuzzy neural network.[3.4]

Fuzzy system is a strong self-adaptation system, which can update automatically and continuously modify the membership functions of the fuzzy subsets. Fuzzy system is defined by the "if–then " rules, based on the R^i. Its fuzzy rule is listed below.

$$R^i : If \quad x_i \text{ is } A_1^i, x_2 \text{ is } A_2^i, \cdots, x_k \text{ is } A_k^i \text{ then } y_i = p_0^i + p_1^i x_1$$
$$+ \cdots + p_k^i x_k \tag{1}$$

Here, A_j^i are the fuzzy subsets of the fuzzy system; i is the number of fuzzy subsets, j is the number of input parameters; P_j^i (j = 1, 2, ..., k) is the parameter of the fuzzy system; y_i is the output based on the fuzzy rules, input part (that is, if...) is fuzzy, output part (that is, then...) is determined. The fuzzy reasoning proves that the output is the linear combination of the input.

If x = $[x_1, x_2, \cdots, x_k]$, the degrees of membership can be calculated according to the fuzzy rules.

$$\mu_{A_j^i} = \exp(-(x_j - c_j^i)^2 / b_j^i), \qquad j = 1, 2, ..., k; i = 1, 2, ..., n \tag{2}$$

In the formula, c_j^i, b_j^i belong to the center and breadth of membership functions; K is the number of input parameters; n is the number of fuzzy subsets. We calculate fuzzily based on the degrees of membership and use fuzzy operator as continually multiplying operator.

$$w^i = \mu_{A_j^1}(x_1) * \mu_{A_j^2}(x_2) * \cdots \mu_{A_j^k}(x_k), \qquad i = 1, 2, ..., n \tag{3}$$

We calculate the output y_i according to the result of fuzzy calculation.

$$y_i = \sum_{i=1}^{n} w^i \left(p_0^i + p_1^i x_1 + \cdots p_k^i x_k \right) / \sum_{i=1}^{n} w^i \tag{4}$$

Fuzzy neural network is divided into four layers: input, fuziness, fuzzy rule calculation, and output. Input layer connects input vector x, and the number of nodes are the same as the dimensions of input vector. Fuziness layer gets fuzzy degree of membership μ by using membershiup function (2) to blur the inputs. Fuzzy rule calculation layer is calculated by formula (3). Output layer is calculated by using formula (4). 【5】

The learning algorithm of fuzzy neural network is explained below.

　　(1)　Error calculation

$$e = \frac{1}{2}(y_d - y_c)^2 \tag{5}$$

In the formula, y_d is network expectation output; y_c is network actual output; e is the error between expectation and actual output.

　　(2)　Coefficient correction

$$q_j^i(k) = q_j^i(k-1) - \alpha \frac{\delta e}{\delta q_j^i} \tag{6}$$

$$\frac{\delta e}{\delta q_j^i} = (y_d - y_c) w^i / \sum_{i=1}^{m} w^i \bullet x_j \tag{7}$$

In the formula, q_j^i is neural network coefficient; α is network learning rate; w^i is the multiplication.

　　(3)　Parameter correction

$$c_j^i(k) = c_j^i(k-1) - \beta \frac{\delta e}{\delta c_j^i} \tag{8}$$

$$b_j^i(k) = b_j^i(k-1) - \beta \frac{\delta e}{\delta b_j^i} \tag{9}$$

In the formula, c_j^i b_j^i are the center and breadth of membership functions.[6]

2.2　Influntial Factors of DO Pond Aquaculture Defense

In aquaculture ponds, the content of DO are affected by multiple factors. The main source of DO in the pond are originated from the disslolving of molecular oxygen in the air and the photosynthesis of aquatic plants. Atmospheric pressure, light, temperature, wind speed and direction are the main outside factors which affect the dissloved oxygen content in crab ponds.

In the ponds, the consumption of DO are mainly from sediment, zooplankton, fish and chemical factors. According to the scientific research, the sediment, zooplankton, and even fish consumption of DO is very small and have little influence on the changes of DO. So, those factors are not taken into the account. However, because of the complex detection, and poor detection method, it's hard to detect the chemical consumption of DO. The water quality parameters, such as water temperature, water depth, electrical conductivity, PH, turbidity are also affect the redox reactions. In the aquaculture water, conductivity change is a significant water quality indicator and has some connection with the change of DO. So, it choose conductivity as one important indicator of measuring DO changes.

Due to the fact that the changing process of the pond aquaculture water quality is a dynamic concecutive change process. The changes of DO have continuities between two consecutive periods. During the prediction of DO, the DO in the previous time period is usually considered as a crucial index to measure the changes of DO.

2.3   DO Prediction Algorithm Based on the Fuzzy Neural Network

With fuzzy neural network, DO prediction is divided by three parts: the foundation of fuzzy neural network, the training of fuzzy neural network and the DO prediction of fuzzy neural network.

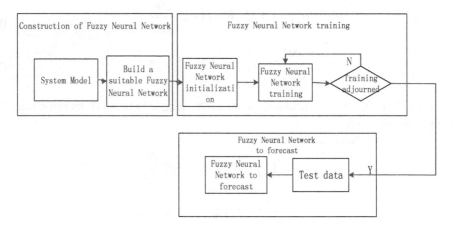

Fig. 1. Process of Fuzzy Network DO Prediction Algorithm

3   Simulation Experiment and Result Analysis

The data used in this study are produced by the Water Quality Monitoring System which Based on Wireless NetworkSystem. When it has been equipped at China Agricultural University-Yixing Aquaculture Internet of Thing research base in Jiangsu province, China, the system has Stable operated more than one year and has obtained. many water quality parameters.

When being used in the water quality monitoring of Crab ponds, the water quality monitoring system is stable and can meet the production need. The Sampling interval is 30 minutes, which means 48 sets of data has been collected per day. we take 350 for training, 50 for prediction. The real-time DO, conductivity, and water tempreture are taken as the inputs of neural network, that is, x_i, in the formula (1). We train the fuzzy neural network with actual DO as the outputs of neural network expectation. The training result is shown below.

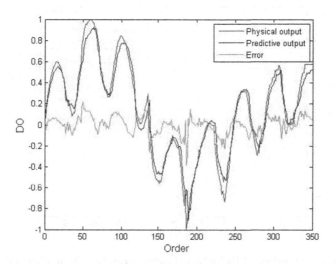

Fig. 2. Resolved Oxygen Training Data Forecast Chart

We can get forecast data by using 50 groups of data for testing to forecast neural network. And the result is showed in Fig 3. From the chart, the error of the result is no more than 0.1, which means that fuzzy neural network has a better impact on the forecast of the pond resolved oxygen.

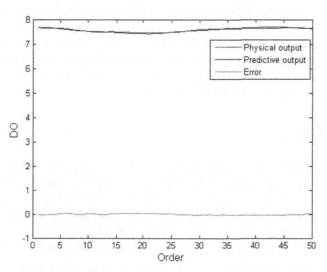

Fig. 3. Resolved Oxygen Testing Data Forecast Chart

In order to compare the adapatability of this method, we use the same data and compare them with the forecast result of grey neural network.The forecast result is illustrated as follows in Fig 4.

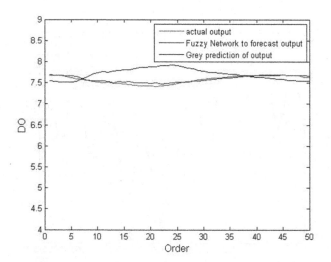

Fig. 4. Comparison of Fuzzy Neural Network and Grey Neural Network

According to the Fig 4, we can see that grey neural network forecast data appears obvious errors, and the forecast results of fuzzy neural network is better compared with that of grey neural network. The experiment shows that fuzzy neural network can better resolve the problem of aquatic dissolved oxygen in ponds. And the forecast results are more accurate, comprehensive and objective.

Based on the analysis of the influential factors of dissolved oxygen, the research, which adopts fuzzy neural network methods and focuses on the forecast issue of aquatic dissolved oxygen in ponds, has resolved the problem of dissolved oxygen forecasts. According to the simulation experiment, it shows that fuzzy neural network prediction is more accurate. Compared with gray neural network, the accuracy of forecasts is higher, especially for DO forecasts.

4   Conclusion

Based on the analysis of the influential factors for dissolved oxygen content in crab ponds, the prediction model which based on fuzzy neural network has been proposed in this paper. It can control the learning performance error value and the total error, the performance of the model can be optimized for water dissolved oxygen prediction. When compared with grey neural network, the The experimental results has shown that the model has a better ability to predict the dissolved oxygen prediction content.

Acknowledgements. This work was supported by the National Agricultural Science and technology achievement transformation project (2012GB2E000330), the state of

12th five-year science and technology support projects (2012BAD35B03) and (2011BAD21B01), and the programs "Agro-scientific Research in the Public Interest" (201203017).

References

1. Faruk, D.O.: A hybrid neural network and ARIMA model for water quality time series prediction. Engineering Applications of Artificial Intelligence 23(4), 586–594 (2010)
2. Ahmed, A.N., et al.: Evaluation the efficiency of radial basis function neural network for prediction of water quality parameters. Engineering Intelligent Systems for Electrical Engineering and Communications 17(4), 221–231 (2009)
3. Han, H.G., et al.: An efficient self-organizing RBF neural network for water quality prediction. Neural Networks 24(7), 717–725 (2011)
4. Skowronska, K.T., et al.: Application of a fuzzy neural network for river water quality prediction. Chemia Analityczna 51(3), 365–375 (2006)
5. Li, Z.B., et al.: An Improved Gray Model for Aquaculture Water Quality Prediction. Intelligent Automation and Soft Computing 18(5), 557–567 (2012)
6. Faruk, D.O.: A hybrid neural network and ARIMA model for water quality time series prediction. Engineering Applications of Artificial Intelligence 23(4), 586–594 (2010)
7. Wu, J., et al.: Application of chaos and fractal models to water quality time series prediction. Environmental Modelling & Software 24(5), 632–636 (2009)
8. Sun, X.: Prediction of fluorite deposit in Yixian based on fuzzy-neural network. Journal of China University of Geosciences 18, 279–281 (2007)
9. Rahman, M.S., et al.: Thermal conductivity prediction of foods by Neural Network and Fuzzy (ANFIS) modeling techniques. Food and Bioproducts Processing 90(C2), 333–340 (2012)
10. Taner, A.: Prediction of moisture dependent some physical properties of wheat using artificial neural network and fuzzy logic. Energy Education Science and Technology Part a-Energy Science and Research 29(1), 395–406 (2012)

Flexible Embedded Telemetry System
for Agriculture and Aquaculture

André Weiskopf[1], Frank Weichert[1], Norbert Fränzel[2], Manuel Schneider[3]

[1] Fraunhofer Advanced System Technology (AST),
Am Vogelherd 50, 98693 Ilmenau, Germany
{andre.weiskopf,frank.weichert}@iosb-ast.fraunhofer.de
[2] Technical University Ilmenau, Ehrenbergstraße 29, 98693 Ilmenau, Germany
norbert.fraenzel@tu-ilmenau.de
[3] University of Applied Science Schmalkalden,
Blechhammer 9, 98674 Schmalkalden, Germany
m.schneider@fh-sm.de

Abstract. This paper describes a system for data acquisition and remote maintenance via wireless communication. In this work the concept of managing time critical tasks via networks is shown. The system consists of a client-server-structure with a microcontroller-based module connected to the user devices and maintenance software, both as clients and a maintenance server for establishing the connection between the clients and to identify users. The special interest in this paper is the implementation of a specialized communication protocol to optimize the communication between the clients.

Keywords: Telemetry, ISOBUS, CAN, remote maintenance, wireless communication, real-time, tracking, time-critical process.

1   Introduction

The options for diagnosis and maintenance by software have raised rapidly with the growing number of embedded systems in all kinds of products. By changing the parameterization or the firmware of embedded systems a lot of adaptions are possible. These tasks can be done for stationary systems and by an increasing availability of wireless communication services also remotely for mobile systems. The use of remote maintenance services reduces downtimes and travel expenses for service personnel.

In this paper a system is described for both the diagnosis and remote maintenance of stationary systems, e.g. greenhouses and fish farming, and additionally fleet management of mobile systems, e.g. agriculture machines like harvester. In modern automation and automotive technology a lot of different digital systems are used e.g. for control of systems or parts thereof. These consist of individual modules which communicate via a common data bus, i.e. CAN, ISOBUS, Ethernet or RS485. To fulfill the different tasks various combinations of input-, output, sensor- and actor modules are possible. By an increasing number of software parameters in these modules the overall systems are becoming increasingly complex. For an optimal

D. Li and Y. Chen (Eds.): CCTA 2013, Part II, IFIP AICT 420, pp. 552–560, 2014.

adaption to the requirements of the task the modules can be parameterized in large scale. At the same time the modules can be used for reading sensor values to make this information available to other components of the system or to the service personnel. By using the internal error history and monitoring feature of the different modules troubleshooting is supported. The diagnosis and maintenance of such systems demand comprehensive knowledge by specially trained personnel to fulfill the tasks.

The technology "FETS" can be used for remote maintenance and diagnosis of such control systems. "FETS" stands for Flexible Embedded Telemetry System. It can be adapted to various stationary and also mobile systems. The monitoring of the individual systems can take place via a service control center. The development of a command pattern, which allows control of time critical processes on the CAN bus side of a system over communication networks without Quality of Service (QoS), was one main challenge to solve. With this technology the service personnel has the opportunity to upload new firmware to the modules, which have to be remotely maintained. In this way new features can easily be implemented and the behavior can be adjusted, so the range of functions increases, e.g. by using new sensor or actor modules.

In modern agriculture the ISOBUS is of growing importance [1, 2, 3]. ISOBUS is defined by 250kbaud CAN-2.0b communication protocol based on the ISO11783 norm [4, 5, 6]. At this various auxiliary equipment will be connected to the agriculture machine, e.g. modern tractor or harvester, via a common data bus. In this way information like driving speed, hydraulic pressures and valve positions are made available by the Electronic Control Unit (ECU) of the agriculture machine. Depending on the number and types of the auxiliary equipment the agriculture will be more efficient by resource-conserving use of plant protection products. In a further expansion phase of the system the remote control of agriculture machines with certain limits is possible. This will be examined at precision farming. In addition to using FETS for mobile systems in agriculture the technology can be applied to stationary systems, i.e. control of greenhouses and aquaculture by monitoring of vital factors like oxygen concentration, pH value and temperature.

Currently the technology FETS is applied as remote maintenance system for modern electrical wheelchairs, as tracking system for pedelec hire and also as maintenance and diagnosis system for solar power plants. For adaption to various user interfaces FETS is based on a modular design principle.

2   Related Work

Mainly in the automotive sector the use of mobile communication networks for maintenance tasks is discussed, e.g. [7]. In [8] an idea to connect a local onboard module to a remote server is presented. In the field of maintenance and monitoring for a variety of purposes numerous reports exist, e.g. in [9, 10] mobile communication is used for monitoring a wheelchair and in [11] for automated alarming if a failure is detected. Bidirectional communication is discussed in [12] in the area of additional services.

With regards to the focus of this paper, the remote maintenance of agriculture machines and monitoring of greenhouse and aquaculture, a few reports exists. In [1, 2, 3] some aspects for embedded systems, which are used for communication with agricultural machines via ISOBUS, are given.

3   System Overview

The developed technology is based on three main components. Figure 1 shows a simplified overview of the system.

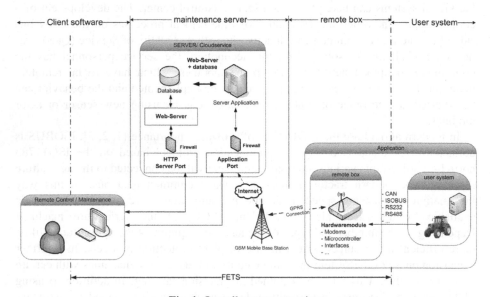

Fig. 1. Overall system overview

The first component, called "remote-box", is a typical embedded system based on a Microcontroller which offers sufficient communication interfaces and hardware resources to fulfill all tasks of remote maintenance. This box realizes the time critical communication with the data bus of the system that has to be maintained. Another key aspect is the opportunity to connect to a server via GPRS-based communication. Within the "FETS" technology modular software was designed. With an individual adaption of the corresponding hardware the firmware can be ported easily onto other Microcontrollers. The expansion phase of the remote box depends on the boundary conditions of the user system. This contains for instance the time behavior of the system, the number of components to be monitored and configured as well as the required calculation time. In addition the box may contain a GPS receiver, so the localization of mobile systems is possible. Furthermore it could be equipped with RF-components, so the box can be used as a network node or transmit data via ISM-Band. In this case the transfer of data can be encrypted by using crypto-algorithms. Figure 2 illustrates a used expansion stage of the hardware of the remote-box.

Fig. 2. Hardware of the remote-box

The server is the central communication relay between the clients (remote-boxes and maintenance software). The server application verifies the authorization status of the actual user. So it can decide which user has access to which remote box. After that the user can connect to a selected remote-box. Now the server establishes a transparent connection between remote-box and maintenance client software. It is important to note that the whole data is transmitted via the server and not point-to-point. So, only the server needs a static, visible internet address. Additionally the server will be used for databases and updates. So the remote-boxes and client-PCs can download the most up to date software. By integrating automatic updates the firmware of the remote-boxes and software of the maintenance client-PCs are always up-to-date. This ensures that no different versions of software will be used, so the system maintenance gets more efficient.

4   Research Issue

The main task is to connect the time-critical communication of the end devices with the latency afflicted communication via GPRS to the server. In picture 3 a histogram of the transfer of data packets with typical latencies of GPRS communication is shown.

The measurements in the histogram are for the transmission of a single data packet from remote-box to server and back. So an average transfer of a single packet is around 750 milliseconds, but it is also possible that this data transfer time is much higher.

In CAN-Bus based systems the data exchange between different nodes is organized by service data objects (SDO) and process data objects (PDO). For example by connecting the remote-box to a CAN-BUS based system, typical timeouts for answering to service data objects are 250 milliseconds. To process time critical commands on the CAN-Bus it is necessary to know exactly the maximal timeouts. If such a timeout is exceeded, it is also possible that warnings, errors in or shut-offs of the user system can be produced.

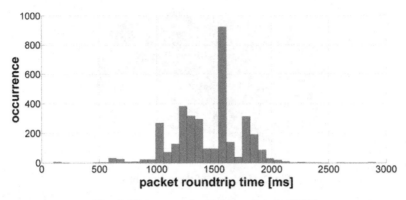

Fig. 3. Histogram of packet transfer via GPRS

Due to different latencies of transfers via GPRS and CAN, a direct transfer from the server to the remote-box is not reasonable.

5   Design of the Communication Process

To solve the problem with the different latencies a specialized macro language was included. With these macros it is possible to execute time critical operations self-contained. Therefore a specialized communication protocol stack is integrated in all 3 parts of the remote maintenance system. Its job is to allocate different services for time-critical processes.

This stack is based on the TCP-Stack. To optimize the communication process a packet manager and action manager are added to the TCP-Stack. Figure 4 shows the schematic of the communication process.

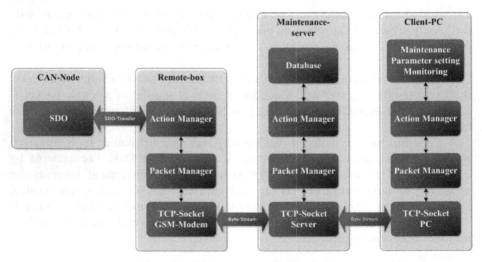

Fig. 4. Structure of communication process

The **action manager** layer executes actions as commands and macros. That means its job is the data processing in PC, server and remote-box. On the remote-box side typical actions are reading out complete list of parameters, measurements or setting new parameters and on PC-side to display measurements or to indicate errors. The most critical part is to keep the user system in a secure state at all time. Therefore it is necessary to transfer all data from the maintenance software to remote-box first and after that to process the data in an appropriate manner. The **TCP socket** is included in the PC's and also in the firmware of embedded modems. Its job is to transfer the data between the 3 system parts. TCP by itself provides a secure byte-stream, but during the process of design there are some restrictions like the loss of the connection in wireless systems or another typical problem is the overflow of buffers for the received data in the modem.

To optimize the time of complete data transfer a **packet manager** is included between TCP socket and action manager. Its job is to divide the complete data stream of the transmitter in single data packets, to include some control bytes in these packets and also to summarize the received single data packets to one data stream.

A whole data packet consists of an info field, an ID-field, the macro command, and the data to be transmitted. The packet manager is also responsible for detecting lost packets. So it can react to this by different actions like packet repeats.

Most processes in user systems are designed to use simple commands. So it is usually enough to send one data packet like readout one measurement or set one parameter. The structure of this process is shown in figure 5a). Some processes require more communication, e.g. firmware updates or the transmission of a whole parameter set. In this case typically two transfer mechanisms are used. The first is to send a confirmation after each packet and send then the second data packet and so on, see figure 5b). The second one is to send the whole byte-stream without confirmation. Only one confirmation is send when the data stream is finished, see 5c). Here it is possible to loose single packets and thus it is necessary to send the whole data stream again.

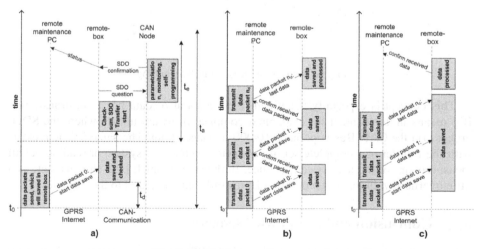

Fig. 5. Standard transmit processes

In the FETS Technology a mixture of both processes is used to optimize the transfer. In the remote maintenance system all the data packets are send in series. And for every single data packet a confirmation with the ID is transmitted backwards. So it is possible to identify lost packets. With this principle it is not necessary to repeat the whole byte-stream again. The only packets to repeat are these packets where no confirmation was transmitted backwards. In figure 6 this new approach is shown.

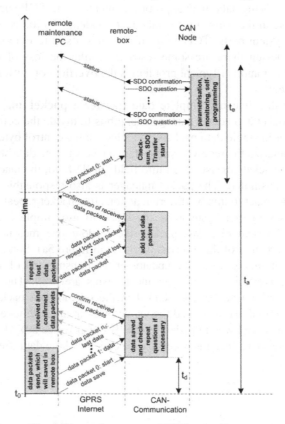

Fig. 6. Transmit process at FETS technology

This concept is especially beneficial if big amount of data has to be transferred. Big means in this context for example a complete firmware update where the SDO-process on the CAN-Bus is perhaps 10.000 messages long. The same problem exists with the transfer of a complete set of parameters to the user system or reading out all measurements of the user system at once. In the case of small byte-streams there is a little overhead with this concept, but it is insignificant compared to the payload.

6   Conclusion and Acknowledgements

A fast, easy and safe transmission of data via GPRS cannot be guaranteed. To prevent the accumulation of long waiting periods the communication dialog should be reduced to realize an efficient communication with user system via remote maintenance technology.

Here, the maintenance system was extended by a special protocol stack. This was realized on the one side in the firmware for the remote-box and on the other side in the application software for the client PC. The task is to summarize the time-critical command sequences (actions) in packets in the remote-box. So a secure time critical communication with the system can be ensured.

With this technology a Client-Server-Structure was realized, where the remote-boxes and Maintenance-PCs communicate via a specific server as clients. A specifically developed application is the basis of the server. This accepts incoming connection requests from the clients, manages the user authentication and transfers data packets between the clients which are connected. Specific data packets can be saved in a database. Although the connection between Maintenance-PC and remote box is established through a server, there is a transparent connection between them. The server is the central exchange.

A special software design is needed for the resulting complexity of the firmware with many parallel processes. A special multitasking system was set up to obtain functional embedded software with the used hardware. This design pattern allows an efficient use of the processor resources with the help of nested interrupts.

The developed technology "FETS" is an approach for the realization of maintenance and remote diagnostic system. By the presented concept a system was developed, which can easily be adapted to a variety of different user systems. The technology is not limited to CAN based systems, it can easily be adapted to other interfaces and communication protocols.

FETS has already been successfully tested in a field test phase in Europe as a localization-, diagnosis- and maintenance module on electrical wheelchairs and pedelecs. A present field of application is the maintenance of solar power plants. It also became obvious, that the technology is very robust and reliable. So the application of this technology is interesting for the maintenance and diagnosis of vehicles and stationary systems.

References

1. Fantuzzi, C., Marzani, S., Secchi, C., Ruggeri, M.: A Distributed Embedded Control System for Agricultural Machines. In: IEEE International Conference on Industrial Informatics, Singapore (2006)
2. Zhang, M., Zhou, Z., Xi, Z.: In: Second International Conference on Intelligent Computation Technology and Automation (ICICTA), Cangsha (2009)
3. Sarker, M.K., Park, D.S., Badarch, L.: Electronic Control Sensors Applications for the Next Generation Tractor Based on Open Source Library. In: Sixth Conference on Sensing Technology (ICST), Kolkata (2012)
4. Tractors and machinery for agriculture and forestry', International Standard ISO 11783-1 to 13, ISO (2007)
5. Felimeth, P.: CAN-based tractor- agricultural implement communication ISO 11783, CAN-Newsletter 9 (2003)

6. Stone, M.L.: ISO 11783 – An Electronic Communications Protocol for Agricultural Equipment. ASEA Distinguished Lecture Series (1999)
7. Herrtwich, R.G.: Automotive Telematics - Road Safety versus IT Security? In: International Conference on Computer Safety, Reliability and Security, Potsdam (2004)
8. Zhang, Y., Salman, M., Subramania, H.S., Edwards, R., Correia, J., Gantt, G.W., Rychlinski, M., Stanford, J.: Remote Vehicle State of Health Monitoring and Its Application to Vehicle No-Start Prediction. In: IEEE Autotestcon, Anaheim (2009)
9. Touati, Y., Ali-Cherif, A., Achili, B.: Smart Wheelchair Design and Monitoring via Wired and Wireless Networks. In: IEEE Symposium on Industrial Electronics and Applications, Kuala Lumpur (2009)
10. Sevillano, J.L., Cascado, D., Vicente, S., Lujan, C.D., del Rio, F.D.: A Real-Time Wireless Sensor Network for Wheelchair Navigation. In: International Conference on Computer Systems and Applications, Rabat (2009)
11. Lobardi, P., Giaconia, C.G., Di Dio, V.: An Embedded Diagnostic System for Wheelchairs Brushless Drives Monitoring. In: International Symposium on Power Electronics, Electrical Drives, Automation and Motion, Taormina (2006)
12. Zhihong, T., Jinsheng, Y., Jianguo, Z.: Location Based Services Applied to an Electrical Wheelchair Based on the GPS and GSM Networks. In: International Workshop on Intelligent Systems and Applications, Wuhan (2009)

Author Index

Printed in the United States
By Bookmasters